Motivation Science

Myint Swe Khine
Editor

Motivation Science

A Machine-Generated Literature Overview

Springer

Editor
Myint Swe Khine
Curtin University
Perth, WA, Australia

ISBN 978-981-97-9246-7 ISBN 978-981-97-9247-4 (eBook)
https://doi.org/10.1007/978-981-97-9247-4

This Springer imprint is published by the registered company Springer Nature Singapore Pte Ltd.
The registered company address is: 152 Beach Road, #21-01/04 Gateway East, Singapore 189721, Singapore

About This Book

Auto-summaries can be generated by either an abstractive or extractive auto-summarization:

- An extraction-based summarizer identifies the most important sentences of a text and uses the original sentences to create the summary.
- An abstraction-based summarizer creates new text based on deep learning. New phrases are created to summarize the content.

The auto-summaries you will find in this book have been generated via an extractive summarization approach.

Each chapter was carefully edited by Dr. Myint Swe Khine. The editor selected the papers which were then auto-summarized. The editor has not edited the auto-summaries due to the extraction-based approach, and have not changed the original sentences. You will find the editor's reviews and guidance on the auto-summaries in their chapter introductions.

In machine-generated books, editors are defined as those who curate the content for the book by selecting the papers to be auto-summarized and by organizing the output into a meaningful order. Next to the thoughtful curation of the papers, editors should guide the readers through the auto-summaries and make transparent why they selected the papers.

The ultimate goal is to provide a current literature review of Springer Nature publications on a given topic in order to support readers in overcoming information overload and to help them dive into a topic faster; to identify interdisciplinary overlaps; and to present papers which might not have been on the readers' radar.

Please note, that the selected papers are not used to train a LLM while the auto-summaries are created.

Contents

About the Editor

Professor Myint Swe Khine has more than 30 years of experience in teacher education. He holds master's degrees from the University of Southern California, USA, the University of Surrey, UK, and a doctorate in education (EdD) from Curtin University, Australia. He has worked at the National Institute of Education, Nanyang Technological University, Singapore, and was a professor at the Emirates College for Advanced Education in the United Arab Emirates. He serves on the editorial boards of several international academic journals. Throughout his career, he has published over 40 edited books. His most recent volumes include *Rhizome Metaphor: Legacy of Deleuze and Guattari in Education and Learning* (Springer, 2023) and *Machine Learning in Educational Sciences: Approaches, Applications and Advances* (Springer, 2024). He currently teaches in the School of Education, Curtin University, Australia.

Motivation Science 1

Introduction by the Editor

Motivation science is a field that delves into the very core of what drives us—why we do what we do. It's a field that bridges several disciplines, including psychology, neuroscience, economics, and even sociology. Researchers in this field are fascinated by the question of what makes people tick. In particular, they focus on understanding what motivates us and how we deal with challenges. Researchers investigate in depth the desires, needs, and goals that propel humans into action, and what processes they go through to translate those motivations into behaviors. They are also looking at how people persevere and overcome obstacles to achieving their goals.

Some of the key areas that motivational science explores are intrinsic vs extrinsic motivation and goal setting. This looks at the difference between doing something because you enjoy it (intrinsic) and doing something for an external reward (extrinsic). Research suggests that intrinsic motivation leads to greater long-term commitment and satisfaction. How setting specific, achievable goals can increase motivation and focus is another area that has been explored. The role of rewards and incentives can influence behavior, but also how the anticipation of a reward can be more powerful than the reward itself. Understanding the science of motivation can be incredibly useful in many areas of life. Educators can use these principles to create engaging learning environments. Businesses can use this knowledge to create motivating work cultures and incentive programs. And on a personal level, you can use this science to understand your own motivations and develop strategies to stay focused and achieve your goals.

M. S. Khine (ed.), *Motivation Science*, https://doi.org/10.1007/978-981-97-9247-4_1

Motivation Science

Motivation is a fundamental aspect of human behavior that drives us to act and achieve our goals. It is a complex phenomenon that has been studied extensively across various fields, including psychology, neuroscience, and organizational behavior. Motivation science is an interdisciplinary field that aims to understand the underlying mechanisms and factors that influence motivation, as well as how it can be harnessed to improve performance, well-being, and overall life satisfaction. One of the earliest and most influential theories in motivation science is the self-determination theory (SDT) proposed by Edward L. Deci and Richard M. Ryan (1985). SDT suggests that three basic psychological needs drive human motivation: autonomy, competence, and relatedness. Autonomy refers to the need for control and self-determination, competence refers to the need for mastery and effectiveness, and relatedness refers to the need for belonging and connection with others. According to SDT, when these needs are satisfied, individuals experience intrinsic motivation, which leads to enhanced performance, creativity, and well-being (Ryan & Deci, 2000).

Another prominent theory in motivation science is the expectancy-value theory, which posits that individuals' motivation is influenced by their expectations of success and the value they place on the task or activity (Wigfield & Eccles, 2000). This theory has been widely applied in educational settings, where researchers have found that student's motivation to engage in academic tasks is influenced by their expectations of success and the value they assign to the task, such as its perceived usefulness or interest (Eccles & Wigfield, 2002).

In recent years, neuroscience has emerged as a crucial contributor to the study of motivation science. Researchers have used advanced neuroimaging techniques, such as functional magnetic resonance imaging (fMRI), to investigate the neural correlates of motivation and reward processing. In addition to these theoretical frameworks, motivation science has also explored various intervention strategies and techniques to enhance motivation. For example, goal-setting theory suggests that setting specific, challenging, and achievable goals can increase motivation and performance (Locke & Latham, 2002). Self-regulation strategies, such as self-monitoring, self-evaluation, and self-reinforcement, have also been found to be effective in promoting motivation and goal attainment (Zimmerman, 2000).

Despite the significant progress that has been made in the science of motivation, there are still many unanswered questions and areas for further research. One area of ongoing investigation is the role of individual differences, such as personality traits, cultural background, and life experiences, in shaping motivation. Researchers are also exploring the interplay between motivation and other psychological processes, such as emotion regulation, decision-making, and cognitive control. Motivation science is a vibrant and multidisciplinary field that explores the complex factors and mechanisms underlying human motivation. Drawing on insights from different theoretical frameworks and empirical studies, researchers in this field aim to develop strategies and interventions to enhance motivation, performance, and well-being in a variety of contexts, ranging from education and work to sports and personal development.

The articles included in this section are, toward a general theory of motivation: problems, challenges, opportunities, and the big picture, motivation and new media, individual variability in motivation intensity, and commitment to change from locomotion motivation during deliberation. Motivated level of construal: how temperature affects the construal level of state-relevant stimuli, Hate: evidence for a motivational conceptualization, and consensual and behavioral validity of a measure of adaptive individual differences dimensions in human motivation are also included in this section.

References

Deci, E. L., & Ryan, R. M. (1985). Intrinsic motivation and self-determination in human behavior. Plenum Press.

Eccles, J. S., & Wigfield, A. (2002). Motivational beliefs, values, and goals. Annual Review of Psychology, 53(1), 109–132.

Locke, E. A., & Latham, G. P. (2002). Building a practically useful theory of goal setting and task motivation: A 35-year odyssey. American Psychologist, 57(9), 705–717.

Ryan, R. M., & Deci, E. L. (2000). Self-determination theory and the facilitation of intrinsic motivation, social development, and well-being. American Psychologist, 55(1), 68–78.

Wigfield, A., & Eccles, J. S. (2000). Expectancy–value theory of achievement motivation. Contemporary Educational Psychology, 25(1), 68–81.

Zimmerman, B. J. (2000). Self-efficacy: An essential motive to learn. Contemporary Educational Psychology, 25(1), 82–91.

Machine Generated Summaries

Disclaimer: The summaries in this chapter were generated from Springer Nature publications using extractive AI auto-summarization: An extraction-based summarizer aims to identify the most important sentences of a text using an algorithm and uses those original sentences to create the auto-summary (unlike generative AI). As the constituted sentences are machine selected, they may not fully reflect the body of the work, so we strongly advise that the original content is read and cited. The auto generated summaries were curated by the editor to meet Springer Nature publication standards. To cite this content, please refer to the original papers.

Machine generated keywords: people, motive, individual difference, personality, event, judgment, goal, article, participant, emotion, person, difference, stimulus, thought, mental

Toward a General Theory of Motivation: Problems, Challenges, Opportunities, and the Big Picture [166]

This is a machine-generated summary of:

Baumeister, Roy F.: Toward a general theory of motivation: Problems, challenges, opportunities, and the big picture [166]

Published in: Motivation and Emotion (2015)

Link to original: https://doi.org/10.1007/s11031-015-9521-y

If you want to cite the papers, please refer to the original.

For technical reasons we could not place the page where the original quote is coming from.

Abstract-Summary

"Motivation theories have tended to focus on specific motivations, leaving open the intellectually and scientifically challenging problem of how to construct a general theory of motivation."

"Both state (impulses) and trait (basic drives) forms of motivation must be explained, and their relationship must be illuminated."

"Not all motivations are the same, and indeed it is necessary to explain how motivation evolved from the simple desires of simple animals into the complex, multifaceted forms of human motivation."

"The relationship between liking and wanting, and the self-regulatory management of motivational conflict, also require explanation by an integrative theory."

Introduction

"Motivation researchers have made impressive progress at advancing a basic theoretical understanding of various specific motivations, such as sexual desire, hunger, achievement striving, and aggression."

"The purpose of this article (like that of the presidential address on which it is based) is to look ahead to the possibility of generating a general theory of motivation: a broad understanding of motivation that would apply across multiple domains of desire and motivated behavior."

"Most of psychology can be analyzed down into the basic processes of motivation and cognition."

"The recent resurgence of interest in motivation may be a sign of the belated recognition of the indispensability of both cognition and motivation for a comprehensive understanding."

"Motivational states and traits must be covered."

"It should address how motivations change, including the possibility that traits and drives change."

Two Meanings of Motivation

"Motivation in this sense refers to recurrent patterns of desire and frequent behavioral tendencies."

"The second concept of motivation refers to a particular desire to perform a particular behavior on a particular occasion."

"It is reasonable to discuss sexual motivation as a general pattern of desiring sex on various occasions, of tending to respond to sexual stimuli, and of being broadly interested in a wide range of sexual activities (and perhaps highly averse to others)."

"Sexual motivation can also take the form of a specific desire to engage in a specific action with a specific partner on a specific occasion."

"For now, I will refer to the broad, general form of motivation as drive and to the specific, instantiated form as impulse."

"People may have a basic drive for sex, which on specific occasions takes the form of targeted impulses to engage in particular sexual activities."

Primacy of Motivation

"The primacy of motivation entails that cognition, emotion, and social interaction emerged largely to serve motivation (e.g., Baumeister [1])."

"Emotion provides subjective feelings that provide feedback about motivationally relevant events, such as feeling happy upon satisfying some basic need, fear of something that jeopardizes one's safety, and anger at something that blocks one's quest for satisfaction."

"To be sure, many emotion theorists suppose that emotion causes motivation, such as when the feeling of fear causes the animal to hide or flee."

"Only if something is motivationally relevant does it evoke emotion."

"This process postulates that the animal is motivated to obtain food."

"In other ways too, cognitive processes facilitate motivated behavior."

"As I said, one does not have emotions if there is no relevant motivation in the first place."

Wanting and Doing

"The difference between plants and animals does not lie in needing things from the environment to sustain life: Both plants and animals have needs."

"For that reason, it needs some kind of action control mechanism that a plant does not (usually a brain and central nervous system)."

"I propose, motivation evolved in step with animals' control of movement and other action."

"A central nervous system may be an essential prerequisite for subjective experience, so that plants do not feel joy and pain in the sense that animals do."

"Having devoted much of my research efforts in recent years to problems of agency—from simple processes of effortful self-regulation to grand questions of free will and moral decision making—I can say with some confidence that theory of agency has sadly underappreciated the role of motivation."

"In whatever sense humans have free will, then, it did not emerge by some odd quirk of causal indeterminacy so much as having evolved to facilitate motivated behavior."

Basic and Advanced Motivations

"No one is likely to dispute that most animals are motivated to obtain food and water, to engage in reproductive activities, and to seek safety from predators."

"Insofar as learning can facilitate survival and reproduction, curiosity and other motivations that promote learning would also be among these basic motivations."

"It seems likely that human motivation has moved beyond mere survival and reproduction."

"Humankind re-invented social interaction in the form of culture as a strategy for achieving survival and reproduction."

"The view of humans as having invented new forms of social life raises the question of whether there are forms of motivation that are quite new."

"My own early survey of research on meaning in life proposed that people have four needs for meaning, which can be considered as components of a basic drive to achieve a meaningful life (Baumeister [2])."

Satiation and Beyond

"Motivation may also adapt to the local opportunity structure, in the sense that the person's general drive may gradually change as a function of whether it meets regular satisfaction or not."

"The idea that motivation adapts to the local opportunity structure invokes change in the trait, not just the state."

"Although the satiation cycle implies continual change, it also suggests a kind of stable equilibrium, as might be illustrated by someone who eats the same meals each day at the same times and with the same quantity of food."

"The well-accepted satiation cycle can be taken as a kind of equilibrium, in which desire and satisfaction follow each other in a steady sequence without any enduring change."

"The idea that getting begets wanting rejects the assumption that each cycle of satiation returns to the same baseline."

Motivational Plasticity

"Freud [3, 4] conceptualized motivation as innate drives requiring regular satisfaction and continuing more or less the same regardless of external factors."

"Sexual desire can also seem like something that emerges from inside (presumably as Freud thought of it), and plasticity pertains mainly to the forms it takes, such as what specific sexual acts one desires."

"The diversity of sexual desires suggests greater plasticity."

"This led to the conclusion that the female sex drive enjoys higher plasticity than the male in adulthood, although early experiences may have a powerful effect on the male sex drive, resulting in such things as paraphilias (perversions) that, once established and perhaps finalized at puberty, show little responsiveness to change, including even targeted therapy."

"One implication of that work was that the greater plasticity of female than male sexuality could possibly stem from a difference in drive strength."

"Plasticity of the underlying drive carries profound implications for understanding motivated behavior."

Addiction

"The process of becoming addicted to a substance that one had until recently never experienced, and hence was perfectly well able to live without, presumably involves acquiring a motivation to experience something."

"The idea that getting begets wanting suggests that addiction resembles other motivations in that its appeal increases across repeated satisfactions."

"Satisfying the addictive motivation reduces it in the short run but increases it in the long run."

"The key to understanding addiction is not a transformation in agency (i.e., volitional control of action) but rather a transformation of motivation."

"One remarkable testament to motivational plasticity in addiction reflects what happens when the addictive substance is no longer available."

"My treatment of addiction here is necessarily much briefer than is warranted for a comprehensive theory of motivation."

Liking Versus Wanting

"Motivation is wanting, but presumably it is sustained by liking."

"Motivation theory will have to address the question of how liking and wanting are related."

"Can wanting even be sustained at all without liking? (Some addiction theorists are fond of asserting that addicts desire the drug despite getting no pleasure from it, but I have not found convincing evidence that this is true.) If wanting and liking are fully orthogonal, there should be plenty of cases in which people continue to want things they dislike."

"These cases would have to go beyond routine problems of addiction and appetite control (I like cake but want to not eat it)—because these are not true instances of this category."

"Satisfaction only reinforces wanting if the person likes it."

"Liking is however the normal and natural result of getting what you want."

Motivational Conflict, Restraint, and Regulation

"To the time frame, the higher and lower motivations differ according to cultural values and enlightened self-interest, with the lower desires reflecting simple, animalistic motivations whereas the higher ones are linked to traits that set humans apart from other animals, including moral principles and rational calculation."

"The fact of motivational conflict requires theory to do more than simply explaining why one desire takes priority over another."

"Motivational conflict was particularly prevalent in connection with desires to do things incompatible with the workplace (such as sleeping, playing games, and having sex)."

"That includes recognition of conflict, prioritization of which motivation should be allowed to guide behavior, and self-regulatory processes to inhibit the disfavored desires."

"When motivation strays from those preferred patterns, people need to have a mechanism to decide whether to act on the actual or the desired desire!"

Conclusion

"A general theory could provide a framework to elaborate such theories, such as by helping them see what general aspects have not yet been fully studied in connection with each particular motivation, and even to highlight ways in which particular motivations depart from the general patterns."

"I hope that in a decade or two, progress will have reached the point at which someone (perhaps another future president of SSM!) can and will write a general theory of motivation."

"Third, I hope that moving toward a general theory of motivation will help psychology as a whole acknowledge and embrace the fundamental importance of motivation in the grand scheme of integrative psychological theory."

Motivation and New Media: An Introduction to the Special Issue [167]

This is a machine-generated summary of:

Small, Ruth V.: Motivation and new media: an introduction to the special issue [167]

Published in: Educational technology research and development (2011)

Link to original: https://doi.org/10.1007/s11423-011-9189-2

If you want to cite the papers, please refer to the original.

For technical reasons we could not place the page where the original quote is coming from.

Abstract-Summary
"This article provides an introduction to the theme of this special issue, "Motivation and New Media," followed by an explanation of the motivation for the issue and a brief description of the process for selecting articles to be included."

Introduction
"Scholars from Marshall McLuhan to Don Ely to Chris Dede, have suggested creating and integrating media, or new media (i.e., computer-based technologies integrated with other, more traditional media such as video and audio), into instructional design in order to more actively engage students in the learning experience."

"Recent research indicates that digital media used in teaching and learning enhance engagement and increase participation in learning activities (Schilling [5]; Sadik [6]) and improve learning motivation (House, [7]; Hsu [8])."

"New media fit well with constructivist learning environments."

"These are all powerful motivators for student learning."

Motivation for this Special Issue

"The motivation for this special issue was triggered by a 2005 National Science Foundation Science of Learning catalyst grant which brought together several leading scholars in the areas of motivation, technology and child development to explore how children are motivated to learn in informal, technology-rich learning contexts."

"Research luminaries, such as Edward Deci, Eliot Solloway, Jacquellyn Eccles, and John Keller, provided inspiration to the cohort of more than 30 researchers who participated in that project, including several contributors to this special issue."

Paper Selection Process

"In spring 2010, a call for submissions was issued to the AECT community from Michael Hannafin, research editor of ETR&D, for abstracts for articles on "Motivation and New Media" for a special issue of the journal."

"The call stated: New media have become important tools for motivating learning performance in the 21st century."

"This special issue will bring together researchers who focus on exploring how motivation influences the use of new media, how the use of new media stimulates motivation, and the methods and complications of studying motivation using new media in a learning context."

"Criteria for selection included a (1) strong foundation of motivation theory, (2) focus on motivation and learning outcomes, and (3) technology-related learning context."

Contributors to this Special Issue

"The articles in this special issue address a variety of motivation theories and models, such as expectancy-value theory, intrinsic motivation, curiosity, self-determination theory, and interest, as the authors explore ways to stimulate and maintain student engagement in learning."

"The authors introduce their article, Curiosity, Interest and Engagement in Technology-Pervasive Learning Environments: A New Research Agenda, with scenarios that demonstrate the differences between how information access in the 20th and 21st centuries influences students' curiosity and their attempts to satisfy that curiosity."

"An article by Amy L. Baylor (Florida State University), entitled The Design of Motivational Agents and Avatars, explores how characteristics of interface agents and avatars can serve as virtual social models and affect student motivation while using Internet-based game and learning systems."

Moxie: Individual Variability in Motivation Intensity [168]

This is a machine-generated summary of:

Curtis, Jessica; Hatvany, Thomas; Barber, Kathryn E.; Burkley, Edward: Moxie: individual variability in motivation intensity [168]

Published in: Current Psychology (2022)

Link to original: https://doi.org/10.1007/s12144-022-03304-7

If you want to cite the papers, please refer to the original.

For technical reasons we could not place the page where the original quote is coming from.

Abstract-Summary
"Motivation intensity differs across individuals."

"We developed a new tool to measure a novel construct we call "moxie" to assess individual differences in motivation intensity."

"Convergent, discriminant, criterion-related, and incremental validity were established by exploring associations between moxie and the Big Five personality traits, as well as related motivational constructs such as grit, self-control, volition and inhibition systems, and procrastination."

"Moxie also predicted intrinsic and extrinsic motivation above other motivational constructs."

"These findings suggest that moxie is an important and distinct motivational construct and offer new insights regarding individual differences in motivational intensity."

Motivational Theories of Effort
"Much of motivation research is concerned with identifying the processes and mechanisms that affect the (1) direction, (2) initiation, (3) persistence, and (4) intensity of this energy (Burkley, [9]; Geen, [10]; Richter and others, [11])."

"The final component motivational energy, intensity, refers to the amount of resources expended to carry out a goal."

"Despite these classic roots, intensity has received less empirical attention than the other four components of the energy process, especially regarding individual differences (Gendolla and others, [12])."

"In terms of modern approaches, research on energy intensity has almost exclusively focused on its situational influences and has fallen under the guidance of Brehm's Motivation Intensity Theory (Brehm & Self, [13]; Brehm and others, [14])."

"When task difficulty is unknown, then the goal's perceived importance or value often determines energy intensity (Richter and others, [11]; Stanek & Richter, [15])."

Moxie and Its Relation to Other Motivational Constructs

"People high in moxie tend to approach tasks with a high level of energy and willingly mobilize and invest resources and effort into achieving their goals."

"On the surface, a number of constructs and measures within the field of motivation may appear to relate to moxie, including conscientiousness, volition and inhibition systems, trait self-control, grit, and trait procrastination."

"Volition and inhibition/self-control are focused on how people direct their motivational energy."

"While moxie captures energy intensity, grit is focused exclusively on the persistence of motivational energy and effort employed over time."

"Although these constructs all play a role in motivation, they are theoretically distinct from moxie because they focus on individual variability in the efficiency (conscientiousness), direction (volition and inhibition, trait self-control), persistence (grit), and initiation (procrastination) of motivational energy, rather than its intensity (moxie)."

Scale Development and Studies Overview

"Studies 1–3 examined the Moxie Scale's factor structure, internal reliability, and test–retest reliability."

"Studies 4–6 investigated the scale's convergent and divergent validity by exploring its relation to other relevant constructs."

"Studies 6 and 7 examined the scale's predictive and incremental validity."

"These studies aimed to establish moxie as a distinct motivational construct, develop a new scale to assess individual differences in moxie, and examine the psychometric properties of this measure."

Study 1

"In Study 1, we conducted an exploratory factor analysis (EFA) to determine if the hypothesized single-factor structure of the Moxie Scale was supported."

"Sample size was determined based on recommendations for factor analysis which suggest 10 participants per item (Everitt, [16]) or a total sample size of at least 100 (Kline, [17]; MacCallum and others, [18])."

"Parallel analysis and scree plots were used to determine the number of factors to retain."

"With consideration of the parallel analysis, factors from the data were retained if their eigenvalues were greater than those obtained from simulating random data (O'Connor, [19])."

Study 2

"Following previous recommendations (Hu & Bentler, [20]; Schreiber and others, [21]), we used the following fit indices to test model fit: Tucker-Lewis-Index (TLI), the comparative fit index (CFI), and the root-mean-square-error of approximation (RMSEA), where good fit is indicated by values greater than or equal to 0.95 for the TLI and CFI and values less than or equal to 0.06 for the RMSEA."

"We also calculated the chi-square and confidence intervals for RMSEA for this CFA."

"This was to be expected, however, as previous work (e.g., Bentler & Bonett, [22]) states that cases with sample sizes larger than 200 often result in significant chi-square statistic ($n = 214$)."

"These fit indices demonstrated excellent fit: TLI = 0.98, CFI = 0.99, and RMSEA = 0.06, 90% Confidence Interval (CI) for RMSEA = 0.04-0.09."

Study 3

"In Study 3, we assessed the Moxie Scale's test–retest reliability to provide support for our assertion that moxie is a stable individual difference variable."

"Participants completed the Moxie Scale using a 1 (not at all true) to 7 (very true) rating scale."

"The Moxie Scale demonstrated excellent internal consistency at Time 1 (Cronbach's $\alpha = 0.93$, 95% CI [0.89,0.96], McDonald's $\omega = 0.94$, 95% CI [0.90,0.97]) and Time 2 (Cronbach's $\alpha = 0.96$, 95% CI [0.93,0.98], McDonald's $\omega = 0.96$, 95% CI [0.94,0.98])."

"This finding provides support for the test–retest reliability of the Moxie Scale."

Study 4

"In Study 4, we investigated the convergent and divergent validity of the Moxie Scale in relation to a common personality measure: the global Big Five personality facets (John, and others, [23]; McCrae & Costa, [24])."

"Of the five traits assessed in this measure, we expected moxie to be most strongly associated with the trait of conscientiousness."

"We expected this relationship with conscientiousness to be moderate in size, given our assertion that moxie is a related but distinct construct."

"Moxie correlated most strongly with conscientiousness ($r = 0.47$) and this correlation was positive and moderate in size."

Study 5

"In Study 5, we examined the convergent and divergent validity of the Moxie Scale by examining its relation to measures associated with the volition and inhibition systems."

"These included measures of promotion and prevention orientation (ELSamen, [25]), and sensitivity to the Behavioral Activation System (BAS) and Behavioral Inhibition System (BIS; Carver & White, [26])."

"Because moxie is concerned with the mobilization of effort toward a goal (and not away from a goal), we expected moxie to be most associated with measures relevant to the volition system (i.e., BAS and promotion subscales)."

"The 10-item General Regulatory Focus Measure-Short (ELSamen, [25]) was utilized to assess promotion ($\alpha = 0.84$) and prevention orientations ($\alpha = 0.78$)."

"Moxie correlated with the volition measures, including the promotion subscale ($r = 0.47$) and the three BAS subscales (r's $= 0.23$ to 0.44)."

"As expected, moxie did not significantly correlate with the inhibition measures."

Study 6

"In Study 6, we examined the convergent and divergent validity of the Moxie Scale by examining its relation to several common measures in the motivational literature, including procrastination, trait self-control, and grit."

"We also examined the predictive and incremental validity of the Moxie Scale by examining its relation to intrinsic and extrinsic motivation."

"Because of the clear relationship between energy mobilization and goal enjoyment (Brehm and others, [14]; Wright & Brehm, [27]), we predicted that moxie would be positively related to intrinsic motivation."

"Moxie may relate only to intrinsic motivation because of the relationship between effort and goal enjoyment (Brehm and others, [14]; Wright & Brehm, [27])."

"Since moxie refers to a high level of motivational intensity regardless of the type of goal, it may be that people high in moxie tend to be more motivated for both intrinsic and extrinsic reasons."

"Regression analyses were utilized to examine the relationship between moxie and intrinsic and extrinsic motivation, controlling for procrastination, grit, and self-control."

Study 7
"In Study 7, we again examined the predictive and incremental validity of the Moxie Scale, this time in regards to goal achievement."

"We predicted that these momentary goal investments would mediate the relationship between moxie and goal achievement."

"As predicted, moxie at Time 1 was a significant predictor of goal achievement at Time 3, after controlling for grit and self-control."

"Moxie at Time 1 predicted greater goal achievement at Time 3, $B = 0.69$, $t(37) = 3.08$, $p = 0.004$, $R^2 = 0.20$, and greater goal investment at Time 2, $B = 0.78$, $t(36) = 3.40$, $p = 0.002$, $R^2 = 0.24$."

"When controlling for goal investments, the direct effect of moxie on achievement became non-significant, $B = 0.19$, $t(34) = 0.77$, $p = 0.44$, suggesting full mediation."

General Discussion
"Across seven studies (each using independent samples), the results provided initial evidence that the Moxie Scale is a reliable, valid, and novel contribution to motivation research."

"Moxie was also found to be correlated but distinct from several other individual difference measures commonly used in motivation research."

"Providing support for the Moxie Scale's criterion-related and incremental validity, Study 6 found that moxie accounted for significant variance in intrinsic and extrinsic motivation, above and beyond that of other individual difference measures related to motivational energy (i.e., procrastination, self-control, grit)."

"The present findings, especially those of Study 7, indicated that moxie predicted achievement within people's self-selected goals, which included a range of different domains such as health, school, finances, and hobbies."

"Upcoming studies could explore if individual variability in moxie also corresponds to differences in cardiovascular reactivity during goal pursuit."

[Section 12]
"Some give it their all, running the entirety of the course and finishing in under 2.5 h. Others take a slow and steady approach, walking the course and concluding their race in 5–6 h. Consequently, marathons illustrate how people differ in the amount of energy they put into achieving a given goal."

"Much of motivation research has been devoted to identifying the factors that influence how much energy or effort people put into a task or goal (e.g., Bargh and others, [28]; Freydefont and others, [29]; Gendolla and others, [30]; Silvestrini & Gendolla, [31])."

"As the marathon example illustrates, the idea that some people put more effort and energy into a task than others is not new."

"No clear, direct individual difference measure exists to globally assess trait variability in energy intensity."

"In appreciation of this fact, the present work sought to develop and validate a measure that assesses individual differences in motivation intensity."

Commitment to Change from Locomotion Motivation During Deliberation [169]

This is a machine-generated summary of:

Scholer, Abigail A.; Tory Higgins, E.: Commitment to change from locomotion motivation during deliberation [169]

Published in: Motivation and Emotion (2011)

Link to original: https://doi.org/10.1007/s11031-011-9239-4

If you want to cite the papers, please refer to the original.

For technical reasons we could not place the page where the original quote is coming from.

Abstract-Summary

"Academic Press, New York, vol 35, pp 293–344, 32), we propose that increased commitment to change can result from increased locomotion motivation in the deliberation phase."

"Three studies provide evidence that increased commitment to change is related to locomotion motivation arising either from a chronic orientation or from a movement-focused deliberation tactic that intensifies that orientation."

"Although locomotion motivation is typically associated with goal pursuit, the current work highlights the impact that locomotion motivation can have on commitment to change in the initial deliberation phase."

Introduction

"The current research explores how an individual's motivational orientation affects the likelihood that people will commit to change in the deliberation phase of self-regulation, particularly when deliberating about difficult, ambivalence-evoking change decisions."

"We propose that: (a) locomotion motivation may play an important role during deliberation when individuals are facing difficult change decisions; and (b) a deliberation approach specifically designed to activate locomotion concerns will be particularly helpful to people who have chronically high locomotion motivation, strengthening commitment to change."

"We have proposed that intensifying locomotion motivation in deliberation may increase commitment to change, regardless of whether it comes from a chronic orientation, a deliberation tactic that promotes locomotion, or through the intensification of an individual's chronic locomotion orientation."

"Unlike locomotion, assessment is unlikely to have a simple relation to commitment to change, whether the motivation arises from one's chronic assessment orientation, a deliberation tactic that promotes assessment, or through the intensification of an individual's assessment orientation."

Study 1
"Participants deliberated about a change decision using either an exhaustive-focused deliberation tactic that serves the assessment system or a movement-focused deliberation tactic that serves the locomotion system."

"Consistent with our predictions, as chronic locomotion increased in the Exhaustive-focused deliberation condition, there was no effect on commitment to change, $\beta = .30$ (CI = −.28, .88), $t(76) = 1.03$, $p = .31$, but as chronic locomotion increased in the Movement-focused deliberation condition, commitment to change significantly increased, $\beta = 1.04$ (CI = .42, 1.66), $t(76) = 3.52$, $p < .001$."

"This suggests that increased locomotion motivation in deliberation, arising both from a chronic orientation and from a deliberative approach which serves that orientation (i.e., regulatory fit), can lead to greater commitment to changing from the current state to a new state."

Study 2
"There was a main effect of deliberation frame, such that the Movement-focused deliberation led to greater commitment ($M = 7.84$, $SD = 1.57$) than the Exhaustive-focused deliberation ($M = 6.32$, $SD = 1.82$), $\beta = .38$ (CI = .14, .62), $t(50) = 3.09$, $p = .003$."

"As detailed below, there was no evidence that the effects on goal commitment of deliberation frame or its interaction with locomotion were mediated by the subjective value of the new state or by self-efficacy beliefs."

"The only significant effect was a marginal effect of deliberation frame, $\beta = .23$ (CI = −.01, .47), $t(70) = 1.93$, $p = .06$, such that individuals in the Movement-focused deliberation reported more positivity about the new state (M = 8.83, CI = 7.33, 8.68) than individuals in the Exhaustive-focused frame (M = 8.18, CI = 5.32, 7.24)."

Study 3

"On the dichotomous choice measure, locomotion motivation significantly predicted switching to the new set of tasks, B = 2.11 (SE = 1.05), odds ratio = 8.26, $p < .05$."

"Locomotion also significantly predicted likelihood of wanting to switch to the new set of tasks on the continuous choice measure, $\beta = .43$ (CI = .10, .75), $t(27) = 2.69$, $p = .01$, whereas assessment significantly predicted wanting to stick with the first set of tasks, $\beta = .48$ (CI = −.80, −.15), $t(27) = 3.02$, $p = .005$."

"They also did not predict any difference in participants' self-efficacy beliefs on task 1 versus task 2: locomotion $\beta = .12$ (CI = −.28, .52), $t(27) = .63$, $p = .53$, assessment $\beta = .00$ (CI = −.40, .40), $t(27) = .003$, $p = .99$; information version $\beta = -.23$ (CI = −1.01, .55), $t(27) = .61$, $p = .55$."

General Discussion

"To suggesting that an individual's motivational orientation may be a predictor of commitment to change, the current work also suggests that locomotion motivation, often associated with the goal pursuit phase, can play an important role in the deliberation phase as well."

"The current studies add to research that has shown that locomotion motivation is related to openness to change more broadly (e.g., Kruglanski and others 33) by demonstrating how it can affect deliberation and goal commitment."

"It will be important to investigate in the future the conditions under which locomotion motivation arising from either a chronic orientation or a situational factor (e.g., deliberation tactic) is sufficient to increase commitment to change versus conditions in which both locomotion factors together are necessary (e.g., regulatory fit)."

Motivated Level of Construal: How Temperature Affects the Construal Level of State-Relevant Stimuli [170]

This is a machine-generated summary of:

Hansen, Jochim; Steinmetz, Janina: Motivated level of construal: How temperature affects the construal level of state-relevant stimuli [170]

Published in: Motivation and Emotion (2019)

Link to original: https://doi.org/10.1007/s11031-018-09750-w

If you want to cite the papers, please refer to the original.

For technical reasons we could not place the page where the original quote is coming from.

Abstract-Summary
"Motivation affects perception, for example when people overestimate the psychological proximity of goal-relevant objects."

"In Experiment 1, people who were warm (versus cold) grouped cold drinks into more and narrower categories, as cold drinks served their goal to cool off."

"In Experiment 2, heat simulations lead people to represent cooling actions (e.g., drinking cold water) in more concrete terms (i.e., as for how instead of why), whereas cold simulations lead people to represent warming actions in more concrete terms."

"In Experiment 3, participants evaluated heaters and ventilators based on concrete (i.e., single reviews) instead of abstract product information (i.e., overall ratings) if the product was motivationally relevant."

Motivated Perception
"Thirsty participants perceived an ambiguous stimulus as more transparent (i.e., water-like) than quenched participants (Changizi and Hall [34])."

"The concept "water" and its associated features—such as transparency—may be more active in thirsty participants than in quenched participants, and may thus be more likely to be applied when interpreting an ambiguous stimulus."

"Of interest for the present research, motivations have been shown to influence distance perception (Balcetis 35; Cole and others 36; Stefanucci e al."

"Money appears closer than financially valueless objects (Cole and Balcetis [37])."

"Thirsty participants estimated a water bottle as being closer than participants who were not thirsty (Balcetis and Dunning [38])."

"These effects can be explained by different motivational directions elicited from the current situation (i.e., the exposure to an affective object, the situational constraints, and the available resources; Balcetis [35])."

Motivated Construal Level

"We argue that concreteness of mental representation (i.e., low-level construal) more broadly allows for preparation of adaptive responding and thereby point out another operation of successful self-regulation."

"This is the case because concrete, low-level construal promotes adaptive responding."

"Identifying a novel path by which concreteness and motivation might be linked, we propose that not only are concrete stimuli more motivating, but people might actively construe any stimulus (i.e., any object or person) on a more concrete level when that stimulus is appropriate to fulfill one's goals and thus is motivationally relevant."

"Adaptive responding to a motivationally relevant object may generally be easier when the object is already construed on a low level."

The Present Research

"The present research tested the idea that motivational states affect the level of construal of stimuli that are relevant to the person in that state."

"To measure level of construal, we used breadth of categorization of motivationally relevant (vs. irrelevant) objects (Exp)."

"1), level of identification of motivationally relevant (vs. irrelevant) actions (Exp)."

"2), and weighting of aggregated versus individualized information when evaluating motivationally relevant (vs. irrelevant) products (Exp)."

Experiment 1: Category Breadth

"Participants were seated in either a warm or a cold room and asked to classify diverse hot or cold beverages into as many categories as they deemed appropriate."

"We expected that participants in a cold (versus warm) room would use more and narrower categories of hot beverages."

"We expected that participants in a warm (versus cold) room would use more and narrower categories of cold beverages."

Method

"They were randomly assigned to a 2 (Room temperature: cold, warm) × 2 (Type of beverages: cold, hot) between-participants design."

"Entering the average of these three effect sizes ($f = 0.26$) and a desired power of $1—\beta = 0.80$, an alpha level of $\alpha = 0.05$ into the program G*Power (Faul and others [39]) with specifying the numerator df = 1 and number of groups = 4 resulted in a total sample size of 119, which means at least 30 participants per experimental condition."

"A thermometer was placed in each room, outside of the visibility of the participants, to constantly monitor the lab temperature."

"New participants were only brought into the lab room when the appropriate temperature was reached."

Results and Discussion

"This main effect was qualified by a significant interaction between room temperature and type of beverages, $F(1,132) = 4.91$, $p = .028$, $\eta_p^2 = 0.04$, 90% CI [0.002; 0.101]."

"Participants in the warm room classified the cold beverages into more (and smaller) categories ($M = 7.31$, $SD = 2.60$) than participants in the cold room ($M = 6.15$, $SD = 2.16$), $F(1,132) = 6.78$, $p = .010$, $\eta_p^2 = 0.05$, 90% CI [0.006; 0.120], for the contrast, indicating that cold beverages were construed on a more concrete, low level when sitting a warm (vs. cold) room."

"In contrast to our hypothesis, room temperature did not affect category breadth of the hot beverages ($M_{\text{warm room}} = 4.21$, $SD = 0.98$; $M_{\text{cold room}} = 4.45$, $SD = 1.15$), $F(1,132) = 0.30$, $p = .586$, $\eta_p^2 = 0.002$, 90% CI [0.000; 0.034]."

Experiment 2: Action Identification

"Simulating being in a warm versus cold environment makes people actually experience physical warmth and cold, respectively similarly to actual experiences, which is why people's preferences and behavior are affected by the simulations (Steinmetz and others 40)."

"Following the format of the behavioral identification form (BIF, Vallacher and Wegner 41) that measures the abstractness level on which actions are identified, we presented participants with two alternative descriptions of the actions (one more concrete and one more abstract) and asked participants to choose which of the two best represented the given action."

"We proposed that the level of action identification of participants primed with the desert movie would be higher (i.e., more abstract) for the warm than for the cold items whereas the level of action identification of participants primed with the winter movie would be higher (i.e., more abstract) for the cold than for the warm items."

Method

"Participants were randomly assigned to a 2 (Movie: desert, winter-scenery) × 3 (Type of actions: cold, hot, neutral) split-plot design with the first factor varied between participants and the second factor varied within participants."

"Participants watched a movie clip that lasted approximately 90 s. For half of the participants, the movie clip depicted a winter landscape with a lot of snow, accompanied by sounds of cold winds."

"Participants were provided with 24 actions (e.g., recycling) and two alternative descriptions of each action."

"8 actions were related to warmth (e.g., sitting at the fire, having a sauna, using a hot-water bottle, running through hot sand), 8 were related to cold (e.g., drinking cold water, walking through cold rain, hiking in the snow, making ice cubes), and 8 items were neutral (e.g., skydiving, making an expensive purchase, recycling, teaching; see Fujita and others 42, Exp."

Results and Discussion

"Depending on the temperature content of the movie, participants chose different numbers of high-level action descriptions of the heat and cold items: Participants who watched the cold movie chose more abstract descriptions for the cold items (M = 6.06, SD = 1.36) than participants who watched the warm movie (M = 5.61, SD = 1.82), whereas participants who watched the warm movie chose more abstract descriptions for the heat items (M = 5.84, SD = 1.60) than participants who watched the cold movie (M = 5.29, SD = 1.76), $F(1,98) = 5.50$, $p = .021$, $\eta_p^2 = 0.05$, 90% CI [0.004; 0.138] for the interaction contrast."

Experiment 3: Weighting of Individualized Versus Aggregated Information

"Low-level, concrete representations enhance the relative influence of product information that is closely tied to a single experience within a specific context compared to broad, decontextualized information which is averaged across multiple individuals and instances (Ledgerwood and others [43])."

"We tested the hypothesis that warmth would cause participants to put more weight on individualized information than on averaged information when the evaluated object could help to reduce the warmth, that is, when participants evaluate ventilators."

"Cold, in contrast, should cause participants to put more weight on individualized information than on averaged information when the evaluated object could help reducing the cold, that is, when participants evaluate heaters."

Method

"Participants were randomly assigned to a 2 (Prime: warm, cold) × 2 (Product: ventilators, heaters) × 2 (Information Type: product favored by average information, product favored by individualized information) mixed design with the second and third factors varied within participants."

"Entering the effect of $f = 0.26$ and a desired power of $1—\beta = 0.80$, an alpha level of $\alpha = 0.05$ into the program G*Power (Faul and others [39]) with specifying the number of groups with 2 and the number of measurements with 4 (since participants rated heaters and ventilators favored by individualized information and heaters and ventilators favored averaged information), with an assumed correlation between the ratings of $r = 0$, resulted in a total sample size of 42, which means at least 21 participants per experimental condition."

"Participants read a story about a person who was waiting for the bus either on a very hot summer day or on a very cold winter day (between participants)."

Results and Discussion

"A 2 (Prime) by 2 (Product) by 2 (Information Type) mixed-design ANOVA on the evaluation revealed a main effect of the product, $F(1,56) = 9.41$, $p = .003$, $\eta_p^2 = 0.14$, 90% CI [0.030; 0.282], indicating that participants preferred ventilators ($M = 4.02$, $SD = 0.83$) over heaters ($M = 3.59$, $SD = 1.07$)."

"When participants were primed with cold, they evaluated the concrete version of the heater (i.e., favored by individualized information) more positively ($M = 5.23$, $SD = 1.51$) than the abstract version of the heater (i.e., favored by aggregate information) ($M = 2.63$, $SD = 1.61$), $F(1,56) = 39.35$, $p < .001$, $\eta_p^2 = 0.41$, 90% CI [0.245; 0.535] for the contrast."

General Discussion

"We argue that experiencing a visceral state, such as cold or warmth, can motivate participants to reduce that state (see also Zhang and Risen [44]), and thus causes concrete construal of actions and objects that help to reduce the state."

"Non-motivational account of the present findings, one could argue that any visceral state per se, such as warmth, would foster an abstract construal of actions and objects which are related to that state (e.g., warm beverages), instead of increasing concrete construal of actions and objects related to the opposite, pursued state (e.g., cold drinks)."

"According to construal level theory, when experiencing warmth, participants should represent actions and objects related to this present state as more concrete (and not as more abstract)."

"The present findings suggest that motivational relevance is a moderator of the general finding that avoidance motivation causes concrete construal."

[Section 14]

"Motivation can influence how people construe their subjective reality."

"Most of this research literature on the "motivated tactician" demonstrates that motivations affect the content of people's construal."

"We extend previous research by showing that motivations can also affect the level of construal, that is, how concretely versus abstractly people represent their reality."

"We suggest that motivations cause people to construe stimuli that are relevant to the current motivation on a more concrete level than stimuli that are not relevant, much like in a cartoon in which a very hungry castaway who floats on a raft through the ocean imagines a yummy steak very concretely with all its visual and gustatory details."

Hate: Evidence for a Motivational Conceptualization [171]

This is a machine-generated summary of:

Rempel, John K.; Burris, Christopher T.; Fathi, Darius: Hate: Evidence for a motivational conceptualization [171]

Published in: Motivation and Emotion (2018)

Link to original: https://doi.org/10.1007/s11031-018-9714-2

If you want to cite the papers, please refer to the original.

For technical reasons we could not place the page where the original quote is coming from.

Abstract-Summary

"A prototype approach showed that statements central to hate focused on the desire for a target to suffer harm (Study 1)."

"Statements including the desire for target harm received higher hate ratings than those in which this desire was absent (Study 2)."

"Hate attributions were higher when harm was a desired outcome rather than unwanted byproduct of goal pursuit (Study 4)."

"Lay attributions of hate consistently link with the perceived desire for a target to experience harm."

The Intersection of Motivation and Emotion

"The emotion system is commonly understood as a network of survival circuits (e.g., LeDoux [45]) designed to provide an organism with extremely rapid, flexible responses that address common evolutionary issues and problems."

"Once the emotion system has been activated by an unexpected change in the environment (e.g., Frijda [46]; Levenson [47]), the organism must identify the personal relevance of the change in terms of opportunities and threats (e.g., Tracy [48])."

"Roseman [49, 50] also suggested that motives proper tend to be more deliberate in nature, whereas emotions tend to have a more impulsive quality."

"Based on Rempel and Burris [51] and Rempel and Sutherland [52], we suggest that hate has a distinctively motivational signature—that is, we predict that whether or not people label an experience as "hate" will primarily hinge on the salience of the desire for a target to experience harm."

"We hypothesize that people will most readily apply the "hate" label when the desire for a target to incur harm is a discernible component of their own or others' experience, irrespective of its emotional content."

The Present Research

"Based on our motivational conceptualization (Rempel and Burris [51]), we expected the desire for harm to be central to the hate prototype."

"In Study 2, we further tested the centrality of motivation in a hate experience by having participants rate statements in which motivational, attitudinal, and emotional elements were either present or absent in a given experience."

"In Study 4, we tested the extent to which a salient desire to harm is necessary and sufficient for increased attributions of hate, even if the goal of harming the target is a means to some other superordinate goal."

"We expected that the clear presence of the desire to harm, whether or not it was the primary goal, would result in higher attributions of hate compared to the condition in which harm was presented as an unfortunate and undesirable byproduct of goal attainment."

Study 1

"A second sample of participants then rated the centrality of 52 distinct statements abstracted from the open-ended definitions."

"The sample that generated the open-ended list of hate definitions consisted of 213 undergraduates (111 men, 102 women) at the University of Waterloo who participated either for payment (approx."

"Participants subsequently rated each of the 52 statements on a scale ranging from 1 (extremely poor example of hate) to 6 (extremely good example of hate)."

"The items rated as most central appeared to be strong, clear motivational statements of desired or intended harm, e.g., "wishing death upon the target," "thought of murder," and "premeditated acts of violence and abuse."""

"When presented with a menu of spontaneously generated features of hate, participants rated the desire for a target to come to harm as significantly more central than even intense emotions."

Study 2
"Although we used a different methodology in our second study, we nevertheless remained focused on showing that harmful intentions (vs. emotional or attitudinal features) are central to people's conception of hate."

"Quotes expressing a clear desire for harm to befall the target were rated as more diagnostic of hate than quotes wherein this expressed desire was absent."

"As in Study 1, the desire for a target to experience harm was seen as more diagnostic of hate than the absence of this desire."

"Being agitated to the point of wanting to smash things, but not wishing harm upon the target, only reached the hate midpoint."

"Harm framed as teaching the target a lesson—and, thus, arguably in the service of the target's ultimate wellbeing—likewise barely reached the hate midpoint."

"Study 2 showed that the expressed desire that a target experience harm is seen as diagnostic of hate."

Study 3
"More participants mentioned harmful intentions when justifying their hate ratings in response to the Nihilistic Hate description (60 of 97, or 62%) compared to the Redress description (15 of 102, or 15%), χ^2 (1) = 47.07, $p < .001$ (CI 0.36, 0.59), $\varphi = .49$."

"There was also a significant main effect of harmful intent, $F(1, 191) = 38.32$, $p < .001$, $\eta_p^2 = .167$, such that hate ratings were higher when harm was mentioned in participants' justifications of their ratings ($M = 6.48$, $SD = 0.92$) versus when it was not ($M = 4.16$, $SD = 1.87$)."

"Participants were also more likely to mention harmful intent spontaneously when justifying their hate ratings in response to Nihilistic Hate versus Redress."

Study 4
"Participants all read six brief vignettes (based on the six subtypes of hate from Rempel and Burris [51]) in which the main character either harms, or expresses a desire to harm, a diverse range of targets in a variety of contexts: (1) a young woman wished that her grandfather would become ill and be hospitalized so that she would have greater freedom; (2) a man wanted to send endangered refugees back to their home country for economic reasons; (3) a young woman who was sexually assaulted at a party reflected on the type of punishment she wanted for her assailant; (4) a man enjoyed humiliating acquaintances with his practical jokes; (5) a man prevented his wife from interacting with other people so that she would not leave him; and (6) a woman criticized and sabotaged her co-workers so that they would not be promoted ahead of her."

"In the first condition, harm took the form of "collateral damage," such that the main characters regarded it as an undesirable but necessary means of achieving their ultimate goals, as in the refugee vignette: Carl knows that refugees from a neighboring country have fled violent situations and moved into his town."

General Discussion

"Given this heterogeneity, we deliberately focused our research on the construal of hate across a broad cross-section of people."

"In all four of our studies, attributions of hate were most clearly and consistently linked to the desire for a target to incur harm."

"The desire to harm was consistently associated with the hate label across an array of diverse stimuli in acceptably large samples from two continents."

"If the experience of hate fundamentally promotes the desire for the other to suffer harm, then to be hated is to be a potential victim, and to hate is to be a potential perpetrator."

"People who feel hated could potentially use the sense of impending peril as a justification for directly or indirectly enacting a "first strike" against the perceived hater in an effort to eliminate the threat (cf. Halevy [53])."

Conclusion

"Hate can be a challenging, risky, and uncomfortable psychological process to study—but all of these deterrents, if anything, underscore the need for hate to be systematically investigated and clearly understood."

"Showing that the motive to harm is central to how people construe hate sets the stage to explore other theoretical and practical issues, such as how hate is triggered and sustained—with a particular focus on the contributing role of negative emotions and attitudes."

"There is also the question of what, if any, specific mechanisms allow hate to dissipate, and whether reduction-focused interventions can facilitate this process."

[Section 9]

"They further suggested that intended harm can function either as an end in itself or as a means to some other end (i.e., as an ultimate or an instrumental goal), with distinct emotions (see below) and attitudes (i.e., negative evaluations of a target) "fueling" the motivational engine and contributing to the distinctive signatures of different types of hate."

"In the decade or so since Rempel and Burris [51], research on hate has remained sparse, but it has yielded some results consistent with a motivational conceptualization in which the core goal is the desire to harm a target."

"Subsequent studies by Halperin and others also suggested that people perceived a close link between hate and behavioral intentions involving the desire for harm."

"In four large Israeli samples, Halperin and others [54] found that hate was more strongly associated with political intolerance of Palestinians than was anger or fear."

You're Too Much for Me: Contagion of Motivation Depends on Perceiver-Model Distance [172]

This is a machine-generated summary of:

Radel, Rémi; Fournier, Marion; de Bressy, Virginie; d'Arripe-Longueville, Fabienne: You're too much for me: Contagion of motivation depends on perceiver-model distance [172]

Published in: Motivation and Emotion (2014)

Link to original: https://doi.org/10.1007/s11031-014-9451-0

If you want to cite the papers, please refer to the original.

For technical reasons we could not place the page where the original quote is coming from.

Abstract-Summary
"Based on assimilation/contrast models in priming research, we hypothesized that motivational contagion should only occur in case of moderate distance between perceiver and model's motivation."

"A first lab-study supported this hypothesis by showing that the effect of the exposure to a model presented as highly intrinsically motivated (compared to a model presented as moderately intrinsically motivated or to a not-presented model) depended on participants' initial intrinsic motivation."

"A second study in a real educational context confirmed our hypothesis by showing that weakly intrinsically motivated students invested more effort during a physical education term when yoked with a peer with a moderate intrinsic motivation than when yoked a highly intrinsically motivated peer."

Introduction
"Since the task of the model was different from the task of the perceiver, it suggests that this motivational contagion effect would rather come from a priming effect than from expectancies about the task."

"As motivational contagion seems to work as a priming influence, we assumed, in line with the models of assimilation and contrast, that the degree of distance between the perceiver's motivation and the model's motivation could moderate the effect of motivational contagion."

"If a learner perceives a slightly more motivated model than her/him, the learner should assimilate this prime as it should trigger a comparison process focusing on similarities."

"While we expected that, for participants with high initial levels of IM, the exposure to the moderately and extremely intrinsically motivated model would both lead to more effort than in the control condition, we predicted that, for those with low initial levels of IM, only the exposure to a moderately intrinsically motivated model would lead to greater effort than in the control condition."

Study 1

"Simple effects tests further revealed that, as predicted, participants with low initial IM score (i.e., at $M - 1SD$) exerted less effort in the task when exposed to the highly motivated model ($M = 3.89$) than those exposed to a moderate IM model ($M = 5.12$), $t(72) = -3.351$, $p = .001$."

"Participants with high initial IM score (i.e., at $M + 1SD$) exerted more effort in the task when exposed to the highly motivated model ($M = 6.28$) than those exposed to a moderate IM model ($M = 5.13$), $t(72) = 3.315$, $p = .001$."

"Tests of simple effects further indicated that participants with low initial IM score (i.e., at $M - 1SD$) tended to perform better in the task when exposed to a moderately motivated model ($M = 5{,}457$ RPM) than those exposed to a highly IM model ($M = 4{,}317$ RPM), $t(72) = -1.702$, $p = .093$."

Study 2

"Each participant was yoked with a model student in the high jump activity for all the remaining sessions on this activity (i.e., six to seven 2-hour sessions depending on the groups of students)."

"In the moderate distance condition, participants were yoked with model students who had moderately distant IM score from their own IM score (the mean distance between participants' and models' IM was 1.38 Likert scale units, $SD = 0.29$ with distance values ranging from 1.03 to 1.98)."

"In the high distance condition, target students were yoked with model students who had very distant IM score from their own IM score (the mean distance between participants' and models' IM was 3.51 Likert scale units, $SD = 0.33$ with distance values ranging from 2.67 to 4.17)."

General Discussion
"Results generally supported the models predicting assimilation and contrast as we showed that a model with a high level of intrinsic motivation only lead to greater effort (in comparison to control participants) for participants who had a high initial level of IM."

"We observed that when weakly intrinsically motivated students were yoked with a highly intrinsically model for the whole physical education term, they reported less effort than weakly intrinsically motivated students yoked with a moderately intrinsically motivated model."

"Concerning the mechanism through which the perceiver-model distance would affect the motivational contagion effect, this is hard to provide a firm response to this question as there are different models explaining the assimilation and contrast effects."

"Since the status has been also identified as a powerful moderator in the domain of emotional contagion (see Hatfield and others [55]), further studies should be conducted to examine how the rapport and the motivational gap together influence the motivational contagion effect."

Consensual and Behavioral Validity of a Measure of Adaptive Individual Differences Dimensions in Human Motivation [173]

This is a machine-generated summary of:

Bernard, Larry C.: Consensual and behavioral validity of a measure of adaptive individual differences dimensions in human motivation [173]

Published in: Motivation and Emotion (2009)

Link to original: https://doi.org/10.1007/s11031-009-9131-7

If you want to cite the papers, please refer to the original.

For technical reasons we could not place the page where the original quote is coming from.

Abstract-Summary
"These results support the utility of the AIM-Q as a new measure of motives in purposeful human behavior."

Introduction
"The present study is part of a research program that combines evolutionary and individual differences psychology in an attempt to understand some of the dimensions of human motivation."

"A few evolutionary psychologists (e.g., Buss [56, 57]; Figueredo and others [58]; MacDonald [59]) have taken the position that individual differences in psychological adaptations—specifically, dispositions and personality traits—could have evolved."

"The personality research form (PRF; Jackson [60]) is another attempt to identify a comprehensive set of independent individual differences dimensions in motivation."

"Given this caveat, the MTMM approach did demonstrate adequate convergent and discriminant validities within the 15 AIM-Q scales, suggesting the motives may be successfully measured as individual differences dimensions (Bernard and others [61])."

"If the AIM2 scales' reliability and validity were supported, they could offer a new option for motivational, personality, and evolutionary psychologists, and behavioral geneticists who want to incorporate a multidimensional measure of individual differences based on evolutionary psychology theory into their research."

Method
"Self, friend, and relative samples were directed to different internet addresses where the wording of AIM2 and behavior report items matched the sample."

"Sex of participant was entered in this step to provide an indication of whether it was statistically significantly related to the criterion and to control for any potential effects when AIM2 scale scores were entered."

"Step 2 (to test explicit hypotheses of convergent validity)—To simplify the multiple regression equations, partial correlation coefficients (with Sex of participant controlled) were calculated between each AIM2 scale and all behavioral variables."

"Only AIM2 scales hypothesized to be related to each criterion behavior and that had statistically significant partial correlations with it were entered in this step."

"Step 3 (to test implicit hypotheses of discriminant validity)—In certain cases, some AIM2 scales not hypothesized to be related to some of the criterion behaviors did have statistically significant partial correlations with them."

Results
"Participants in the friend and relative samples were asked how many years they were acquainted with the person they were rating."

"The mean values on Degree of Acquaintance with the person in the self sample across different circumstances suggested that both friends and relatives knew their target least well in a work setting, while the friend sample reported knowing their target best in a social setting, and the relative sample reported knowing their target best in a home setting."

"Participants in the friend and relative samples were also asked to indicate their relationship to the person they were rating."

"Relevant was the change in multiple R from Step 2 to 3, which indicated whether the AIM2 predictors that were not hypothesized to be related to the behavior criterion, but were found to have statistically significant partial correlations with it, accounted for a statistically significant portion of additional variance in the behavior after the influence of the hypothesized AIM2 predictors."

Discussion

"These analyses supported almost all of the 36 hypothesized relationships between AIM2 scales the behavior reports included in this study."

"These results were obtained with the effect of Sex of participant controlled, so that they indicate relationships between AIM2 motive dimensions and behavior without regard to participant Sex."

"The AIM2 motives can be viewed as accounting for some of the variance in these reported behaviors and the results do provide evidence of the AIM2 scales' validity through support of hypothesized relationships to meaningful behaviors."

"While considerable work remains to be done, the present results suggest that the AIM2 provides a relatively reliable, relatively valid, standard method of researching multiple individual differences dimensions in motivation."

"With most of the AIM2 scales demonstrating adequate reliability and validity, and able to predict self-reported behavior, the next step is to proceed with studies of the evolutionary theory of motivation on which the motives are based."

The Impact of Method, Motivation, and Empathy on Diversity Training Effectiveness [174]

This is a machine-generated summary of:

Lindsey, Alex; King, Eden; Hebl, Michelle; Levine, Noah: The Impact of Method, Motivation, and Empathy on Diversity Training Effectiveness [174]

Published in: Journal of Business and Psychology (2014)

Link to original: https://doi.org/10.1007/s10869-014-9384-3

If you want to cite the papers, please refer to the original.
For technical reasons we could not place the page where the original quote is coming from.

Abstract-Summary

"The purpose of this paper is to examine method, motivation, and individual difference variables as they impact the effectiveness of a diversity training program in a field setting."

"We conducted a longitudinal field experiment in which participants (N = 118) were randomly assigned to participate in one of three diversity training methods (perspective taking vs. goal setting vs. stereotype discrediting)."

"Results suggest the effectiveness of diversity training can be enhanced by increasing motivation in carefully framed and designed programs."

"Training effects were mediated by internal motivation to respond without prejudice, and the model was moderated by trainee empathy."

"This study indicates that taking the perspective of others may have a lasting positive effect on diversity-related outcomes by increasing individuals' internal motivation to respond without prejudice."

"We identify an important mediator (internal motivation to respond without prejudice) and boundary condition (trainee empathy) for examining diversity training effectiveness."

Introduction

"The aim of this paper will be to investigate the often asked but seldom answered question: How can diversity training programs be designed effectively?"

"We investigate whether certain diversity training methods are more effective than others."

"We will compare diversity-related perspective taking, goal setting, and stereotype discrediting methods as they relate to diversity training effectiveness."

"We will investigate internal motivation to respond without prejudice as a potential mediator that explains the effect of diversity training methods on diversity-related outcomes."

"Decidedly few studies have been able to identify mediators that can explain the relative effectiveness of a given diversity training exercise."

"In conducting this study, we hope to unveil not only the relative effectiveness of diversity training methods, but also why and for whom they are effective when they do indeed work as intended."

Comparison of Diversity Training Methods

"Of the aforementioned studies and theory surrounding goal setting, we expect diversity-related goal setting to be a relatively effective method (when compared to other training methods) associated with favorable diversity-related attitudes and behaviors."

"One might conclude that actively discrediting stereotypes via diversity training may be an effective method of reducing prejudiced attitudes against stigmatized groups."

"Although it is possible that stereotype discrediting may lead to a decrease in prejudice in the short-term due to engagement in active argument, the fact that this strategy relies on prejudice suppression suggests that this method will not be as effective as time passes or when compared to perspective taking and goal setting."

"Training methods involving perspective taking and goal setting will be associated with more favorable diversity-related attitudes and behaviors when compared to the stereotype discrediting method."

How Diversity Training Methods Impact Outcomes

"An individual who sets a goal to respond with less prejudice toward a given stigmatized group should theoretically be internally motivated to respond without prejudice to meet their own goal, which should then be associated with favorable diversity-related attitudes and behaviors."

"Previous work has also shown that internal motivation to respond without prejudice is related to more positive diversity-related attitudes (Ratcliff and others [62])."

"This internal motivation to respond without prejudice should be associated with favorable diversity-related attitudes and behaviors by motivating people to respond to stigmatized individuals in an egalitarian manner."

"Based on the rationale above, we anticipate that internal motivation to respond without prejudice will serve as a partial mediator between successful diversity training methods and diversity-related outcomes."

"More formally, we hypothesize the following: Internal motivation to respond without prejudice will partially mediate the relationships between diversity training methods and diversity-related attitudes and behaviors."

Individual Differences in Responsiveness to Diversity Training

"The final question this paper will address is whether individual differences may alter the effectiveness of the different diversity training methods discussed above."

"Individuals who are low in empathy may need to be prompted by diversity training to enhance understanding and awareness of the experiences of diverse populations."

"These individuals should show the previously hypothesized mediation effect in that they would need to have their internal motivation to respond without prejudice promoted by diversity training before this motivation could show a beneficial association with diversity-related attitudes and behaviors."

"We believe individuals who are low in dispositional empathy stand to benefit the most from diversity training methods aimed at improving their diversity-related attitudes and behaviors through their influence on internal motivation to respond without prejudice."

"We test this model using diversity training sessions focused on African Americans and LGB individuals in order to examine any differences and/or crossover effects."

Method

"To assess the generalizability of effects across targets of prejudice, the marginalized group which was the target of the diversity training (African Americans vs. LGB individuals) varied across participants."

"We used this same scale to measure supportive behaviors exhibited toward African Americans by adapting the items such that the focus of each item was on African Americans rather than LGB individuals (sample item: "Used derogatory terms to refer to African Americans"; reverse coded; $\alpha = 0.67$)."

"We used this same scale to measure internal motivation to respond without prejudice toward LGB individuals by adapting the items such that the focus of each item was LGB individuals rather than ethnic minorities (sample item: "Being nonprejudiced toward gays and lesbians is important to my self concept"; $\alpha = 0.82$)."

Results

"The indirect effect (0.13) was also significant for perspective taking in predicting LGB supportive behavior [0.01, 0.31], but non-significant for all other outcomes and training methods (note that a confidence interval excluding zero is indicative of a significant mediation effect)."

"The index of moderated mediation was significant for this model [0.04, 0.80].When running this model with LGB supportive behavior as our dependent variable, again the indirect effect was significant for individuals low in empathy [0.03, 0.58] but not for individuals high in empathy [−0.35, 0.10]."

"When running this model with African American supportive behaviors as our dependent variable, the indirect effect was yet again significant for individuals low in empathy [0.03, 0.47], but not for individuals high in empathy [−0.58, 0.02]."

Discussion

"Speaking, this study indicates that taking the perspective of others may have a lasting positive effect on diversity-related outcomes, regardless of the focus of training, when compared to setting diversity-related goals or discrediting commonly held stereotypes about stigmatized groups."

"Our second hypothesis was partially supported in that internal motivation to respond without prejudice served as a partial mediator between the perspective taking training method and diversity-related attitudes and behaviors toward LGB individuals."

"Future work should seek to replicate these findings while also investigating other individual difference characteristics that may impact the effectiveness of diversity training programs."

"Although this makes the effects that we did discover even more impressive, we would encourage future researchers to continue testing the effectiveness of the perspective taking and goal setting diversity training exercises with larger samples."

Conclusion

"The current study provides novel evidence regarding the relative effectiveness of three diversity training methods on diversity-related attitudes and behaviors."

"While the perspective taking method was most successful overall, we showed that diversity trainers and researchers may need to consider the empathy levels of the individuals in their sample before deciding which training method to use."

"Appealing to individuals who are low in empathy may be a fruitful opportunity for future research and practice."

Beyond Perspective-Taking: Mind-Reading Motivation [175]

This is a machine-generated summary of:

Carpenter, Jordan M.; Green, Melanie C.; Vacharkulksemsuk, Tanya: Beyond perspective-taking: Mind-reading motivation [175]

Published in: Motivation and Emotion (2016)

Link to original: https://doi.org/10.1007/s11031-016-9544-z

If you want to cite the papers, please refer to the original.

For technical reasons we could not place the page where the original quote is coming from.

Abstract-Summary

"Mind-reading motivation (MRM) is an individual difference in individuals' willingness to effortfully engage with other people's perspectives and mental states, a tendency which has consequences for processes ranging from persuasion to teamwork."

"Study 1 demonstrated that levels of mind reading motivation are stable over time and distinct from mind-reading ability."

"Study 3 demonstrated effects of MRM on persuasion: matching the source (salient mind versus not) to individuals' levels of MRM increased elaboration."

"Study 4 demonstrated that MRM directly increases individuals' perceptions of co-leadership during a dyadic task, in turn affecting joint performance as a pair."

Introduction

"We propose a new construct which could explain this difference: mind-reading motivation (MRM), an individual difference in the extent to which a person is willing to effortfully engage in understanding the perspectives and mental states of other people."

"MRM describes a general orientation towards others: it should predict not only the depth of people's attempted perspective-taking, but also the breadth of situations in which they might be inclined to engage with other people's mental states (i.e., even when another person is not physically present; Study 3)."

"People high in MRM may have a chronic goal of figuring out the perspectives of others for its own sake, and so they would be expected to engage in recreational mind-reading and to actively ponder other people's mental-state information even outside an immediate social interaction."

Mind-Reading Motivation and Social Behavior

"An inherent aspect of MRM is its influence on interpersonal interactions: a dispositional curiosity about others' mental states should have an enduring influence on a person's social behaviors."

"In addition to its effects in personal relationships, MRM should predict a general attunement toward mental information even outside face-to-face interactions."

"Individuals high in MRM should spontaneously take any opportunity to engage with information about others' mental states even when others are not physically present; meanwhile, people low in MRM should be less likely to engage with mental-state information."

"These benefits do not necessarily depend on accurate perspective-taking; rather, the mere fact of focusing one's attention on the perspective of the other may signal interest in the other person, which may contribute to positive relationship feelings."

"Perspective-taking encourages pro-relationship behaviors across a variety of circumstances, and individuals high in MRM should spontaneously attempt to take the perspective of others."

Measuring Mind Reading Motivation

"To explore the factor structure of the scale, we performed maximum likelihood exploratory factor analysis on MRM responses in a larger sample (N = 462)."

"Although differentiating between aspects of MRM may be useful in some contexts (e.g., Carpenter and others [63]), in the studies reported here, the three factors/subscales generally did not differentially predict relevant outcomes."

"The correlations between the factors, the scree plot pattern, and the good Cronbach's alphas for the scale as a whole across studies provided further support for a single-scale approach."

"We conducted four studies to examine the effects of MRM across a variety of interpersonal contexts."

"Study 1 explores the relationship between MRM and perspective-taking ability, as well as examining the stability of MRM over time."

"Study 3 examines whether MRM affects how people respond to persuasive messages that are explicitly connected to other minds."

Study 1

"At Time 1, participants completed the MRM scale, the Interpersonal Perception Task (the IPT-15; Archer and Costanzo [64]), the Interpersonal Reactivity Index (Davis [65]) and demographics."

"We predicted that participants' MRM scores would be highly correlated across the two waves, indicating reliability across time."

"There was no difference between those who completed both surveys and those who dropped out on Time 1 MRM, $t(266) = -0.28$, $p = .78$ or on the IPT-15, $t(257) = -.80$, $p = .42$."

"When MRM, Empathic Concern, and Perspective Taking were all entered into a regression predicting IPT performance, $R^2 = .13$, $F(3, 230) = 11.60$, $p < .01$, MRM's impact remained significant, $b = .19$, $p < .05$, as did the effect of empathic concern, $b = .19$, $p < .05$."

"Study 1 showed that individuals' Motivation for Mind-Reading was stable across time and provided evidence for the distinction between MRM and the separate concept mind-reading ability."

Study 2

"As predicted, Personal Approach scores were significantly predicted by MRM, $r(94) = .36$, $p < .01$."

"When MRM, perspective-taking, empathic concern, and the Big Five were all used to simultaneously predict Personal Approach score ($R^2 = .15$, $F(8, 82) = 3.02$, $p < .01$), only MRM emerged as a significant predictor ($\beta = .40$, $p < .01$)."

"Study 2 demonstrated the effects of motivation to mind-read within personal relationships: participants' MRM was positively related to how deeply they discussed a close other's mental state when spontaneously describing that person."

"Personal Approach scores were not significantly correlated with either perspective-taking or empathic concern above and beyond MRM, indicating that the motivational aspect of MRM allows it to demonstrate unique predictive effects compared to extant empathy instruments."

"MRM predicted Personal Approach score above and beyond the effect of openness to experience and agreeableness, indicating that MRM's predictive effect is distinct from those of general trait curiosity or a trait desire to please others."

Study 3

"Using multiple regression, we tested the effect of mental state, argument quality, and individual MRM on product liking, $R^2 = .31$, $F(9, 131) = 6.65$, $p < .01$."

"A significant two-way interaction between MRM and argument quality emerged in the no mental state condition, $b = .89$, $p < .05$."

"Individuals high in MRM elaborated more on the arguments (that is, showed a stronger differentiation between strong and weak arguments) in the mental state condition, whereas individuals low in MRM showed greater elaboration in the no mental-state condition."

"We then tested the effects of the mental state manipulation, argument quality, and MRM on intention to buy the product, $R^2 = .29$, $F(9, 131) = 6.17$, $p < .01$."

"There were no significant effects in the weak argument condition, but in the strong argument condition, there was a significant MRM by mental state interaction, $b = -1.16$, $p < .01$."

Study 4

"Because this orientation may be tacitly communicated to partners, we also predicted that individuals who work closely with another person who is high in MRM would be more likely to perceive co-leadership."

"To examine task performance across pairs, we averaged MRM and co-leadership scores to the pair-level in our model of task performance."

"We then tested the hypothesis that pairs with higher MRM report higher perceived co-leadership, which in turn predicts faster time completing the task."

"Consistent with our hypotheses, MRM predicted lower scores of self-reported leadership on an interdependent task wherein both participants had the potential to emerge as leader."

"There was no significant association between one's MRM and perception that their partner led the task; rather, MRM predicted higher scores of mutual leadership."

"Working with a partner high in MRM was associated with participants attributing lower levels of leadership to their partner and a greater sense of both partners leading together."

General Discussion

"In Study 4, we also demonstrated that people high in trait MRM were less likely to spontaneously adopt positions of power in ambiguous social situations."

"The four studies provided evidence that MRM is a general cognitive motivational tendency with consequences for a range of social behaviors, and thus is useful to study as an important individual difference."

"The current studies did not directly examine MRM's effects on egocentrism, but some of the current results combined with those from other lines of work imply a connection: in Study 2, MRM was positively correlated with participants' willingness to engage in effortful thought about hypothetical people's perspectives, implying less reliance on automatic assumptions."

"Even if high MRM individuals' attempts at mind-reading are not always accurate, the mere fact of developing sophisticated understandings of a partner's perspectives may be taken as an indication of caring about the partner, and thus may lead to positive relationship outcomes."

Conclusion

"Mind-reading has implications for a variety of cognitive and interpersonal outcomes, and the current research highlights the key role of individual motivation in this process."

"Motivation for mind-reading is an important and useful construct that illuminates the ways mind-reading may occur in real life, separating those who seek out and enjoy theorizing about other people's mental states from those who find the process annoying, unimportant, or dull."

Motivation in Behavior Analysis: A Critique [176]

This is a machine-generated summary of:

Hayes, Linda J.; Fryling, Mitch J.: Motivation in Behavior Analysis: A Critique [176]

Published in: The Psychological Record (2014)

Link to original: https://doi.org/10.1007/s40732-014-0025-z

If you want to cite the papers, please refer to the original.

For technical reasons we could not place the page where the original quote is coming from.

Abstract-Summary

"We consider the concept of motivation in behavior analysis from a unique, radical monistic perspective derived from interbehaviorism and interbehavioral psychology."

"We describe an alternative means by which the issues typically addressed under the concept of motivation might be approached and reconciled with our monistic, interbehavioral perspective."

Radical Monism

"The subject matter is then articulated as a single event, continuously ongoing in the present moment, wherein space and time may be seen to be illusory, causality is superfluous, and reality is gratuitous."

"Radical monism is radical enough to deny the legitimacy of its own description: Radical monism takes as its task the conceptualization of an event in the moment, assuming this event to constitute the all, while, at the same time, it recognizes that the conceptualization achieved is something other than the event conceptualized, thereby violating its own premise of "the event in this moment" being the all."

"While our monistic position is an extension of interbehaviorism, we have also found interbehaviorism and interbehavioral psychology to be the only way to continue speaking about the world in a way that is consistent with radical monism."

Interbehaviorism and Interbehavioral Psychology
"Interbehaviorists conceptualize the subject matter of psychology as an integrated field event, whereby the participating factors are considered interdependent."

"There is no actual distinction between the factors participating in the psychological event; rather, the participating factors are constructs developed for the purposes of analyzing the psychological event."

"From the perspective of an interbehaviorist, setting factors might have their primary impact on (a) stimulus objects, (b) the reacting individual, or (c) the entire interaction (see Kantor and Smith [66], p. 47), and participate in psychological events as all participants do, by altering the entire event."

"The presence or absence of specific setting factors alters the entire psychological event, whereby stimulus and response functions are impacted, various interbehavioral histories are made present, and more."

"That we have articulated the relevant features of interbehaviorism and interbehavioral psychology, let us continue to discuss events from our radical monistic, interbehavioral perspective."

Motivation
"As Michael [67] pointed out, the topic of motivation has historically received relatively little attention in the behavior-analytic literature, and what has been discussed in this context has been articulated in categorical concepts appropriate to sciences other than psychology, namely physiology and ecology (p. 192)."

"Michael cites Skinner's recognition of the need for a motivation-like concept in his identification of deprivation and satiation as something other than stimulus variables, and in his treatment of these concepts and aversive stimulation in the analysis of verbal behavior in particular (pp."

"While the MO refinement of the EO concept has been welcomed and certainly extends upon the multiple ways in which motivational variables impact behavior (specifically, by distinguishing between establishing operations [EOs] and abolishing operations [AOs] as types of MOs), our analysis pertains to features of the EO concept that are unchanged by the MO refinement."

Establishing Operations
"Despite bearing responsibility for the occurrence of a type of response at a particular point in time, as does a discriminative stimulus, an EO is not thereby taken to be a discriminative stimulus, the distinction between them having to do with certain dynamic characteristics of a forthcoming event."

"An EO, by contrast, is responsible for the occurrence of a response at a particular time due to an increase in the reinforcing effectiveness of a forthcoming event, should it take place (Michael [68])."

"An EO has two momentary effects: First, it increases the reinforcing effectiveness of future events, to be observed when and if they occur (called the reinforcer establishing effect); second, it alters the frequency of occurrence of behaviors previously consequated by those events, observed in the present situation (called the evocative effect; Michael [67], p. 192)."

Eliminating the Reinforcer-Establishing Function

"Recall Michael's [67] conceptualization of the evocative effect, the altering of "the frequency of occurrence of that part of the organism's repertoire relevant to those events as consequences" (p. 192)."

"This evocative effect is further understood as the result of "1) a direct effect of the establishing operation on such behavior; 2) an indirect effect of an increase in the evocative effect of all discriminative stimuli for behavior that has previously been followed by food; and 3) an increase in the frequency of behavior that has been followed by conditioned reinforcers whose effectiveness depends on food deprivation" (Michael 67, pp."

"We will concentrate on the other two means by which the evocative effect is assumed to take place, namely, as a direct effect on behavior and indirectly by increasing the evocative effect of discriminative stimuli for behavior."

Concerns About the Direct Effect

"It is not clear what it means to suggest that an event, such as food deprivation, which is not regarded as a stimulus event, has a direct effect on behavior."

"While the case has sometimes been made that operations such as food deprivation and aversive stimulation increase the "general activity" of organisms (e.g., Finger [69]), this is not a common view among behavior analysts."

"Little is said about possible direct effects, such as "general activity" in behavior-analytic circles and, at least to our way of thinking, for good reason."

"Do we not regard specific actions as standing in relation to specific stimuli from a behavior-analytic perspective?"

"Given this, it is not clear why we would argue that EOs alter the frequency of behavior by direct action."

Salvaging the Indirect Effect

"When food deprivation is present, the discriminative effect of the lever on pressing responses may be increased."

"It is also very similar to Kantor's [70, 71]; Kantor and Smith [66] views as to the nature and participation of setting factors in event fields."

"From Kantor's standpoint, setting factors of an organismic sort, such as food deprivation, as well as those having more of an environmental character participate in psychological events, and thus are interrelated with particular functions of stimulating and responding at specific points in time."

"It may be worth elaborating on the conceptual differences between the traditional behavior-analytic approach to motivation (i.e., the EO and related constructs) and Kantor's conceptualization of the setting factor."

"From the perspective of the psychological event, setting factors are participants in this multifactored field and therefore alter the entire event of interest, including both stimulus and response functions."

"Setting factors operate as constituents of the present event."

Summary

"Michael [67] has articulated two effects of EOs: a reinforcer-establishing effect and an evocative effect."

"We have argued that the effect of the EO be limited to the evocative effect for this reason."

"Upon closer scrutiny of the evocative effect, however, we have further suggested that only one of the means by which this effect of an EO is articulated from within the general framework of a behavior-analytic account, namely the indirect means of increasing the effectiveness of discriminative stimuli, is plausible."

"We have salvaged this indirect effect and rejected the account of direct effects of EOs."

Distinguishing Motivated From Nonmotivated Events

"We think, in fact, that there are such factors and that they involve implicit actions with respect to the consequences of behavior, occurring by way of the past consequences of behavior being present substitutively in currently available stimulus objects."

"Central to the distinction between motivated and nonmotivated events is the interbehavioral construct of stimulus substitution."

"As we have mentioned, motivated events involve responding to "future consequences" of behavior in the present moment, and this necessarily involves stimulus substitution because future consequences of behavior do not actually exist."

"Rather, from the perspective of the event, motivated events appear to constitute specific setting factors, substitute stimulus functions, and implicit responses with respect to those stimulus functions."

"The difference between motivated and nonmotivated behavior may then be understood as the degree to which these substitutional or implicit functions of stimuli are operating in event fields, and this, in turn, may be massively influenced by verbal activity and its stimulus products."

Return to Radical Monism

"In behavior analysis, the principal causal variable is reinforcement, and the category of reinforcement has undergone some changes since it was first introduced."

"What continues to exist or occur needs no explanation, and for this reason reinforcement interpreted as selection may be said to be simply a descriptive, as opposed to an explanatory, concept."

"The attraction to causality is common in behavior analysis, and the EO construct is by no means alone in this regard (see Hayes and others [72])."

"When it is held that there is only one event—one event with no parts—as claimed by radical monism, there is no need for a concept of causality."

"The concept of EOs as setting factors need not imply causality."

"It was our goal to consider the concept of motivation, especially as it is conceptualized in behavior analysis, from our radical monistic, interbehavioral perspective."

[Section 11]

"Most philosophies in psychology attempt to deal with the concept of motivation."

"Because of several key scholarly works on the topic (e.g., see Klatt and Morris [73]; Michael [68, 67]), behavior analysts now regularly consider the concept of motivation."

"We critique the concept of motivation, and in particular the concept of establishing operations in behavior analysis, from a radical monistic perspective, largely derived from interbehavioral thinking."

"We will begin by describing our radical monistic perspective, including its relationship to interbehaviorism and interbehavioral psychology, before considering the topic of motivation more specifically."

Emotion Regulation Strategy Selection in Daily Life: The Role of Social Context and Goals [177]

This is a machine-generated summary of:

English, Tammy; Lee, Ihno A.; John, Oliver P.; Gross, James J.: Emotion regulation strategy selection in daily life: The role of social context and goals [177]

Published in: Motivation and Emotion (2016)

Link to original: https://doi.org/10.1007/s11031-016-9597-z

If you want to cite the papers, please refer to the original.

For technical reasons we could not place the page where the original quote is coming from.

Abstract-Summary
"Suppression was particularly tied to social features of context: it was used more when others were present, especially non-close partners, and when people had instrumental goals, especially more interpersonal ones (e.g., avoid conflict)."

"Distraction and reappraisal were used more when regulating for hedonic reasons (e.g., to feel better), but these strategies were also linked to certain instrumental goals (e.g., getting work done)."

"When contra-hedonic regulation occurred, it primarily took the form of dampening positive emotion during high points."

"These findings highlight the social nature of emotion regulation and underscore the importance of examining regulation in both positive and negative contexts."

Introduction
"We test the idea that social aspects of the context (i.e., who is present) and situation-specific regulation goals play an important role in predicting the emotion regulation strategies that people deploy in daily life."

"Most experimental research on emotion regulation has instructed people to use a certain strategy in the lab in order to determine the effects of the strategy on various outcomes, such as emotional experience and expression, physiology, cognition, and social interaction quality (e.g., Butler and others [74]; Gross [75], Richards and Gross [76])."

"We tested the role of two factors in predicting strategy use: (1) The social context (i.e., who is present), and (2) the emotion regulation goals being pursed (i.e., hedonic or instrumental goals, pro-hedonic or contra-hedonic regulation)."

Method
"Each day participants were asked to describe the most positive event (i.e., high point) and most negative event (i.e., low point) of the day."

"Participants indicated the reason(s) for regulating their emotions from a list that included instrumental goals ("To avoid conflict with others," "To keep up appearances," "To make others feel better/happy," "To get work done,") and a purely hedonic goal ("To change mood")."

"To examine the effects of social context and emotion regulation goals on strategy use, we conducted a series of MLM analyses (see Hox 77; Snijders and Bosker 78, for details regarding general MLM specification) on each of the three strategies (suppression, reappraisal, distraction)."

"For emotion regulation goals, dummy variables were used to test effects for pro-hedonic/contra-hedonic, any instrumental goal, and each specific goal (e.g., avoid conflict/not)."

"Analyses of regulation targets, goals, and strategies were based on 753 events (286 high points, 467 low points) where participants reported regulating their emotions."

Results
"Unlike suppression, reappraisal was used more when hedonic goals were endorsed ($\gamma = 1.594$, SE = .228, $p < .001$, 95% CI [1.147, 2.041]; semi-partial $R^2 = 0.07$), and this effect was weaker in low points ($\gamma = 1.027$, SE = .184, $p < .001$, 95% CI [0.665, 1.390]; semi-partial $R^2 = 0.07$) than high points ($\gamma = 1.615$, SE = .235, $p < .001$, 95% CI [1.151, 2.078]; semi-partial $R^2 = 0.15$; event valence x hedonic goal: $\gamma = -0.622$, SE = .285, $p < .05$, 95% CI [−1.181, −0.063]; semi-partial $R^2 = 0.01$)."

Discussion
"This finding suggests that social features of the context can play an important role in determining which strategy people use to regulate their emotions in daily life."

"Suppression was more likely to be used when individuals were regulating for instrumental reasons or engaging in contra-hedonic regulation (i.e., dampening positive emotion), whereas antecedent-focused strategies were used more when regulating simply for the sake of changing ones mood or engaging in pro-hedonic regulation (i.e., elevating positive emotion or dampening negative emotion)."

"Detecting the relatively rare instances of contra-hedonic regulation (e.g., trying to feel more anger or less joy) might require a different event sampling technique, such as frequent experience sampling or explicitly asking participants whether they tried to increase negative or decrease positive emotion at any point in the day."

Why Some Choose the Vegetarian Option: Are All Ethical Motivations the Same? [178]

This is a machine-generated summary of:

Rosenfeld, Daniel L.: Why some choose the vegetarian option: Are all ethical motivations the same? [178]

Published in: Motivation and Emotion (2018)

Link to original: https://doi.org/10.1007/s11031-018-9747-6

If you want to cite the papers, please refer to the original.

For technical reasons we could not place the page where the original quote is coming from.

Abstract-Summary
"Adopting an ethical-health dichotomy in studying dietary motivation, however, may overlook meaningful variance between vegetarians motivated by different types of ethical concerns—namely, those related to animals and the environment."

"Through two preregistered studies, I compared dietary goal orientations, disgust toward meat, and dietary adherence between vegetarians motivated by animals, health, and the environment."

"In Study 1 ($N = 361$), I found that environmentally motivated vegetarians resembled animal-motivated vegetarians in construing their diets as a means of achieving more prosocial and moral, but less personal, goals than did health-motivated vegetarians."

"In Study 2 ($N = 562$), however, I found that environmentally motivated vegetarians resembled health-motivated vegetarians in reporting less disgust toward meat and lower dietary adherence than did animal-motivated vegetarians."

Prosocial, personal, and moral goal orientations
"Rooted in the Unified Model of Vegetarian Identity (Rosenfeld and Burrow [79]), Rosenfeld and Burrow's [80] dietarian identity framework posits that, when it comes to eschewing animal products, people construe their dietary patterns as a means of achieving prosocial, personal, and moral goals to varying extents."

"Whereas prosocial goals embody aims of following one's dietary pattern in order to benefit something beyond oneself, personal goals constitute one's aims to improve oneself in some way."

"Reviewing qualitative investigations into vegetarian motivation (e.g., Fox and Ward [81]; Greenebaum [82]), Rosenfeld and Burrow [79, 83] have posited that animal-motivated and environmentally motivated vegetarians are likely to construe their food choices as prosocial, whereas health-motivated vegetarians are likely to draw upon a personal aim."

"The current study is needed to clarify how people construe specific animal, health, and environmental motivations as related to these three goal orientations."

Vegetarian motivation and dietary adherence

"Having an ethical aim thus might promote greater dietary adherence, as Rozin and others [84] found that morally motivated, but not health-motivated, vegetarians construe meat avoidance as an "ought"—that is, a moral obligation."

"Variations in the extents to which ethically motivated and health-motivated vegetarians feel disgusted by meat may explain differences in their levels of dietary strictness."

"Although disgust should predict affective and behavioral facets of dietary adherence among both animal-motivated and environmentally motivated vegetarians, I nevertheless aimed to examine both dietary strictness and imagined emotional upset from eating meat in order to test for such a possible divergence in behavioral and affective components of adherence between these groups."

"Rozin and others [84] speculate that unintended dietary violations would disturb ethically motivated vegetarians more than they would health-motivated vegetarians because ethically motivated vegetarians are more disgusted by meat."

The current research: aims and hypotheses

"In Study 1, I investigated the extents to which vegetarians with animal, health, and environmental motivations differ in their dietary goal orientations."

"I hypothesized that both animal-motivated and environmentally motivated vegetarians would have higher prosocial and moral goal orientations than would health-motivated vegetarians and that health-motivated vegetarians would have a higher personal goal orientation than would either animal-motivated or environmentally motivated vegetarians."

"In the current studies, I set to advance the current literature on vegetarianism and dietary motivation through implementing three procedures: first, by distinguishing animal and environmental motivations from one another, as opposed to grouping them together into a single category of ethical motivation; second, by documenting the relationship between animal, health, and environmental motivations and the dietarian identity framework's prosocial, personal, and moral goal orientations; and third, by examining the mediating role of meat disgust in the link between vegetarian motivation and dietary adherence."

Study 1

"Of these participants, 403 indicated that their main reason for being a vegetarian was for animals, health, or the environment and were retained in this study."

"Of these participants, 140 indicated that their main reason for being a vegetarian was for animals, 199 for health, and 22 for the environment."

"Following this dietary pattern item were the prosocial, personal, and moral orientations scales."

"An example item for personal orientation ($\alpha = 0.84$) included "I follow my dietary pattern because I am concerned about the effects of my food choices on my own well-being.""

"An example item for moral orientation ($\alpha = 0.91$) included "I feel that I have a moral obligation to follow my dietary pattern.""

Results

"Animal-motivated vegetarians reported higher levels of prosocial orientation, $t(358) = 8.40$, $p < .001$, $d = 0.94$, and moral orientation, $t(356) = 9.48$, $p < .001$, $d = 1.06$, than did health-motivated vegetarians."

"Environmentally motivated vegetarians reported higher levels of prosocial orientation, $t(358) = 4.26$, $p < .001$, $d = 0.98$, and moral orientation, $t(356) = 3.16$, $p = .002$, $d = 0.73$, than did health-motivated vegetarians."

"Exploratory analyses indicated that animal-motivated and environmentally motivated vegetarians did not report different levels of prosocial orientation, $t(358) = 0.13$, $p = .89$, $d = 0.04$, personal orientation, $t(358) = 1.31$, $p = .19$, $d = 0.27$, or moral orientation, $t(356) = 1.47$, $p = .14$, $d = 0.41$."

Discussion

"These results suggest that vegetarians motivated by concerns about animals and the environment exhibit different goal orientations than do vegetarians motivated by health."

"Both animal-motivated and environmentally motivated vegetarians exhibited stronger prosocial and moral goal orientations than did health-motivated vegetarians, whereas health-motivated vegetarians exhibited a stronger personal goal orientation than did either animal-motivated or environmentally motivated vegetarians."

"Despite having lower personal goal orientation than health-motivated vegetarians, both animal-motivated and environmentally motivated vegetarians exhibited levels of personal orientation that were similarly as strong as their prosocial and moral orientations."

"Vegetarians who are motivated by concerns about animals or the environment view their diets as encompassing strong prosocial, personal, and moral aims, whereas vegetarians motivated by health view their diets as a particularly self-focused pursuit."

Study 2

"Taking this estimation into account, a power analysis indicated that 760 participants would be needed to detect medium effect sizes between animal-motivated vegetarians and either health-motivated or environmentally motivated vegetarians with at least 80% power at a significance threshold of $p = .05$."

"Of these participants, 626 indicated that their main reason for being a vegetarian was for animals, health, or the environment and were retained in this study."

"Of these participants, 195 indicated that their main reason for being a vegetarian was for animals, 319 for health, and 48 for the environment."

"Of assessing the extent to which participants follow their idiosyncratic dietary pattern (i.e., avoiding some combination of red meat, poultry, fish, egg, and dairy), as the DIQ strictness scale does conventionally, the strictness scale in the current study assessed the extent to which participants refrain from eating meat."

Results

"I tested my prediction that animal-motivated vegetarians' greater feelings of disgust toward meat would explain why they would be more upset about eating meat accidentally than would health-motivated vegetarians."

"Distribution-of-the-product confidence limits for the indirect effect of motivation on upset through meat disgust did not include zero, 95% CI [0.57, 0.94], further suggesting that meat disgust fully mediated the link between vegetarian motivation (animal vs. health) and imagined degree of upset from eating meat accidentally."

"I found support for my predictions that animal-motivated vegetarians' greater feelings of disgust toward meat would explain why they follow their diets more strictly and would be more upset if they were to eat meat accidentally than would either health-motivated or environmentally motivated vegetarians."

General Discussion

"In Study 2, I found that environmentally motivated vegetarians resembled health-motivated vegetarians in reporting lower levels of disgust toward meat and dietary adherence than did animal-motivated vegetarians."

"Stronger feelings of disgust toward meat, moreover, explained why animal-motivated vegetarians exhibited greater dietary adherence than did either health-motivated or environmentally motivated vegetarians."

"Environmentally motivated vegetarians resemble animal-motivated vegetarians in terms of their goal orientations—how they see their diets as encompassing prosocial, personal, and moral goals—but resemble health-motivated vegetarians with respect to meat disgust and dietary adherence."

"Vegetarians motivated by a concern for animals, relative to vegetarians motivated by either health or the environment, appear to exercise a particularly high level of dietary adherence because they feel disgusted by the thought of eating meat."

"The current findings highlight value in studying ethical motivation with greater nuance, identifying meat disgust and dietary adherence as two factors that set animal-motivated and environmentally motivated vegetarians apart from one another."

[Section 10]
"Compared to health-motivated vegetarians, ethically motivated vegetarians hold stronger attitudes toward animal rights, attribute greater mind to animals, feel greater disgust toward meat, and set higher standards of dietary adherence for themselves (Jabs and others [85]; Janssen and others [86]; Rothgerber [87, 88]; Rozin and others [84])."

"In reviewing conceptual frameworks for studying vegetarian motivations, Rosenfeld and Burrow [83] highlight that the preponderance of existing research comparing vegetarians with different motivations has categorized individuals as either ethically motivated or health-motivated, with much less research categorizing those reporting ethical concerns by their specific types of ethical motivation."

"In Study 1, I investigated the extents to which vegetarians with animal, health, and environmental motivations differ in their prosocial, personal, and moral dietary goal orientations."

"Across these two studies, I predicted that whereas both animal-motivated and environmentally motivated vegetarians would exhibit a stronger moral goal orientation than would health-motivated vegetarians, this increased moralization would promote particularly strong disgust toward meat and dietary adherence for animal-motivated vegetarians—that is, I predicted that animal-motivated vegetarians would be more disgusted by meat and adhere to their diets more strictly than would either health-motivated or environmentally motivated vegetarians."

The Will to Fight: Aversion-Induced Aggression and the Role of Motivation in Intergroup Conflicts [179]

This is a machine-generated summary of:

Lewon, Matthew; Houmanfar, Ramona A.; Hayes, Linda J.: The Will to Fight: Aversion-Induced Aggression and the Role of Motivation in Intergroup Conflicts [179]

Published in: Perspectives on Behavior Science (2019)

Link to original: https://doi.org/10.1007/s40614-019-00221-2

If you want to cite the papers, please refer to the original.

For technical reasons we could not place the page where the original quote is coming from.

Abstract-Summary

"Aggressive behavior is a source of many significant human problems, most notably the catastrophic loss of life and resources that can result from violent conflicts between groups."

"Aggressive behavior is particularly likely to arise from aversive conditions that function as motivating operations (MOs) that establish the stimulation produced by aggressive acts as reinforcing."

"We describe the behavior that arises from these circumstances as aversion-induced aggression (AIA) and argue that the MOs associated with AIA are important factors in initiating and sustaining violent conflicts between groups."

"We describe how aversive conditions may be exploited by leaders to establish support for aggression against another group in the pursuit of their objectives."

"We advocate against the use of aggression as a tool for resolving conflicts between groups and consider how behavior science may contribute to the development and evaluation of alternative nonviolent practices."

Aversion-Induced Aggression (AIA): Basic Research and Theoretical Framework

"Aversive stimulation, extinction, and the immediate postreinforcement period of certain schedules of reinforcement share a number of functional properties associated with aversive stimuli: they reduce the frequency of previously reinforced behavior, they occasion aggression, and they motivate escape/avoidance behavior (Azrin, [89]; Azrin, Hake, Holz, & Hutchinson, [90]; Thompson [91, 92])."

"Even if there is some evidence to suggest that species-specific attack responses may be elicited the first time an animal of a certain species is exposed to an aggression-motivating aversive event (Azrin, Hutchinson, & Hake, [93]; Bolles, [94]), this behavior is presumably shaped across subsequent occasions via the stimulation it produces, and topographies that produce the most reinforcing stimulation are differentially reinforced."

"With respect to aggression, this means that aversive verbal stimuli are capable of establishing aggression as reinforcing in the same way as the events from which they derive this function (i.e., a description or depiction of some event may be as effective in motivating aggression as experiencing the event itself)."

Motivation for Aggression Within Groups

"The motivational circumstances under which group members may be compelled to support or contribute to acts of violence are determined in part by their environmental circumstances (e.g., political, social, legal, and economic events/conditions) and in part through the actions of group leaders."

"Although leaders often use multiple motivational tools to establish support for group aggression, the MOs associated with AIA, both verbal and nonverbal and often described as "emotional" (Halperin, [95]), are among the most common and powerful."

"When many members of a group are exposed to similar aversive circumstances, and those circumstances are deemed to be due to the actions of another group (either through direct observation or by verbal means), aggression may be motivated across a sufficient number of individuals such that leaders of the "aggrieved" group may exploit these motivational conditions to orchestrate coordinated violence against what is perceived as the "offending" group."

Effectiveness of Group Aggression

"The success of a group in achieving its objectives through aggression and aversive control depends on a) the extent to which it can incapacitate an opposing group's ability to retaliate (e.g., by killing or imprisoning its members and/or taking or destroying resources necessary for retaliation) and/or b) the extent to which it can change the motivational conditions that give rise to the opposing group's use of aversive tactics, either in retaliation or offensive action."

"In a conflict between two groups in which one has a significantly greater capacity for aggression, the stronger group can ostensibly achieve its objectives by quickly incapacitating the weaker group's ability to retaliate in a meaningful way."

"As we have argued above, the aggressive acts of the stronger group serve as MOs that strengthen motivation for retaliation within the weaker group, especially if the aggressive act also significantly worsens the weaker group's social and economic conditions."

Contributions of Behavior Science

"When the members or actions of another group function as abstractly aversive in this way, aggression against the group may be highly motivated and it is therefore easier for leaders to establish support for discriminatory practices, direct attacks on the group, or even genocide (Halperin, [95])."

"The methodological features of behavior science have much to contribute in evaluating group practices in intergroup conflict and bringing the consequences of leaders' actions to bear on their behavior (Mattaini & Holtschneider, [96])."

"The power of this methodology is that it can lead to the development of more effective practices regardless of philosophical orientation: it allows for the identification of what works, thereby making it more likely that leaders adopt courses of action that contribute to the long-term well-being of the group over and above its immediate motivational circumstances."

Conclusion

"The use of violence in intergroup conflict is a vestige of a prescientific time, a time when there was perhaps an excuse for its use to achieve group aims: we simply did not know any better."

"Over the course of human history, knowledge has advanced dramatically in many domains, but the fact that violent action continues to be regarded as anything other than a last resort in conflicts between groups highlights our continued shortcomings in the understanding of human behavior (Skinner, [97])."

"As groups are increasingly dependent upon each other for mutual welfare and we are coming to more fully understand the consequences of violent conflict, aggression is an increasingly indefensible solution to intergroup conflict."

[Section 6]

"Aggressive behavior (i.e., behavior which results in the physical harm of another individual) is at the root of many of the most obvious and troublesome problems facing all human societies, including violent crime, domestic abuse, and intergroup conflict."

"Violent intergroup conflict involves the coordinated aggressive actions of many people and often results in outcomes that are undesirable or even fatal for large numbers of individuals, many of whom may not be directly involved in the conflict."

"Despite the tragic outcomes inevitably associated with coordinated aggressive action of this nature, it appears that the dominant view among scholars and would-be combatants alike is that violent action is still the most effective means of achieving the objectives of a group (Howes, [98]; Pape, [99]; Sprinzak, [100]; Stephan & Chenoweth, [101])."

"We then consider how the MOs associated with AIA contribute to establishing support for coordinated aggressive action within a group."

When Motivation Backfires: Optimal Levels of Motivation as a Function of Cognitive Capacity in Information Relevance Perception and Social Judgment [180]

This is a machine-generated summary of:

Roets, Arne; Van Hiel, Alain; Kruglanski, Arie W.: When motivation backfires: Optimal levels of motivation as a function of cognitive capacity in information relevance perception and social judgment [180]

Published in: Motivation and Emotion (2012)

Link to original: https://doi.org/10.1007/s11031-012-9299-0

If you want to cite the papers, please refer to the original.

For technical reasons we could not place the page where the original quote is coming from.

Abstract-Summary

"Motivation and cognitive capacity are central variables in major models of social judgment and persuasion."

"In Study 1 we demonstrated a cross-over pattern of means for the interaction effect of motivation and cognitive capacity on judgment, with high motivation being beneficial under high cognitive capacity but detrimental under low cognitive capacity."

"In Study 2, the role of information relevance was further investigated, showing that for highly motivated participants, judgment was differently affected by information with low versus high relevance when cognitive capacity was high but not when it was low."

"In the discussion, we elaborate on these effects, advocating a more dynamic perspective on (social) judgment, acknowledging the conditional (opposite) effects of motivation, in function of cognitive capacity."

Introduction

"According to the unimodel, judgment occurs through a generic processing mode that relies on the individual's perception of the information's relevance (i.e., perceived informativeness) as determined by situational and personal characteristics (parameters) such as the individual's cognitive capacity and motivation (see also Kruglanski and Gigerenzer [102])."

"More general account, we propose that cognitive capacity simply refers to whether or not resources are sufficient for explicit and deliberate processing of information and its relevance, whereas motivation refers to whether or not people feel the need or willingness to engage in such information and relevance processing to reach judgments."

"Two studies on social judgment were conducted, investigating whether the positive impact of high levels of motivation under high cognitive capacity can indeed reverse when cognitive capacity is low."

"Study 2 investigated the joint impact of motivation and cognitive capacity on judgment contrasting a condition in which the potential relevance of information was high with one in which the potential relevance was low."

Study 1

"Planned contrasts demonstrated that when cognitive capacity was high, participants in the high motivation condition showed larger judgmental changes ($M = 4.26$, $SD = 2.56$) after receiving additional incriminating information than in the low motivation condition ($M = 3.11$, $SD = 1.88$), $F(1, 70) = 2.73$, $p = .10$."

"Participants in the low cognitive capacity condition considered the additional information less relevant in the high motivation condition ($M = 3.75$, $SD = 1.19$) compared to the low motivation condition ($M = 4.32$, $SD = 0.47$), $F(1, 68) = 5.17$, $p = .03$."

"The remaining direct effect was not significant; Direct effect estimate $= .29$, $SE = .19$, $p = .18$, indicating that the interaction effect between motivation and cognitive capacity on judgment change was mediated by perceived information relevance."

Study 2

"Based on the findings of Study 1, we hypothesize that under conditions that showed to facilitate accurate perception of information relevance (i.e., motivation and cognitive capacity being both either high or low), the manipulation of the potential information relevance (high vs. low) will have the strongest effects on judgment."

"Planned comparisons showed that in the high motivation conditions, participants high in cognitive capacity changed their judgment to a greater extent when the additional information was framed as highly relevant ($M = 24.71$, $SD = 14.02$) versus less relevant ($M = 10.82$, $SD = 14.04$); $F(1, 94) = 6.36$, $p = .01$."

"When cognitive capacity was low, high motivation yielded no more judgment change in response to information with high relevance ($M = 15.86$, $SD = 12.23$) compared to information with low relevance ($M = 15.08$, $SD = 14.37$), $F(1, 94) = 0.02$, $p = .88$."

General Discussion

"Applying this perspective to information sampling quantity in judgment tasks, Roets and others (103) reported that low levels of relative cognitive capacity indeed resulted in a substantial increase in information sampling when motivation was high, but not when motivation was low."

"Although high motivation might increase the amount of information sampled when cognitive capacity is low, it was found to hamper the perception of the relevance of this information, which accounts for the effects on judgment outcome."

"Account, we proposed that cognitive capacity simply refers to whether or not resources are sufficient for explicit and deliberate processing of the information and its relevance, whereas motivation refers to whether or not people are willing to engage in such information processing to reach judgments."

"Low levels of motivation under impaired cognitive capacity allow for judgments to be more strongly guided by implicit value accounts, which are less dependent on the available cognitive resources, and may actually yield more adequate judgments than disorganized deliberate information processing."

Conclusion

"Models of judgment and persuasion in social psychology have generally assumed that high levels of motivation are better than low levels."

"It was shown that when cognitive resources are taxed, high levels of motivation may backfire and judgment may be better served by lower levels of motivation."

"The present studies encourage a more dynamic perspective on the role of motivation and cognitive capacity, acknowledging that also in social judgment and persuasion, the effects of motivation are conditional."

Motivation and Interpersonal Sensitivity: Does it Matter How Hard You Try? [181]

This is a machine-generated summary of:

Hall, Judith A.; Blanch, Danielle C.; Horgan, Terrence G.; Murphy, Nora A.; Rosip, Janelle C.; Schmid Mast, Marianne: Motivation and interpersonal sensitivity: Does it matter how hard you try? [181]

Published in: Motivation and Emotion (2009)

Link to original: https://doi.org/10.1007/s11031-009-9128-2

If you want to cite the papers, please refer to the original.

For technical reasons we could not place the page where the original quote is coming from.

Abstract-Summary

"Across 11 experiments, motivation to be accurate on a test of interpersonal sensitivity was manipulated using five methods for increasing motivation (monetary incentive, ego motive, forewarning that accuracy would be tested, exhortation to try hard, and framing the interpersonal sensitivity test description to suggest that performance was gender relevant)."

"Participants were then given an interpersonal sensitivity test consisting of interpreting the meanings of cues or recalling a target person's appearance, nonverbal cues, or spoken utterances."

"Neither the individual studies, nor a meta-analysis of the 11 studies, found that the motivation manipulations improved participants' accuracy on interpersonal sensitivity tests that involved the processing of nonverbal cues."

Introduction

"Ickes and others [104] proposed that women's typical advantage on tests of interpersonal sensitivity might be due to motivational differences stemming from the stereotypically female nature of such tasks."

"Whether self-evaluation should be considered a motivational manipulation is unclear, but if it is, these studies are consistent with the notion that female-relevant motivation can positively influence women's performance on an interpersonal sensitivity test."

"Female-relevant manipulations have improved women's accuracy, but it needs to be stressed again that the empathic accuracy paradigm used in these studies predominantly measures sensitivity to verbal cues."

"Koenig and Eagly [105] framed an interpersonal sensitivity test (one based predominantly on nonverbal cues) as female relevant with a strongly worded manipulation that included, among other things, the statement that women typically perform better than men."

"Manipulations designed to appeal to female stereotypes had positive, though qualified, impact on women's interpersonal sensitivity in studies that primarily measured sensitivity to verbal cues."

Methods

"Ego relevance was the basis of motivation versus standard task instructions (control), with the interpersonal sensitivity task consisting of accuracy in judging the trait dominance of target persons."

"Participants also received accuracy scores for their ability to recall what the target person had said."

"The study was identical to Study 8 in testing the effect of forewarning on accuracy in recalling a target person's behavior, but it used different stimuli (four different interview responses from the Study 8 target person), also ~4 min in duration."

"Interpersonal sensitivity was defined as how much of a target person's nonverbal and verbal behavior the participants remembered."

"The motivation manipulation was exhortation (vs. control), as in Study 10, and interpersonal sensitivity consisted of recalling nonverbal and verbal behavior."

Results

"Within the individual studies, there was only one instance of manipulated motivation significantly influencing performance (in Study 4, a female-relevant manipulation hurt women's performance on a nonverbal task)."

"The effects for gender-neutral manipulations were calculated for men and women together."

"To see whether there were any differential effects for men and women, the gender × manipulation interaction was calculated within each of the relevant studies."

"When individual studies show small and nonsignificant effects, it is still possible that meaningful trends can be detected through meta-analysis."

"There was no motivation effect, either for the individual motivation manipulations or for all 13 of the gender-neutral manipulations considered together, as indicated by the small and nonsignificant combined Zs."

"There was a positive but nonsignificant effect for forewarning and a significant effect for exhortation as well as a significant effect across all four of the studies (with no significant heterogeneity)."

Discussion

"The possibility that nonverbal accuracy may be based more on people's knowledge of the domain in question than on their immediate motivation is consistent with one of the present studies (Study 2), which did not find a motivational effect but did find a significant positive correlation between a paper-and-pencil test of nonverbal cue knowledge and performance on the interpersonal sensitivity test (Rosip and Hall [106])."

"This result fits with those previous studies that found positive motivation effects on interpersonal sensitivity tasks based on Ickes' empathic accuracy paradigm, in which verbal cues contribute much more than nonverbal cues to the attainment of accuracy."

"In the present studies one might argue that men started out at a relatively low motivational level yet manipulations were still not successful in raising their accuracy on the nonverbal cue based tests."

Seventy-Five Years of Motivation Measures (1930–2005): A Descriptive Analysis [182]

This is a machine-generated summary of:

Mayer, John D.; Faber, Michael A.; Xu, Xiaoyan: Seventy-five years of motivation measures (1930–2005): A descriptive analysis [182]

Published in: Motivation and Emotion (2007)

Link to original: https://doi.org/10.1007/s11031-007-9060-2

If you want to cite the papers, please refer to the original.

For technical reasons we could not place the page where the original quote is coming from.

Abstract-Summary
"The literature on motivational measures from 1930 to 2005 is reviewed."

"A descriptive taxonomy of measures in the field of motivation is suggested in order to synthesize ideas about measurement scales."

"Suggestions are offered for further research in motivational measurement."

Introduction
"The third section of the article, "Motivational Measures: A Descriptive Analysis," summarizes findings from the survey and provides a descriptive view of measures of motivation that may facilitate thinking about approaches to measurement."

"We think of this article as providing a first statement in a renewed conversation about measurement in motivation."

"Measurement in the motivation area is important not only for motivation researchers themselves, but also because of the issues that the area's findings raise for the different ways of assessing any part of personality."

"The careful study of both individual motives (e.g., n Achievement) and multiple motives, as well as the various ways that they have been measured (e.g., thematic and self-judgment), and the integration of those motives within personality (e.g., intrinsic motivation), make a rich area for the consideration of how such a key personality attribute should be assessed."

The Evolving Understanding of Motivation and Its Measure
"Their study included some of the first measures of motivation, including the Thematic Apperception Test (TAT; Murray [107])."

"The Personality Research Form (PRF; Jackson [108]) is a self-judgment test of 20 motives based on Murray's needs theory, and also includes scales to monitor social desirability and infrequency of response."

"Another group of researchers departed from Murray's motive-by-motive measurement approach to examine a person's sources of motives, incentives acting upon the person, and the expectancy of a reward (Deci and Ryan [109]; Rotter [110]; Vroom [111])."

"If there is a consistent theme underlying the motivational concepts and measures discussed so far, it is that motivation involves the organization of needs and goals within the personality system; that is, it concerns the organized pursuits of the individual (Koch [112])."

Measures of Motivation and Trends in Measuring Motives
"In our e-mail message we explained that we were preparing a review of measures of motivation and asked: What do you consider to be the most important measures (i.e., tests and scales) of motivation, either historically, or in use today?"

"General measures of motivation: We first examined the most widely studied scales from PsycINFO and divided them into several categories."

"A number of more recent tests now include scales of intrinsic and extrinsic motivation as well, such as the Motivated Strategies for Learning Questionnaire (MSLQ; Pintrich et al. [113]), the Work Preference Inventory (WPI; Amabile et al. [114]), and the Motivation Sources Inventory (MSI; Barbuto and Scholl [115])."

"A second group of scales involves "biological bases of motives," i.e., scales that address a person's general responses to rewards and punishments; these include the Regulatory Focus Questionnaire, which concerns styles of approaching pleasure and avoiding pain, and the Behavioral Inhibition System/Behavioral Activation System Scale (RFQ; Cesario et al. [116]; Higgins et al. [117]; BIS/BAS scales; Carver and White [118])."

Motivational Measures: A Descriptive Analysis
"Some scales measure motivational dynamics; others, the self as a motivator, and still others, para-motivational features such as values."

"The second content area of measurement concerns self-related motives."

"A third area of motivational measurement concerns motivational dynamics, or how motives are integrated within an individual's mental life."

"An early example of such a scale was Allport and Ross's [119] measure of intrinsic motivation for religion."

"The Academic Motivation Scale measures an intrinsic–extrinsic motivational locus dynamic, but does so specifically in a school context (Vallerand et al. [120])."

"Psychologists increasingly understood that scales that measured the same motives with different approaches often failed to converge (e.g., McClelland [121])."

"When a person's achievement motives are congruent across measures such as the TAT and self-judgment scales, that individual's need for achievement appears more self-determined and better predicts certain outcomes (Biernat [122]; Thrash and Elliot [123])."

Discussion

"Just over 70 years ago, Murray [124] wrote of measurement practices in motivation: Some use physiological techniques, others present batteries of questionnaires."

"Speaking, the use of measures of motives increased through the 1960s and then leveled off."

"Measures of motives that focus on specific social contexts such as work and education are increasing in relation to the use of general measures of motives."

"Among approaches to measurement, some measures employ self-report, some attempt to tap a person's implicit models of his or her self and world, and others examine motives in specific psychological and social contexts."

"We surveyed motivational measures from PsycINFO and polled members of the SPSP Listserv to identify scales of significance."

"It is our hope that this revisiting of the area of motivational measures, after an absence of 30 or so years, can promote two directions already present in the field."

Motivational and Personality Variables Distinguish Academic Underachievers from High Achievers, Low Achievers, and Overachievers [183]

This is a machine-generated summary of:

Dings, Alexander; Spinath, Frank M.: Motivational and personality variables distinguish academic underachievers from high achievers, low achievers, and overachievers [183]

Published in: Social Psychology of Education (2021)

Link to original: https://doi.org/10.1007/s11218-021-09659-2

If you want to cite the papers, please refer to the original.

For technical reasons we could not place the page where the original quote is coming from.

Abstract-Summary
"Previous studies have associated several variables concerning motivation and other domains with underachievement, i. e. a student's academic achievement falling short of what their cognitive abilities, as the best predictor of academic performance, would indicate."

"Using discriminant analysis, underachievers identified in a German twin family study were compared not only to achievers with comparable IQ scores, but also with students of lower aptitude, but comparably low grades, as well as overachieving students."

"Results confirm previous findings that compared to successful students, underachievers report lower levels of motivation and parental support; beyond this comparison, underachievers also differed from other low achievers, mostly in terms of their personality."

Introduction
"Its reverse case of overachievement has also been studied, albeit to a much lesser extent (Sparfeldt, Buch & Rost, [125])–it might simply be a less pressing issue; also note that if a good IQ test result is used as a benchmark which the academic achievement of underachieving students is said to fall short of, you might just as well ask the reverse question for apparent "overachievers": why did their IQ test performance fall short of what they achieve in school?"

"As Hanses and Rost [126] point out, this is questionable when using the regression approach, as it subsumes very different cases of partially unclear validity under the label of underachievement: it will „diagnose" underachievement in many cases where the actual grades are just fine–but the student would have been expected to achieve even better results."

Materials and Methods

"The distribution of grades is skewed towards good grades so strongly that a few students were classified as underachievers or low achievers with a cutoff of 0.5 SD while receiving a 3 ("satisfactory") in both subjects, despite the fact that this selection was already much more restrictive than the usual regression approach."

"As far as our German sample can be generalized, and as far as readers follow our interpretation of „serious" underachievement, it thus appears to be a remarkably rare phenomenon: In total, there were 352 different students who fell either into the high IQ group or the low achievement group, or both; the latter case, the overlap between both categories, represents underachievement, but happened only in 21 cases out of the 352."

"That the usual approach of using a difference of 1 SD between standardized grades and IQ scores results in about 15% selected cases in either direction, but, in our view, will include many cases whose classification as practically meaningful underachievement is doubtful."

Results

"The position of the group centroids on the first function suggest that it almost exclusively separated successful students from unsuccessful ones: Underachievers (group centroid at − 0.79) were grouped together with low achievers (− 0.75), while overachievers (0.44) and high achievers (0.45) were indistinguishable at the other end."

"The group centroids of the second function indicate that, after separating successful and unsuccessful students, classification can mostly be improved by contrasting low-IQ and high-students, and in particular by contrasting underachieving students (group centroid at − 0.73) to overachieving students (0.77)."

"The mean Z-scores in the variables that most closely resembled the AOM components (learning motivation, self-efficacy, academic self-concept and parental support as the closest proxy to environmental support) correlated substantially with the underachievement score ($r = 0.37$, $p < 0.001$; the coefficient is positive because higher scores in the regression approach denote overachievement)."

Discussion

"This was in part a replication of established differences between underachievers and high achievers; adding the simultaneous comparison to low achievers (and overachievers) as well as, in terms of variables used, the Big Five, were the novel parts of the study."

"With these restrictions in mind, our study adds to previous research in justifying the construct itself: It appears that underachievers do in fact differ not only from other highly intelligent students, as previous research had already demonstrated, but that some differences to other low achievers are also discernible, albeit smaller."

"Besides not being identical to either other low achievers or other highly intelligent students, detailed results of the discriminant analysis indicated that in terms of the motivational and personality variables used here, underachievers are hardly even closer to one group than the other – they cannot clearly be put into either box."

Conclusions

"That they differ from other highly intelligent students in various personal and environmental aspects was already established; besides replicating these differences, the patterns found in our sample of 51 under- or overachievers and 176 comparison students suggest that underachievers also differ from fellow unsuccessful students in terms of their personality."

"Explaining the specific phenomenon of underachievement, as opposed to low achievement in general, through motivational problems therefore might miss the point; this is somewhat at odds with previous research and calls for future studies clarifying this issue."

"In practical terms, this research suggests that while targeting motivation in unsuccessful students is not necessarily wrong, it is not specifically underachievers who are in need of such an intervention."

"The differences in the Big Five variables do not immediately suggest practical steps, but one point of interest could be the unusually low conscientiousness reported by underachievers; rather than lacking motivation, some of the might rather be in need of being taught skills and strategies to help them organize their academic endeavors effectively."

You Can Do it if You Really Try: The Effects of Motivation on Thinking for Pleasure [184]

This is a machine-generated summary of:

Alahmadi, Sarah; Buttrick, Nicholas R.; Gilbert, Daniel T.; Hardin, Amber M.; Westgate, Erin C.; Wilson, Timothy D.: You can do it if you really try: The effects of motivation on thinking for pleasure [184]

Published in: Motivation and Emotion (2017)

Link to original: https://doi.org/10.1007/s11031-017-9625-7

Copyright of the summarized publication:

If you want to cite the papers, please refer to the original.

For technical reasons we could not place the page where the original quote is coming from.

Abstract-Summary

"People find it difficult to enjoy their own thoughts when asked to do so, but what happens when they are asked to think about whatever they want?"

"Evidence in support of four reasons why this is: without instructions people do not realize how enjoyable it will be to think for pleasure, they do not realize how personally meaningful it will be to do so, they believe that thinking for pleasure will be effortful, and they believe it would be more worthwhile to engage in planning than to try to enjoy their thoughts."

Introduction

"The answer to this question is unknown, because to date there have not been any studies that have compared instructions to enjoy one's thoughts with no instructions, that is, with a condition in which people are not given the goal of enjoyment."

"These possibilities are not mutually exclusive; e.g., people could underestimate how much they would enjoy thinking for pleasure and prioritize other goals."

"The present studies examined the role of possible moderators of thought enjoyment, such as people's initial mood and level of physical activity while thinking."

"In the interest of full disclosure, we note that the initial purpose of some of the studies was to test these potential moderators of the conditions under which people enjoy thinking."

"What emerged from these studies was the powerful effect of motivating people to try to enjoy their thoughts."

Study 1

"Participants in the enjoy condition rated the extent to which they thought about the eight topics listed at the beginning of the study versus other topics (1 = only about other topics, 9 = only about the 8 topics), and all participants rated how surprised they were by the thoughts that came to mind; the extent to which their goal had been to make plans for what they would do later; the extent to which their goal had been to think about things that were pleasant or entertaining; the extent to which they were letting their thoughts flow; the extent to which they were trying to control the direction of their thoughts; how interesting the thinking period was, all on 9-point scales with appropriate labels."

Study 2

"In the enjoy condition, participants were asked to spend the baseline period "entertaining yourself with your thoughts as best you can."

"The experimenter, who was unaware of whether participants were in the enjoy or no instructions condition, then told participants that the baseline period would begin."

"We performed a 2 (Instructions: enjoy vs. no instructions) x 2 (Activity: treadmill vs. chair) analysis of variance (ANOVA) on these ratings to test our hypothesis that participants in the enjoy condition would have the highest scores on the enjoyability index (which would result in a main effect of the instructions manipulation) and that it would be easier to think for pleasure while walking than sitting (which would result in a main effect of the Activity manipulation)."

"As in Study 1, participants in the enjoy condition reported less of a tendency to let their thoughts flow."

Study 3

"A 2 (Instructions: enjoy vs. none) x 2 (Hassles: hassles vs. none) ANOVA revealed a significant main effect of the mood manipulation on negative affect, $F(1, 147) = 42.87$, $p < .001$, $\eta_p{}^2 = 0.23$, reflecting the fact that participants in the hassles condition were in a more negative mood than participants in the no hassles condition, Ms = 2.56 versus 1.75 (SDs = 0.89, 0.61)."

"As in Study 1, there was no effect of the instructions manipulation on people's preference to engage in more thinking versus proof reading, $Fs(1, 146) = 0.04$, $p = .84$; again, this may be due to the fact that proofreading was an unpopular task: 71% of participants preferred thinking or had no preference."

"In Studies 1 and 3, on the other hand, participants in the no instructions conditions reported more of a goal to engage in planning than did participants in the enjoy conditions, and reported lower enjoyment of the thinking periods."

Studies 4a and 4b

"Participants then read that the study was about "people's thought processes in everyday life while they do common activities," that they would be playing the RatMaze videogame, that the game would be interrupted with a "time out period," and that they would receive further instructions about what to do at that time."

"Participants randomly assigned to the enjoy condition were instructed to spend the time out period "entertaining yourself with your thoughts as best you can" while remaining in their chair."

"As predicted, participants in the enjoy condition reported greater enjoyment of the time out periods than did participants in the no instructions or planning conditions."

"Participants in the enjoy condition also reported that they enjoyed the video game more, both right after the time out period and at the end of the study, ts(201) > 2.88, p < .005, suggesting that the improved experience of the time out period extended to their game play."

Study 5

"Forecasters were then asked to predict their responses if they were instructed to spend the thinking period in each of three ways: entertaining themselves with their thoughts, thinking about whatever they wanted, and planning what they would be doing over the next 48 h, in random order (there were no significant effects of order on participants' responses)."

"Because the confidence interval of d for the predicted difference did not overlap with the confidence interval for the actual effect of enjoyment instructions in Studies 1–4 (95% CI = 0.56, 0.88), we can conclude that forecasters significantly underestimated how enjoyable thinking for pleasure would be."

"Participants in the enjoy condition reported that this experience was significantly more personally meaningful than did participants in the planning or normal activity conditions, suggesting that not only were forecasters in Study 5 incorrect in their predictions about how meaningful planning would be, but also that these predictions were in the opposite direction to the true effect."

General Discussion

"Studies 1–4 found that people enjoyed such times more if they were given the goal to entertain themselves with their thoughts than if they were given no instructions."

"In many ways, the most striking result of Studies 1–4 is how little people enjoyed themselves when given the freedom to think about whatever they wanted."

"It is important to note that these ratings are lower than the enjoyment people get from engaging in everyday activities such as playing with their phones, reading, or watching television (Buttrick and others [127]; Smith and Frank [128]; Wilson and others [129], Study 8)."

"The present studies suggest that thinking for pleasure may be undervalued, however, and may be a viable alternative to "device obsession," especially if people's goal is to find meaning as well as pleasure."

Priming Autonomous and Controlling Motivation and Effects on Persistence [185]

This is a machine-generated summary of:

Murphy, Stephen L.; Taylor, Ian M.: Priming autonomous and controlling motivation and effects on persistence [185]

Published in: Current Psychology (2020)

Link to original: https://doi.org/10.1007/s12144-020-00921-y

If you want to cite the papers, please refer to the original.

For technical reasons we could not place the page where the original quote is coming from.

Abstract-Summary
"The present studies examined whether priming distinct motivational states influenced persistence at a task designed to promote repeated failure, and post-task plans for engaging in self-regulatory activity."

"Two double-blind, between-subject experiments (Study 1: N = 58; Study 2: N = 92) involved participants being randomized to Autonomous Motivation, Controlling Motivation, or Neutral prime conditions using a scrambled-sentence test."

"Using frequentist and Bayesian analyses, Study 1, Study 2, and an internal meta-analysis showed no differences in persistence or planned exercise across priming conditions, thus contrasting with previous research."

Introduction
"The aim of the present research is to contribute in this regard and examine whether priming autonomous motivation promotes persistence relative to a neutral prime, and whether priming controlling motivation hinders persistence, while ensuring experimenters remain blind to participant condition."

"While research confirms more autonomous involvement in a persistence-requiring activity aids performance in a subsequent activity that requires self-regulation (i.e., a handgrip endurance task; Muraven and others 130), it has yet to be established whether priming distinct motivational states may elicit similar effects."

"A secondary aim of the present research is to examine whether, relative to a neutral prime, an autonomous motivation prime enhances the plans participants make to engage in activity following a persistence-requiring task, and whether a controlling motivation prime attenuates these plans."

Study 1

"Bayesian analyses support these findings, with data 6.63 (mood), 5.13 (SST duration), and 5.97 (typical exercise behavior), times more credible under the null than the alternative hypothesis."

"Post-hoc analysis revealed participants in the neutral condition persisted more than participants in the autonomous motivation condition (28.6 vs. 21.6 min, $p = .05$, $d = .63$), but Bayesian analyses contrastingly demonstrated weak support for the null hypothesis (data 1.25 times more credible under the null than the alternative hypothesis)."

"From a Bayesian perspective, data revealed weak support in favor of a difference between the neutral and controlling motivation prime conditions (data 1.10 times more credible under the alternative hypothesis than the null)."

"No significant differences were found between controlling and autonomous conditions (21.8 vs. 21.6 min, $p = .96$, $d = .05$), with Bayesian analyses indicating data to be 5.46 times more credible under the null than the alternative hypothesis."

Study 2

"Study 1 found weak evidence in support of our hypothesis that priming participants to experience controlling motivation may reduce persistence relative to a neutral prime."

"We therefore hypothesized that participants in the autonomous/controlling condition would report greater/less frustration during the tracing task relative to the neutral condition (hypothesis 3)."

"We hypothesized participants in the autonomous/controlling condition would experience greater/lower increases in HRV (hypothesis 4) and report higher/lower mental exertion (hypothesis 5) during the tracing task relative to the neutral condition."

"After the tracing task, participants completed measures of mental exertion, task motivation, frustration, and planned exercise behavior."

"The Situational Motivation Scale (SIMS; Guay and others [131]) measured the quality of motivation participants had during the impossible tracing task."

"Bayesian analyses support this interpretation, with data 9.11 (impossible tracing task duration), 4.32 (planned exercise behavior), 7.80 (frustration), 5.36 (HRV during the impossible tracing task), and 6.90 (mental effort) times more credible under the null compared to the alternative hypothesis."

Internal Meta-Analysis

"The autonomous motivation primes influence on tracing task duration was not significantly different to the controlling motivation prime, p = .85, d = .04, 95% CI (−.36, .45), or neutral prime, p = .15, d = .30, 95% CI (−.11, .72)."

"The controlling motivation primes influence on tracing task duration was not significantly different to the neutral prime, p = .10, d = .35, 95% CI (−.06, .77)."

"The autonomous motivation primes influence on planned exercise was not significantly different to the controlling motivation prime, p = .19, d = .27, 95% CI (−.14, .69), or neutral prime, p = .44, d = .16, 95% CI (−.25, .57)."

"The controlling motivation primes influence on planned exercise was also not significantly different to the neutral prime, p = .19, d = .27, 95% CI (−.13, .69)."

General Discussion

"The present research shares many similarities with literature demonstrating motivational priming effects, and cast a large net to capture initially unanticipated manifestations of motivational differences (i.e., investigating quality of engagement), which reduces concerns key methodological decisions explain our null findings."

"There is some support for the idea that contextual and/or population characteristics may specifically reverse or eliminate motivational priming effects, for instance, a controlling motivation prime (compared to an autonomous motivation and neutral prime) was found to promote persistence during a problem-solving task (the reverse effect to that hypothesized in the present research), but only for individuals high in attachment anxiety (Milyavskaya and others [132])."

"The motivational primes used in the present research were expected to exert a large effect due to prior literature demonstrating similar effect magnitudes (e.g., Keatley and others [133])."

Conclusion

"This research provides evidence that priming autonomous/controlling motivation exerts no influence on persistence or plans to engage in exercise (relative to a neutral prime), thus conflicting with extant theory and evidence."

"Numerous explanations exist for these null findings."

"These include unanticipated moderating factors unique to the present research, and that motivational priming effects are smaller than anticipated (thus preventing the ability of our research to identify these effects)."

"Future research should seek to implement pre-registration and conduct research capable of identifying small motivational priming effects."

The Emotions in Cultural-Historical Activity Theory: Personality, Emotion and Motivation in Social Relations and Activity [186]

This is a machine-generated summary of:

Burkitt, Ian: The Emotions in Cultural-Historical Activity Theory: Personality, Emotion and Motivation in Social Relations and Activity [186]

Published in: Integrative Psychological and Behavioral Science (2021)

Link to original: https://doi.org/10.1007/s12124-021-09615-x

If you want to cite the papers, please refer to the original.

For technical reasons we could not place the page where the original quote is coming from.

Abstract-Summary
"This article conceptualizes human emotion as formed within the cultural and historical development of each individual within their society, using theories of emotion created within cultural-historical activity theory (CHAT)."

"Moving forward, I draw on more recent work on emotions in the field of CHAT, as well as on the work of M. M. Bakhtin, to create a more embodied understanding of the cultural-historical formation of the emotions as part of the dynamics of personality as a whole, including its conscious and unconscious elements."

Vygotskyan Approaches to the Emotions
"For Vygotsky the term also indicated how we experience life from the centre of our own personality as a unity or whole, so that our active relationship to others and to the social and cultural world is not primarily cognitive or emotional, based on intellect or perception, or motivated by need or will, but is all these things at once."

"The problem in this approach from a Vygotskyan perspective in terms of a theory of emotions, is that if we follow the main body of Vygotsky's [134, 135] work on child development, we have to assume an early period of ontogeny where the lower mental functions are governed by biological processes that only begin their social development with the gradual mastery of speech, at which point they become subject to cultural laws of development through various different stages."

Activity Theory: Emotion, Motivation and Need

"At times, Leontyev [136] means to use the term 'object' in this way, as when he breaks activity down into three macrostructures; the first being the patterns of social activity we participate in as individuals and which form the basis of our motivations: the second are the specific actions of persons and the goals at which they are aimed: and the third are operations, the means and methods of action, or the way in which actions are performed."

"We find the broader notion of an object of activity in Leontyev's work, which I feel allows us to create a more fully formed view of motive, need and, also, of emotion."

"This leads to a deeper critique of Leontyev's work on emotions: that he pays too little attention to social relations and joint activities, and how the object of an action is not simply a goal, but another person who is its addressee (Sève [137])."

Social Relations, Personality and Emotions

"What this demonstrates is that in social relations and activities we are always both the object and the subject of emotional evaluation; it is not just that, as Bakhtin said, we experience emotion subjectively in the way we find the object of our love as loveable, we also find we are the object of emotional-evaluative acts by others, so we feel ourselves to be loveable from the experience of being loved by another or experience the pain of neglect and rejection."

"Emotion is like all other psychological phenomena – a product of symbolic and linguistic meaning that, as such, has both a social and historical existence and a private sense for me in terms of the objects that I love or dislike from within my own emotional experience."

Discussion and Conclusions

"The emotions we feel are part of the social relations of society, which are manifested in our own personal experience, and these relations are actively made meaningful in the symbols, signs, metaphors and everyday concepts of a culture."

"The objects of emotional experience are crystallized in relationships and joint activities and they answer to deep seated personal and emotional needs that have developed in our own life history, located in the specific place and time of the culture in which we are raised and create our life trajectory."

"This then emerges into consciousness as imagination and fantasy driven by deep-seated emotional needs that are the product of cultural history and personal experience within our own unfolding biography."

"In this productive activity, we transform something of the social world along with some aspect of ourselves, including our own emotional experience, and we set ourselves new goals and find new objects that create for us new needs and desires."

[Section 5]

"The subject of emotions has become more prominent amongst those who work in the field of cultural historical psychology and activity theory."

"In the history of the social relations of production, humans have transformed the natural environment into a socially produced mode of life (Marx & Engels [138]) and, in the process, have transformed their own nature to become individuals who are created through social and historical development rather than in biological maturation alone."

"For Vygotsky [134] and Leontyev [136, 139, 140], although humans are biological organisms, each human personality is the product of social development within the historical mode of life, involving the mastery of tool and sign mediated activity through which they appropriate the social and cultural heritage of their society."

"I also want to suggest ways forward towards a more fully formed understanding of emotion as a cultural and historical phenomenon, and as part of the socially developed personality considered as a whole."

Acting in Our Interests: Relational Self-Construal and Goal Motivation Across Cultures [187]

This is a machine-generated summary of:

Gore, Jonathan S.; Cross, Susan E.; Kanagawa, Chie: Acting in our interests: Relational self-construal and goal motivation across cultures [187]

Published in: Motivation and Emotion (2008)

Link to original: https://doi.org/10.1007/s11031-008-9113-1

If you want to cite the papers, please refer to the original.

For technical reasons we could not place the page where the original quote is coming from.

Abstract-Summary

"A cross-cultural model tested the hypotheses that (a) people with a highly relational self-construal will pursue their goals for RARs, and (b) RARs will predict positive goal outcomes after controlling for variance explained by personally-autonomous reasons (PARs) and social support."

"One hundred seventy Americans and 219 Japanese completed a well-being and self questionnaire then generated and rated seven goals on several attributes."

"Results showed that relational self-construal was associated with RARs for goals."

"Americans and Japanese differed in the types of goals they pursued and the degree to which social support predicted effort."

Personally- and Relationally-Autonomous Reasons for Goals

"Self-concordance theory identifies two types of reasons for goals: reasons that that "express enduring interests and values" (Sheldon and Elliott 141, p. 482; termed autonomous reasons), and reasons that are demanded by the situation or by others (termed controlled reasons)."

"We represent self-concordant goals as motivated by personally-autonomous reasons (PARs), because the rationale for their pursuit derives from a person's individual or personal endorsement."

"In their original work, self-concordance was measured by combining ratings of autonomous reasons and subtracting controlled reasons for goals, so that the score reflected the relative autonomy of personal reasons for goal pursuit (what we term personally-autonomous reasons, or PARs)."

"We defined RARs as the relative autonomy of relational reasons for goal pursuit, where the volition stems from the relationship rather than from solely individual or personal sources."

Relational Self-construal and Relational Reasons

"They found that RARs for goals are likely to be valued or preferred by people who strongly define themselves in terms of their close relationships (termed relational-interdependent self-construal by Cross et al. 142; see also Sedikides and Brewer 143, for reviews; this construct will be referred to as relational self-construal in the remainder of this paper)."

"Highly relational individuals were no more likely than others to pursue their goals for relationally-controlled reasons (RCRs)."

"Cross et al. [142] found that individuals with a highly relational self-construal were more likely than other people to consider the needs and wishes of close others when making an important decision."

"Individuals whose culture emphasizes the importance of close others' interests may be especially likely to employ RARs while pursuing their goals in comparison to individuals whose culture emphasizes expressing one's individuality."

Culture and Motivation

"Pursuing goals for PARs represents a form of disjoint agency because it emphasizes personal endorsement of one's goals in contrast to the influence of other sources, including the influence of close relationships."

"Research with East Asian populations shows that important relationships, social obligations, and other relational concerns are much more important in directing behavior than individual preferences and wishes (Heine and Lehman 144; see Miller 145 for a review)."

"Oishi and Diener [146] found that Asian-Americans' subjective well-being was enhanced when they pursued goals that reflected important relationships (e.g., to please parents and friends), whereas European-Americans' subjective well-being was enhanced when they pursued goals that reflected personal interests alone (e.g., for my fun and enjoyment)."

"Pursuing goals for RARs allows an individual to engage in conjoint agency ("acting in our interests"), which is particularly emphasized in East Asian cultures."

Relationally-Autonomous Reasons and Social Support

"Social support is well-recognized as an interpersonal influence that predicts goal effort (e.g., Hektner [147]; Wentzel [148]; Zaleski [149]), and is associated with positive well-being in the context of goal pursuit (Ruehlman and Wolchik [150])."

"Support is defined in the current study as perceived encouragement from close others to pursue a goal, which can be expressed through verbal, monetary, and other forms of assurance or assistance."

"Although RARs and social support both involve close others, they are distinct constructs."

"Social support as we have defined it (i.e., encouragement for the pursuit of goals) also differs from what others have called autonomy support."

"We also expect that individuals with a highly relational self-construal will perceive others as providing more support for their goals than will low relationals."

"By attending to close others' interests, high relationals are also likely to receive some encouragement from those close others while pursuing their goals."

Goal Content, Duration, and Attributes

"Findings from cross-cultural research suggest that some differences in content and duration may exist between Americans and Japanese, particularly regarding self-improvement goals (Heine et al. [151]) or goals that involve changing one's personality (Heine et al. [152])."

"Although both effort and progress are indicative of a person's motivation to achieve a goal, and the two constructs tend to be highly correlated, they are not necessarily identical."

"Previous research on goal pursuit and well-being has shown that the personal meaning of goals is associated with global feelings of purpose (Gore and Cross [153]; McGregor and Little [154]; Ryff and Singer [155])."

"Gore and Cross [153] found that people pursuing their goals for RARs tended to exert more effort toward their goals than did others; effort, in turn, was associated with perceived progress toward the goal and feelings of purpose."

"PARs predicted feelings of purpose directly, but not indirectly (through effort or progress made toward the goal)."

The Proposed Model

"The purpose of this study is to examine the role of RARs in goal pursuit for members of Eastern and Western cultures, with a focus on the degree to which these motives predict positive goal outcomes."

"We seek to replicate the Gore and Cross [153] finding that RARs will predict perceived effort toward goals, controlling for PARs."

"We expect that RARs will predict positive goal outcomes controlling for both PARs and social support."

"Based on the findings from Gore and Cross [153], we propose a model that examines the associations among self-construal, motives, and goal outcomes."

"This model consists of Relational Self-Construal, two motivational variables (RARs, PARs), Social Support, two goal attributes (Effort and Progress), and a well-being component (Purpose in Life)."

Method

"Participants then completed the first packet of scales, which assessed a variety of personality and well-being constructs, including the measures of relational self-construal and purpose in life."

"Participants rated all items measuring reasons, support, effort, and progress multiple times, but each time was in reference to a different goal."

"For all goal-related measures (i.e., reasons, support, effort, and progress), reliability coefficients were calculated across all items and across all goals."

"The 3-item Support scale was used to measure the amount of perceived social support and encouragement received from close others while working toward goals."

"Effort was measured using five items that assessed the degree to which the individual had worked toward each goal."

"Progress was measured using three items that assessed the degree of perceived advancement for each goal."

Results

"To test for mean differences between men and women, between Americans and Japanese, and Gender × Country interaction effects, a MANOVA was conducted using Gender and Country as the independent variables and all of the variables in the model as the dependent variables."

"Conventional fit assessments use a non-significant chi-square statistic as an assessment of a good fit, but often multiple variable models with large samples typically result in significant chi-square statistics, as was the case noted below."

"Follow-up analyses examining individual paths were conducted; the path was considered to be moderated by culture if an equality constraint across the two samples resulted in a model fit that was significantly different from the fit of the model using no constraints."

"The final model, with all of the factor loadings and all but the aforementioned path constrained to be equal across groups, fit the data well, $\chi^2(376, N = 389) = 561.73$, $p < .01$; CFI = 0.97; RMSEA = 0.04."

Discussion

"Replicating previous research, relationally-autonomous reasons for goals (RARs) predicted goal effort directly, and progress and purpose indirectly, when controlling for PARs and support for both Americans and Japanese."

"Although the items used to measure RARs and PARs are similar, this research demonstrated that the two constructs serve distinct functions in the goal pursuit process, replicating the findings of Gore and Cross [153] and extending them to include individuals in a collectivistic culture."

"Replicating Gore and Cross' [153] findings, the degree to which a person pursued a goal for RARs directly predicted the amount of effort that person perceived expending on his or her goal, whereas PARs did not predict effort when controlling for RARs."

"This study expands upon Gore and Cross' [153] findings, because we found that these patterns were evident in both American and Japanese samples, which indicates that the two sources of motivation have unique roles in the goal pursuit process and in predicting well-being for both individualistic and collectivistic cultures."

[Section 10]
"Individual preferences, beliefs, and values are often argued to be the optimal sources of goal motivation, but this assertion has been recently called into question by cultural psychologists (Markus and Kiatayama [156]) and with the investigation of relational reasons for goals."

"The degree to which people consider the interests of close others in their reasons for pursuing their goals can be a powerful impetus for action."

"The purpose of the current study is to investigate the degree to which reasons that integrate close others' interests with one's own (termed relationally-autonomous reasons, or RARs) motivate people across cultures, and the differences between cultures in the effectiveness of these motives."

Academic Self-efficacy: From Educational Theory to Instructional Practice [188]

This is a machine-generated summary of:

Artino, Anthony R.: Academic self-efficacy: from educational theory to instructional practice [188]

Published in: Perspectives on Medical Education (2012)

Link to original: https://doi.org/10.1007/s40037-012-0012-5

If you want to cite the papers, please refer to the original.

For technical reasons we could not place the page where the original quote is coming from.

Abstract-Summary
"Over the last 34 years, educational researchers from diverse fields of inquiry have used the notion of self-efficacy to predict and explain a wide range of human functioning, from athletic skill to academic achievement."

"This article is not a systematic review of the empirical research on self-efficacy; instead, its purpose is to describe the nature and structure of self-efficacy and provide a brief overview of several instructional implications for medical education."

"This article is meant to encourage medical educators to consider and explicitly address their students' academic self-efficacy beliefs in an effort to provide more engaging and effective instruction."

The Nature and Structure of Self-efficacy

"A second important aspect of Bandura's definition of self-efficacy is the idea that individuals make use of their efficacy judgments in reference to some goal ('attain designated types of performances'), which reflects both the task- and situation-specific nature of efficacy beliefs."

"Self-efficacy theory postulates that people acquire information to evaluate efficacy beliefs from four primary sources: (a) enactive mastery experiences (actual performances); (b) observation of others (vicarious experiences); (c) forms of persuasion, both verbal and otherwise; and (d) 'physiological and affective states from which people partly judge their capableness, strength, and vulnerability to dysfunction' [157]."

"Although it is clear that task and domain-specific measures of perceived efficacy have greater predictive power than global measures of the construct, Bandura [157] warned that it is incorrect to believe that self-efficacy is concerned solely with 'specific behaviors in specific situations.'"

Self-efficacy in Medical Education: Instructional Implications

"The evidence has shown that students with high self-efficacy in various academic domains choose to engage in tasks that foster the development of their knowledge, skills, and abilities in those areas; exert effort in the face of difficulty; and persist longer at challenging tasks [158, 159]."

"Much of the research suggests that students' perceptions of confidence (i.e., their self-efficacy beliefs) may more accurately predict their motivation and future academic choices than actual competence."

"Instructional strategies focused on providing students with opportunities for performance success align well with Bandura's [157, 160, 161] emphasis on enactive attainment as the most influential source of self-efficacy information."

"Several techniques have been suggested for helping students accurately calibrate their self-efficacy beliefs."

"Feedback not only helps students grow their self-efficacy beliefs, as described above, but it can play a prominent role in calibration."

Conclusions

"The self-efficacy component of Bandura's social-cognitive theory has had a profound impact on the study of motivation and achievement in academic settings."

"Results from a meta-analysis of more than 100 empirical studies conducted over the last 20 years found that of nine commonly researched psychosocial constructs, academic self-efficacy was the strongest single predictor of college students' academic achievement and performance [162]."

"It seems, then, that cultivating students' academic self-efficacy is a worthwhile goal for any educator."

Essentials

"Self-efficacy is a personal belief in one's capability to organize and execute courses of action required to attain designated types of performances."

"Individuals develop self-efficacy beliefs in relation to specific goals, such as the ability to accurately assess heart sounds in a patient."

"Instructional practices designed to develop students' self-efficacy beliefs and improve learning include the following: encourage students to set clear, specific, and challenging proximal goals; provide students with honest and explicit feedback; facilitate accurate calibration of self-efficacy; and use peer modeling."

[Section 5]

"Since the publication of Albert Bandura's seminal article entitled 'Self-Efficacy: Toward a Unifying Theory of Behavioral Change,'[160] countless researchers in the social and behavioral sciences have used self-efficacy to predict and explain a wide range of human functioning."

"The purpose of this article is to describe the nature and structure of self-efficacy, a key component of social-cognitive theory, and to provide a brief overview of several potential instructional implications for medical education."

"By explicating Bandura's theory of self-efficacy, this article encourages medical educators to consider and explicitly address their students' academic self-efficacy beliefs in an effort to provide more engaging and effective instruction."

Assessing Smallholder Farmers' Motivation to Adopt Agroforestry Using a Multi-group Structural Equation Modeling Approach [189]

This is a machine-generated summary of:

Buyinza, Joel; Nuberg, Ian K.; Muthuri, Catherine W.; Denton, Matthew D.: Assessing smallholder farmers' motivation to adopt agroforestry using a multi-group structural equation modeling approach [189]

Published in: Agroforestry Systems (2020)

Link to original: https://doi.org/10.1007/s10457-020-00541-2

If you want to cite the papers, please refer to the original.

For technical reasons we could not place the page where the original quote is coming from.

Abstract-Summary

"We used the theory of planned behaviour framework to assess the adoption behaviour of these farmer categories resulting from project interventions."

"About 40% of the variation in farmer motivation to integrate trees in their coffee plantations was explained by the significant variables of 'attitude' and 'perceived behavioural control' among farmers actively participating in the T4FS project from phase 1."

"The neighbors of participating farmers and farmers who had never interacted with the project were only motivated by 'attitude' and 'social norms' respectively."

"Farmer motivation resulting from social pressure was strongest among farmers who had never interacted with the project, and in the absence of project interventions, rely on existing social structures to drive change in their community."

"Farmers' perceived behavioural control to overcome tree planting barriers and their attitude to the economic benefits of shaded coffee were significantly different among the four farmer categories ($p < 0.05$)."

"The adoption behaviour of smallholder farmers is mainly shaped by existing community social norms and beliefs that tend to promote knowledge exchange, as opposed to the conventional knowledge transfer extension approaches."

Introduction

"This study applied the multi-group structural equation modeling technique to identify the differences in farmer psychological drivers to adopt shaded coffee across four farmer categories in the Mt. Elgon region of Uganda."

"The cheaper alternative available to smallholder coffee farmers is the integration of shade trees, facilitated by the T4FS project, to sustain their coffee production."

"The study addresses the extent to which project interventions influenced smallholder farmers' motivations to integrate trees in their coffee farming systems."

"To investigate smallholder farmers' motivation to adopt shaded coffee on their farms, this study employed the Theory of Planned Behaviour (TPB), which suggests that behavioural intentions are shaped by attitude, subjective norms and perceived behavioural control (Ajzen [163])."

"The components of the TPB (attitude, subjective norm and perceived behavioural control) relate to the key aspects influencing smallholder farmers' decision-making on integration of trees in their coffee plantations in the Mt. Elgon region of Uganda (Buyinza and others [164])."

Methods

"The farmers actively participating in the project since its inception and those neighbouring active project participants, were selected from Manafwa district, the only area where the T4FS project has been operating since 2014."

"Farmers actively participating in the second phase of the project and those who had never participated in the T4FS project were selected from Bududa and Sironko districts respectively."

"While a random sample of farmers actively participating in the project since its inception and farmers actively participating in phase 2 of the project was obtained from the list of project beneficiaries, farmers who had never participated in the project were randomly selected from a list of households from the local council leaders."

Data Analysis

"Multi-group SEM provides a simultaneous estimation of different interdependent multiple regressions (Hair and others [165]) that allows analysis of several groups of data from a population."

"Multi-group SEM aims to identify the differences in farmer psychological motivations to adopt agroforestry practices across these farmer categories in the Mt. Elgon region of Uganda."

"Using the validated structural model, two structural models were estimated as meeting the requirements of the multi-group SEM technique."

"The first was the unconstrained structural model in which all parameters were to be estimated for each of the four groups."

"The two models were then assessed using the Chi-square test to decide which model was better to address factors that influenced the adaptation intentions across the four farmer categories at different levels of interaction with the T4FS project in the Mt. Elgon region of Uganda."

Results

"Farmers generally displayed a high intention to plant shade trees in their coffee plantations."

"Farmers evaluated planting of trees in coffee as being highly favorable, expressed by means above 6 out of 7 for all variables measuring attitude across all farmer categories."

"Farmers displayed a moderate perception of control on planting scattered trees on their farms."

"Farmers' perceived own capability (perceived behavioural control) to overcome tree planting barriers and their evaluation of the economic benefits of shaded coffee (attitude) were significantly different among the four farmer categories ($p < 0.05$)."

Multi-group Analysis Model Parameter Outputs

"The intention of farmers actively participating in the T4FS project from phase 1 to plant trees in coffee is significantly influenced by their attitude and perceived behavioural control."

"Only the attitude construct influences farmers neighbouring those actively participating in the T4FS project to plant trees in coffee plantations, and none of the TPB constructs influences the intentions of the farmers who have never participated in the T4FS project and who are living far from the farmers participating in the project, to plant trees in coffee plantations."

"There was also an insignificant negative influence of perceived behavioural control to plant trees in coffee among farmers who had never participated in the T4FS project and living far from project participating farmers."

Discussion

"This implies that the motivation of project participating farmers (phase 1 and 2) to adopt shaded coffee was a result of their positive evaluation of shaded coffee as being more favorable and their own perceived capability to implement the practice of integrating trees in their coffee plantations."

"The neighbors of project participating farmers and farmers that had never interacted with the project were only motivated by 'attitude' and 'social norms' respectively."

"Early research in the agricultural context found that the process of diffusion exerted social pressure on farmers to adopt innovative farming practices championed by early adopters in the neighbourhood."

"This could explain why social norms are predominant in Sironko, where farmers have never interacted with the T4FS research project."

"The expression of social norms as drivers towards integration of trees in coffee systems among farmers who had never interacted with the project and lived far from project beneficiaries seems to demonstrate this theory."

Conclusions

"Sustainable agricultural technology adoption requires that researchers and development agencies pay more attention to the internal decision-making processes, and look beyond the mere characteristics of agricultural innovations and the household to include psychological factors in technology uptake."

"The Theory of Planned Behaviour provides a useful model for exploring the psychological factors that influence smallholder farmers' tree planting decisions."

"The findings indicate that psychological factors are key drivers to the farmers' internal decision-making processes in agroforestry technology adoption."

"The psychological factors vary among different groups of farmers, usually shaped by the existing community social norms and beliefs."

Motivation Gap and Achievement Gap Between Public and Private High Schools in the Philippines [190]

This is a machine-generated summary of:

Bernardo, Allan B. I.; Ganotice, Fraide A.; King, Ronnel B.: Motivation Gap and Achievement Gap Between Public and Private High Schools in the Philippines [190]

Published in: The Asia-Pacific Education Researcher (2014)

Link to original: https://doi.org/10.1007/s40299-014-0213-2

If you want to cite the papers, please refer to the original.

For technical reasons we could not place the page where the original quote is coming from.

Abstract-Summary

"As in many countries, public school students in the Philippines have lower levels of achievement compared to private school students."

"Filipino students (N = 1,694) enrolled in high school Chemistry from private and public high schools participated in the study."

"Multivariate Analysis of Variance indicated that public school students reported less support for schooling from their social groups, lower academic related self-concept, and lower achievement goals compared to private school students."

Introduction

"There is some indication that factors related to students' motivation may be related to the achievement gaps between private and public schools in the Philippines."

"Applying these perspectives to the students in the private and public education sector in the Philippines, we propose that these two groups of students are likely to also have distinct motivations that emerge from their different school environments and experiences, and that are associated with varying levels of achievement and other important school outcomes."

"At the outset we should assert that we are not assuming that motivation can completely account for the achievement gap between students in private and public schools."

"How similar or different are the perceived facilitating conditions, sense of self, and achievement goals of Filipino students from private and public schools?"

Method

""; α = .93); and Positive Self-Concept refers to the degree to which a student holds positive feelings about his/her general intellectual ability in school (5 items, e.g., "I think I am as good as everybody else at school."

"To measure student engagement in learning, three scales developed by McInerney and colleagues (2005) were used: (1) intention for further education scale: the plan of a student to continue his/her education in college or university (5 items, e.g., "I intend to go on to college or university"; α = .82); (2) school valuing scale: the degree to which the a student values education (9 items, e.g., "Education is important for me to get a job"; α = .75); and (3) positive affect for schooling scale: the degree to which the student appears to like school (3 items, e.g., "Subjects at school interest me"; α = .96)."

Results

"The MANOVA conducted involving the eight facilitating conditions indicated significant differences between the public and private school groups; $F(8, 1685) = 203.09$, $p = .0001$."

"An overall MANOVA was conducted across the three self-report school engagement measures (i.e., university intention, affect to school, and school valuing) and indicated significant overall differences between the public and private high school students, $F(3, 1{,}690) = 111.27$, $p = .0001$."

"The sense of self variables seemed to explain more of the variance of the engagement outcomes for the private school students (the R^2s were significant, although small) compared to the public school students (the R^2s were mostly not significant)."

"Although sense of self explained a significant amount of variance in the achievement scores of both groups, sense of purpose and self-reliance predict achievement in the private school students, whereas positive self-concept predicts achievement in the public school students."

Discussion

"The public schools students also reported lower levels of all achievement goals compared to private school students, which may be a reflection the low levels of social support and personal self-beliefs related to learning."

"The public school students may not set as high achievement goals as those from private schools because there is not much social support for it and they do not enough positive self-beliefs to do so."

"We realize that there is a complex array of interaction organizational, political, economic, social, and even historical factors that have shaped and that will continue to shape the achievement gap between students in public and private schools in the Philippines, but we hope that our study calls attention to an important gap in the educational experiences of Filipino students in the public schools as part of the broader collective efforts at improving the educational experiences and outcomes of the disadvantaged groups in the country."

Bibliography

1. Baumeister, R. F. (2005). The cultural animal: Human nature, meaning, and social life. New York: Oxford University Press.
2. Baumeister, R. F. (1991). Meanings of life. New York: Guilford Press.
3. Freud, S. (1920/1961). Beyond the pleasure principle (standard ed.). London: Norton.
4. Freud, S. (1933). New introductory lectures on psycho-analysis. New York: Norton.
5. Schilling, K. (2009). The impact of multimedia course enhancements on student learning outcomes. Journal of Education for Library and Information Science, 50(4), 214–225.
6. Sadik, A. (2008). Digital storytelling: A meaningful technology-integrated approach for engaged student learning. Education Technology Research and Development, 56, 487–506.
7. House, J. D. (2009). The effects of instructional and computer activities on interest in science learning for students in the United States and Korea: Results from the TIMSS 2003 assessment. International Journal of Instructional Media, 36(1), 119–131.
8. Hsu, Y.-S. (2008). Learning about seasons in a technologically enhanced environment: The impact of teacher-guided and student centered instructional approaches on the process of student's conceptual change. Science Education, 92(2), 320–344.
9. Burkley, E. (2016). The MP3 model of effort mobilization: Motivation as promptness, perseverance and power. Unpublished manuscript.
10. Geen, R. G. (1995). Human motivation: A social psychological approach. Thomson Brooks/Cole Publishing Co.
11. Richter, M., Gendolla, G. H. E., & Wright, R. A. (2016). Three decades of research on motivational intensity theory: What we have learned about effort and what we still don't know. In A. Elliot (Ed.), Advances in motivation science. Elsevier Press.
12. Gendolla, G. H. E., Wright, R. A., & Richter, M. (2012). Effort intensity: Some insights from the cardiovascular system. In R. M. Ryan (Ed.), The Oxford handbook of human motivation (pp. 420–438). Oxford University Press.
13. Brehm, J. W., & Self, E. A. (1989). The intensity of motivation. Annual Review of Psychology, 40, 109–131.
14. Brehm, J. W., Wright, R. A., Solomon, S., Silka, L., & Greenberg, J. (1983). Perceived difficulty, energization, and the magnitude of goal valence. Journal of Experimental Social Psychology, 19, 21–48. https://doi.org/10.1016/00221031(83)90003-3

15. Stanek, J. C., & Richter, M. (2021). Energy investment and motivation: The additive impact of task demand and reward value on exerted force in hand grip tasks. Motivation and Emotion. https://doi.org/10.1007/s11031-020-09862-2
16. Everitt, B. S. (1975). Multivariate analysis: The need for data, and other problems. The British Journal of Psychiatry, 126, 237–240. https://doi.org/10.1192/bjp.126.3.237
17. Kline, P. (1994). An easy guide to factor analysis. Routledge.
18. MacCallum, R. C., Widaman, K. F., Zhang, S., & Hong, S. (1999). Sample size in factor analysis. Psychological Methods, 4(1), 84–99. https://doi.org/10.1037/1082-989X.4.1.84
19. O'Connor, B. P. (2000). SPSS and SAS programs for determining the number of components using parallel analysis and velicer's MAP test. Behavior Research Methods, Instruments & Computers, 32(3), 396–402.
20. Hu, L., & Bentler, P. M. (1999). Cutoff criteria for fit indexes in covariance structure analysis: Conventional criteria versus new alternatives. Structural Equation Modeling, 6(1), 1–55.
21. Schreiber, J. B., Stage, F. K., King, J., Nora, A., & Barlow, E. A. (2006). Reporting structural equation modeling and confirmatory factor analysis results: A review. The Journal of Educational Research, 99(6), 323–337. https://doi.org/10.3200/JOER.99.6.323-338
22. Bentler, P. M., & Bonett, D. G. (1980). Significance tests and goodness of fit in the analysis of covariance structures. Psychological Bulletin, 88(3), 588–606. https://doi.org/10.1037/00332909.88.3.588
23. John, O. P., Naumann, L. P., & Soto, C. J. (2008). Paradigm shift to the integrative Big Five trait taxonomy: History, measurement, and conceptual issues. In O. P. John, R. W. Robins, & L. A. Pervin (Eds.), Handbook of personality: Theory and research (pp. 114–158). Guilford Press.
24. McCrae, R. R., & Costa, P. T. (1987). Validation of the five-factor model of personality across instruments and observers. Journal of Personality and Social Psychology, 52(1), 81–90. https://doi.org/10.1037/00223514.52.1.81
25. ELSamen, A. A. (2011). Examining the construct validity of the Lockwood goal orientation scale using the general hierarchal model: An exploratory study. Journal of Management Policy and Practice, 12(4), 81–93.
26. Carver, C. S., & White, T. L. (1994). Behavioral inhibition, behavioral activation, and affective responses to impending reward and punishment: The BIS/BAS scales. Journal of Personality and Social Psychology, 67(2), 319–333. https://doi.org/10.1037/00223514.67.2.319
27. Wright, R. A., & Brehm, J. W. (1989). Energization and goal attractiveness. In P. A. Lawrence (Ed.), Goal concepts in personality and social psychology (pp. 169–210). Lawrence Erlbaum Associates Inc.
28. Bargh, J. A., Gollwitzer, P. M., & Oettingen, G. (2010). Motivation. In S. Fiske, D. T. Gilbert, & G. Lindzay (Eds.), Handbook of social psychology (pp. 268–316). Wiley. https://doi.org/10.1002/9780470561119.socpsy001008
29. Freydefont, L., Gollwitzer, P. M., & Oettingen, G. (2016). Goal striving strategies and effort mobilization: When implementation intentions reduce effort-related cardiac activity during task performance. International Journal of Psychophysiology, 107, 44–53. https://doi.org/10.1016/j.ijpsycho.2016.06.013
30. Gendolla, G. H. E., Wright, R. A., & Richter, M. (2019). Advancing issues in motivation intensity research: Updated insights from the cardiovascular system. In R. M. Ryan (Ed.), The Oxford handbook of human motivation (pp. 373–392). Oxford University Press.
31. Silvestrini, N., & Gendolla, G. H. E. (2019). Affect and cognitive control: Insights from research on effort mobilization. International Journal of Psychophysiology, 143, 116–125. https://doi.org/10.1016/j.ijpsycho.2019.07.003
32. Higgins, E. T., Kruglanski, A. W., & Pierro, A. (2003). Regulatory mode: Locomotion and assessment as distinct orientations. In M. P. Zanna (Ed.), Advances in experimental social psychology (Vol. 35, pp. 293–344). New York: Academic Press.
33. Kruglanski, A. W., Pierro, A., Higgins, E. T., & Capozza, D. (2007). "On the move" or "staying put": Locomotion, need for closure and reactions to organizational change. Journal of Applied Social Psychology, 37, 1305–1340.

34. Changizi, M. A., & Hall, W. G. (2001). Thirst modulates a perception. Perception, 30, 1489–1497. https://doi.org/10.1068/p3266 .
35. Balcetis, E. (2016). Approach and avoidance as organizing structures for motivated distance perception. Emotion Review, 8, 115–128. https://doi.org/10.1177/1754073915586225 .
36. Cole, S., Balcetis, E., & Dunning, D. (2013). Affective signals of threat increase perceived proximity. Psychological Science, 24, 34–40. https://doi.org/10.1177/0956797612446953 .
37. Cole, S., & Balcetis, E. (2013). Sources of resources: Bioenergetic and psychoenergetic resources influence distance perception. Social Cognition, 31, 721–732. https://doi.org/10.1521/soco.2013.31.6.721 .
38. Balcetis, E., & Dunning, D. (2010). Wishful seeing: Desired objects are seen as closer. Psychological Science, 21, 147–152. https://doi.org/10.1177/0956797609356283 .
39. Faul, F., Erdfelder, E., Buchner, A., & Lang, A.-G. (2009). Statistical power analyses using G*Power 3.1: Tests for correlation and regression analyses. Behavior Research Methods, 41, 1149–1160. https://doi.org/10.3758/BRM.41.4.1149 .
40. Steinmetz, J., Tausen, B. M., & Risen, J. L. (2018). Mental simulation of visceral states affects preferences and behavior. Personality and Social Psychology Bulletin, 44, 406–416. https://doi.org/10.1177/0146167217741315 .
41. Vallacher, R. R., & Wegner, D. M. (1989). Levels of personal agency: Individual variation in action identification. Journal of Personality and Social Psychology, 57, 660–671. https://doi.org/10.1037/0022-3514.57.4.660 .
42. Fujita, K., Trope, Y., Liberman, N., & Levin-Sagi, M. (2006). Construal level and self-control. Journal of Personality and Social Psychology, 90, 351–367. https://doi.org/10.1037/0022-3514.90.3.351 .
43. Ledgerwood, A., Wakslak, C. J., & Wang, M. A. (2010). Differential information use for near and distant decisions. Journal of Experimental Social Psychology, 46, 638–642. https://doi.org/10.1016/j.jesp.2010.03.001 .
44. Zhang, Y., & Risen, J. L. (2014). Embodied motivation: Using a goal systems framework to understand the preference for social and physical warmth. Journal of Personality and Social Psychology, 107, 965–977. https://doi.org/10.1037/a0038153 .
45. LeDoux, J. E. (2014). Comment: What's basic about the brain mechanisms of emotion? Emotion Review, 6, 318–320. https://doi.org/10.1177/1754073914534506 .
46. Frijda, N. H. (2013). Comment: The why, when, and how of appraisal. Emotion Review, 5, 169–170. https://doi.org/10.1177/1754073912468905 .
47. Levenson, R. W. (2014). The autonomic nervous system and emotion. Emotion Review, 6, 100–112. https://doi.org/10.1177/1754073913512003 .
48. Tracy, J. L. (2014). An evolutionary approach to understanding distinct emotions. Emotion Review, 5, 308–312. https://doi.org/10.1177/1754073914534478 .
49. Roseman, I. J. (2008). Motivations and emotions: Approach, avoidance, and other tendencies in motivated and emotional behavior. In A. J. Elliot (Ed.), Handbook of approach and avoidance motivation (pp. 343–366). New York: Psychology Press.
50. Roseman, I. J. (2013). Appraisal in the emotion system: Coherence in strategies for coping. Emotion Review, 5, 141–149. https://doi.org/10.1177/1754073912469591 .
51. Rempel, J. K., & Burris, C. T. (2005). Let me count the ways: An integrative theory of love and hate. Personal Relationships, 12, 297–313. https://doi.org/10.1111/j.1350-4126.2005.00116.x .
52. Rempel, J. K., & Sutherland, S. (2016). Hate: Theory and implications for intimate relationships. In K. Aumer (Ed.), The psychology of love and hate in intimate relationships (pp. 105–129). Cham: Springer. https://doi.org/10.1007/978-3-319-39277-6_7 .
53. Halevy, N. (2017). Preemptive strikes: Fear, hope, and defensive aggression. Journal of Personality and Social Psychology, 112, 224–237. https://doi.org/10.1037/pspi0000077
54. Halperin, E., Canetti-Nisim, D., & Hirsch-Hoefler, S. (2009). The central role of group-based hatred as an emotional antecedent of political intolerance: Evidence from Israel. Political Psychology, 30, 93–123. https://doi.org/10.1111/j.1467-9221.2008.00682.x

55. Hatfield, E., Cacioppo, J. T., & Rapson, R. L. (1994). Emotional contagion. Paris: Cambridge University Press.
56. Buss, D. M. (1991). Evolutionary personality psychology. Annual Review of Psychology, 42, 459–491. https://doi.org/10.1146/annurev.ps.42.020191.002331 .
57. Buss, D. M. (1999b). Adaptive individual differences revisited. Journal of Personality, 67, 259–264. https://doi.org/10.1111/1467-6494.00055 .
58. Figueredo, A. J., Sefcek, J. A., Vasquez, G., Brumbach, B. H., King, J. E., & Jacobs, W. J. (2005). Evolutionary personality psychology. In D. M. Buss (Ed.), The handbook of evolutionary psychology (pp. 851–877). Hoboken, NJ: Wiley.
59. MacDonald, K. (1998). Evolution, culture, and the five-factor model. Journal of Cross-Cultural Psychology, 29, 119–149. https://doi.org/10.1177/0022022198291007 .
60. Jackson, D. N. (1999). Personality research form manual (3rd ed.). Port Huron, MI: Sigma Assessment Systems.
61. Bernard, L. C., Mills, M., Swenson, L., & Walsh, R. P. (2008). Measuring motivation multidimensionally: Development of the assessment of individual motives-questionnaire (AIM-Q). Assessment, 15, 16–35. https://doi.org/10.1177/1073191107306131 .
62. Ratcliff, J. J., Lassiter, G. D., Markman, K. D., & Snyder, C. J. (2006). Gender differences in attitudes toward gay men and lesbians: The role of motivation to respond without prejudice. Personality and Social Psychology Bulletin, 32, 1325–1338.
63. Carpenter, J. M., Green, M. C. , & LaFlam, J. (2011). People or profiles? individual differences in Facebook use. Personality and Individual Differences, 50, 538–541.
64. Archer, D., & Costanzo, M. (1993). The Interpersonal perception task (IPT-15). Berkeley, CA: University of California, Extension Center for Media and Independent Learning.
65. Davis, M. H. (1980). A multidimensional approach to individual differences in empathy. JSAS Catalog of Selected Documents in Psychology, 10, 85.
66. Kantor, J. R., & Smith, N. W. (1975). The science of psychology: An interbehavioral survey. Bloomington, IN: The Principia Press.
67. Michael, J. L. (1993). Establishing operations. The Behavior Analyst, 16, 191–206.
68. Michael, J. L. (1982). Distinguishing between discriminative and motivational functions of stimuli. Journal of the Experimental Analysis of Behavior, 37, 149–155.
69. Finger, F. W. (1951). The effect of food deprivation and subsequent satiation upon general activity in the rat. Journal of Comparative and Physiological Psychology, 44(6), 557–564.
70. Kantor, J. R. (1938). The nature of psychology as a natural science. Acta PsychologIA, 4, 1–61.
71. Kantor, J. R. (1958). Interbehavioral psychology. Chicago, IL: The Principia Press.
72. Hayes, L. J., Adams, M. A., & Dixon, M. R. (1997). Causal constructs and conceptual confusions. The Psychological Record, 46, 97–111.
73. Klatt, K. P., & Morris, E. K. (2001). The Premack principle, response deprivation, and establishing operations. The Behavior Analyst, 24, 173–180.
74. Butler, E. A., Egloff, B., Wilhelm, F. W., Smith, N. C., Erickson, E. A., & Gross, J. J. (2003). The social consequences of expressive suppression. Emotion, 3, 48–67.
75. Gross, J. J. (1998). Antecedent- and response-focused emotion regulation: Divergent consequences for experience, expression, and physiology. Journal of Personality and Social Psychology, 74, 224–237.
76. Richards, J. M., & Gross, J. J. (2000). Emotion regulation and memory: The cognitive costs of keeping one's cool. Journal of Personality and Social Psychology, 79, 410–424.
77. Hox, J. (2002). Multilevel analysis: Techniques and applications. Mahwah, NJ: Erlbaum.
78. Snijders, T., & Bosker, R. (1999). Multilevel analysis. London: Sage Publications.
79. Rosenfeld, D. L., & Burrow, A. L. (2017a). The unified model of vegetarian identity: A conceptual framework for understanding plant-based food choices. Appetite, 112, 78–95.
80. Rosenfeld, D. L., & Burrow, A. L. (2018). Development and validation of the Dietarian Identity Questionnaire: Assessing self-perceptions of animal-product consumption. Appetite, 127, 182–194.

81. Fox, N., & Ward, K. J. (2008). Health, ethics and environment: A qualitative study of vegetarian motivations. Appetite, 50, 422–429.
82. Greenebaum, J. (2012). Veganism, identity and the quest for authenticity. Food, Culture & Society, 15, 129–144.
83. Rosenfeld, D. L., & Burrow, A. L. (2017b). Vegetarian on purpose: Understanding the motivations of plant-based dieters. Appetite, 116, 456–463.
84. Rozin, P., Markwith, M., & Stoess, C. (1997). Moralization and becoming a vegetarian: The transformation of preferences into values and the recruitment of disgust. Psychological Science, 8, 67–73.
85. Jabs, J., Devine, C. M., & Sobal, J. (1998). Model of the process of adopting vegetarian diets: Health vegetarians and ethical vegetarians. Journal of Nutrition Education, 30, 196–202.
86. Janssen, M., Busch, C., Rödiger, M., & Hamm, U. (2016). Motives of consumers following a vegan diet and their attitudes towards animal agriculture. Appetite, 105, 643–651.
87. Rothgerber, H. (2013). A meaty matter. Pet diet and the vegetarian's dilemma. Appetite, 68, 76–82.
88. Rothgerber, H. (2014a). A comparison of attitudes toward meat and animals among strict and semi-vegetarians. Appetite, 72, 98–105.
89. Azrin, N. H. (1961). Time-out from positive reinforcement. Science, 133, 382–383. https://doi.org/10.1126/science.133.3450.382 .
90. Azrin, N. H., Hake, D. F., Holz, W. C., & Hutchinson, R. R. (1967). Motivational aspects of escape from punishment. Journal of the Experimental Analysis of Behavior, 8, 31–44. https://doi.org/10.1901/jeab.1965.8-31 .
91. Thompson, D. M. (1964). Escape from SD associated with fixed-ratio reinforcement. Journal of the Experimental Analysis of Behavior, 7, 1–8. https://doi.org/10.1901/jeab.1964.7-1 .
92. Thompson, D. M. (1965). Punishment by SD associated with fixed-ratio reinforcement. Journal of the Experimental Analysis of Behavior, 8, 189–194. https://doi.org/10.1901/jeab.1965.8-189 .
93. Azrin, N. H., Hutchinson, R. R., & Hake, D. H. (1967). Attack, avoidance, and escape reactions to aversive shock. Journal of the Experimental Analysis of Behavior, 10, 131–148. https://doi.org/10.1901/jeab.1967.10-131 .
94. Bolles, R. C. (1970). Species-specific defense reactions and avoidance learning. Psychological Review, 77, 32–48. https://doi.org/10.1037/h0028589 .
95. Halperin, E. (2011). The emotional roots of intergroup aggression: The distinct roles of anger and hatred. In P. R. Shaver & M. Mikulincer (Eds.), Herzilya series on personality and social psychology: Human aggression and violence. Washington, DC: American Psychological Association.
96. Mattaini, M., & Holtschneider, C. (2017). Collective leadership and circles: Not invented here. Journal of Organizational Behavior Management, 37, 126–141. https://doi.org/10.1080/01608061.2017.1309334 .
97. Skinner, B. F. (1971). Beyond freedom & dignity. Indianapolis, IN: Hackett.
98. Howes, D. E. (2013). The failure of pacifism and the success of nonviolence. Perspectives on Politics, 11, 427–446. https://doi.org/10.1017/S1537592713001059 .
99. Pape, R. A. (1997). Why economic sanctions do not work. International Security, 22, 90–136. https://doi.org/10.1162/isec.22.2.90 .
100. Sprinzak, E. (2000). Rational fanatics. Foreign Policy, 120, 66–73.
101. Stephan, M. J., & Chenoweth, E. (2008). Why civil resistance works: The strategic logic of nonviolent conflict. International Security, 33, 7–44. https://doi.org/10.1162/isec.2008.33.1.7 .
102. Kruglanski, A. W., & Gigerenzer, G. (2010). Intuitive and deliberative judgments are based on common processes. Unpublished manuscript: University of Maryland.
103. Roets, A., Van Hiel, A., Cornelis, I., & Soetens, B. (2008). Determinants of performance and invested effort. A need for closure by cognitive capacity interaction analysis. Personality and Social Psychology Bulletin, 34, 779–792.

104. Ickes, W., Gesn, P. R., & Graham, T. (2000). Gender differences in empathic accuracy: Differential ability or differential motivation? Personal Relationships, 7, 95–109. https://doi.org/10.1111/j.1475-6811.2000.tb00006.x .
105. Koenig, A. M., & Eagly, A. H. (2005). Stereotype threat in men on a test of social sensitivity. Sex Roles, 52, 489–496. https://doi.org/10.1007/s11199-005-3714-x .
106. Rosip, J. C., & Hall, J. A. (2004). Knowledge of nonverbal cues, gender, and nonverbal decoding accuracy. Journal of Nonverbal Behavior, 28, 267–286. https://doi.org/10.1007/s10919-004-4159-6 .
107. Murray, H. A. (1943). Thematic Apperception Test manual. Cambridge, MA: Harvard University Press.
108. Jackson, D. N. (1987). Personality Research Form—Form E. Port Huron, MI: Sigma Assessment Systems, Inc.
109. Deci, E. L., & Ryan, R. M. (2000). The "what" and "why" of goal pursuits: Human needs and the self-determination of behavior. Psychological Inquiry, 11, 227–268.
110. Rotter, J. B. (1990). Internal versus external control of reinforcement: A case history of a variable. American Psychologist, 45, 489–493.
111. Vroom, V. H. (1964). Work and motivation. Oxford: Wiley.
112. Koch, S. (1951). The current status of motivational psychology. Psychological Review, 58, 147–154.
113. Pintrich, P. R., Smith, D. A., Garcia, T., & McKeachie, W. (1993). Reliability and predictive validity of the Motivated Strategies for Learning Questionnaire (MSLQ). Educational and Psychological Measurement, 53, 801–813.
114. Amabile, T. M., Hill, K. G., Hennessey, B. A., & Tighe, E. M. (1994). The Work Preference Inventory: Assessing intrinsic and extrinsic motivational orientations. Journal of Personality and Social Psychology, 66, 950–967.
115. Barbuto, J. E. Jr., & Scholl, R. W. (1998). Motivational Sources Inventory: Development and validation of new scales to measure an integrative taxonomy of motivation. Psychological Reports, 82(3, Pt 1), 1011–1022.
116. Cesario, J., Grant, H., & Higgins, E. T. (2004). Regulatory fit and persuasion: Transfer from "feeling right". Journal of Personality and Social Psychology, 86, 388–404.
117. Higgins, E. T., Friedman, R. S., Harlow, R. E., Idson, L. C., Ayduk, O. N., & Taylor, A. (2001). Achievement orientations from subjective histories of success: Promotion pride versus prevention pride. European Journal of Social Psychology, 31, 3–23.
118. Carver, C. S., & White, T. L. (1994). Behavioral inhibition, behavioral activation, and affective responses to impending reward and punishment: The BIS/BAS scales. Journal of Personality and Social Psychology, 67, 319–333.
119. Allport, G. W., & Ross, J. M. (1967). Personal religious orientation and prejudice. Journal of Personality and Social Psychology, 5, 432–443.
120. Vallerand, R. J., Pelletier, L. G., Blais, M. R., & Brière, N. M. (1992). The Academic Motivation Scale: A measure of intrinsic, extrinsic, and amotivation in education. Educational and Psychological Measurement, 52, 1003–1017.
121. McClelland, D. C. (1985). How motives, skills, and values determine what people do. American Psychologist, 40, 812–825.
122. Biernat, M. (1989). Motives and values to achieve: Different constructs with different effects. Journal of Personality, 57, 69–95.
123. Thrash, T. M., & Elliot, A. J. (2002). Implicit and self-attributed achievement motives: Concordance and predictive validity. Journal of Personality, 70, 729–756.
124. Murray, H. A. (1938). Explorations in personality. New York: Oxford University Press.
125. Sparfeldt, J. R., Buch, S. R., & Rost, D. H. (2010). Klassenprimus bei durchschnittlicher Intelligenz – Overachiever auf dem Gymnasium [Excellent Scholastic Achievement with Average Intelligence – Overachievement in Secondary School]. Zeitschrift für pädagogische Psychologie, 24(2), 147–155. https://doi.org/10.1024/1010-0652/a000012
126. Hanses, P., & Rost, D. H. (1998). Das „Drama der hochbegabten Underachiever": "gewöhnliche" oder "außergewöhnliche" Underachiever? [The "drama" of the gifted underachievers

- "ordinary" or "extraordinarty" underachievers?]. Zeitschrift für Pädadogische Psychologie, 12(1), 53–71.
127. Buttrick, N. R., Choi, H., Wilson, T. D., Oishi, S., Boker, S. M., Wilks, D. C. (2017). Thinking versus doing for enjoyment: A cross-cultural replication. Under editorial review.
128. Smith, E. N., & Frank, M. C. (2015). Replication of "Just think: The challenges of the disengaged mind; Study 8" by Wilson et al. (2014, Science). Unpublished manuscript, Stanford University.
129. Wilson, T. D., Reinhard, D., Westgate, E. C., Gilbert, D. T., Ellerbeck, N., Hahn, C., Brown, C. L., & Shaked, A. (2014). Just think: The challenges of the disengaged mind. Science, 345, 75–77.
130. Muraven, M., Gagné, M., & Rosman, H. (2008). Helpful self-control: Autonomy support, vitality, and depletion. Journal of Experimental Social Psychology, 44, 573–585. https://doi.org/10.1016/j.jesp.2007.10.008 .
131. Guay, F., Vallerand, R. J., & Blanchard, C. (2000). On the assessment of situational intrinsic and extrinsic motivation: The situational motivation scale (SIMS). Motivation and Emotion, 24, 175–213. https://doi.org/10.1023/A:1005614228250 .
132. Milyavskaya, M., McClure, M. J., Ma, D., Koestner, R., & Lydon, J. (2012). Attachment moderates the effects of autonomy-supportive and controlling interpersonal primes on intrinsic motivation. Canadian Journal of Behavioural Science/Revue canadienne des sciences du comportement, 44, 278–287. https://doi.org/10.1037/a0025828 .
133. Keatley, D. A., Clarke, D. D., Ferguson, E., & Hagger, M. S. (2014). Effects of pretesting implicit self-determined motivation on behavioral engagement: Evidence for the mere measurement effect at the implicit level. Frontiers in Psychology, 5, 125. https://doi.org/10.3389/fpsyg.2014.00125 .
134. Vygotsky, L. S. (1987a). Thinking and speech. In R. W. Rieber & A. S. Carton (Eds.), The collected works of L. S. Vygotsky: Volume 1, Problems of general psychology (N. Minick, Trans., pp. 39–285). New York, NY: Plenum Press. (Original work published 1934).
135. Vygotsky, L. S. (1994). The problem of the environment. In R. van der Veer & J. Valsiner (Eds.), The Vygotsky reader (pp. 338–354). (Original work published 1935).
136. Leontyev, A. N. (1974). The problem of activity in psychology. Soviet Psychology, 13(2), 4–33. https://doi.org/10.2753/RPO1061-040513024
137. Sève, L. (in press). Thinking with Marx today, volume 2: "Man?" (J. H. Membrez, Trans.). Leiden, NL: Brill. (Original work published 2008).
138. Marx, K. & Engels, F. (1970). The German ideology. (C. J. Arthur, Ed.) London, UK: Lawrence & Wishart. (Original work published 1846).
139. Leontyev, A. N. (1978). Activity, consciousness and personality. (M. J. Hall, Trans.). Englewood Cliffs, NJ: Prentice Hall.
140. Leontyev, A. N. (1981). Problems of the development of the mind. Progress Publishers.
141. Sheldon, K. M., & Elliot, A. J. (1999). Goal striving, need satisfaction, and longitudinal well-being: The self-concordance model. Journal of Personality and Social Psychology, 76, 482–497. https://doi.org/10.1037/0022-3514.76.3.482 .
142. Cross, S. E., Bacon, P. L., & Morris, M. L. (2000). The relational-interdependent self and relationships. Journal of Personality and Social Psychology, 78, 791–808. https://doi.org/10.1037/0022-3514.78.4.791 .
143. Sedikides, C., & Brewer, M. B. (2001). Individual self, relational self, collective self. New York: Psychology Press.
144. Heine, S. J., & Lehman, D. R. (1997). The cultural construction of self-enhancement: An examination of group-serving biases. Journal of Personality and Social Psychology, 72, 1268–1283. https://doi.org/10.1037/0022-3514.72.6.1268 .
145. Miller, J. G. (1997). Culture and self: Uncovering the cultural grounding of psychological theory. In J. G. Snodgrass & R. L. Thompson (Eds.), The self across psychology: Self-recognition, self-awareness, and the self concept. Annals of the New York Academy of Sciences (Vol. 818, pp. 217–231). New York: New York Academy of Sciences

146. Oishi, S., & Diener, E. (2001). Goals, culture, and subjective well-being. Personality and Social Psychology Bulletin, 27, 1674–1682. https://doi.org/10.1177/01461672012712010 .
147. Hektner, J. M. (2001). Family, school, and community predictors of adolescent growth-conducive experiences: Global and specific approaches. Applied Developmental Science, 5, 172–183. https://doi.org/10.1207/S1532480XADS0503_5 .
148. Wentzel, K. R. (1998). Social relationships and motivation in middle school: The role of parents, teachers, and peers. Journal of Educational Psychology, 90, 202–209. https://doi.org/10.1037/0022-0663.90.2.202 .
149. Zaleski, Z. (1988). Close relationships and acting for self-set goals. European Journal of Social Psychology, 18, 191–194. https://doi.org/10.1002/ejsp.2420180210 .
150. Ruehlman, L. S., & Wolchik, S. A. (1988). Personal goals and interpersonal support and hindrance as factors in psychological distress and well-being. Journal of Personality and Social Psychology, 55, 293–301. https://doi.org/10.1037/0022-3514.55.2.293 .
151. Heine, S. J., Lehman, D. R., Markus, H. R., & Kitayama, S. (1999). Is there a universal need for positive self-regard? Psychological Review, 106, 766–794. https://doi.org/10.1037/0033-295X.106.4.766 .
152. Heine, S. J., Kitayama, S., & Lehman, D. R. (2001). Divergent consequences of success and failure in Japan and North America: An investigation of self-improving motivations and malleable selves. Journal of Personality and Social Psychology, 81, 599–615. https://doi.org/10.1037/0022-3514.81.4.599 .
153. Gore, J. S., & Cross, S. E. (2006). Pursuing goal for us: Relationally-autonomous reasons in long-term goal pursuit. Journal of Personality and Social Psychology, 90, 848–861. https://doi.org/10.1037/0022-3514.90.5.848 .
154. McGregor, I., & Little, B. R. (1998). Personal projects, happiness, and meaning: On doing well and being yourself. Journal of Personality and Social Psychology, 74, 494–512. https://doi.org/10.1037/0022-3514.74.2.494 .
155. Ryff, C. D., & Singer, B. (1998). The role of purpose in life and personal growth in positive human health. In P. T. P. Wong & P. S. Fry (Eds.), The human quest for meaning: A handbook of psychological research and clinical applications (pp. 213–235). Mahwah, NJ: Lawrence Erlbaum Associates.
156. Markus, H. R., & Kiatayama, S. (2003). Models of agency: Sociocultural diversity in the construction of action. In G. Berman & J. Berman (Eds.), The Nebraska symposium on motivation: Cross-cultural differences in perspectives on self (Vol. 49, pp. 1–57). Lincoln, NE: University of Nebraska Press.
157. Bandura A. Self-efficacy: the exercise of control. New York: W. H. Freeman and Company; 1997.
158. Schunk DH, Pintrich PR, Meece JL. Motivation in education: theory, research, and applications. 3rd ed. Upper Saddle River: Pearson Education; 2008.
159. Schunk DH. Self-efficacy and academic motivation. Educ Psychol. 1991;26:207–31.
160. Bandura A. Self-efficacy: toward a unifying theory of behavioral change. Psychol Rev. 1977;84:191–215.
161. Bandura A. Social foundations of thought and action: a social cognitive theory. Englewood Cliffs: Prentice Hall; 1986.
162. Robbins SB, Lauver K, Le H, Davis D, Langley R, Carlstrom K. Do psychosocial and study skill factors predict college outcomes? A meta-analysis. Psychol Bull. 2004;130:261–88.
163. Ajzen I (2011) The theory of planned behaviour: reactions and reflections. Psychol Health 26(9):1113–1127. https://doi.org/10.1080/08870446.2011.613995
164. Buyinza J, Nuberg I, Muthuri C, Denton M (2020) Psychological factors influencing farmers' intention to adopt agroforestry: A structural equation modeling approach. J Sustain For. https://doi.org/10.1080/10549811.2020.1738948
165. Hair JF, Black WC, Babin BJ, Anderson RE (2010) Multivariate data analysis, 7th edn. Prentice Hall, Upper Saddle River

166. Baumeister, Roy F. Toward a general theory of motivation: Problems, challenges, opportunities, and the big picture. Motivation and Emotion (2015). https://doi.org/10.1007/s11031-015-9521-y
167. Small, Ruth V. Motivation and new media: an introduction to the special issue. Educational technology research and development (2011). https://doi.org/10.1007/s11423-011-9189-2
168. Curtis, Jessica; Hatvany, Thomas; Barber, Kathryn E.; Burkley, Edward Moxie: individual variability in motivation intensity. Current Psychology (2022). https://doi.org/10.1007/s12144-022-03304-7
169. Scholer, Abigail A.; Tory Higgins, E. Commitment to change from locomotion motivation during deliberation. Motivation and Emotion (2011). https://doi.org/10.1007/s11031-011-9239-4
170. Hansen, Jochim; Steinmetz, Janina Motivated level of construal: How temperature affects the construal level of state-relevant stimuli. Motivation and Emotion (2019). https://doi.org/10.1007/s11031-018-09750-w
171. Rempel, John K.; Burris, Christopher T.; Fathi, Darius Hate: Evidence for a motivational conceptualization. Motivation and Emotion (2018). https://doi.org/10.1007/s11031-018-9714-2
172. Radel, Rémi; Fournier, Marion; de Bressy, Virginie; d'Arripe-Longueville, Fabienne You're too much for me: Contagion of motivation depends on perceiver-model distance. Motivation and Emotion (2014). https://doi.org/10.1007/s11031-014-9451-0
173. Bernard, Larry C. Consensual and behavioral validity of a measure of adaptive individual differences dimensions in human motivation. Motivation and Emotion (2009). https://doi.org/10.1007/s11031-009-9131-7
174. Lindsey, Alex; King, Eden; Hebl, Michelle; Levine, Noah The Impact of Method, Motivation, and Empathy on Diversity Training Effectiveness. Journal of Business and Psychology (2014). https://doi.org/10.1007/s10869-014-9384-3
175. Carpenter, Jordan M.; Green, Melanie C.; Vacharkulksemsuk, Tanya Beyond perspective-taking: Mind-reading motivation. Motivation and Emotion (2016). https://doi.org/10.1007/s11031-016-9544-z
176. Hayes, Linda J.; Fryling, Mitch J. Motivation in Behavior Analysis: A Critique. The Psychological Record (2014). https://doi.org/10.1007/s40732-014-0025-z
177. English, Tammy; Lee, Ihno A.; John, Oliver P.; Gross, James J. Emotion regulation strategy selection in daily life: The role of social context and goals. Motivation and Emotion (2016). https://doi.org/10.1007/s11031-016-9597-z
178. Rosenfeld, Daniel L. Why some choose the vegetarian option: Are all ethical motivations the same?. Motivation and Emotion (2018). https://doi.org/10.1007/s11031-018-9747-6
179. Lewon, Matthew; Houmanfar, Ramona A.; Hayes, Linda J. The Will to Fight: Aversion-Induced Aggression and the Role of Motivation in Intergroup Conflicts. Perspectives on Behavior Science (2019). https://doi.org/10.1007/s40614-019-00221-2
180. Roets, Arne; Van Hiel, Alain; Kruglanski, Arie W. When motivation backfires: Optimal levels of motivation as a function of cognitive capacity in information relevance perception and social judgment. Motivation and Emotion (2012). https://doi.org/10.1007/s11031-012-9299-0
181. Hall, Judith A.; Blanch, Danielle C.; Horgan, Terrence G.; Murphy, Nora A.; Rosip, Janelle C.; Schmid Mast, Marianne Motivation and interpersonal sensitivity: Does it matter how hard you try?. Motivation and Emotion (2009). https://doi.org/10.1007/s11031-009-9128-2
182. Mayer, John D.; Faber, Michael A.; Xu, Xiaoyan Seventy-five years of motivation measures (1930–2005): A descriptive analysis. Motivation and Emotion (2007). https://doi.org/10.1007/s11031-007-9060-2
183. Dings, Alexander; Spinath, Frank M. Motivational and personality variables distinguish academic underachievers from high achievers, low achievers, and overachievers. Social Psychology of Education (2021). https://doi.org/10.1007/s11218-021-09659-2
184. Alahmadi, Sarah; Buttrick, Nicholas R.; Gilbert, Daniel T.; Hardin, Amber M.; Westgate, Erin C.; Wilson, Timothy D. You can do it if you really try: The effects of motivation on thinking for pleasure. Motivation and Emotion (2017). https://doi.org/10.1007/s11031-017-9625-7

185. Murphy, Stephen L.; Taylor, Ian M. Priming autonomous and controlling motivation and effects on persistence. Current Psychology (2020). https://doi.org/10.1007/s12144-020-00921-y
186. Burkitt, Ian The Emotions in Cultural-Historical Activity Theory: Personality, Emotion and Motivation in Social Relations and Activity. Integrative Psychological and Behavioral Science (2021). https://doi.org/10.1007/s12124-021-09615-x
187. Gore, Jonathan S.; Cross, Susan E.; Kanagawa, Chie Acting in our interests: Relational self-construal and goal motivation across cultures. Motivation and Emotion (2008). https://doi.org/10.1007/s11031-008-9113-1
188. Artino, Anthony R. Academic self-efficacy: from educational theory to instructional practice. Perspectives on Medical Education (2012). https://doi.org/10.1007/s40037-012-0012-5
189. Buyinza, Joel; Nuberg, Ian K.; Muthuri, Catherine W.; Denton, Matthew D. Assessing smallholder farmers' motivation to adopt agroforestry using a multi-group structural equation modeling approach. Agroforestry Systems (2020). https://doi.org/10.1007/s10457-020-00541-2
190. Bernardo, Allan B. I.; Ganotice, Fraide A.; King, Ronnel B. Motivation Gap and Achievement Gap Between Public and Private High Schools in the Philippines. The Asia-Pacific Education Researcher (2014). https://doi.org/10.1007/s40299-014-0213-2

Emotion Based Motivation 2

Introduction by the Editor

Emotion-based motivation is a concept that bridges the realms of psychology and neuroscience, examining how emotional states influence motivational processes. Emotions are complex psychological states that involve physiological arousal, expressive behaviors, and conscious experience. They play a critical role in shaping our motivations, guiding us towards goals, and influencing our actions. This essay explores the intricate relationship between emotions and motivation, examining theoretical frameworks, neural mechanisms, and practical applications in various domains such as education, mental health, and organizational behavior.

Emotion-Based Motivation

Emotion-based motivation refers to the idea that our emotions significantly impact our motivations and drive our behavior. Motivation is a multifaceted phenomenon that drives human behavior. It is influenced by various factors, including biological, psychological, and social elements. One crucial aspect of motivation that has garnered significant attention in psychological research is emotion-based motivation. Emotions, which are complex psychological states involving physiological arousal, expressive behaviors, and conscious experience, play a pivotal role in motivating actions and decisions. Articles in this chapter explore the concept of emotion-based motivation, examining the theoretical frameworks, empirical evidence, and practical implications in various contexts such as education, workplace, and personal development.

Empirical research provides robust evidence supporting the influence of emotions on motivation. Studies have shown that emotions can significantly impact goal setting, persistence, and performance in various domains. For example, positive emotions, such as joy, excitement, and contentment, have been found to enhance motivation. According to Fredrickson's broaden-and-build theory, positive emotions broaden individuals' thought-action repertoires and build enduring

M. S. Khine (ed.), *Motivation Science*, https://doi.org/10.1007/978-981-97-9247-4_2

personal resources (Sriwidharmanely et al., 2021). Several other studies that found how positive affect increased intrinsic motivation and creative problem-solving in participants are also reported in the literature. Similarly, research by Hirschi and Spurk (2021) reported that individuals experiencing positive emotions are more likely to set ambitious goals, persist in the face of challenges, and achieve higher levels of success.

While positive emotions generally enhance motivation, negative emotions can have complex effects. Fear, anger, and anxiety can either undermine or enhance motivation, depending on the context and individual differences. For instance, fear of failure can lead to avoidance behavior and decreased motivation, but it can also spur individuals to work harder to avoid negative outcomes. A study by Liu et al. (2020) found the multiplicative effect between intrinsic and extrinsic motivations showed that extrinsic motivation was detrimental to academic performance for students with high intrinsic motivation. For students with low intrinsic motivation, however, the extrinsic motivation helped to improve academic performance. Worth noting, intrinsic motivation also had a moderate positive effect on academic performance.

Emotional intelligence, which involves the ability to perceive, understand, manage, and regulate emotions, plays a crucial role in emotion-based motivation. Individuals with high emotional intelligence are better equipped to harness their emotions for motivational purposes. Research by Drigas and Papoutsi (2020) highlighted that emotionally intelligent individuals are more adept at managing stress, staying motivated, and achieving their goals. Moreover, a meta-analysis by MacCann et al. (2020) confirmed the positive relationship between emotional intelligence and academic achievement, underscoring the importance of emotional skills in motivating behavior. Understanding emotion-based motivation has significant practical implications in various contexts, including education, workplace, and personal development.

In educational settings, fostering positive emotions can increase students' motivation to learn and achieve. Teachers and educators can create a supportive and engaging learning environment that promotes positive emotions such as curiosity, excitement and a sense of accomplishment. Techniques such as providing constructive feedback, setting achievable goals and incorporating enjoyable activities can increase students' intrinsic motivation. In addition, addressing negative emotions such as anxiety and fear through counselling and emotional support can help students stay motivated and perform better academically. At individual level, understanding and harnessing emotions can have a significant impact on goal achievement and overall well-being. Individuals can use techniques such as mindfulness, emotional regulation strategies and positive self-talk to manage their emotions and stay motivated. Setting emotionally meaningful goals that align with personal values and interests can increase intrinsic motivation and persistence. For example, a study by Gonzalez Olivares et al. (2020) showed that individuals who pursued goals driven by intrinsic motivation reported higher levels of well-being and satisfaction than those motivated by extrinsic rewards.

Emotion-based motivation is a complex phenomenon that intersects with various disciplines, including psychology, neuroscience, and sociology. Integrating insights from these fields can provide a more comprehensive understanding of how emotions influence motivation. For instance, advances in neuroimaging techniques can shed light on the neural mechanisms underlying emotion-based motivation, while sociological research can explore the role of social and cultural factors. Emotion-based motivation is a critical aspect of human behavior, influencing actions and decisions across various domains. Theoretical frameworks such as the James-Lange theory, Cannon-Bard theory, Schachter-Singer two-factor theory, and appraisal theory provide valuable insights into the complex interplay between emotions and motivation. Empirical evidence highlights the significant impact of both positive and negative emotions on motivation, as well as the crucial role of emotional intelligence. Practical applications in education, workplace, and personal development underscore the importance of fostering positive emotions and managing negative ones to enhance motivation. However, challenges such as individual differences, the need for longitudinal studies, and the integration of multidisciplinary perspectives call for further research. Understanding and leveraging emotion-based motivation holds the potential to improve well-being, performance, and overall quality of life.

In this chapter, the effects of an emotion control treatment on academic emotions, motivation and achievement in an online mathematics course, visual design as a holistic experience: how students' emotional responses to the visual design of instructional materials are formed, antecedents and consequences of emotion regulation in STEM degree programs, and influence of personality and motivation on oral presentation performance among university students are presented. Studies related to motivation across domains: Do goals and attributions change with subject matter for Grades 4 and 5 students, putting effort into emotion regulation: manipulating desirability and motivational strength, initial examination of state and longitudinal effects of loving-kindness practice on affective and motivational states at work, and leaning forward to increase approach motivation: the role of joy, exercise, and posture in achieving goals are also included in this chapter.

References

Drigas, A., & Papoutsi, C. (2020). The Need for Emotional Intelligence Training Education in Critical and Stressful Situations: The Case of Covid-19. *Int. J. Recent Contributions Eng. Sci. IT*, *8*(3), 20–36.

Gonzalez Olivares, A. L., Navarro, Ó., Sánchez-Verdejo, F. J., & Muelas, Á. (2020). Psychological well-being and intrinsic motivation: relationship in students who begin university studies at the School of Education in Ciudad Real. *Frontiers in Psychology*, *11*, 2054.

Hirschi, A., & Spurk, D. (2021). Ambitious employees: Why and when ambition relates to performance and organizational commitment. *Journal of Vocational Behavior*, *127*, 103,576.

Liu, Y., Hau, K. T., Liu, H., Wu, J., Wang, X., & Zheng, X. (2020). Multiplicative effect of intrinsic and extrinsic motivation on academic performance: A longitudinal study of Chinese students. *Journal of personality*, *88*(3), 584–595.

MacCann, C., Jiang, Y., Brown, L. E., Double, K. S., Bucich, M., & Minbashian, A. (2020). Emotional intelligence predicts academic performance: A meta-analysis. *Psychological bulletin*, *146*(2), 150.

Sriwidharmanely, S., Sumiyana, S., Mustakini, J. H., & Nahartyo, E. (2022). Encouraging positive emotions to cope with technostress's adverse effects: insights into the broaden-and-build theory. *Behaviour & Information Technology*, *41*(10), 2201–2214.

Machine Generated Summaries

Disclaimer: The summaries in this chapter were generated from Springer Nature publications using extractive AI auto-summarization: An extraction-based summarizer aims to identify the most important sentences of a text using an algorithm and uses those original sentences to create the auto-summary (unlike generative AI). As the constituted sentences are machine selected, they may not fully reflect the body of the work, so we strongly advise that the original content is read and cited. The auto generated summaries were curated by the editor to meet Springer Nature publication standards. To cite this content, please refer to the original papers.

Machine generated keywords: emotion, student, pekrun, regulation, mathematic, goal, emotion regulation, elementary, negative emotion, contextual, academic, school, anxiety, affective, life.

Effects of an Emotion Control Treatment on Academic Emotions, Motivation and Achievement in an Online Mathematics Course [293]

This is a machine-generated summary of:

Kim, ChanMin; Hodges, Charles B.: Effects of an emotion control treatment on academic emotions, motivation and achievement in an online mathematics course [293].

Published in: Instructional Science (2011).

Link to original: https://doi.org/10.1007/s11251-011-9165-6.

If you want to cite the papers, please refer to the original.

For technical reasons we could not place the page where the original quote is coming from.

Abstract-Summary

"We designed and developed an emotion control treatment and investigated its effects on college students' academic emotions, motivation, and achievement in an online remedial mathematics course."

"The treatment group showed more positive emotions of enjoyment and pride than the control group."

Role of Academic Emotions

"Positive emotions, such as enjoyment and pride, impact motivation as well; for instance, the joy of a discovery experience (e.g., "oh, now I get it") can motivate students to work on mathematics problem-solving (Liljedahl [1]; McLeod [2])."

"Despite the aforementioned influence of academic emotions on motivation, much research is done with how motivation can be a factor for academic emotions."

"Academic emotions and motivation have bidirectional influence (e.g., via a learner's goals; Linnenbrink and Pintrich [3])."

"Academic emotions interact with motivation and play a crucial role in mathematics achievement (McLeod [2]; Hannula [4]; Op 't Eynde and others [5])."

"A holistic view is needed to understand both motivation and academic emotions (Hannula [4]; Meyer and Turner [6]; Op 't Eynde and Turner [7]; Turner and Patrick [8])."

"Our intent is to help researchers and practitioners understand how to approach academic emotions as well as "how to induce a desired motivational state in students" throughout one potential approach to emotions (Hannula [4], p. 166)."

Design of Support for Emotion Control

"Some interventions were designed to reduce students' motivational problems in mathematics education (Croft and Ward [9]; Hodges [10]) but few attempts to design and develop interventions to facilitate emotion control have been reported, especially in the context of online remedial mathematics courses."

"If students perceive the controllability of their emotions in a math learning context, they could suppress negative emotions and maintain positive emotions."

"Our emotion control treatment aimed to help initiate students' conscious awareness of their emotions and deliberate appraisals that could lead to automation later on."

"The treatment also intended to facilitate students' deliberate efforts (a) to shift their attention from debilitating emotions to facilitative emotions (e.g., pay attention to a positive side of anxiety), (b) to become aware of the situation and their emotions as well as to conduct conscious appraisals (e.g., recognize the feeling of enjoyment), and (c) to provoke or suppress emotional responses (e.g., suppress anger)."

Research Questions

"What is the impact of the emotion control treatment on students' academic emotions?"

"What is the impact of the emotion control treatment on students' motivation toward the mathematics course?"

"What is the impact of the emotion control treatment on students' achievement?"

Method

"The participants were randomly assigned to either a treatment group or a control group."

"Students could choose to work on the course in a mathematics lab with computers and tutors available; but no group work was required."

"The participants were randomly assigned to either a treatment group or a control group for the study."

"The participants were randomly assigned to either a treatment group ($n = 50$) or a control group ($n = 51$) for the study."

"In the third week, all participants were asked to complete an online survey containing questions to measure their academic emotions and motivation and then they took the first course exam."

"The AEQ-M measures emotions that are linked to achievement activities and their outcomes and contains 60 items measuring seven discrete emotions relating to mathematics: enjoyment, pride, anger, anxiety, shame, hopelessness, and boredom."

Results

"Academic emotions were analyzed through a one-factor MANOVA, with enjoyment, pride, anger, anxiety, shame, hopelessness, and boredom, and with condition (Treatment, $n = 44$; Control, $n = 51$) as the between-subject factor."

"Follow-up ANOVAs indicated that significant differences occurred only for the emotions of enjoyment, $F(1, 93) = 4.117$, $p = 0.045$, and pride, $F(1, 93) = 4.232$, $p = 0.042$."

"The effect size estimates were $d = 0.41$ for enjoyment and $d = 0.43$ for pride, indicating a medium effect (Cohen [11])."

"Levene's test revealed appropriate homogeneity of the posttest scores of boredom, $F(1, 93) = 0.051$, $p = 0.937$, enjoyment, $F(1, 93) = 0.035$, $p = 0.852$, anger, $F(1, 93) = 0.761$, $p = 0.551$, shame, $F(1, 93) = 0.358$, $p = 0.551$, pride, $F(1, 93) = 0.063$, $p = 0.803$, and hopelessness, $F(1, 93) = 0.547$, $p = 0.461$."

Discussion

"We investigated the effects of an emotion control treatment on students' academic emotions, motivation, and achievement (RQs 1, 2, and 3)."

"Results indicate that the treatment group showed more positive emotions of enjoyment and pride as well as higher level of motivation than the control group but there was no difference between the two groups in achievement."

RQ1: What Is the Impact of the Emotion Control Treatment on Students' Academic Emotions?

"The presence of the emotion treatment affected participants' emotions of enjoyment and pride."

"The total time students actually were in contact with the emotion control treatment video (6 min) was short, yet the video produced statistically significant medium effects on the emotions of enjoyment and pride."

"Our treatment aimed to help students control not only negative emotions but also promoted positive emotions, which distinguishes our research from the previous studies whose main focus was only on negative emotions such as anxiety (McLeod [2]; Goetz and others [12]; Pekrun and others [13])."

"The impact of the treatment on positive emotions provides further evidence that the controllability emphasized in the treatment may have the potential to help students reappraise the situation as enjoyable and feel pride (Schallert and others [14])."

"If there was no negative emotion recognized, the controllability encouraged in the treatment would have not been processed in students."

RQ2: What Is the Impact of the Emotion Control Treatment on Students' Motivation?

"This finding suggests a possibility of improving motivation through emotion control."

"Our finding illustrates that emotions can be "an empowering source of information about how to influence motivational patterns" (Ford [15], p. 145), rather than simply variables produced by motivation (Meyer and Turner [6])."

"Our study is important in that it shows that promoting positive emotions can positively influence motivation."

"Our findings further support the theory of the interdependence among emotions, motivation, and cognition (Ainley [16])."

RQ3: What Is the Impact of the Emotion Control Treatment on Students' Achievement?
"A statistically significant impact of the improved emotions and motivation on achievement was not found."

"Our expectation was not that the treatment will directly affect students' mathematical knowledge and skills during the short-time study."

"A study for an entire semester with series of emotion control treatments would explain if positive changes in students' emotions and motivation could influence their achievement as well."

"Previous studies found that emotions impacted cognitive processes of learning and performance (Pekrun and others [13]; Goetz and others [17])."

RQ4: How Do Students' Academic Emotions Relate to Each Other and to Motivation?
"The correlations among emotions were significant."

"The emotion of pride was not significantly correlated with the emotions of boredom, enjoyment, anger, and shame."

"This could suggest that there is a need for a specific examination of mastery-oriented and performance-oriented motivation when examining academic emotions."

"Considering the correlations between participants' emotions and motivation were significant, further research should consider an integrative view of academic emotions and motivation in order to facilitate students' emotion control."

Re-design of Support for Emotion Control
"This focus was to examine a way of approaching academic emotions to facilitate students' emotion control and thereby their motivation as well."

"Without the effort to enhance motivation at the same time, the emotion control treatment might have not been optimized."

"If a motivation enhancement treatment is given to help students recognize the usefulness of a given math task by showing a practical example use of math (e.g., using algebra in grocery shopping), the value of the task can be reevaluated and the boredom can be reduced also by the use of controllability supported by the emotion control treatment."

"Considering conceptually overlapping components between motivation and emotions such as value and goals (Schutz and others [18]; Weiner [19]), an integrative view of emotions and motivation is needed in the design of support for either emotion control or motivation enhancement in math."

Limitations and Suggestions for Future Research

"For further research, the emotion control treatment should be provided while students encounter obvious struggles with content, homework, exams, and so forth."

"If more math-specific emotions were addressed in the emotion control treatment when the strategies were embedded in the scenarios and strategies, the study findings might have been different."

"Future studies should take this aspect into considerations during the process of an emotion control treatment design."

"In order to address this social desirability issue as well as to better understand academic emotions and design support, future studies should consider the use of a multi-method approach (Meyer and Turner [20]) including the examination of online interactions and discourse analysis (discussion board postings, email conversations, synchronous chats) combined with self-reported data sources (surveys, interviews)."

"Forth, our emotion control treatment was given to the participants only once."

Implications

"The framework used for the emotion control treatment design should be further revised, validated, and refined throughout investigations of direct and indirect effects of the treatment on academic emotions, motivation, and achievement (e.g., by using structural equation modeling)."

"It is meaningful that our emotion control treatment design involved theoretical foundations not only because theories can be a source of verified strategies (Ertmer and Newby [21]) but also because a sound theory-based instructional design is considered critical to help students with academic emotion control (Goetz and others [17]; Astleitner [22])."

"Although the emotion control treatment was a six-minute video implemented only once, we expected that the strategies embedded in the video would be effective."

"Our emotion control treatment as well as our conceptual framework in this study provide implications for research on and design and development of support for students' academic emotions and motivation in domains typically regarded as difficult such as mathematics."

"The emotion control treatment can be redesigned so as to support middle and high school mathematics instruction."

[Section 14]

"Too often, however, college mathematics courses have become barriers to students' pursuit of undergraduate degrees (NMAP [23])."

"Pervasive and persistent mathematics deficits have been reported among entering college students across the nation."

"In response, remedial mathematics courses have been offered at 71% of universities and community colleges (National Center for Education Statistics [24])."

"Students reluctantly enroll in college remedial mathematics courses due to college requirements, and few embrace or engage in mathematical concepts and procedures (Hagedorn and others [25]; Stage and Kloosterman [26])."

Visual Design as a Holistic Experience: How Students' Emotional Responses to the Visual Design of Instructional Materials Are Formed [294]

This is a machine-generated summary of:

Tomita, Kei: Visual design as a holistic experience: how students' emotional responses to the visual design of instructional materials are formed [294].

Published in: Educational technology research and development (2022).

Link to original: https://doi.org/10.1007/s11423-022-10088-x.

If you want to cite the papers, please refer to the original.

For technical reasons we could not place the page where the original quote is coming from.

Abstract-Summary
"To understand the complex phenomenon of how students' emotional responses to instructional materials are formed, this qualitative thematic study explored factors thought to affect their visual perceptions of instructional materials by utilizing the general framework of approach and avoidance motivation."

"Four sets of printed instructional materials were prepared, each with the same content drawn from finite mathematics, but with different visual designs."

"The results showed that students selected instructional materials based on their holistic impression of the materials and on their individual expectations as shaped by previous experiences."

"Students who had math anxiety selected materials that did not look like math textbooks, although opinions regarding which materials did or did not look like textbooks were diverse due to different material experiences."

"While existing studies tend to be deterministic about the attractiveness of visual materials (e.g., Plass and others in Learn Instruct 29:128–140, 2014), the present study confirms that there is no universal design that elicits comfortable experiences for everyone."

Introduction

"The present study aims to bring new perspectives to the field of instructional design about how students' emotional responses to the visual design of instructional materials are formed, which will aid in the creation of engaging and motivating materials."

"It is important to understand how the state of approach-avoidance can be triggered by the visual design of instructional visual materials so that students will not be discouraged but motivated to engage in the learning."

"Embracing such individual differences or student factors in the design of instructional materials could enhance students' positive emotions and motivation (Kim & Pekrun [30]; Park & Lee [33]), which existing research does not always successfully observe (e.g., Brom and others [34])."

"To successfully embrace individual differences in order to design engaging and motivating instructional visual materials, it would be critical to understand the factors influencing students' diverse visual perceptions and motivational needs."

"What are the factors affecting the college students' perception and experience of instructional materials with different visual designs?"

Theoretical Framework

"To analyze the factors affecting students' emotional response to instructional visual materials, the present study utilizes the general framework of approach and avoidance motivation as defined by Feltman and Elliot [35]: Approach and avoidance motivation is composed of three conceptually distinct components."

"Motivation is defined as the energization and direction of behavior. (p. 1) Elliot [36] explains that positive stimuli cause approach motivation, in which people want to approach the stimuli such as objects or events."

"Negative stimuli cause avoidance motivation, in which people want to stay away from the stimuli."

"The author of the present study found the framework of approach and avoidance motivation helpful for analyzing factors affecting students' emotional responses to the instructional materials."

"The fact that the approach and avoidance motivation theory considers both external and internal stimuli, and explains why people engage or disengage from some objects, made the concept an appropriate framework for the present study."

Methodological Framework

"Since the present study investigates how the visual design of instructional materials impacts student emotions, it is inevitable to touch on the issue of visual appeal or aesthetics."

"Theories on how people recognize and form opinions about objects suggest that the concept of beauty or attractiveness is not quantifiable and should be described by qualitative investigation (Frick [37]; Maccia [38]; Steiner [39])."

"Barriball and While [40] argue that the semi-structured interview method allows probing and clarification, which makes it an effective strategy to understand the participants' intentions, despite the fact that each participant uses words differently."

"Such qualities of the data collection method seemed to be particularly important for the present study, which deals with the complex phenomenon of how students' visual perceptions and emotional responses are formed."

Method

"The quiz was intended simply to enhance the students' attention to and engagement with the material; the quiz score was not of interest in the present study."

"For the sake of the research, the number of students who study the materials is not equally distributed."

"The duration of the interviews varied from 43 min to one hour, depending on the time that students took to select and study the material and take the quiz."

"To increase the study's relevance to the students, the instructional materials were based on a textbook from a finite mathematics course that is often required at the university."

"As the interview data revealed that students' visual perceptions seem to be influenced by their perspectives on what constitutes the 'standard look' of learning materials and other types of media, the present study adopted Spradley's [41] approach to interpretation."

Results and Discussion

"While many students did not like the thin font of the Simple material, Taylor liked it because she looks at music scores a lot."

"The three code relationships under this theme are about how student expectations impact their emotional reactions to materials."

"While students like Vanessa preferred a serious look, most students selected materials that did not look academic."

"Katherine, one of only four students who indicated "Quite a bit of Anxiety," avoided materials that looked like regular textbooks because of her math anxiety, which led to the selection of the Vivid material."

"While students like Sophia selected the Cool material because it did not look like a textbook, Taylor avoided it since it looked like an advanced math textbook."

"The fourth theme summarizes the code relationships on how the materials' visual design features influence students' emotions."

Summary of Themes

"That students developed expectations for materials through their various experiences with materials."

"If the materials met their expectations, they developed positive emotional reactions to them."

"If students developed a negative feeling for typical textbooks, they tended to develop a resistance to materials that looked like a textbook."

"Students' design beliefs also impacted their expectations for the materials."

Implications

"Based on the findings of the present research, the following section will discuss the theoretical and practical implications for the design of instructional materials in terms of three major issues: Impacts of visual design on students' emotions."

"The students in the present study did develop resistance to some visual designs, even declining to look at the inside of the materials."

"The findings of the present study suggest issues to consider when designing instructional materials to create positive learning experiences: Reading Habits: If the students look for information quickly, visual cues that emphasize essential information would be beneficial."

"Since the findings of the present research suggest that students develop positive emotions as they interact with the materials, allowing students to customize the visual design could enhance students' positive emotions toward the materials."

"Designers could prepare instructional materials with the same content but different visual designs, allowing students to select the design they want to use."

Conclusion

"To understand the complex phenomenon of how students' emotional responses to instructional materials are formed, this qualitative thematic study explored factors thought to affect college students' visual perceptions of instructional materials."

"Although existing visual design studies tend to focus on the reporting of visual design factors, the results of the present study suggest that student factors, or individual differences, significantly affect their perceptions of instructional materials."

"Each student brings a different background to the materials, and these diverse backgrounds resulted in their differing visual perceptions of learning materials; the same design can be perceived negatively or positively by different students."

"Recognizing the diversity of students' visual perceptions and seeking the factors that contribute to these differences will allow us to better predict ways to design instructional materials that provide positive learning experiences."

Examining Motivations to Walk the Camino De Santiago: A Typology of Pilgrims [295]

This is a machine-generated summary of:

Brumec, Snežana; Lavrič, Miran; Naterer, Andrej: Examining Motivations to Walk the Camino de Santiago: A Typology of Pilgrims [295].

Published in: Pastoral Psychology (2023).

Link to original: https://doi.org/10.1007/s11089-023-01071-1.

If you want to cite the papers, please refer to the original.

For technical reasons we could not place the page where the original quote is coming from.

Abstract-Summary

"Three dimensions of motivation were identified using the following factor analyses: secular, spiritual, and religious."

"The results revealed that pilgrims are not homogeneous in their motivations and that, except for two types with a total frequency of 4%, secular motivation is more or less present in all other types of prospective pilgrims."

"Religious motivation always appears together with secular, spiritual, or, most often, both these forms of motivation."

"The findings suggest that secular motivation may be related to a need for exploration, the core motive underlying self-actualization."

Introduction

"Since 1987, when the Council of Europe declared the cultural importance of routes crossing all of Europe with the destination of Santiago de Compostela, the number of pilgrims on the Camino de Santiago increased from fewer than 3,000 to more than 347,000 in 2019 (Pilgrim's Reception Office [42])."

"The Pilgrim's Reception Office collects data on reasons for walking the Camino."

"About 18% walked the Camino because of their religious beliefs, while 17% of the pilgrims said it was the natural heritage of the route that had attracted them."

"There is a large discrepancy in the data on motivation between these two centers in Spain; according to the Pilgrim's Reception Office, 11% of pilgrims walked the Camino because of nonreligious motives, whereas according to the CETUR that percentage is 82%."

Pilgrims' Motivations

"Considering the complexity of the pilgrims' motives, Oviedo and others [43] suggest that individuals with "various, often contrasting, motivations and expectations walk side by side" on the Camino de Santiago pilgrimage route."

"They found that most pilgrims reported a strengthened commitment to self-actualization after the Camino de Santiago pilgrimage; additionally, according to the research, these changes occurred separately from the motivation for pilgrimage."

"The author, who had walked the route several times, claimed that contemporary pilgrims went to Santiago for a variety of motives that might change throughout the Camino."

"Frey claimed that pilgrims noted the difference between religious and spiritual motives as well as the difference between orthodoxy and personal devotion."

"A qualitative study by Chemin [44] also explored the motives of pilgrims on the Camino de Santiago."

The Present Study

"To bridge qualitative and quantitative studies on the Camino de Santiago pilgrims' motivation, we divided the present study into two stages."

"In line with our chosen approach, we based the study on a qualitative analysis of 32 travelogues about the Camino de Santiago pilgrimage, written by the pilgrims themselves, published in book form in the Slovene language and recorded in the Slovenian library information system COBISS (Co-operative Online Bibliographic System & Services)."

"Since certain qualitative studies mentioned earlier reported that pilgrims' motives may change over the course of the Camino, we targeted the survey at prospective pilgrims, those who had made the firm decision to walk at least 300 km of the Camino de Santiago for the first time."

"We chose the 27-item scale from an empirical study of pilgrims' motivations along the Camino de Santiago by Farias and others ([45], p. 33)."

Findings and Results

"I am not religious, I am skeptical of so-called new spiritual movements and as far as hiking is concerned, I must admit that until the Camino I was more a Sunday walker than an enthusiast who uses every free hour for a walk in nature. (p. 13) Nevertheless, in his narrative we noticed that his personal motives were (1) to test himself: "I think the fact that I had never done anything like this contributed to the final decision."

"After we analyzed the sections codded as personal motive, we identified seven reasons to walk the Camino: (1) to experience something new, (2) to rethink my past, (3) to test myself, (4) to think about the future, (5) to take time just for myself, (6) to reflect on my faith and practice it, and (7) to get to know another country and its people."

"Prospective pilgrims who scored higher on secular motivation decided to walk the Camino to get to know another country and its people."

The Seven Types of Camino de Santiago Pilgrims

"We ranked respondents with a high average ($M \geq 4$) on the scale of secular motivation and low average values on the scale of spiritual ($M < 4$) and religious ($M < 4$) motivations as being the prospective pilgrims with secular motivation."

"Approximately 28% of our sample are above-average secular-spiritual and below-average religiously motivated prospective pilgrims who want to get to know another country and its people better."

"Our sample consists of nearly 20% of mostly secularly motivated prospective pilgrims (above-average secular and below-average religious and spiritual motivation), the majority of whom are driven by sensation-seeking motivations to enjoy the trekking experience."

"They crave new knowledge about and experiences in another country and with its people. (4) Of the sample, 8% are secular-religious motivated pilgrims (above-average secular and religious motivation and below-average spiritual motivation)."

Discussion

"Except for two types with a total frequency of 4%, secular motivation is more or less present in all other types of prospective pilgrims."

"We suggest that the Camino de Santiago pilgrims are driven by an irreducible and fundamental need for exploration, which is the core motive underlying self-actualization."

"Our suggestion that pilgrims are driven by the need for exploration aligns with that of Kim and others [46]; internally driven motivation emerges as a common element across religious, spiritual, and secular travelers, represents a greater motivational variation, and prevails as a unifying force across different groups, mainly or partially only for secular motivated pilgrims."

"According to the Pilgrim's Reception Office [42], in 2019 around 40% of Camino pilgrims reported having a religious motive and 11% a nonreligious motive (in our sample, there are no "purely religious" pilgrims, and around 50% are nonreligious pilgrims)."

Conclusion and Implications

"Secular motivation is by far the most common motivation; moreover, highly motivated pilgrims represent the largest group."

"The drive for exploration emerged as a common element across all types of pilgrims who were mainly or partially secular motivated."

"Exploration is a specific need underlying the self-actualization that helps people grow as a whole person (Kaufman [47]), our findings have important theoretical and practical implications."

"Studies measuring the drive for exploration and self-actualization before and after the journey are needed to validate and potentially modify the findings of this study."

Antecedents and Consequences of Emotion Regulation in STEM Degree Programs [296]

This is a machine-generated summary of:

Jarrell, Amanda; Lajoie, Susanne P.; Hall, Nathan C.; Horrocks, Phoenix T. M.: Antecedents and Consequences of Emotion Regulation in STEM Degree Programs [296].

Published in: Innovative Higher Education (2022).

Link to original: https://doi.org/10.1007/s10755-021-09587-1.

If you want to cite the papers, please refer to the original.

For technical reasons we could not place the page where the original quote is coming from.

Abstract-Summary
"The purpose of this study was to address this gap by examining how students in STEM (Science, Technology, Engineering, Math) degree programs (N = 174) use emotion regulation strategies related to their achievement emotions, approaches to learning, and exam performance."

"Findings also showed cognitive reappraisal to correspond with more positive emotions, less negative emotions, more complex approaches to learning, and better exam performance."

"Results additionally showed that associations between cognitive reappraisal and emotions were stronger during exams than while studying."

"These findings indicate that how post-secondary students choose to regulate their emotions in STEM degree programs has important implications for how they feel, learn, and perform."

Challenges and Persistence in STEM Degree Programs
"Many factors have been posited as contributing to these low graduation rates in STEM programs including large time commitments, competitive and unsupportive environments, excessive course loads during the first two years, and gatekeeping courses specifically designed to "weed out" students (Chen [48]; Hunter [49]; Larcombe and others [50])."

"To the substantial time commitment, many STEM programs have deliberately difficult first-year courses meant to discourage all but the most talented students from continuing (Gasiewski and others [51]; Weston and others [52])."

"How students handle their emotions in these situations can have a substantial impact on their decision to persist in STEM programs and careers, their academic success more generally, as well as their emotional well-being (Ben-Eliyahu & Linnenbrink-Garcia [53]; Pelch [54]; Simon and others [55])."

"With some scholars further suggesting that students' emotions may play an even larger role in their persistence than specific academic challenges (Szulecka and others [56]), emotion regulation is thus of potentially crucial importance to the success of undergraduate STEM students."

Regulation of Academic Achievement Emotions

"According to Pekrun's control-value theory, achievement emotions are proposed to arise from students' appraisals of controllability and perceived value concerning their academic environment and challenges."

"To appraisals of control and value, achievement emotions can be shaped by emotion regulation (Jacobs & Gross [57])."

"While the control-value theory of achievement emotions acknowledges that emotion regulation can influence achievement emotions, it is unclear how appraisals and emotion regulation jointly explain student achievement emotion experiences."

"The joint contribution of students' control and value appraisals and emotion regulation in predicting their emotions has yet to be tested, with this research being particularly lacking in a post-secondary STEM context."

Academic Consequences of Emotion Regulation

"Previous research has primarily examined associations between two emotion regulation strategies (cognitive reappraisal and suppression) and emotions."

"Students' spontaneous use of cognitive reappraisal and suppression when studying were measured in relation to their emotional experiences in their favorite and least favorite courses (Ben-Eliyahu & Linnenbrink-Garcia [53])."

"Research on achievement emotions suggests that positive activating emotions are adaptive because they conserve cognitive resources, promote motivation, facilitate flexible and complex learning strategies, self-regulation, and achievement (e.g., Ahmed and others [58]; Frenzel and others [59]; Pekrun and others [60, 61])."

"Whereas negative activating emotions generally reduce cognitive resources and achievement, specific emotions like anxiety and shame can correspond with greater motivation to avoid failure and use of analytical strategies (e.g., Artino & Jones [62]; Frenzel and others [59]; Naismith & Lajoie [63]; Pekrun and others [60, 61])."

"It is possible that emotion regulation might influence learning and performance because of the cognitive demands associated with using these strategies and by shaping achievement emotion experiences."

Current Study

"This study also sought to examine the direct and indirect effects of emotion regulation on approaches to learning and exam performance, as mediated by achievement emotions."

"To address this research gap, the following four primary hypotheses were tested: Hypothesis 1: Appraisals and cognitive reappraisal should positively predict positive achievement emotions and negatively predict negative achievement emotion, with suppression being unrelated to negative emotion and negatively predicting positive achievement emotions."

"Hypothesis 3: Exam performance should be positively predicted by reappraisal and negatively predicted by suppression."

"Hypothesis 4: The relationships between emotion regulation and learning as well as exam performance could be direct or indirect through achievement emotions."

Method

"Students' perceptions of course-specific control (eight self-efficacy items; e.g., "I expect to do well in this class") and value (six items; e.g., "Understanding the subject matter of this course is very important to me) were measured with the Motivated Strategies for Learning Questionnaire (MSLQ; Pintrich and others [64])."

"These measures assessed students' use of cognitive reappraisal during studying (six items; e.g., "When I want to feel a more positive emotion (such as enjoyment or pride) while studying for this class, I change what I'm thinking about") and exams (six items; e.g., "When I was faced with a stressful situation while taking this exam, I made myself think about it in a way that helped me stay calm"), and expressive suppression during studying (four items; e.g., "I control my emotions while studying for this class by not expressing them") and exams (four items; e.g., "I kept my emotions to myself while taking this exam")."

Results

"The results also showed that reappraisal predicted less frustration ($B = -0.188$, $p < 0.05$, $SE = 0.089$) that, in turn, predicted less critical thinking ($B = -0.573$, $p < 0.05$, $SE = 0.251$)."

"The results showed that reappraisal positively predicted hope ($B = 0.391$, $p < 0.001$, $SE = 0.071$) that, in turn, predicted higher exam grades ($B = 7.896$, $p < 0.01$, $SE = 2.970$)."

"The results also indicated that reappraisal predicted less shame ($B = -0.230$, $p < 0.001$, $SE = 0.069$) that, in turn, predicted lower exam grades ($B = -5.556$, $p < 0.05$, $SE = 2.462$)."

Discussion

"This finding indicates that students' cognitive appraisals are likely more important than their emotion regulation strategies for predicting their emotions while studying; a finding largely consistent with the primary assumption of control-value theory of achievement emotions (Pekrun & Perry [65])."

"The present findings support existing research showing cognitive reappraisal to be one such adaptive emotion regulation strategy, and extend prior studies by demonstrating how emotion regulation may be more important than control and value appraisals for students' academic development depending on the type of task (e.g., exams vs. studying) and type of emotion involved (e.g., anger vs. enjoyment)."

"The findings from this study indicate that not only should adaptive appraisals of control and value be promoted in STEM degree programs, but that students should be explicitly encouraged to use cognitive reappraisal strategies in specific academic contexts."

[Section 8]

"Possibility, students' use and the consequences of emotion regulation in higher education STEM programs remains an underexplored topic of research."

"We address these gaps by examining undergraduate STEM students' spontaneous use of emotion regulation during studying and exam taking in a required STEM course."

"We examine how appraisals of control and value, and emotion regulation strategies, jointly predict students' emotional experiences to gain insight into the processes of emotion generation and regulation in post-secondary STEM degree programs."

"We also considered how students' use of different emotion regulation strategies directly or indirectly impact their approaches to learning and exam performance."

Influence of Personality and Motivation on Oral Presentation Performance [297]

This is a machine-generated summary of:

Liang, Hsin-Yi; Kelsen, Brent: Influence of Personality and Motivation on Oral Presentation Performance [297].

Published in: Journal of Psycholinguistic Research (2018).

Link to original: https://doi.org/10.1007/s10936-017-9551-6.

If you want to cite the papers, please refer to the original.

For technical reasons we could not place the page where the original quote is coming from.

Abstract-Summary

"Personality and motivation have been identified as influential variables associated with foreign language learning; however, few studies have investigated their effect on oral presentations."

"This study addresses the importance of both personality and motivation in students' collaborative oral presentation performance."

"A Big Five personality trait questionnaire measuring Extraversion, Agreeableness, Conscientiousness, Neuroticism and Openness to Experience, together with the Collaborative Inquiry-based Project Questionnaire measuring Task, Project Work, Reinforcement, Social Learning and Social Pressure motivational constructs were employed to evaluate 257 university students."

"The results showed that Extraversion, Project Work and Social Pressure were significant correlates of oral presentation scores."

"These findings extend previous research by highlighting the contextual relationships between these affective variables and performance in collaborative oral presentation contexts."

Introduction

"Language learning success has been associated with individual personality and motivation differences (Dörnyei [66]; Dörnyei and Ushioda [67]; Dweck and Leggett [68]; Ehrman et al. [69])."

"This study extends the existent EFL literature by investigating the effects of personality traits and motivational constructs on student performance in a collaborative oral presentation context."

"Surmising that these factors may affect extraverts' speech production rather than their entire language learning process, Dewaele and Furnham [70] assert that: …

some researchers found links between extraversion scores and linguistic variables, depending on the type of linguistic material they used."

"Task (motivation derived from satisfaction or interest in the activity) and Reinforcement (motivation arising from external rewards or avoidance of negative consequences) factors relate to the traditional intrinsic/extrinsic dichotomy; Project Work connects with the project-related activities inherent in inquiry-based learning contexts conducted in teams; and Social Learning and Social Pressure signify the social aspects of working together in groups and how this shapes collective learning motivation (pp."

"There has been little research conducted on the relationship between personality and motivation in inquiry-based learning contexts where collaborative presentations were the means of assessment."

Materials and Methods

"EFL students enrolled in an ESP program at a university in northern Taiwan participated in this study, where collaborative oral presentation scores served as the dependent variable (DV)."

"Presentation scores were calculated as follows: i) the instructor rated each student and team, and ii) each team rated each other team."

"Ratings for presentations were based on a scoring rubric (see Appendix), which was decided upon by the participating students and posted on the class e-learning website so that each team could use it as a guide while preparing their presentation."

"Due to the collaborative nature of the presentations under investigation and the integration of peer (team) and instructor (individual and team) assessment in the present study, a composite rating (Score) incorporating linguistic (speech), metalinguistic (delivery) and paralinguistic (verbal and nonverbal) elements as well as visual communication was preferred for use as the DV."

Results

"To gain further understanding of the data and identify which explanatory variables made contributions in explaining the DV Score, regression analysis was employed."

"Factors with significant correlations with the DV Score were selected as predictor variables and sequentially entered into regression equations."

"Extraversion was the sole selected personality variable, explaining 6% of the variation; motivational variables Project Work and Social Pressure were selected, explaining a combined 4% of the variance."

"When TOEIC was selected as an additional explanatory variable along with personality and the motivation predictor variable, the model accounted for 15% of the variability in Score."

"The model including gender with the aforementioned predictors led to a model explaining 16% of the variability in Score, suggesting that the inclusion of gender added little explanatory power to the regression model."

"Additional regressions were performed using motivational variables as the DVs and entering personality variables as predictors based on the significant correlations revealed above."

Discussion

"Within the Extraversion domain, five items displayed significant correlations, indicating that talkative, energetic, enthusiastic, assertive and outgoing/sociable students were more likely to perform well in collaborative oral presentations."

"Further examination of the data showed that lower ability students' performance in the oral presentation positively correlated with almost all of the extraversion items, while higher ability students' scores correlated primarily with being assertive."

"These results suggest that lower ability students benefit from the presence of the extraversion characteristics associated with collaborative oral presentation scenarios, where an outgoing personality can compensate for insufficient language ability, particularly when this personality trait results in the ability to overcome a lack of confidence."

"The selection and significance of Social pressure as a predictor of collaborative oral presentation performance adds an interesting contrast to this consideration of intrinsic motivation and shares similarity with the results of Flowers [71] who found friendship and reciprocity as motivators for Japanese tertiary students engaged in a computer-supported collaborative learning context."

Conclusion

"The present study offers evidence for the role personality traits and collaborative inquiry-based motivation play in explaining achievement in oral presentations conducted by teams."

"These findings provide support for the argument that extraverts are likely to perform better in spoken tasks, potentially pointing to their superior verbal processing capacity and lower language speaking anxiety as they interact with peers in collaborative projects and engage in oral presentations."

"It highlights the importance of both intrinsic and extrinsic motivational factors concurrently playing a role in collaborative inquiry-based projects as teams of students prepare and deliver oral presentations."

"The authors acknowledge that measures based solely on verbal assessment may offer different results to those presented in this research."

"Investigating the hypothesis that extraverts prefer fluency while introverts favor lexical richness, further examination of students' academic discourse in EFL oral presentations, and consideration of different measures of communication ability offer potential directions for future research."

Motivation Across Domains: Do Goals and Attributions Change with Subject Matter for Grades 4 and 5 Students? [298]

This is a machine-generated summary of:

Vogler, Jane S.; Bakken, Linda: Motivation across domains: Do goals and attributions change with subject matter for Grades 4 and 5 students? [298].

Published in: Learning Environments Research (2007).

Link to original: https://doi.org/10.1007/s10984-007-9021-4.

If you want to cite the papers, please refer to the original.

For technical reasons we could not place the page where the original quote is coming from.

Abstract-Summary
"Grades 4 and 5 students from a large, metropolitan school district were surveyed regarding these motivational variables across the academic domains of reading and mathematics."

"The relationships among these motivational variables were explored, as well as differences across domains."

"Results indicate that children could have a generalised notion of motivation that becomes differentiated when students are asked to reflect on these variables within specified domains."

"The existence of few differences across ethnic and socioeconomic groups suggest that instructional style could be a more powerful influence than either of these variables."

Introduction
"To further confuse educators' attempts at intervention, students who demonstrate learned helplessness in one academic area might not exhibit the same behaviour in another, leading one to wonder if students have different motivational patterns across academic domains (e.g. reading and mathematics)."

"Although Dweck and her colleagues have looked at how her theory of motivation translates across domains of academic achievement and social behaviour (e.g. Benenson & Dweck [72]), little has been done to examine students' beliefs about intelligence or the goals that they set for themselves across specific academic domains, such as reading and mathematics."

"The purpose of this study was to investigate what motivates students to persist and seek challenges in their learning through the examination of students' beliefs about the stability and controllability of intelligence, goals and behaviours."

Methods

"The subscales utilised for measuring students' goal orientation included the original version for both mastery (learning) goal orientation ($\alpha = 0.86$; e.g. "I like reading work that I'll learn from even if I make a lot of mistakes") and performance-approach goal orientation ($\alpha = 0.86$; e.g. "I'd like to show my teacher that I'm smarter than the other students in reading"), but the revised version for performance-avoid goal orientation ($\alpha = 0.74$; e.g. "One of my goals is to keep others from thinking I'm not smart in reading")."

"These statements constitute the Avoiding Novelty sub-scale ($\alpha = 0.78$; e.g. "I would prefer to do reading work that is familiar to me, rather than work I would have to learn how to do") and refer to students' preference for avoiding unfamiliar tasks."

Results

"As predicted, entity beliefs negatively correlated with a mastery (learning) goal orientation; however a significant correlation was only found within the domain of reading ($r = -0.13$, $p < 0.05$)."

"In order to determine if the significant differences found across domains within the subscales of entity beliefs, mastery (learning) goals and approach goals were influenced by teacher variability, a 2×2 factorial repeated-measures analysis of variance was used."

"Using this design, significant differences were found within subjects (regarding domain specificity) but were not found between subjects (regarding the number of teachers): entity beliefs ($F[1, 250] = 2.97$, $p > 0.05$), mastery (learning) goals ($F[1, 251] = 0.86$, $p > 0.05$), and performance-approach goals ($F[1, 249] = 3.04$, $p > 0.05$)."

"In reading, gender differences were found on the mastery (learning) goals subscale, $t(266) = -2.12$, $p < 0.05$, indicating that females score higher on the mastery (learning) goal subscale than males."

Discussion

"While not all motivational variables were different across domains, the subscales used to measure entity beliefs, mastery (learning) goals and performance-approach goals showed significant differences in reading and mathematics."

"Because the goals that students set stem from the beliefs that they have about intelligence (according to Dweck's theory), it makes sense that learning (mastery) goals and performance-approach goals would also be domain dependent."

"With differences across domains evident within the entity beliefs, mastery (learning) goal and performance-approach goal subscales, an obvious question was whether these differences were the result of having different teachers."

"Dweck [73] included in her theory the element of student expectations, explaining that students who hold entity beliefs and set performance goals will exhibit mastery-oriented behaviour if they have high expectations for success, but will exhibit learned helplessness if they have low expectations for success."

Conclusion

"As long as some students continue to demonstrate learned helpless behaviour, researchers will continue to study the complex nature of motivation."

"Dweck's [73] unique motivational goal theory provides a sound basis from which to conduct such research."

"This study not only corroborated earlier findings consistent with Dweck's theory, but attempted to add to the existing body of knowledge regarding group differences and domain specificity."

Putting Effort into Emotion Regulation: Manipulating Desirability and Motivational Strength [299]

This is a machine-generated summary of:

Gutentag, Tony; Tamir, Maya: Putting Effort into Emotion Regulation: Manipulating Desirability and Motivational Strength [299].

Published in: Affective Science (2022).

Link to original: https://doi.org/10.1007/s42761-022-00155-0.

If you want to cite the papers, please refer to the original.

For technical reasons we could not place the page where the original quote is coming from.

Abstract-Summary

"We propose that people could be motivated to invest effort in emotion regulation, by rendering the target emotional state more desirable."

"Rendering an emotion goal more desirable can motivate people to invest effort in emotion regulation, ultimately facilitating successful emotion regulation."

"In three studies (N = 452), we show that both inside and outside the lab, rendering calmness more desirable, boosted the motivational strength to increase calmness, increasing the effort people invested to increase calmness, and ultimately made people calmer."

Motivational Strength in Emotion Regulation

"The model highlights the critical role of motivation in emotion regulation, but it focuses specifically on motivational content (i.e., what people want to feel)."

"We suggest, however, that motivational strength in emotion regulation (i.e., how strongly motivated people are to feel what they want to feel) is similarly critical for success."

"Boosting motivational strength may entice people to engage in emotion regulation, even when it is effortful."

"A third alternative, therefore, is that people may be more motivated to regulate emotions, when they find the emotion goal sufficiently desirable."

"Increasing the desirability of emotion goals by rendering their instrumental value more salient can change motivational content in emotion regulation (e.g., Tamir and others [74])."

"We tested whether rendering an emotion goal more desirable by highlighting its instrumental value could also boost motivational strength, increasing effort, and ultimately success in emotion regulation."

Studies 1a and 1b

"We also found an Emotion × Condition interaction, Wilks' $\Lambda = 0.95$, $F(1,98) = 4.92$, $p = 0.025$, $\eta_p^2 = 0.05$, such that in the high desirability of calmness condition participants exerted more effort to change calmness ($M = 131.28$, $SD = 173.64$) than excitement ($M = 110.20$, $SD = 157.88$), Wilks' $\Lambda = 0.95$, $F(1,98) = 5.16$, $p = 0.025$, $\eta_p^2 = 0.05$."

"Emotion (calmness, excitement) and Time (before the manipulation, and after the effortful regulation paradigm in Study 1a, before the manipulation, after the effortful regulation paradigm, and after the music exposure task in Study 1b) were entered as within-subject factors, and Condition (high vs. low desirability of calmness in Study 1, high desirability of calmness vs. neutral control in Study 1b) as a between-subjects factor."

Study 2

"Before the effortful music paradigm, people experienced more calmness in the high desirability of calmness vs. excitement condition, $F(1,198) = 13.60$, $p < 0.001$, $\eta_p^2 = 0.06$."

"When looking at emotions across time, in the high desirability of calmness condition, there was a significant change in calmness across time, $F(2,197) = 5.75$, $p < 0.001$, $\eta_p^2 = 0.06$, such that calmness increased from baseline to before the effortful music paradigm ($p = 0.021$), and from before to after the effortful music paradigm ($p = 0.022$), whereas excitement didn't change across time ($F < 1.19$)."

"There was also a significant change in calmness across time in the high desirability of excitement condition, $F(2,197) = 9.44$, $p < 0.001$, $\eta_p^2 = 0.09$, such that calmness increased from baseline to before the effortful music paradigm ($p < 0.001$), and from before to after the effortful music paradigm ($p = 0.028$)."

General Discussion

"Building on research on motivational strength in other domains (e.g., Gollwitzer [75]), our findings show that the more people consider the target emotion desirable, the more motivated they are to invest effort to achieve it, and the more successful they ultimately are in achieving it."

"Although research on emotion regulation has typically focused on effects of motivational content (Tamir [76]), our findings demonstrate the additional and independent contribution of motivational strength."

"From an applied perspective, our findings raise the possibility that deficits in motivational strength may explain why people refrain from regulating their emotions, even when they have the means to do it, both in healthy (e.g., Suri and others [77]) and clinical (e.g., Liu & Thompson [78]; Yoon & Rottenberg [79]) populations."

"Future research could examine when motivational strength is (or is not) likely to promote success in emotion regulation, and what might moderate such effects."

[Section 5]

"Success in emotion regulation is important for adaptive functioning (e.g., Gross and others [80]); yet, people are often reluctant to invest the necessary effort for regulating emotions."

"How can we motivate people to invest effort in emotion regulation?"

"We propose that to engage and succeed in emotion regulation people must be sufficiently motivated to do so, and that the strength of their motivation is informed by the desirability of their target emotional state."

Splitting the Affective Atom: Divergence of Valence and Approach-Avoidance Motivation During a Dynamic Emotional Experience [300]

This is a machine-generated summary of:

Kaczmarek, Lukasz D.; Behnke, Maciej; Enko, Jolanta; Kosakowski, Michał; Guzik, Przemysław; Hughes, Brian M.: Splitting the affective atom: Divergence of valence and approach-avoidance motivation during a dynamic emotional experience [300].

Published in: Current Psychology (2019).

Link to original: https://doi.org/10.1007/s12144-019-00264-3.

If you want to cite the papers, please refer to the original.

For technical reasons we could not place the page where the original quote is coming from.

Abstract-Summary
"Valence and approach-avoidance motivation are two distinct but closely related components of affect."

"We also expected that high-approach positive emotions (e.g., desire) would differ from low-approach positive emotions (e.g., amusement) producing a stronger link between valence and approach-avoidance motivation."

"We also explored other dynamic properties of discrete emotions such as the difference between approach-avoidance motivation and valence as a marker of balance within affective components."

"We asked 69 participants to provide continuous ratings of valence and approach-avoidance motivation for eight standardized clips representing different discrete emotions."

"Using multilevel modeling, we established a significant relationship between valence and approach-avoidance motivation with high-approach emotions producing a stronger link between valence and approach-avoidance motivation compared to neutral states and low-approach emotions."

"We found that awe was a distinct positive emotion where approach motivation dominated over valence."

Introduction

"We focused on the relationship between valence and approach-avoidance motivation because the general link between positive valence and approach-avoidance motivation observed across the majority of affective responses (Marchewka and others [81]) is different for some specific emotions (Shiota and others [82])."

"Individuals are motivated to approach objects of anger despite their negative valence, e.g., with the intention to put an end to a situation that aroused anger or to punish the anger-provoking agents (Carver and Harmon-Jones [83])."

"Accounting for the interplay between valence and approach-avoidance motivation is essential to explain the primary affective differences between discrete emotions that stem from different evolutionary processes and engage different biological systems (Harmon-Jones [84]; Shiota and others [82])."

"Given that the majority of literature has been focused on happiness as the most representative positive emotion (e.g., Lench and others [85]), it is essential to account for more numerous discrete positive emotions in testing how the valence and approach-avoidance motivation produce different experiential qualities."

Elementary Components of Affect

"MDMA posits that besides valence and arousal that have been often studied together, comprehensive models of affect should account for approach-avoidance motivational intensity because this component provides additional explanatory power (Harmon-Jones and others [86])."

"MDMA has been increasingly focused on the fact that emotions of the same valence can vary in approach-avoidance motivation (Gable and Harmon-Jones [87])."

"Research on MDMA limited this general claim indicating that the broadening effect does not occur for high-approach positive emotion, e.g., desire (Gable and Harmon-Jones [87])."

"Despite the abundance of experimental studies on the MDMA where the stimuli were carefully selected to fit the MDMA assumptions (Harmon-Jones [84]), we know little whether the strength of valence and approach-avoidance motivation is the same across a variety of high-approach vs. low-approach positive emotions."

The Present Study

"We aimed to examine the relationship between valence and approach-avoidance motivation in several discrete emotions as well as affective categories, i.e., high-approach and low-approach emotions."

"We also expected that approach-avoidance motivation would be stronger for high-approach positive emotions (such as awe, desire, enthusiasm, tenderness) than compared to other low-approach emotions (H3)."

"As explorative aims, we examined differences between approach-avoidance motivation and valence to provide more information about the balance between the components of affect (intraaffective balance)."

"We also aimed to present descriptive data with the dynamics of valence and approach-avoidance motivation."

Method

"We tested whether the relationship between approach-avoidance motivation and valence is significant and differs for discrete emotions (Model 4) for affective categories among all emotions (Model 5), and only for positive emotions (Model 6)."

"To test which emotions are over-motivated, balanced or undermotivated, we examined the difference between approach-avoidance motivation and valence for discrete emotions (Model 7), affective categories among all emotions (Model 8) and only for positive emotions (Model 9)."

"Level 3 Level 4 We regressed the outcome (difference between approach-avoidance motivation and valence) on discrete emotions (Model 7), affective categories for all emotions (Model 8), and positive emotions only (Model 9) as predictors with the following structure: Level 5 Level 6".

Results

"We found that awe, desire, enthusiasm, and tenderness produced higher levels of approach-avoidance motivation relative to the neutral conditions."

"Amusement, anger, and fear produced lower levels of approach-avoidance motivation relative to the neutral conditions."

"High-approach clips produced higher levels of approach-avoidance motivation, and low-approach clips produced lower levels of approach-avoidance motivation relative to the neutral clip."

"We found that the relationship between valence and approach-avoidance motivation was stronger for anger, desire, enthusiasm, fear, and tenderness relative to the neutral conditions."

"Amusement and awe the relationship between valence and approach-avoidance motivation did not differ relative to the neutral conditions."

"Supporting H3, the relationship between valence and approach-avoidance motivation was stronger for high-approach positive emotions relative to low-approach emotions, the neutral conditions, and low-approach positive emotion."

"In relative terms, amusement, and anger tended to be undermotivated (the negative difference between approach-avoidance affect and valence) compared to neutral conditions."

Discussion

"The present study examined absolute levels, relationships, and relative differences between valence and approach-avoidance motivation intensity in dynamic affective experience."

"We demonstrated that the relationship between valence and approach-avoidance motivation differed between discrete emotions and groups of emotions."

"For individuals who experienced high-approach positive emotions, the congruence between valence and approach motivation was higher compared to low-approach emotions, low-approach positive emotions, and neutral conditions."

"The relationship between valence and the approach-avoidance motivation seemed to be enhanced for high-approach emotions."

"This indicated that it was easier to predict approach-avoidance motivation from valence when individuals experienced high-approach emotions."

"The examination approach-avoidance motivation levels in discrete emotions revealed that individuals who experienced awe, enthusiasm, desire, and tenderness (or high-approach positive affect in general) increased their approach-avoidance motivation compared to the neutral condition."

An Initial Examination of State and Longitudinal Effects of Loving-Kindness Practice on Affective and Motivational States at Work [301]

This is a machine-generated summary of:

Masters-Waage, Theodore C.; Reb, Jochen; Tov, William; Bandara, Udana: An Initial Examination of State and Longitudinal Effects of Loving-Kindness Practice on Affective and Motivational States at Work [301].

Published in: Mindfulness (2021).

Link to original: https://doi.org/10.1007/s12671-021-01781-8.

If you want to cite the papers, please refer to the original.

For technical reasons we could not place the page where the original quote is coming from.

Abstract-Summary

"We investigated the state and longitudinal effects of this training on employees' affective and motivational states at work in two studies."

"Study 1 tested this training program in a randomized controlled trial, comparing the effects of loving-kindness practice on employee affect and motivation with an active (mindfulness) and a passive (waitlist) control condition."

"Analyses focused on both longitudinal effects (increases in affect and motivation over the training period) and state effects (effects of practice on daily affect and motivation)."

"Study 2 conducted a 1-week study to further probe the state effects of loving-kindness and the effectiveness of formal vs. informal practice."

"Results indicated mixed support for longitudinal effects, with individuals in the loving-kindness condition showing increases in work motivation, affective valence, and activation over time but the majority of these increases not being statistically different from trends in the two control conditions."

"Analysis of state (day-level) effects found consistent support for a beneficial effect of loving-kindness practice on daily affective valence and motivation."

Study 1

"After completing the pre-survey, participants were randomly assigned to one of three conditions: loving-kindness practice, active (mindfulness) control, or passive (waitlist) control."

"Analyses controlled for participants' mean level of formal and informal practice over the intervention period and the total number of daily surveys they completed in order to take into account potential individual differences in motivation to practice and/or complete the surveys."

"We also tested whether the effects of daily formal and informal practice differed between loving-kindness and mindfulness conditions."

"As above, participants' mean level of formal and informal practice over the intervention period and the total number of daily surveys they completed were entered as controls."

"In the loving-kindness condition, participants showed increased valence, activation, and motivation over the course of the six-week intervention."

Study 2

"Participants were randomly assigned to one of two conditions, (i) loving-kindness formal practice or (ii) loving-kindness informal practice."

"After the information session, participants were asked to complete these practices over the next seven days."

"In study 2, participants were randomly assigned to either formal or informal practices."

"For participants assigned to the formal loving-kindness group, we predicted daily outcomes based on whether or not they engaged in formal practices on a particular day."

"While these results are different from study 1 in that in study 2 it was informal (and not formal) practice that was associated with increases in valence, they nonetheless provide consistent support of an overall effect of loving-kindness practice on affective valence."

"Consistent with study 1, neither formal practice ($b = 0.729$, $SE = 0.37$, $p = 0.051$) nor informal practice ($b = 0.006$, $SE = 0.17$, $p = 0.970$) was significantly related to activation."

General Discussion

"Further, within-person analyses of the daily practice effects provided greater support for loving-kindness being of "help" to employees."

"Replicated results across two studies found that on days in which individuals practiced loving-kindness they also experienced greater work motivation and positive valence (but not activation)."

"By investigating loving-kindness as a secular workplace practice, this paper advances contemplative research in organizations."

"This research suggests the possibility—to be confirmed in future research—that loving-kindness practice might be more impactful for increasing motivation than mindfulness practice."

"It would be intriguing to find that communal loving-kindness practice within an organization leads to changes in company culture or climate; we encourage future multilevel research in this direction."

"While we uncovered several noteworthy effects of loving-kindness practice on affective and motivational states at work, this research only stands at the very beginning of the systematic study of the role of loving-kindness in the workplace, hopefully encouraging future research."

[Section 4]
"The consideration of loving-kindness practice in the workplace as a tool for cultivating a virtuous ethical stance has clear parallels with the voluminous research on the benefits of mindfulness at work (Reb and others [88]; Reb & Atkins [89])."

"Beyond the theoretical link between loving-kindness and affective and motivational states at work, there is also a practical question concerning how the potential benefits of loving-kindness manifest themselves (i.e., longitudinal vs. short-term effects) and how they are best elicited (i.e., formal vs. informal practices)."

"It is plausible and consistent with work on informal mindfulness practice (Birtwell and others [90]), that practicing loving-kindness during employees' interactions with others is an important driver of the practice's salubrious effects."

"To address the above questions, we examined the effects of loving-kindness practice on affective and motivational states, from both a longitudinal (i.e., trend over the training period) and a state (i.e., day-level effects) perspective, and considered both formal and informal practice."

Leaning Forward to Increase Approach Motivation! the Role of Joy, Exercise, and Posture in Achieving Goals [302]

This is a machine-generated summary of:

Miragall, Marta; Vara, M. Dolores; Cebolla, Ausiàs; Etchemendy, Ernestina; Baños, Rosa M.: Leaning forward to increase approach motivation! The role of joy, exercise, and posture in achieving goals [302].

Published in: Current Psychology (2019).

Link to original: https://doi.org/10.1007/s12144-019-00175-3.

If you want to cite the papers, please refer to the original.

For technical reasons we could not place the page where the original quote is coming from.

Abstract-Summary
"The aim of this study is to analyze whether the manipulation of the body (via posture) and behavior (via exercise) increases approach motivation to achieve a personal goal while individuals are joy-induced through a virtual reality mood-induction procedure (VR-MIP)."

"This study highlights that adopting a leaning forward posture while pedaling during joy induction increases approach motivation to achieve a personal goal, and the role of joy as a moderator in this relationship."

Introduction

"Of achieving goals, several studies have pointed out that specific affective and behavioral factors, such as positive affect or exercise, may increase approach motivation (Custers and Aarts [91]; May and others [92]; Orehek and others [93])."

"While the affective and behavioral factors have received great attention in the process of achieving goals, the role of the body has largely been neglected in this process."

"The scientific literature increasingly emphasizes the influence of the body in shaping emotional and cognitive processes (e.g., Cuddy and others [94])."

"The effects of bodily factors – along with affective and behavioral factors – in changing the approach motivation will be the focus of this study."

"Subsections, specific bodily, affective and behavioral factors that have been linked to the increase in approach motivation will be provided."

Posture: A Body Factor to Increase Approach Motivation

"One of the most interesting and novel findings related to the embodied cognition theories is the effect of body posture on approach motivation."

"The manipulation of the body posture has been related to changes in approach motivation intensity."

"Harmon-Jones and others [95] found that a supine body posture (vs. an upright posture), which is associated with reducing approach motivation, reduced the tendency that facilitates effective actions toward goal achievement (e.g., valuing one action over another)."

Positive Affect and Exercise: Affective and Behavioral Factors that Increase Approach Motivation

"Affective and behavioral factors, such as positive affect and exercise, have been related to the increase in approach motivation."

"Studies have shown that positive affective states like joy may be relevant motivators in the goal achievement process."

"Custers and Aarts [91] found that priming behavioral goal states (e.g., going for a walk) that were implicitly associated with positive affect increased the motivation to achieve these states to a greater extent."

"This study showed that moderate-intensity aerobic exercise increased implicit approach motivation for appetitive stimuli, even beyond its natural state."

Current Study
"The main objective of this study is to test whether the manipulation of the body (via body posture) and behavior (via exercise) increases approach motivation and other motivational variables related to achieving a personal goal while individuals are joy-induced."

"The secondary objective of this study is to explore the role of joy as a moderator of the effects of the manipulations of the body (via body posture) and behavior (via exercise) on approach motivation, that is, to explore whether a higher level of joy helps to increase approach motivation to a greater extent."

"Participants in the conditions that involve the manipulation of the body (via body posture) and behavior (via exercise) are expected to achieve a greater increase in approach motivation when they have high levels of joy, both before and after the experimental procedure."

Method
"Participants were assigned to one of three conditions using the Random Allocation Software 2.0 (Saghaei [96]): (a) a high-approach motivation condition (high condition) ($n = 26$), where participants adopted a leaning forward posture while they pedaled on a vertical stationary bicycle during a joy VR-MIP; (b) a medium-approach motivation condition (medium condition) ($n = 34$), where participants adopted a reclining backward posture while they pedaled on a horizontal stationary bicycle during a joy VR-MIP; and (c) a low-approach motivation condition (low condition) ($n = 25$), where participants adopted a reclining backward posture and did not have to pedal while they received the VR-MIP."

"Once the goal had been determined, participants answered four items on a 7-point graduated horizontal scale (1 = not at all; 2 = very little; 3 = a little; 4 = some; 5 = quite; 6 = a lot; 7 = completely) about their 'approach motivation intensity' ('I am motivated to do it'), their 'perceived behavioral control' ('I should be able to get it' and 'I am able to do it'), and the 'planning' ('I clearly see the steps I have to follow to achieve my goal') related to achieving their goal."

Results
"To analyze whether the relationship between the independent variable (condition) and the dependent variable (change in approach motivation intensity) was moderated by the joy scores before (model 1) and/or after (model 2) the experimental procedure; or more specifically, to analyze whether participants in medium and high conditions (vs. low condition) achieved a greater increase in approach motivation when they had high levels of joy, both before and after the experimental procedure."

"Tests of equality of conditional means at each value of the moderator were performed, showing that there were differences between conditions in the change in approach motivation at 'medium', $F(2,77) = 3.52$, $p = 0.034$, $R^2 = 0.086$, and 'high' levels of joy before the experimental procedure, $F(2,77) = 5.77$, $p = 0.005$, $R^2 = 0.125$."

Discussion
"The objective of this study was to examine whether the manipulation of the body (via body posture) and behavior (via exercise) increased the approach motivation and other motivational variables related to achieving a personal goal while individuals received a joy induction."

"This study addressed the links between positive affect (joy), exercise (pedaling), approach-oriented body postures (a leaning forward posture), and motivation variables related to achieving goals (e.g., approach motivation intensity), as well as the role of joy as a moderator of these relationships."

"This study shows the role of joy, exercise, and body posture in changing the approach motivation intensity to achieve a personal goal."

"This study also highlights the need to incorporate body posture manipulations (e.g., leaning forward posture) during exercise when people want to achieve a goal, as well as the need to experience relatively high levels of joy to increase approach motivation."

Exploring the Roles of Emotions, Motivations, Self-Efficacy, and Secondary Control Following Critical Unexpected Life Events [303]

This is a machine-generated summary of:

Turner, Jeannine E.; Goodin, Joel B.; Lokey, Cathryn: Exploring the Roles of Emotions, Motivations, Self-Efficacy, and Secondary Control Following Critical Unexpected Life Events [303].

Published in: Journal of Adult Development (2012).

Link to original: https://doi.org/10.1007/s10804-012-9148-0.

If you want to cite the papers, please refer to the original.

For technical reasons we could not place the page where the original quote is coming from.

Abstract-Summary

"Analyses from interviewed subjects who had encountered significant life-changing events revealed three major themes that influenced their decisions, abilities for self-regulation, and life course paths: (1) types of unexpected events, (2) types and intensity of emotional responses to the critical event, and (3) beliefs about primary control (i.e., personal agency and self-efficacy) and mediated control (i.e., external sources of influence)."

Introduction

"Research has shown that holding a belief of personal control is a key aspect in cultivating and sustaining an individual's motivation for ultimate goal attainment (e.g., Weiner [19]; Bandura [97]; Rotter [98]; Shell and Husman [99])."

"While appraisal theories provide a broad foundation for interpreting individuals' responses to life events, Pekrun's [100] control-value theory provides more specific cognitive-emotional linkages for understanding individuals' emotions, decisions, and motivated behaviors."

"Building upon existing expectancy-value theories of motivation (e.g., Eccles [101]; Pekrun [102]; Pintrich [103, 104]; Wigfield and Eccles [105]), Pekrun's control-value theory (100) particularly underscores the importance of control beliefs and personal values as the vehicle for linking individuals' unique cognitions, emotions, and motivations."

"Individuals' interpretations of mediated control (e.g., Berrenberg [106]; Ferguson and Cox [107]; Hall and others [108]; Rothbaum and others [109]) could provide an important conceptual understanding of the ways that individuals respond to critical life events."

"We were particularly interested in exploring differences in ways individuals' perceived a critical life event, their emotions about the event, their perceptions of control (both primary and mediated), and elements that influenced their abilities to manage their life disruptions."

Method

"We identified those who might be appropriate to the study (e.g., had life experiences that took them on a different path than what they had originally intended) and requested interviews."

"The research participants took part in a semi-structured interview during which they were asked the primary research question, "Have you ever had something unexpected (positive or negative) happen in your life that caused you to reevaluate your course of action for what you were trying to accomplish?"."

"During the interview, participants were asked questions over four major domains: (a) the unexpected event that changed their life course, (b) the feelings they had at the time, (c) the ways in which the events were resolved (or not), and (d) their perceptions of control in relation to the event and its outcome(s)."

Results

"The major themes that emerged from participants' interviews were (1) types of unexpected events, (2) types and intensity of emotional responses, and (3) perceptions of self-efficacy and secondary control."

"Participants' emotional reactions to unexpected events were related to their current level of life satisfaction."

"For some subjects, their self-reactions and self-evaluations appeared to require thought and time to integrate within their broader life context, especially when participants experienced negative unexpected events with accompanying negative emotions."

"Following their unexpected events, the extent to which participants experienced control seemed to fall into one of three categories: external attributions of chance, attributions to self, and attributions to a mediating control (i.e., forces, God, and/or fate)."

"The greatest amount of overall sense of control that individuals felt seemed to occur when they had perceptions of a benevolent, mediating control and they mentioned personal participation in the event (i.e., self-efficacy)."

Discussion

"The resilient participants in our study talked about finding meaning from their critical life event."

"Because our participants were finding meaning for critical life events, their meaning-making most often involved accommodative cognitive changes, but the meaning of the event was assimilated into their general life story."

"Participants who could not find meaning, who could not make accommodative cognitive changes, and who could not assimilate the event into their life stories seemed "stuck" and unable to emotionally move forward in life."

"While we did not find contamination sequences (McAdams and others [110])—whereby individuals describe moving from a positive to a negative state following adversity—some of our participants did describe being emotionally and motivationally "stuck" following their critical life event."

Predicting Students' Negative Emotions in College Remedial Mathematics Courses [304]

This is a machine-generated summary of:

Park, Seung Won; Cho, Moon-Heum; Lim, Seongmi; Hwang, Seokwon: Predicting students' negative emotions in college remedial mathematics courses [304].

Published in: European Journal of Psychology of Education (2023).

Link to original: https://doi.org/10.1007/s10212-023-00697-2.

If you want to cite the papers, please refer to the original.

For technical reasons we could not place the page where the original quote is coming from.

Abstract-Summary
"Although one of the challenges in college remedial mathematics courses involves dealing with negative emotions adversely impacting student achievement, little empirical research has been conducted to examine the factors contributing to the formation of such emotions in these settings."

"The purpose of this study was to examine whether background variables, motivation, and self-regulation predicted students' negative emotions, measured with boredom, frustration, and test anxiety."

"Results indicate that students' age (a background variable) and motivation as shown in extrinsic goal orientation, task value, and self-efficacy were significantly associated with boredom, frustration, and test anxiety."

"Students' self-regulation, specifically effort regulation and metacognitive regulation, was related to negative emotions."

Expectancy–Value Theory of Motivation
"The theoretical framework that undergirds the study that we designed to examine the predictors of negative emotions in the context of remedial mathematics course

is the expectancy–value theory of motivation and self-regulation (Duncan & McKeachie [111]; Pintrich & De Groot [112]; Pintrich and others [113]; Wigfield & Eccles [114]).”

“According to this theory of motivation, achievement-related choices like self-regulation (e.g., paying attention to the content or persistence in learning even if the content is challenging) derive from a combination of student expectations for success and subjective value placed on a task in a particular domain like mathematics.”

“In numerous studies Pintrich and others have examined the links among these three motivational components, self-regulation, and performance in diverse content domains (e.g., mathematics, English) with students of diverse backgrounds (e.g., K–12 and college) both domestically and internationally (Duncan & McKeachie [111]).”

“Expanding Pintrich’s model, we examined the way background variables, motivation, and self-regulation relate to students’ negative emotions in remedial mathematics courses.”

Negative Emotions

“We were particularly interested in three negative emotions: boredom, frustration, and test anxiety, considering the research contexts in which students may experience high levels of stress while taking remedial mathematics courses.”

“The boredom students often experience in their mathematics courses has been shown to negatively influence their achievement (Pekrun and others [115]).”

“Because learning mathematics concepts in a remedial course can be challenging and students may feel that they will be blocked from achieving their goals (e.g., pursuit of an academic degree) if they fail, perceived frustration can run very high, especially for underprepared students.”

“The pursuit of a college degree for many students is dependent upon passing remedial mathematics courses; thus, the requirement to achieve a certain number of points on exams to pass the course can result in high test anxiety.”

Background Variables and Negative Emotions

“Regarding gender, researchers of academic emotions have consistently reported that female students tend to exhibit higher negative emotions about mathematics than male students (Holm and others [116]), compared to female students, who have reported lower enjoyment and pride but more anxiety and hopelessness with regard to learning mathematics (Arroyo and others [117]; Frenzel and others [118]).”

“We assumed that female students may also have higher negative emotions than male students in remedial mathematics.”

"Age also seemed to be associated with negative emotions in remedial mathematics courses (Krinzinger and others [119])."

"We assumed older students may report fewer negative emotions than younger students in remedial mathematics courses."

"Besides gender and age the number of hours university students spend at work may also relate to the negative emotions they experience while taking remedial mathematics courses."

"If students worked too many hours per week, they may not have had enough time for learning, thus increasing their negative emotions in remedial mathematics courses."

Motivation and Negative Emotions

"On the MSLQ expectancy was measured with self-efficacy about learning, which produces positive achievement emotions like enjoyment and pride; whereas low perceived self-efficacy produces negative achievement emotions like test anxiety (Ahmed and others [120]; Pintrich and others [113])."

"Research has shown that intrinsic goal orientation concerned with the development of competence (e.g., mastery goals) was positively related to enjoyment and hope, whereas extrinsic goal orientation concerned with the demonstration of competence (e.g., performance goals) was positively related to negative emotions like anxiety and hopelessness (Pekrun and others [121]; Putwain and others [122]."

"Task value (i.e., recognizing the importance of academic tasks) related with positive achievement emotions, whereas negative value placed on the activities related to negative emotion (e.g., test anxiety Ahmed and others [120]; Pintrich and others [113])."

"We assumed students who have high self-efficacy and value (high intrinsic goal orientation, low extrinsic goal orientation, and high task value) report negative emotions less often than their counterparts."

Self-regulation and Negative Emotions

"Self-regulation refers to students' systematic use of metacognitive strategies and behaviors to manage their learning process to achieve personal goals (Pintrich & De Groot [112]; Pintrich and others [113]; Zimmerman & Schunk [123])."

"Effort regulation is very important because even if students felt boredom or disliked learning mathematics concepts, they still needed to exert effort to learn them in order to continue to pursue their college coursework."

"Metacognitive regulation refers to students' use of metacognitive strategies for deep learning in a challenging learning task (Pintrich & De Groot [112])."

"Pausing mathematics problem solving and reflecting upon the problem-solving process is an essential regulation process that students must master while learning mathematics concepts in remedial mathematic courses."

Research Questions

"The main goal of the study was to investigate factors that predict students' experience of negative emotions in remedial mathematics courses."

"To what extent do background variables, motivation, and self-regulation predict boredom in remedial mathematics courses?"

"To what extent do background variables, motivation, and self-regulation predict frustration in remedial mathematics courses?"

"To what extent do background variables, motivation, and self-regulation predict test anxiety in remedial mathematics courses?"

Methods

"ALEKS individually guides students to master mathematical concepts by providing course materials and feedback, assessing their progress at a certain point to update student states, and prescribing a new path."

"ALEKS was used as the main learning material in remedial mathematics courses offered in an emporium teaching model, in which students attended two lab sessions per week and learned mathematics concepts with ALEKS at their own pace."

"The data were collected from those who took basic remedial mathematics courses in ALEKS, in which students learn core elements of mathematics, including fractions, functions, quadratics, factoring, absolute value equations, and inequalities."

Results

"Among the predictors extrinsic goal orientation ($\beta = 0.20$, $p < 0.01$), self-efficacy ($\beta = -0.55$, $p < 0.001$), and effort regulation ($\beta = -0.23$, $p < 0.01$) made significant individual contributions to explaining frustration."

"The variables entered in Step 1 together explained 6.3% of the variance, $F(4, 196) = 3.27$, $p < 0.01$, and pretest scores were the only variable that significantly predicted test anxiety ($\beta = —0.17$, $p < 0.05$)."

"Among the variables extrinsic. goal orientation ($\beta = 0.30$, $p < 0.001$), self-efficacy ($\beta = -0.67$, $p < 0.001$), and effort regulation ($\beta = -0.20$, $p < 0.01$) made significant individual contributions to explaining the variance in test anxiety."

Discussion

"The purpose of this study was to examine significant predictors of negative emotions that students experienced in remedial mathematics courses."

"The results are somewhat encouraging because gender and personal situations in which students had to work at jobs in addition to schoolwork were not related to the negative emotions they experienced in the remedial mathematics courses."

"Emotion researchers viewed motivation, self-regulation, and emotions as interrelated and reciprocal (Pekrun [124]; Pintrich & De Groot [112]), but very few empirical studies have been conducted to investigate how motivation and self-regulation related to multiple negative emotions in remedial mathematics courses."

"In consideration of the remedial mathematics course contexts, in which students had high levels of stress and external goals (e.g., passing the course), we examined negative emotions from the perspectives of expectancy–value of motivation and self-regulation."

[Section 10]
"Students who must take or retake remedial mathematics courses experience negative emotions while doing so, concerned about possible failure."

"Linked to poor performance, negative emotions are known to negatively influence student motivation, competencies, and the use of deep learning strategies (Pekrun and others [115, 125])."

"Many researchers have demonstrated that negative emotions like test anxiety and boredom negatively related to students' intrinsic goal orientation, self-efficacy, and use of active learning strategies (Pekrun and others [115]; Pintrich & De Groot [112]); however, few empirical studies have been conducted in remedial mathematics courses to examine possible factors that influence those negative emotions."

"The purpose of this study was to investigate factors related to students' negative emotions in remedial mathematics courses from the perspectives of Pintrich's expectancy–value theory of motivation and self-regulation (Duncan & McKeachie [111]; Pintrich & De Groot [112]; Pintrich and others [113])."

The Role of Epistemic Emotions in Undergraduate Students' Proof Construction [305]

This is a machine-generated summary of:

Schubert, Sandra; Pekrun, Reinhard; Ufer, Stefan: The role of epistemic emotions in undergraduate students' proof construction [305].

Published in: ZDM – Mathematics Education (2022).

Link to original: https://doi.org/10.1007/s11858-022-01413-y.

If you want to cite the papers, please refer to the original.

For technical reasons we could not place the page where the original quote is coming from.

Abstract-Summary
"Considering the challenge that proof construction poses to learners of all ages, prior research has investigated its cognitive determinants, but the impact of affective-motivational experiences on proof construction has been insufficiently investigated."

"We investigated how the performance of 80 mathematics undergraduate students in a geometric proof construction task relates to the epistemic emotions experienced during proof construction."

"Controlling for geometry knowledge, we included control and value appraisals as antecedents in our investigation of epistemic emotions, and attention and motivation as mediators of their effects on proof construction performance."

"These two markers of the proof construction process, in turn, mediate the effects of enjoyment and boredom on proof construction performance."

"We investigated systematically the role of epistemic emotions in geometric proof construction and we offer insights that complement the existing research on the cognitive determinants of proof performance."

"This study extended research on epistemic emotions into the area of proof construction, an epistemic process central to mathematics."

Introduction
"We examined the epistemic emotions that arise when students are constructing a multistep geometry proof, and how these emotions shape markers of the proof

construction process, such as cognitive resources and motivation, as described in Pekrun's [124] cognitive-motivational model of emotions and learning."

"We investigated whether differences in university students' epistemic emotions during a geometric proof construction task are attributable to differences in their control and value appraisals, and whether epistemic emotions mediated the links among cognitive appraisals, the markers of epistemic processes (attention and motivation), and proof construction performance."

"We investigated whether interindividual differences in students' experiences of epistemic emotions during proof construction were attributable to control and value appraisals."

"We expected that epistemic emotions, including attention and motivation, would predict markers of proof construction processes."

Method

"The students' subjective competencies of proof construction were assessed with a 12-item scale based on the Self-Description Questionnaire (SDQ; Marsh [128]) (e.g., "I have always done well in such geometric problems.") with answer options from 1 (strongly disagree) to 5 (strongly agree)."

"Perceived values for geometric proof construction were measured using six items from Pekrun and others's [129] task value scale, covering intrinsic task value (e.g., "I like solving such geometric problems"), utility value, and attainment value."

"The 6-item scale (Wild & Schiefele [130]) to measure students' attention during the proof construction task assessed the students' lack of concentration, distractibility, and task-irrelevant thinking during task completion (e.g., "It was hard for me to stay focused during the task") on a scale ranging from 1 (not at all) to 5 (completely)."

"The students' knowledge of the mathematical framing theory was included as a covariate of proof construction performance."

Results

"Indirect effects of appraisals on attention and motivation via epistemic emotions (Q5a)."

"Enjoyment and curiosity mediated the effects of control and value on motivation, but only enjoyment mediated relationships between participants' control and value appraisals and their attention."

"The more the control and value are attributed to the task, the more students enjoy constructing a proof, which, in turn, enhances their attention and motivation."

"Contrary to the prominent role of curiosity as a mediator between appraisals and the students' attention and motivation, the indirect effect of curiosity on proof construction performance failed to reach statistical significance."

"Enjoyment showed a significant indirect effect on performance, that primarily stemmed from an indirect effect via students' attention."

"Mediation via attention was slightly stronger than via motivation, but both specific indirect effects failed to reach significance."

"The indirect effects of frustration and confusion on proof construction performance were not significant."

Discussion

"We examined the relationship between university students' subjective appraisals of a proof construction task, their emotional epistemic experiences, attention allocated to and motivation invested in the task, and proof construction performance, controlling for knowledge of the framing theory."

"Some emotions are experienced at a low intensity, on average. (2) Primarily positive epistemic emotions depend on students' control and value appraisals of the task. (3) Epistemic emotions function as mediators between these appraisals and markers of the proof construction process (attention and motivation). (4) We observed partial support for the assumption that value moderates the relationship between control appraisals and emotional experiences. (5) Positive and negative emotions function as predictors of proof construction performance indirectly, via the markers of the proof construction process."

"The findings of this study suggest that students' epistemic emotions and cognitive as well as motivational factors impact proof construction performance."

Conclusions

"It shows that proof construction is an affectively charged experience, as the participating students reported a range of positive and negative epistemic emotions."

"In considering the effects of epistemic emotions, the findings suggest that curiosity and enjoyment enhance students' attention and motivation invested in proof construction, while boredom hinders their concentration and motivation."

"This study's findings also suggest that students' appraisals are linked to their attention and motivation indirectly through positive epistemic emotions, and that positive as well as negative epistemic emotions are linked with the students' proof construction performance through the emotions' effects on attention and motivation."

"Evoking curiosity and fostering enjoyment regarding proof construction is a means of promoting adaptive epistemic processes among students that will ultimately benefit their proof construction performance."

Capturing Teacher Students' Emotional Experiences in Context: Does Inquiry-Based Learning Make a Difference? [306]

This is a machine-generated summary of:

Litmanen, Topi; Lonka, Kirsti; Inkinen, Mikko; Lipponen, Lasse; Hakkarainen, Kai: Capturing teacher students' emotional experiences in context: does inquiry-based learning make a difference? [306].

Published in: Instructional Science (2012).

Link to original: https://doi.org/10.1007/s11251-011-9203-4.

If you want to cite the papers, please refer to the original.

For technical reasons we could not place the page where the original quote is coming from.

Abstract-Summary
"Teacher students' contextual learning experiences were examined longitudinally in authentic study environments using the contextual activity sampling system, a means of mobile-supported experience sampling."

"The students' (n = 9) experiences were first recorded during a 2 week period in their first year of study."

"The first year of study consisted mostly of lectures and ordinary small-group work, whereas the second measurement period ran parallel to the completion of an intensive inquiry-based project, which was the focus of the present study."

"A multivariate analysis of variance revealed that studying during the inquiry-based period produced stronger experiences of being challenged as well as negative affects than the teacher-centered period."

"Interview data indicated that the participants enjoyed the inquiry-based period and that the work was intensive."

"Contextual data and interviews were also used to describe students' experiences during one particular study session during the inquiry-based project."

Introduction
"The study follows how a group of students experienced shifting from teacher-centered learning to inquiry-based learning over 2 years."

"Studying and learning are not merely cognitive activities; rather, students experience a wide range of emotions while studying (Pekrun and others [13])."

"An ideal learning environment would, instead, provide shared control, where teachers facilitate and promote learning by using activating and student-centered methods, in order to help all types of students develop their study strategies and self-regulatory skills."

"We wanted to determine whether a variation in emotions and motivational experiences would play a role when the same teacher students would be introduced to increasingly student-centered ways of learning."

"To explore how students experience teacher-centered instructional arrangements as compared to inquiry-based learning."

"We anticipated that the students would experience intense affects when moving from teacher-centered to the inquiry-based study."

Methods

"Data collection took place when the students were in their first and second year of a 5 year Master's program."

"The students study intensively as a small group for 3 years, applying progressive inquiry-based learning as one of their main approaches."

"The present study focused on comparing teacher students' contextual learning experiences across two 14 day follow-ups, the first (year 1) consisting mostly of lectures, and the second (year 2) focusing on an intensive inquiry-based project."

"During the first-year follow-up, the students studied various subject-matter based courses in the teacher-training program, including physical education, drama education, scientific writing, introductory methodology and educational psychology."

"While the teachers had given the schedule during the first-year follow-up, during the second year follow-up the students were responsible for arranging the study schedule."

"To the CASS process, each participant underwent a semi-structured interview before and after the inquiry-based period in the second year."

Statistical Procedures

"To see whether the students' study-related experiences differed during the teacher-centered and inquiry-based periods, a 2×2 MANOVA was performed on five dependent variables."

"The statistical analyses were completed in five stages: missing value analysis, creation of sum variables, outlier analysis followed by a MANOVA, and a time series analysis."

"In the 16 observations having only one missing value concerning dependent variables, the missing values were replaced using the EM method in the SPSS."

"Following the missing value analysis, three sum variables were created."

"Following the missing value and outlier analyses, the number of observations included in the MANOVA analysis was 575."

"Students' experiences during the inquiry-based study period were analyzed using time-series analysis at a group level."

"This resulted in having one value for every hour of every day of the research period, the value being the mean of the observations during that hour."

Results

"The participants experienced study situations during the inquiry-based period as more challenging and affect-provoking than during the teacher-centered period."

"The MANOVA showed that participants' experiences differed between the two periods (teacher-centered and inquiry-based) and in the different situations (studying and not-studying)."

"With Wilks' criterion, the main effects for dependent variables (positive affects, negative affects, importance, challenge and competence) were significantly effected by both study period (teacher-centered vs. inquiry-based) [$F(5, 571) = 7.52$, $p < 0.001$] and situation (studying vs. not studying) [$F(5, 571) = 39.52$, $p < 0.001$]."

"While the CASS data revealed only higher negative experiences and challenges, the interviews revealed that the students also enjoyed working on the project."

"The variation of positive and negative affects of all nine participants was examined during the second period, when the students were collaboratively involved in the inquiry-based study."

Discussion

"The aim of this study was to compare teacher students' experiences during two different periods (teacher-centered and inquiry-based) and to explore how experience-sampling data could be used to describe students' experiences during the inquiry-based period."

"The results indicated that the students' studying experiences differed during the two periods."

"Further, the students reported being challenged more while studying during the inquiry-based period."

"The students reported both positive and negative emotional experiences, along with those of being challenged, and all of these mostly related to collaboration and co-regulating the learning process."

"It is interesting that the CASS-data and the interviews convey a somewhat different picture of the students' experiences during the inquiry-based period."

"The students did experience higher levels of negative affects and challenge during the inquiry-based period."

"Although methodologically challenging, focusing on students' experiences in authentic learning situations may open fascinating opportunities for educational research."

Motivational Mindsets, Mindset Churn and Academic Performance: The Role of a Goal-Setting Intervention and Purpose in Life [307]

This is a machine-generated summary of:

Hudig, Job; Scheepers, Ad W. A.; Schippers, Michaéla C.; Smeets, Guus: Motivational mindsets, mindset churn and academic performance: The role of a goal-setting intervention and purpose in life [307].

Published in: Current Psychology (2022).

Link to original: https://doi.org/10.1007/s12144-022-03462-8.

If you want to cite the papers, please refer to the original.

For technical reasons we could not place the page where the original quote is coming from.

Abstract-Summary

"The motivational mindset model (MMM) is a new student-centered, multidimensional perspective on motivation in higher education and aims to better explain differences in wellbeing, study success and intervention effectiveness."

"The present study expands the MMM by (1) examining the relationship between the mindsets and academic performance, (2) observing the mindset churn in the first year, and (3) exploring the role of the mindsets in the effectiveness of an online, narrative goal-setting intervention."

"Results further show that students with a low-impact mindset at follow-up were more likely to drop out of the first year compared to the other three mindsets."

"A group of low-impact mindset students show an increased sense of purpose after participating in the goal-setting intervention and moved to a social-impact mindset during the year."

"This pattern provides preliminary support that the goal-setting intervention is a purpose-fostering intervention for students entering higher education with a low-impact mindset."

Motivational Mindset Model

"Students with a high-impact mindset are confident about their reasons for studying and aim to positively affect every aspect of their lives through their study program: earning high grades, self-development, helping others, having a rich social life, and earning money."

"A follow-up study showed that sense of purpose and study engagement were meaningfully related to the motivational mindsets (Hudig and others [131])."

"Students with a high-impact mindset and social-impact mindset showed the same and most optimal pattern in sense of purpose and study engagement."

"The self-impact mindset had a weaker sense of purpose than the high-impact and social-impact mindset, while these students reported equal feelings of engagement toward their studies as the social-impact mindset students."

"The low-impact mindset students revealed the least optimal pattern in the two student qualities: the weakest sense of purpose and lowest level of study engagement (Hudig and others [131])."

Goalsetting and the Goal-Setting Intervention

"In a randomized controlled trial among struggling college students, Morisano and others [132] showed that the experimental goal-setting group increased significantly in academic performance compared to the control group."

"Schippers and others [133] studied the effects of the intervention in a quasi-experimental design among 2928 students, showing a 22% increase in academic performance in the intervention cohorts compared to control cohorts, and remarked increases in performance for students that participated in all stages of the 3-stage goal-setting intervention, wrote more words in the exercise, and showed higher specificity of their goal-achievement plans (GAP)."

"The first stage prompts students with guided questions to reflect on their ideal life, an expressive writing exercise similar to the best-possible-future-self intervention."

"Students write down life goals, clarify the importance of their goals, and specify plans, obstacles, and strategies to overcome these obstacles."

"The underlying mechanisms of this goal-setting intervention among first-year students has not yet been empirically demonstrated."

Goalsetting and Purpose

"A fundamental positive outcome that the goal-setting intervention seems to have an impact on is students' sense of purpose (Schippers & Ziegler [134])."

"One of the opportunities is also to find out which students are more likely to find a sense of purpose through goal-setting activities (Bronk and others [135])."

"To internalize the roadmap, the goal-setting intervention connects goal planning to the goal creation process."

"We therefore propose that the goal-setting intervention is a purpose-fostering intervention."

Purpose and Study Performance

"We have proposed that the goal-setting intervention impacts the purpose of students."

"Research suggests that an increase in sense of purpose is associated with higher study engagement and that higher engagement is related to better study performance, thus a potential mechanism of the goal-setting intervention on performance might be via both purpose and study engagement."

"In a study by Dekker and others [136], self-regulated learning, resilience, grit, and engagement did not appear as mediators between the goal-setting intervention and performance."

"The current study therefore follows the student-oriented, multidimensional approach of the MMM to explore how the goal-setting intervention works."

"We extend our proposition and propose that the goal-setting intervention works differently for students with different pre-intervention levels of purpose."

"The current study examines whether the goal-setting intervention has a different effect on the purpose of the four types of motivational mindsets."

Research Questions and Aims

"The first aim of the present study is to determine whether students' motivational mindset fluctuates or remains stable throughout the first year."

"The second aim of this study is to examine the differences in study performance between the varying types of motivational mindsets."

"In terms of intrinsic, interest-driven motives, research has shown consistently that intrinsic motivation is positively related to academic performance (Cerasoli and others [137]; D'Lima and others [138]; Kaufman and others [139]; Richardson and others [140])."

"For self-transcendent motives, most studies show a positive association with academic performance."

"Research Question 2 is: Are there differences in first-year academic performance between the four types of motivational mindsets?"

"Bearing in mind that purpose is positively related to study engagement, and study engagement is linked to study performance, we also analyze what happens to the study engagement of the mindsets after participation in the intervention."

"Research Question 3 is: Does the goal-setting intervention impact the sense of purpose of the various mindset groups differently?"

Method

"To explore whether the goal-setting intervention impacts students' sense of purpose (RQ3), the over-time developments in purpose across the motivational mindsets were examined."

"We focused on students in the four stable motivational mindset groups and students in the four prominent changed mindset groups."

"To developments in purpose, we tested and compared developments in study engagement across these student groups."

"Paired t-tests were conducted within each stable mindset group both for purpose and study engagement to check for differences between timepoints (T0–T1 and T1–T2)."

"Paired t-tests were conducted within each prominent changed mindset group to analyze developments in purpose and study engagement between timepoints (T0–T1 and T1–T2)."

"Students with a postponed BSA ($n = 50$) were excluded from the analyses on first-year academic performance in the second research question."

"When we specifically tested and compared the four motivational mindsets, we excluded this group of students from the analyses."

Results

"The chi-square test between baseline mindsets was non-significant, indicating that students with a particular motivational mindset at baseline were not more or less likely to drop out later in the research, χ^2 (4) = 5.68, p = 0.224, Cramer's V = 0.09."

"A Kruskal–Wallis H test among the stable motivational mindsets ($n = 117$) indicated no significant differences in first-year credits obtained, χ^2 (3) = 5.11, p = 0.164."

"Fisher's exact test among the stable motivational mindsets was non-significant ($p = 0.067$, Cramer's $V = 0.22$), indicating that these mindset groups were not different in retention rates [$\text{Retention}_{\text{high-impact}} = 100\%$, $\text{Retention}_{\text{self-impact}} = 100\%$, $\text{Retention}_{\text{social-impact}} = 92.3\%$, $\text{Retention}_{\text{low-impact}} = 88.9\%$]."

"Paired samples t-tests were again conducted to check for study engagement differences across timepoints within each stable motivational mindset."

"Paired samples t-tests were conducted to check for study engagement differences between timepoints within the four prominent changed mindset groups."

Discussion

"Low-impact mindset students at follow-up were less likely to advance to the second year of the study program compared to the other three motivational mindsets."

"Future research could test the mindset-performance relationship in a student sample without the goal-setting intervention and measure the study motives more frequently during the academic year."

"Unstable low-impact mindset students can develop self-transcendent and/or self-oriented, intrinsic motives for studying through the intervention."

"A group of students with a self-impact mindset at the beginning of their studies changed to a low-impact mindset, while they reported strong declines in purpose and engagement over the course of the academic year."

"The findings here imply that when students' motives for studying are stable, even without intrinsic motives in a low-impact mindset, their study success can be similar."

Conclusions

"Students who enter higher education with a stable low-impact mindset can be just as successful in the first academic year as the other three motivational mindsets."

"Students who have an unstable mindset and endorse a low-impact mindset during the study program have a higher risk of dropping out."

"For unstable low-impact mindset students does the intensive reflection exercises positively change their mindset, foster purpose, and protect the level of study engagement."

[Section 10]

"Students are most likely to drop out by the end of the first year of the bachelor program (Christie and others [141]; Van Rooij and others [142]; Willcoxson and others [143])."

"In Dutch universities, which is the focus of the current study, 26% of students drop out in the first year, of which 19% switch to another program and 7% leave higher education altogether (Inspectie van het Onderwijs [Dutch Inspectorate of Education] [144]; Van den Broek and others [145])."

"From the students who progress to the second year, most of them (90% to 95%) earn at least their bachelor's degree within five years (Van den Broek and others [145])."

"While research notes that students are motivated to go to university for multiple reasons simulteneously to varying degrees (Côté & Levine [146]; Henderson-King & Smith [147]), most studies have followed a variable-centered, unidimensional approach, largely ignoring the effect that multiple motives may have on study related behavior and outcomes."

The Role of Motivational Components in Metamotivational Monitoring in Medical Students: A Mixed Method Study [308]

This is a machine-generated summary of:

Norouzi, Ali; Parmelee, Dean; Shariati, Mohammad; Norouzi, Saiideh; Alizadeh, Maryam: The role of motivational components in metamotivational monitoring in medical students: a mixed method study [308].

Published in: BMC Medical Education (2023).

Link to original: https://doi.org/10.1186/s12909-023-04081-y.

If you want to cite the papers, please refer to the original.

For technical reasons we could not place the page where the original quote is coming from.

Abstract-Summary

"In metamotivational monitoring, students try to identify the declined motivational component in order to regulate their motivation."

"There is scarcity of evidence on which motivational components are targeted by the medical students when they use each motivational regulation strategies."

"This study aims were identifying motivational components in motivational regulation process, developing a measurement tool and, testing the predictive relationship between the motivational components and motivational regulation strategies."

"In a qualitative study conducting a semi-structured in-depth interview, the motivational components were explored."

"Predictive relation between the motivational regulation strategies and motivational components was assessed utilising Structural Equation Modelling."

"In the Exploratory Factor Analysis of Motivational Components Questionnaire (MCQ), 6 factors were discovered that explained 74% of the total variance."

"In examining the predictive relationships, each of the four components of self-efficacy, intrinsic value, self-relevant value and promotion value were specifically predicted by two motivational regulation strategies."

"Health Profession Educators can improve the academic motivation of students by identifying one or more declined motivational component and teaching specific motivational regulation strategies."

"It is recommended to hold training courses on motivational regulation strategies for medical school faculty, study-skills advisors, and students."

Background

"We know from the literature, outside the field of HPE, students are able to regulate their motivational components utilizing motivation regulation strategies [148]."

"Point of view, Norouzi and others found that medical students use seven motivational regulation strategies, including regulation of value, regulation of situational interest, self-consequating, environmental structuring, promotional situational awareness, preventional situational awareness, and regulation of relatedness to monitor and control their academic motivation [149]."

"The identification of motivational regulation strategies is a significant step in understanding the metamotivational process, but another step, is identifying motivational components which are targeted by these motivational strategies [150]."

"Norouzi and others explained the motivational regulation strategies and designed Metamotivational Strategies in Medical Students Questionnaire (MSMQ) to use in the medical education context [149]."

"First, we identified the motivational components in motivational regulation process in medical students and developed a measurement tool."

Methods

"In a qualitative study, the motivational regulation strategies used by the medical students were explored."

"Taking into account the number of experts ($N = 15$) and the values in the Lawshe table, the items with $CVR < 0.49$ were eliminated from the questionnaire (N: total number of experts; Ne: number of experts who selected the item "necessary") In this step, six medical students were interviewed."

"MCQ and Metamotivational Strategies in Medical Students Questionnaires (MSMQ) [149] were used."

"The students were asked to answer the questions related to each motivational strategy i.e., determine how much they use that strategy and then, determine their reasons (by answering MCQ questions) when using that specific strategy."

"Seven online questionnaires were designed separately and sent to seven different groups of medical students (One questionnaire for each strategy)."

Results

"In the items of factor 3, the students tried to provide reasons for using the motivational regulation strategies, where they referred to the prevention of problematic situations."

"In the items of factor 4, the students tried to provide reasons for using the motivational regulation strategies, where they referred to the academic achievements."

"The reasons referred to in the items of factor 5 were conceptually related to achieving academic values and benefits."

"Divergent validity assessment showed that only in the model of "environmental structuring" "motivational components", item 4 of the subscale of "prevention value" had a factor load of 0.45, and 0.55 on Self-efficacy subscale."

"Investigation of the significance of path coefficients and loading factors indicated that the T-Value of all paths in all seven models was above 1.96."

Discussion

"The present study identifies the motivational components targeted by motivation regulation strategies, creates a tool to measure them, and determines a predictive relation between the motivational components and motivational regulation strategies."

"In his study, Wolters concluded that the students could affect the self-efficacy through their motivational regulation strategies [151]."

"This study demonstrates that these two strategies of regulation of value and promotional situational awareness could correctly predict the self-efficacy."

"The present study results indicate that the two strategies of regulation of situational interest and promotional situational awareness can strongly predict this component."

"The external value was not adequately predicted by any motivational regulation strategies."

"Despite their significant relationship with all motivational components, the three strategies of regulation of relatedness, preventional situational awareness, and self-consequating could not predict a substantial share of one of the motivational components alone."

Conclusion and Recommendations

"The identification of motivational components is a very fundamental idea in the Miele and Scholer's model."

"The results of this study indicate that the six motivational components of the Miele and Scholer's model are the key and essential components in this process."

"Although the expectations of the metamotivational model as to the predictive relation between the strategies and components were not fully met, in this study, the fundamental assumption of the metamotivational model, i.e., the direct relationship between one or two motivational regulation strategies and the motivational components was proven."

"It is expected that future studies investigate the effect of simultaneous use of several motivational regulation strategies on the motivational components."

Motivational Conflict Influences the Timing of Emotions and Their Regulation [309]

This is a machine-generated summary of:

Aldao, Amelia; Wisco, Blair E.: Motivational conflict influences the timing of emotions and their regulation [309].

Published in: Motivation and Emotion (2015).

Link to original: https://doi.org/10.1007/s11031-015-9496-8.

If you want to cite the papers, please refer to the original.

For technical reasons we could not place the page where the original quote is coming from.

Abstract-Summary
"Little is known about how motivational context (e.g., motivational conflict) can affect the timing of emotional experiences (i.e., affective chronometry)."

"The influence of motivational context on the utilization of emotion regulation strategies has been largely unexplored."

"We found that motivational conflict moderated the time course of anxiety."

"These findings highlight the importance of examining motivational processes when seeking to understand how and when individuals experience and regulate their emotions."

Introduction
"ER also unfolds over time (e.g., Sheppes and Gross [152]; Yoon and Joormann [153]; Urry [154]) it is important to examine the influence of motivational conflict on the time course of the use of regulation strategies."

"In this investigation, we sought to advance the literature on motivational conflict and emotions and their regulation by examining the role of sequential motivational conflict on the time course of anxiety and the use of ER strategies."

"We expected that participants in the conflict condition (disgust + food) would experience greater anxiety than those in the control conditions, consistent with prior research that motivational conflict is associated with anxiety (e.g., Aupperle and Paulus [155]; Fauerbach and others [156]; Gray and McNaughton [157]; Mennin and Fresco [158])."

"We expected that participants in the high motivational conflict condition (disgust + food) would experience two increases in anxiety: (1) at the beginning of the disgust film clip, coinciding with the elicitation of disgust (e.g., Aldao and Nolen-Hoeksema [159]), and (2) after the disgust film clip, in preparation for the task that would be at odds with their current emotional state (e.g., Corr [160]; Gray and McNaughton [157]; Mennin and Fresco [158]; Robinson and others [161])."

Method

"Those assigned to the disgust + food (conflict) and craving + food (no conflict) conditions read that after watching the film clip, they would have to participate in a food tasting task, in which they would have to smell and taste a number of foods and rate them on their appeal."

"Following this induction, all participants rated the emotions they were experiencing and the ER strategies they were using (pre-film assessment)."

"Participants then watched either the disgust- or craving-eliciting film clips (see below) and provided additional ratings of their emotions and ER strategies after 100 s (early-film assessment), 200 s (mid-film assessment), and 300 s (post-film assessment)."

"At each time point, participants rated the intensity of their emotions (disgust, desire to eat, anxiety), and the extent to which they used ER strategies (experiential avoidance, perseverative cognition)."

Results

"Comparisons of each time point across the three conditions indicate that there were no differences at pre-film clip, p's > 0.880, and that disgust was higher in the disgust + food (conflict) and the disgust + toy block (no conflict) conditions than in the craving + food (no conflict) condition during the early, mid, post-film clip and pre-task assessments, all p's < 0.001."

"Between condition comparisons Comparisons of each time point across the three conditions indicate that there were no differences at pre-film clip, p's = 1, and that the desire to eat was lower in the disgust + food (conflict) and the disgust + toy block (no conflict) conditions than in the craving + food (no conflict) condition during the early, mid, post-film clip and pre-task assessments, all p's < 0.001."

Discussion

"Anticipating a task that posed a motivational conflict with one's current emotional state resulted in (1) an increase in anxiety immediately prior to the incompatible task, and (2) no changes in avoidance or perseverative cognition."

"A second possibility is that participants in the conflict condition did generate a strong anxiety response to the disgust-eliciting film clip, but they also experienced a stronger motivation to downregulate it because the "worst was yet to come"."

"It is also important to mention that the vast majority of changes in emotional experiences (disgust, desire to eat, anxiety) and the implementation of ER strategies (experiential avoidance, perseverative cognition) primarily took place towards the early phases of the film clip."

"In all, this study suggests that motivational context can have an important influence on the experience of emotions and the spontaneous use of ER strategies."

The Contributions and Prospects of Goal Orientation Theory [310]

This is a machine-generated summary of:

Kaplan, Avi; Maehr, Martin L.: The Contributions and Prospects of Goal Orientation Theory [310].

Published in: Educational Psychology Review (2006).

Link to original: https://doi.org/10.1007/s10648-006-9012-5.

If you want to cite the papers, please refer to the original.

For technical reasons we could not place the page where the original quote is coming from.

Abstract-Summary
"As research in the theory has proliferated, the use of multiple methods to assess goal orientations seems to have contributed to theoretical vagueness, especially with regard to the origin, development, and stability of these orientations."

"This review article starts with a critique of methods used in goal orientation research."

"The article then suggests six possible theoretical models of goal orientations that seem to be suggested by the literature, including the perspectives of goal orientations as emerging from: situation-schemas, self-schemas, self-prime, needs, values, and situated meaning-making processes."

Introduction

"Couple of decades, goal orientation theory (Ames [162]; Dweck [73]; Maehr [163]; Nicholls [164]) emerged as an important theoretical perspective on students' motivation in school (Anderman & Wolters [165]; Elliot [166]; Meece, Anderman, & Anderman [167]; Pintrich & Schunk [168])."

"This theory provided a framework for extensive research on motivational orientations that contributes to students' adaptive and maladaptive patterns of engagement."

"The theory highlights environmental characteristics that foster these motivational orientations (see Midgley [169]; Meece and others [167]; Pintrich & Schunk [168])."

"This article provides a critical review of the concept of goal orientations and of methods used to operationalize them, and follows by suggesting several theoretical models for conceiving these orientations, all with the purpose of clarifying theory and invigorating future research."

Goal Orientation Theory

"Mastery goals orientation can be said to refer to a purpose of personal development and growth that guides achievement-related behavior and task-engagement."

"Some correlational studies, conducted mostly in K-12 settings, found positive relations between mastery goals orientation and classroom achievement (e.g., Brookhart, Walsh, & Zientarski [170]; Kaplan & Maehr [171]), whereas other studies, conducted mainly in college settings (although see Cury and others [172]; Grant & Dweck [173]), did not find such relations (e.g., Harackiewicz, Barron, Carter, Lehto, & Elliot [174, 175])."

"When pursuing performance-avoidance goals, the person is oriented towards the undesired possibility of demonstrating low ability and engages in the task with the purpose of avoiding such a demonstration (Elliot [176])."

"Early conceptualizations of goal orientation theory suggested other goal orientations, in addition to mastery and performance goals, as possible purposes for achievement behavior (e.g., Maehr & Braskamp [177]; Maehr & Nicholls [178])."

"The concept of social goals to date has a distinctly different research history from the work on mastery and performance goals orientations."

The Assessment of Goal Orientations

"Goal orientation theory, which partially emerged from research that investigated the different meanings of achievement in different cultures (e.g., Maehr & Nicholls [178]), would benefit at this point from complementing survey methods, and a cross-cultural approach to motivation in different cultural groups, with research that focuses on the socio-cultural meaning of action in achievement settings within cultural as well as other types of groups (e.g., Bempechat & Boulay [179]; Dowson & McInerney [180]; MacCallum [181])."

"Several researchers used different types of interviews to tap students' goal orientations."

"She conducted two sets of semi-structured interviews with students, one that focused on their general goal orientations in school, and one that employed videotaped scenes from lessons in six subject matter areas and asked about goal orientations in these activities."

"They then gave the students goal orientation rating scales (modified versions of Nicholls and others [182]) on which students had to rate their endorsement of task (mastery), ego (performance), and work-avoidance goals."

"Qualitative studies identified nuances in the manifestation of goal orientations among students within specific contexts, emphasizing the situated nature of the meaning-construction processes that contribute to the elicitation of these orientations (e.g., MacCallum [181]; Lemos [183])."

The Nature of "Goal Orientations"

"One important perspective in goal orientation theory suggests that goal orientations may be defined as purpose-schemas that are associated with certain achievement situations or contexts."

"A fair number of studies have proceeded to investigate directly the perceptions that individuals have of achievement settings (largely in mastery and performance goals terms), and to relate these to the goal orientations that individuals report that they adopt as well as to achievement behavior they exhibit."

"These characteristics of the self-schema suggest the possibility that, whereas individuals may indeed have a disposition to a certain goal orientation, they have in their repertoire the various schemas (e.g., public and private aspects of self; differentiated and undifferentiated conceptions of ability; entity and incremental theories of intelligence) that would make it possible for them to adopt any goal orientation, and it is rather the situated attention to a certain aspect of the self-schema that is the significant defining characteristic eliciting different goal orientations."

Alternative Perspectives on Goal Orientations

"The terminology of goal orientation theory—particularly those terms that concern the ego (e.g., Nicholls [184])—may raise the possibility that what is at stake here is not so much a particular schema (e.g., a certain theory of intelligence, a certain conception of ability), but rather a differential awareness of self."

"The self-awareness perspective suggests that rather than being the manifestations of different schemas, different goal orientations and their associated patterns of behavior are the result of differential attention to the self-schema—regardless of the specific content of the self-schema that is activated."

"It should at least be noted that much of what is discussed and researched in the goal orientation theory literature to date is strikingly similar to the work on self-awareness, and that the relations between the goal orientation literature and literatures on self and identity pose a promising domain of inquiry (e.g., Baumeister [185]; Higgins [186]; Flum & Kaplan [187]; Kaplan & Flum [188]; Kashima, Foddy, & Platow [189]; Kluger & DeNisi [190]; Schlenker, Britt, & Pennington [191])."

Conclusion

"Research within goal orientation theory constitutes a substantial and informative body of knowledge concerning motivational processes in achievement settings and particularly in educational contexts."

"While obviously building on early work on achievement motivation, goal orientation theory is currently firmly positioned within a social-cognitive framework, interlocking perceptions of "meaning," "purpose," and "self" in guiding and framing action, thought and emotion (Maehr [192])."

"Experimental and correlational research suggests that mastery goals orientation is not only related to learning and thinking processes in achievement situations, but also appears to be associated with adaptive orientation toward life more broadly: encourage appropriate social behavior, positive feeling about self and others, and a sense of well-being."

"Working to change educational environments to increase emphasis on mastery goals orientation seems to have the potential to enhance students' motivation, attitudes, and achievement."

Using Achievement Goal-Based Personalized Motivational Feedback to Enhance Online Learning [311]

This is a machine-generated summary of:

Wang, Huanhuan; Lehman, James D.: Using achievement goal-based personalized motivational feedback to enhance online learning [311].

Published in: Educational technology research and development (2021).

Link to original: https://doi.org/10.1007/s11423-021-09940-3.

If you want to cite the papers, please refer to the original.

For technical reasons we could not place the page where the original quote is coming from.

Abstract-Summary

"To mitigate these shortcomings, we implemented a set of principles for designing personalized motivational feedback based on students' achievement goals."

"The results indicated students who received this feedback demonstrated significantly higher levels of motivation and satisfaction."

"Post-interviews showed this personalized feedback helped students regulate their learning goals and behaviors and thus improved their learning."

"The evidence suggests this personalized feedback may benefit learning by improving motivation."

Introduction

"These concerns may be due to the fact that online learning often utilizes a "one-size-fits-all" approach (Brusilovsky [194]), lacks adequate design to enhance motivation (Bawa [195]), and sometimes offers low levels of interactivity (Sunar and others [196]), which may result in low levels of learning engagement and satisfaction leading to high dropout rates (Bawa [195])."

"Various adaptations have been made to customize learning based on learner characteristics (Aleven and others [197])."

"Online personalized learning falls short of considering higher-level learner characteristics such as learning goals and motivational features (Nakic and others [198])."

"Targeting the unresolved drawbacks above, we tested principles of personalized motivational feedback based on students' achievement goals to enhance motivation and learning outcomes."

"Achievement goal is a little-utilized learner characteristic in the design of personalized online learning feedback."

Literature Review

"The differences in students' achievement goals were used as a factor in this study to differentiate learners and guide personalized feedback design."

"In a comparison study in which students received personalized feedback based on their response correctness and response certainty level, Mory [199] found no significant effect of such feedback."

"Some studies have considered achievement goals in designing personalized feedback."

"It is feasible to design personalized feedback based on achievement goals to optimize learning."

"There is little literature on the development and study of personalized feedback based on achievement goals."

"Learners who exhibit one of the four types of achievement goals, HH, HL, LH, and LL, were provided with different feedback."

"For HL students, self-referenced feedback was provided when they were facing an advantageous situation via visuals to show their achievement over time."

Methods

"We had two hypotheses: (1) After delivering the feedback, scores on assignments in the test group will be higher than those in the control group; (2) The post-measurement scores for motivation in the test group will be higher than those in the control group."

"The first phase was a quasi-experimental study (Privitera [200]) in which the effectiveness of the personalized feedback was examined by comparing the assignment scores and motivation scores in the test group (T) and the control group (C)."

"Students in the test group received both the regular instructor's feedback and personalized feedback, while students in the control group received only the regular instructor's feedback."

"The instructor logged into Blackboard, reviewed the assignment submissions, audited the initial assessment scores and the grading comments posted by the researcher, and added his or her feedback."

"Students completed assignments 1 and 2 before they received any feedback."

Results

"A perceived positive impact emerged in both groups since the feedback addressed specific assignment issues and motivated students."

"The personalized feedback in the test group further uniquely motivated students to adjust goals and activate self-regulation processes, "it (self-referential feedback chart) was like a goal setter for me, like I could manage my goals for improving in the course through those charts."

"All test group participants indicated they perceived positive effects, which were caused by regular instructor feedback that addressed specific assignment issues and personalized motivational feedback."

"The test group students further suggested using personalized feedback in future classes, such as "you made the chart in the grading section... and I love it... I wish that every class had that kind of charting..." and motivating them to improve their performance."

Discussion
"The test group students consistently perceived the feedback had a positive impact on their motivation through helping them regulate their goals, activate their self-regulation processes, and take steps to learn more."

"Since the personalized feedback was designed based on learners' achievement goals, this positive effect supports that achievement goal is capable of driving learning from the motivational dimension (Martinez [201])."

"The designed feedback helped learners regulate their goals, which further activated self-regulation processes and motivated students to study, which is consistent with the claims of Pintrich [202] that self-regulation meditates the influence of achievement goals on learning."

"To capture the desired positive results of the feedback intervention, we suggest future researchers consider achievement goal mediating factors, such as learners' self-regulation when classifying students and designing personalization principles."

"Students commented on the positive function of the personalized feedback as it enhanced their learning through goal setting and self-regulation."

Conclusion
"This study brought achievement goals into considerations for designing online personalized feedback."

"It attempted to optimize online learning from a motivational perspective."

"The working mechanism underlying achievement goals is complex, and various factors (e.g., learners' self-regulation skills, the relationship between learners and instructors, etc.) need to be considered for feedback design and examining the effects of feedback."

"We recommend that instructional designers and online instructors consider achievement goals and the personalization principles to enhance the power of the instructional feedback and optimize personalized online learning."

Motivational Factors in Makerspaces: A Mixed Methods Study of Elementary School Students' Situational Interest, Self-efficacy, and Achievement Emotions [312]

This is a machine-generated summary of:

Vongkulluksn, Vanessa W.; Matewos, Ananya M.; Sinatra, Gale M.; Marsh, Julie A.: Motivational factors in makerspaces: a mixed methods study of elementary school students' situational interest, self-efficacy, and achievement emotions [312].

Published in: International Journal of STEM Education (2018).

Link to original: https://doi.org/10.1186/s40594-018-0129-0.

If you want to cite the papers, please refer to the original.

For technical reasons we could not place the page where the original quote is coming from.

Abstract-Summary

"Design-based learning and makerspace programs have been shown to be effective in increasing student motivation for STEM learning."

"The purpose of this study was to understand how elementary school students' (grades 3–6) self-efficacy changed throughout the semester of a design-based makerspace course, and how these changes are associated with experiences of positive and negative achievement emotions."

"This study investigated how self-efficacy and achievement emotions are related to students' interest development in the makerspace course."

"Results of hierarchical growth modeling showed that although students' self-efficacy and situational interest remained moderately high during the course, both declined over the makerspace semester."

"Further, self-efficacy, as well as experiences of excitement and frustration with project tasks were found to be associated with students' situational interest."

"Design-based makerspaces have the potential to trigger elementary school students' interest in STEM activities."

Design-Based Learning and Makerspaces

"Design-based instruction aims to increase students' reasoning skill and transfer of content knowledge by engaging students in designing artifacts that solve real world problems (Doppelt and others [203])."

"The design process within DBL and related programs generally consists of six stages: (1) learning about the design context, identifying needs, and defining the problem; (2) collecting information and gaining knowledge needed for design; (3) brainstorming and creating multiple alternative solutions; (4) choosing the optimal solution based on needs and contextual limitations; (5) constructing the prototype of design ideas; and (6) sharing prototypes with others (including intended users) for evaluation and modifying the design based on feedback (Brown [204]; Doppelt and Schunn [203]; Mehalik and others [205]) DBL is a unique instructional method because, unlike most academic tasks, it requires students to solve ill-structured problems."

"This new term is also useful as makerspaces are increasingly considered in K-12 settings, prompting the need for a more well-defined making and designing process compared to informal makerspaces."

DBL and Learning Outcomes

"At the high school and undergraduate levels, students in DBL programs were successful in transferring content knowledge from fields of engineering (Gerber and others [206]), science (Kolodner and others [207]; Doppelt [208]), and graphic design (Fontaine [209])."

"Students in these programs showed gains in content knowledge according to analyses of their pre- and post-program scores in knowledge domains related to their design projects."

"These gains in content knowledge via DBL instruction were greater when compared to control students in the scripted inquiry condition (Mehalik and others [205])."

"One important component of DBL is the provision of student autonomy in terms of deciding which design project to take on, determining the content knowledge needed to complete design tasks, and establishing a work plan that feasibly allows for on-time completion (Doppelt [208]; Doppelt and others [203])."

"DBL programs are postulated to increase student motivation for STEM learning because of the explicit application of content knowledge to real-world situations and the sense of personal ownership students feel throughout the design process."

DBL, Self-efficacy, and Achievement Emotions: A Social Cognitive Theory Perspective

"The source of self-efficacy our study focused on is the emotional reaction towards a task, (i.e., the makerspace project), which in a learning context is closely related to achievement emotions (Pekrun [124])."

"Scherer (Scherer and others [210]), Pekrun [124], Goetz (Goetz and others [211]) and their associates further described that the relationship between achievement emotion and self-efficacy is related to students' perception of academic control."

"The theoretical relationship between achievement emotions and self-efficacy have found numerous empirical support, with positive emotions such as excitement and pride found to be positively related and negative emotions such as frustration and anger negatively related to ability beliefs (Goetz and others [12]; Goetz and others [17]; Pekrun and others [121]; Gumora and Arsenio [212])."

"Applied to the context of our study, we could expect that students' self-efficacy as appraised within a context-specific makerspace setting would be more strongly associated with activity-related achievement emotions such as excitement and frustration, rather than to outcome-related emotions such as pride or disappointment."

DBL and Interest Development

"Following social cognitive theory, students's self-efficacy is posited to be closely associated with their interest for specific academic activities."

"According to Bandura and Schunk [213], a requisite level of self-efficacy is needed for students to develop intrinsic interest in an academic task or task domain."

"They also suggested that students need to build up a series of mastery experiences that increase their self-efficacy before interest towards an activity can be developed."

"We focused on how students develop triggered and maintained situational interest based on experiences with their design projects, as well as how achievement emotions were related to their developing interest."

"Since design-based makerspaces aim to increase personal relevance and task value by asking students to apply STEM concepts to solve self-identified problems, we expect that students will develop both triggered and maintained situational interest in STEM content related to their design projects."

The Present Study

"Our objective was to gain insights into how participation in a design-based makerspace course influenced students' situational interest, self-efficacy, and achievement emotions."

"We sought to understand how students' self-efficacy changes throughout the semester of the design-based makerspace course, and how these changes are associated with experiences of positive and negative achievement emotions."

"We aimed to investigate how situational interest changed throughout the makerspace course, and how these changes are associated with self-efficacy and achievement emotions."

"Does students' self-efficacy change throughout a design-based makerspace course situated in an elementary school context?"

"a. Is students' self-efficacy associated with achievement emotions experienced during the design process?"

"b. Is students' self-efficacy trajectory associated with achievement emotions?"

"a. Is students' situational interest associated with their self-efficacy and achievement emotions?"

"b. Is students' situational interest trajectory associated with their self-efficacy and achievement emotions?"

Methods

"Students were given surveys during their makerspace class at three time points in the semester."

"Of the semester, students took the self-efficacy and demographic survey before beginning their makerspace project."

"The third, full model added achievement emotion variables to examine how experiences of positive (excitement, curiosity) and negative (frustration, confusion) emotions during the makerspace program are associated with levels of self-efficacy at each time point (Question 1a)."

"In order to directly address the research questions, we used matrices (Miles and others [214]) to sort the coded data by participants' makerspace class level (low vs. high), and interview time point (early, middle, or end) to both account for differences in self-efficacy, emotions, and interest across the two groups, as well as to better see how the participant experiences shifted over the course of the semester."

Results

"Descriptive results showed that self-efficacy and situational interest appeared to decrease as the semester progressed, and this decline is especially apparent for higher-level class students."

"Echoing descriptive trends, students in the higher-level class had significantly lower self-efficacy scores at each time point."

"The graph also shows that older students experienced a steeper decline in self-efficacy between the beginning and middle of the semester, which is related to these students experiencing less excitement on average compared to students in the lower-level class."

"In the early part of the semester, students in both class levels typically expressed high self-efficacy for successfully completing their design projects."

"This result echoed self-efficacy trends in which students in the lower-level class had higher self-efficacy scores at each time point, as well as a less steep decline in self-efficacy overall."

Discussion

"Our objective was to explore elementary school students' self-efficacy trajectory through a semester-long makerspace course, and how changes in self-efficacy are associated with experiences of positive and negative achievement emotions."

"The present study also suggests that students' overall positive emotional reactions towards makerspace activities are likely associated with relatively high self-efficacy and interest in the design-based makerspace program."

"These results point to the need to support students' self-efficacy in design-based makerspaces, particularly in the beginning phases as students start their projects."

"Context-specific factors play a role in students' emotional reactions to design-based makerspace tasks, as well as how their self-efficacy and interest change throughout the course."

"While our study suggests that supporting students' self-efficacy and emotional regulation need to occur across makerspace settings, the types of instructional support that are effective in each context may differ."

Limitations and Future Directions

"One limitation of this study is the lack of a comparison group to discern whether the declining self-efficacy and interest trajectories we found were typical of students participating in makerspaces."

"We cannot infer that the noted decline in these motivational attributes are due to students' participation in the design-based makerspace program, as opposed to a reaction to participating in novel learning tasks in general."

"It is likely the case that students participating in "traditional" STEM instruction begin the course with even lower self-efficacy and interest, and/or have a higher rate of decline in these motivational attributes."

"We note that the focus of our research is on motivational variables, which could only be gathered via self-report."

"Understanding complex motivational reactions to STEM-related activities like design-based makerspaces requires integrating results from multiple studies with different methodological lens, suggesting a need for more varied future research in this area."

Conclusions

"Despite these limitations, our study was one of a few that have examined motivational implications of a STEM-focused, design-based makerspace program."

"The study was also unique because we sought to understand the ways in which makerspace programs influence elementary school students' motivation, compared to previous studies which have focused on older students' motivational responses."

"Results from this study point to the potential for design-based makerspaces to support elementary students' learning in STEM, albeit with a cautionary note that such experiences should include efficacy- and emotion-related scaffolding."

"With this motivational support system in place, design-based makerspace programs have the potential to be an effective instructional medium for developing both STEM-related knowledge and interest."

[Section 11]
"Education scholars and practitioners have sought avenues to increase students' interest in STEM amidst concerns for the inadequacy of the STEM workforce to support growing US needs (NRC [215]; Sadler and others [216])."

"Design-based learning (DBL) has emerged as a possible instructional strategy that could be the solution to both of these current educational issues, offering the potential to support the development of problem-solving skills and STEM interest."

"The purpose of this study is to fulfill the aforementioned research gap by examining motivational variables related to students' situational interest in a DBL course."

"We sought to understand how students' self-efficacy and situational interest changed throughout the semester during which a DBL course took place."

"We situate our mixed-methods research within a design-based makerspace course in an elementary school setting, with students grades 3 through 6."

The Interaction Between Student Motivation and the Instructional Environment on Academic Outcome: A Hierarchical Linear Model [313]

This is a machine-generated summary of:

Los, Ryan; Schweinle, Amy: The interaction between student motivation and the instructional environment on academic outcome: a hierarchical linear model [313].

Published in: Social Psychology of Education (2019).

Link to original: https://doi.org/10.1007/s11218-019-09487-5.

If you want to cite the papers, please refer to the original.

For technical reasons we could not place the page where the original quote is coming from.

Abstract-Summary

"Midwestern university students and instructors responded to a survey consisting of motivation, learning approach, responsibility for learning, and academic success measures."

"HLM coefficients indicated that instructors' use of performance approach goal orientation suggested growth in student self-efficacy for SRL and increase in academic outcome."

"Practical applications of this research include instructors adopting mastery and performance approach goal orientation in the classroom, emphasizing the process and enjoyment of learning and de-emphasizing performance avoid goal orientation, which can increase student academic outcome."

Introduction

"Past research on academic success has largely focused on singular factors and has not observed the complete picture between the student and the instructional environment."

"Hattie found 138 different factors contributing to student success that were addressed by the studies, ranging from personal influences, such as student prior achievement, to contextual factors, such as teaching clarity."

"Few of these studies address personal and contextual influences at the same time."

"The goal of the present study is to create a single encompassing model of academic success in which personal factors interact with contextual factors to predict academic success."

"Urdan and Schoenfelder further state that contextual factors, such as characteristics of the school and classroom, as well as the instructor's method of shaping the instructional environment, have an effect on students' social-cognitive practices, such as their goal orientation."

Theoretical Framework

"This essentially places the responsibility for learning on a spectrum from personal responsibility to their instructor's responsibility and although self-efficacy and self-regulation can affect students' perceptions of their personal responsibility for learning (Zimmerman [217]), it uniquely adds to the prediction of academic outcome."

"This study examined contextual factors similar to student personal factors but from the instructors' perspective, i.e., instructor self-efficacy, teaching goal orientation, teaching approach, and instructors' perceived responsibility for learning."

"It is predicted that self-efficacy, self-regulation, self-efficacy for self-regulated learning, mastery goal orientation, a deep learning strategy, and student perceived responsibility for their own learning, would positively relate to academic success."

"It is predicted that positive aspect of the environment (highly efficacious instructors, encouraging mastery goal orientation, and instructor responsibility for learning) will relate to more positive student motivation and academic success."

Method

"The R-SPQ-2F utilized 20 self-report prompts, ten prompts assess each of the following subscales: deep ($\alpha = 0.67$) and surface ($\alpha = 0.71$) learning approaches, measured via a 5-point Likert-type scale from 'this item is never or only rarely true of me' to 'this item is always or almost always true of me.'."

"The perceived responsibility for learning scale (PRL; Zimmerman and Kitsantas [218]) utilized 19 self-report prompts measured via a 7-point Likert-type scale, with 1 'Mainly the teacher' to 7 'Mainly the student' to assess the degree to which students placed responsibility on themselves or their instructor ($\alpha = 0.92$)."

"The Perceived Responsibility for Learning Scale (PRL; Zimmerman and Kitsantas [218]) measured instructors' perceived responsibility for learning ($\alpha = 0.88$) with 19 self-report prompts measured via a 7-point Likert-type scale, with 1 'Mainly the teacher' to 7 'Mainly the student'."

Results

"Level 1 variables were centered around their group means and included the following 9 student-level variables: self-efficacy, goal orientation (performance avoid, performance approach, and mastery), self-efficacy for SRL, learning approach (deep approach and surface approach), perceived responsibility for learning, and self-regulation."

"Of greatest importance, there were three notable cross-level interactions between students and instructors: student performance avoid goal orientation and instructor mastery goal orientation, student performance avoid goal orientation and instructor perceived responsibility for learning, and student self-efficacy for SRL and instructor performance approach goal orientation."

"Student self-efficacy for SRL and instructor performance approach goal orientation was positive, meaning that an increase in the value of instructor performance approach goal orientation would bring an increase in the relationship between student self-efficacy for SRL and academic outcome."

Discussion

"Results of the level 1 model confirmed that, together, the overall student personal factors, which included: self-efficacy, goal orientation (performance avoid, performance approach, and mastery), self-efficacy for SRL, learning approach (deep approach and surface approach), perceived responsibility for learning, and self-regulation, predict student's self-reported academic outcome."

"When an instructor adopts a performance approach goal orientation, the relationship between student self-efficacy for SRL and academic outcome is increased."

"Students that already have a good sense of self-efficacy for SRL and a mastery goal orientation are able to use adaptive learning strategies that can be adjusted based on the method of instruction (Ames and Archer [219]) to still ensure positive academic outcome regardless of the instructor's goal orientation."

"When instructors adopt both mastery (wherein students are likely to decrease adopting performance-avoid goal orientation) and performance approach (wherein students can experience an increase in self-efficacy for SRL) goal orientations in the classroom there can be corresponding positive effects on their students' academic outcome."

Students' Motivation in Biology Lessons—Can Student Autonomy Reduce the Gender Gap? [314]

This is a machine-generated summary of:

Großmann, Nadine; Hofferber, Natalia; Wilde, Matthias; Basten, Melanie: Students' motivation in biology lessons—can student autonomy reduce the gender gap? [314].

Published in: European Journal of Psychology of Education (2022).

Link to original: https://doi.org/10.1007/s10212-022-00604-1.

If you want to cite the papers, please refer to the original.

For technical reasons we could not place the page where the original quote is coming from.

Abstract-Summary

"To investigate gender-related effects in biology education, a sample of 303 sixth-grade students (M_{age} = 11.31 years, SD_{age} = 0.58 years) participated in a 3-h teaching unit on harvest mice that was taught in either an autonomy-supportive or controlling manner."

"In the treatment with controlling teaching behavior, gender-related differences in self-determined motivation became apparent."

"The gender gap was mainly smaller in the treatment with autonomy-supportive teaching behavior."

"The results suggest that satisfying the need for autonomy appears to be an effective means to help bridge the gender gap in biology lessons."

Introduction

"Studies show that girls have a stronger motivational orientation toward and a higher interest in biology and are assumed to perform better in this subject than boys (Budde [225]; Dietze and others [226]; Hannover & Kessels [227]; Kahle and others [228]; Quaiser-Pohl [229])."

"The current state of research suggests that autonomy-supportive teaching behavior (ASTB) can affect the way in which boys and girls experience autonomy in school and their quality of motivation differently (Budde [225]; Desch and others [230]; Lietaert and others [231]; see also Patall and others [232])."

"Understanding such gender differences and being able to address them in the classroom is important to enabling both boys and girls to experience self-determined motivation and facilitating successful learning processes (see Ryan & Deci [224])."

"We investigated ASTB as a means to reduce possible gender gaps in biology by helping boys to be motivated in a more self-determined manner in this subject (see Desch and others [230]; Lietaert and others [231]; Lüftenegger and others [233]; Schweder & Raufelder [234])."

Theory

"Extrinsically motivated actions can differ in their degree of perceived self-determination (Ryan & Deci [224]; Vallerand & Ratelle [235])."

"Perceived competence is considered to be a crucial prerequisite for the perception of autonomy and the development of self-determined qualities of motivation (Ryan & Deci [224])."

"If autonomy is restricted in controlling learning environments, these self-determined qualities of motivation can be undermined (see Reeve [223]; Ryan & Deci [224]; Taylor and others [236])."

"CTB is believed to hinder the experience of intrinsic motivation and flow, as it favors an external locus of causality and undermines a voluntary execution of actions (Reeve [223])."

"The directives and commands communicated in controlling learning environments might therefore not impair girls' perception of autonomy and motivation to the same extent as that of boys."

"Girls might consequently cope better with controlling learning environments and experience more autonomy and self-determined motivation than boys."

Hypotheses

"Based on previous findings on the effects of ASTB on students' need satisfaction and motivation (Froiland and others [237]; Haerens and others [238]; Hofferber and others [222]; Tessier and others [239]), we assumed that this type of teaching behavior fosters the students' perception of autonomy as well as their intrinsic motivation and flow experience in class."

"CTB was assumed to have a negative effect on students' perceived degree of autonomy, intrinsic motivation, and flow experience (Assor and others [221]; De Meyer and others [240]; Hofferber and others [222])."

"Some studies and discussions suggest that boys and girls may perceive their degree of autonomy in school differently (Budde [225]; Desch and others [230]; Guggenbühl [241]; Lindsey and others [242]; Maccoby [243]; Patall and others [232]), which could affect their quality of motivation (Reeve [223]; Ryan & Deci [224])."

"Gender-specific differences regarding the following predictors of intrinsic motivation are less severe in lessons taught with ASTB than in those taught with CTB: H1a) Interest/enjoyment H1b) Perceived choice H1c) Perceived competence H1d) Pressure/tension H2) Gender-specific differences regarding flow experience are less severe in lessons taught with ASTB than in those taught with CTB."

Method

"One week after the teaching unit, the students' perceived autonomy was assessed using the perceived self-determination questionnaire (Reeve and others [244])."

"Two student teachers (one male and one female) in advanced semesters of university studies conducted the lessons in the treatments with ASTB or CTB equally often."

"Eight translated items of the perceived self-determination questionnaire (Reeve and others [244]) were ultimately used to assess the students' perceived degree of autonomy and thus check the implementation of the respective teaching behavior."

"We calculated an ANOVA to investigate the students' perceived degree of autonomy in both treatments to control the implementation of the teaching behavior."

"For students' intrinsic motivation, we applied a 2 (treatment) × 2 (gender)-multivariate analysis of covariance (MANCOVA), whereas student flow experience was investigated with a 2 (treatment) × 2 (gender)-analysis of covariance (ANCOVA)."

Results

"In the comparison of the ASTB ($M = 3.26$, $SD = 2.91$) and CTB treatments ($M = 2.65$, $SD = 3.06$), no significant difference in the students' motivational regulation was found ($F(1, 301) = 3.22$, $p = 0.074$, $\eta^2 = 0.01$)."

"Simple effects analysis showed that the effect of the treatment is significant for boys ($F(1, 299) = 37.83$, $p = 0.000$, $\eta^2 = 0.11$) and for girls ($F(1, 299) = 7.01$, $p = 0.009$, $\eta^2 = 0.02$)."

"Simple effects analysis showed a significant impact of the treatment on the flow experience for the boys ($F(1, 299) = 16.46$, $p = 0.000$, $\eta^2 = 0.05$), but not for the girls ($F(1, 299) = 0.74$, $p = 0.389$, $\eta^2 = 0.00$)."

Discussion

"We investigated students' quality of motivation in biology lessons with ASTB and CTB."

"Against the background of gender-related theoretical and empirical findings, we assumed these teaching behaviors would likely yield different effects on boys' and girls' quality of motivation."

"Regarding the students' flow experience, our theory-based assumptions about gender-related effects of ASTB and CTB on students' motivation were confirmed (H2)."

"The more destructive effects of CTB on boys' experiences in class and, as a result, a smaller gender gap in the ASTB treatment was also identified concerning this quality of motivation."

"Due to the interrelationships between intrinsic motivation and flow experience, the assumptions about the causes of the gender-related effects of ASTB and CTB on students' interest/enjoyment apply to their flow experience as well."

"Teachers can implement ASTB and CTB differently due to their own orientations toward autonomy and control (Reeve [223]), which can affect students' motivation."

Women's Interest Development and Motivations to Persist as College Students in STEM: A Mixed Methods Analysis of Views and Voices from a Hispanic-Serving Institution [315]

This is a machine-generated summary of:

Talley, Kimberly Grau; Martinez Ortiz, Araceli: Women's interest development and motivations to persist as college students in STEM: a mixed methods analysis of views and voices from a Hispanic-Serving Institution [315].

Published in: International Journal of STEM Education (2017).

Link to original: https://doi.org/10.1186/s40594-017-0059-2.

If you want to cite the papers, please refer to the original.

For technical reasons we could not place the page where the original quote is coming from.

Abstract-Summary

"The constructs of interest and motivation are often identified as factors that contribute to the persistence of undergraduate college women in physics, mathematics, engineering, engineering technology, and computer science."

"A review of the literature regarding interest development and motivation as related to women as science, technology, engineering and mathematics (STEM) learners in diverse communities is presented."

"Results indicated that students identify early participation in STEM activities and family socializing behavior as ones that contributed the most towards influencing their interest in STEM and motivated them to persist in their studies and pathways as future STEM professionals."

"This study is unique in that the participant groups included a substantial representation of Latina and African American women's voices as relayed through the collected quantitative data as well as through the use of focus groups that encouraged women to freely identify experiences they felt contributed to their persistence success."

Background

"This study was undertaken to gain a deeper understanding of what experiences women credited for influencing the development of their career interest goals and their motivation to succeed in academic courses related to studying in a STEM field."

"After following over 1000 students for 3 years, monitoring both the individual and environmental factors and the influence that these factors had upon students' individual and longitudinal regulation of goals and resulting academic performance, the findings in this particular study supported the hypothesis that students do indeed align their goals to the context of STEM academic learning, and this adaptation influences their performance and ultimately their success and persistence in STEM."

"The goal of this study was to gain a deeper understanding of what experiences women credited for influencing their career interest goals and their motivations to succeed in academic courses of their chosen science, technology, engineering and mathematics (STEM) fields."

Methods

"In order to address the research question(s) defined, the survey methodology used incorporated a questionnaire containing closed-ended questions coupled with focus group methodology using open-ended questions to collect the narrative, or qualitative, data."

"A mixed methods research study was used that combined the use of an online questionnaire for quantitative analysis and five focus groups for qualitative analysis."

"Edzie developed a 15-question survey instrument based on the "self-efficacy for learning and performance" motivation subscale of the Motivated Student Learning Questionnaire (Pintrich and DeGroot [245], Pintrich and others [113]) and additional qualitative research findings."

"Further, the groupings of questions focused on students' motivations to persist and factors in choosing their majors, which echo portions of the validated Motivational Sources Inventory (MSI) (Barbuto and Scholl [246])."

"The first three questions of the focus group were developed to explore the constructs of interest and motivation and are presented below, along with their derivative question from the validated MSI instrument."

Results

"As the correlations within this grouping are only sporadically above 0.3, some of these items relating to what different factors motivated the students to pursue their major are likely independent and can therefore be evaluated with a significance level of 0.05 rather than the corrected significance level of 0.008 to control for type I error when comparing the composite scores."

"For most of these questions, the minority and White students had very similar responses, and the overall level of motivation to pursue their STEM majors as measured by the composite scores for these groups were nearly identical (M = 5.58 and 5.58, SD = 1.06 and 0.81, and 95% CI [3.49, 7.66] and [3.99, 7.16] for minority and White students, respectively)."

Discussion

"Other factors with medium effect sizes showed White students to be more likely than minority students to identify their personal ambition as a motivation to persist and their self-confidence in understanding basic concepts in STEM."

"The external motivation factor of family support was also cited many times be the students in the focus groups, and this family support took on many forms as it sparked and drew students' interest in STEM fields."

"On the questionnaire, the students often ranked internal self-concept and instrumental motivation factors more highly in all of the question groupings as the reason behind their choice to select and then persist in their STEM major."

"The students in this study frequently selected internal self-concept and instrumental motivation factors in their decisions to choose and persist in STEM majors."

Conclusions

"This study contributes to the field by revealing a more in-depth account of what experiences women credit for influencing the development of their career interests and their motivation to persist in studying science, technology, engineering and mathematics (STEM)."

"The first research question that drove this study was to examine what factors contribute to the interest development and motivational drive to persist of college women in STEM."

"Quantitative analysis of survey data revealed that the factors of "career options" and "family support" emerged as significant sources of motivation to persistence in STEM majors for the minority-rich student population in this study, as it was seen as a reflection of culture for the minority students in the study."

"Focus groups revealed the particular importance of family expectation as a motivation to pursue and persist in a STEM career; a difference that was also shown with statistical significance in the survey results."

Employing a Case Study Approach to Capture Motivation and Self-regulation of Young Students with Learning Disabilities in Authentic Educational Contexts [316]

This is a machine-generated summary of:

Lichtinger, Einat; Kaplan, Avi: Employing a case study approach to capture motivation and self-regulation of young students with learning disabilities in authentic educational contexts [316].

Published in: Metacognition and Learning (2015).

Link to original: https://doi.org/10.1007/s11409-014-9131-1.

If you want to cite the papers, please refer to the original.

For technical reasons we could not place the page where the original quote is coming from.

Abstract-Summary
"We describe a situated case study approach to investigate the motivational orientation and self-regulation strategies that an elementary school student with learning disabilities employs as he or she engages in a teacher-assigned academic task in an authentic educational context."

"The paper demonstrates the utility of the situated case study approach for the investigation of motivation and self-regulation in one elaborated case of an elementary school student with learning disabilities."

"The paper describes more briefly analyses of seven additional cases of elementary school students with learning disabilities who have different characteristics and who engaged in tasks in different contexts."

"The paper highlights general insights about motivation and self-regulation of elementary students with learning disabilities in integrated classrooms from the comparison of these different cases."

Self-regulation in Academic Settings
"Self-regulation in academic settings has been defined as the "active, constructive process whereby learners set goals for their learning and then attempt to monitor, regulate, and control their cognition, motivation, and behavior, guided and constrained by their goals and the contextual features in the environment" (Pintrich [247], p. 453)."

"Models of self-regulation conceptualize the self-regulating student as a motivated proactive agent, and depict the self-regulatory process as progressing along phases that include assessment of task conditions and relevant personal resources; goal-setting and selection of strategies to pursue these goals; application of the strategies and metacognitive monitoring of this application; and evaluation of the products and metacognitive control of the continued use of the strategies (i.e., whether to maintain the strategies or change them)."

"These phases are thought to operate cyclically, with the evaluation phase leading back to the planning phase of goal setting and selection of strategies, and so on (cf. Winne and Hadwin [248]; Zimmerman [249])."

The Role of Context in Self-regulation

"More recent conceptualizations and research highlight the variability of self-regulation strategies and the important role of the specific contexts, and more particularly, the features of the academic task, in the type of self-regulation strategies that the student applies (Perry and Winne [184])."

"Whereas contextual characteristics and task features frame the kind of self-regulation that would lead to success in the task, it is the student's subjective interpretation of the purpose of engaging in this task that guides her application of self-regulation strategies (Butler and Cartier [250]), which may be more or less compatible with those that would lead to success as defined by the teacher (Kaplan and others [251]; Butler and Cartier [250])."

"Recent recommendations for research on self-regulation highlight the importance of attending to the situated flow of a student's use of self-regulation strategies during task engagement while attending to the interplay between the contextual conditions, primarily task features (Johnson and others [252]; Winne [253]; Zimmerman [254]), the student's subjective interpretation of the purpose for task-engagement (Butler and Cartier [250]), and the student's goals and efficacy to pursue these goals in light of the purpose—or in other words, his or her motivation."

Purpose of Engagement, Motivation, and Self-regulation

"Adopting a mastery-approach goal-orientation focuses students on learning and understanding and guides them to set goals, monitor cues, and select the particular learning strategies that would promote their purpose of developing competencies."

"Adopting a performance-approach goal-orientation would focus students on appearing smart and guide them to engage in self-regulation strategies that would promote their purpose of demonstrating competencies."

"The study found that the student's purpose for engagement guided her type of self-regulation—planning, goal setting, and selection of cognitive and writing strategies and their monitoring and evaluation—and that as goals shifted along phases of task engagement (e.g., from developing understandings, to expressing thoughts in writing, to finishing), the focus and type of her self-regulation strategies shifted accordingly (cf. Winne and Hadwin [255])."

Motivation and Self-regulation Among Young Students with Learning Disabilities

"Another domain of challenge for students with learning disabilities is self-regulation."

"These motivational characteristics among students with learning disabilities have been associated with a sense of alienation from school, avoidance of challenging tasks, and self-handicapping strategies (Fulk and others [256]; Garcıa-Sanchez and de Caso-Fuertes [257])."

"The motivational context seems to play an important role in the motivation and self-regulation of students with learning disabilities."

"There is some indication that the self-regulation of students with learning disabilities may be moderated by their motivation and by task features (e.g., well-structured vs. ill-structured) (Graham and others [258])."

"We propose the case-study approach (cf. Kaplan and others [251]; Butler and others [259]; Yin [260]) as better suited for the investigation of motivation and self-regulation of task engagement among young students with learning disabilities, particularly in authentic educational environments."

The Current Study

"We aimed to demonstrate the utility of the case study approach for investigating how a situated student's purpose of engagement frames her application of self-regulation strategies."

"We sought to demonstrate the utility of the approach with elementary school students with learning disabilities integrated into regular classrooms who engaged in different tasks in authentic educational environments."

"The perspective guiding the present study suggests that a student's self-regulation of engagement would correspond with the purpose of engagement that she adopts."

"The case study approach, employed in the current study, is particularly well suited to investigate this interplay of purpose of engagement and self-regulation during task engagement that is situated in an authentic setting (Kaplan and others [251])."

"The case study approach provides data about the purpose of engagement that students adopted for the task as well as about the learning and self-regulation strategies that he or she employed during task engagement."

Method

"The procedure and data collection followed those described in Kaplan and others [251] and included: (a) identifying traces of strategies in the student's academic product; (b) a direct observation of the engagement process; (c) a stimulated-recall interview (SRI) on the engagement process using the observation as a memory trigger; and (d) a more general interview about the experience of engagement in the task, self-processes related to learning and school, and self-regulation aptitude."

"The research assistant conducted the observation, and upon completion of the assignment, either asked the student to accompany her to another room, explaining that she wanted to learn more about how the student performed the task, or conducted the interview in the individualized instruction setting."

"The research assistant asked questions that aimed at the student's general experiences before, during, and after engagement: "What did you think and feel when you started the assignment?"."

Results

"When she was asked to describe the way she answered the first question, Sapir described in detail the procedure of scanning the text for the location of the required details, monitoring her success in locating the details and writing them down, and correcting her answer upon evaluation of an error: "The title of the article was written down for me, I looked at the second sheet of the text, I looked at the title and wrote it down."

"Her lack of monitoring and evaluation in responding to the last two questions, and the indication from the observation that she made an effort to complete the assignment at the same time as the rest of the class, did not ask for help at any stage, and did not pay attention to the teacher's review, suggest that any goals of correctness were superseded by the goals of simply completing the task."

Cross-Case Comparison and Motivational-Self-Regulation Constructions

"Each case was unique, in the context, task, the student's construction of the purpose of engagement, and the self-regulation strategies used."

"A cross-case comparison did point to some general insights about patterns in the students' construction of purpose of task engagement and employment of self-regulation strategies."

"It seems that despite some contextual differences, both of these students' construction of the purpose of engagement as merely task completion framed their use of strategies of attention regulation and task management (e.g., skipping difficult questions) and avoidance of help-seeking, monitoring for accuracy, and evaluation of understanding."

"It seems that construing the purpose of task engagement as task completion with accuracy, and the context of an individualized session that involved a challenging task, framed these students' perceptions of low efficacy and attention to use of strategies, employment of strategies such as monitoring for accuracy and understanding, attention regulation, and engaging in help-seeking."

Discussion

"The findings corroborate previous findings (Kaplan and others [251]) that suggest that the use of self-regulation strategies during task engagement is framed by the situated purpose and goals that a student adopts for engagement in the task."

"Ample research demonstrates that students who perceive such an emphasis on mastery purposes are more likely to adopt mastery purposes for their own engagement, feel more efficacious, use more adaptive self-regulation strategies that aim at learning and quality of product, and ask for help when they need it (Kaplan and Maehr [261]; Meece and others [262])."

"Whereas our purpose here was to demonstrate the contextual construction of purpose of engagement and self-regulation in a task, research that aims to more holistically understand the patterns of a student's motivational orientation and self-regulation may do well to collect data on the same student's engagement in different tasks and contexts."

Conclusion

"The current study highlights the utility of a situated case study as a methodology for investigating the flow of self-regulated action during task engagement in authentic educational contexts."

"To methodology that employs self-report summated scales and correlational analyses, the situated case study allows for an in-depth view of the dynamics that integrate context, student, and task."

"This integrated approach yields insight into purpose-strategies orientations that constitute a meaningful unit for investigating and understanding students' self-regulation during task-engagement."

"The current study highlights the potentially crucial role of students' subjective purposes for engagement in their employment of self-regulation strategies."

"Such open dialogue would reveal to educators the students' level of knowledge of strategies, as well as the contingencies that students make between adopting certain purposes and employing certain strategies in the particular classroom context and academic task."

[Section 11]

"Much of the research on self-regulation has viewed it as an individual-differences construct: a set of cognitive, metacognitive, motivational, and behavioral strategies that students, who have the strategic knowledge and the motivation to use it, employ during engagement."

"In the case of young students with learning disabilities, who may be challenged in reporting on their use of self-regulation strategies due to difficulties in attention, awareness, or verbal expression, it is imperative that such methods gather data that go beyond students' self-report."

"We demonstrate the use of a situated case study approach in which data from multiple sources—traces in an academic product, direct observation, stimulated-recall interview, and a general interview—are triangulated in order to capture and investigate the motivation and self-regulation strategies of an elementary school student with learning disabilities in a regular classroom as she engaged in an authentic academic task."

Motivational, Identity-Based, and Self-regulatory Factors Associated with Academic Achievement of US Collegiate Student-Athletes: A Meta-analytic Investigation [317]

This is a machine-generated summary of:

García, Agustín J.; Fong, Carlton J.; Regalado, Yvette M.: Motivational, Identity-Based, and Self-Regulatory Factors Associated with Academic Achievement of US Collegiate Student-Athletes: A Meta-Analytic Investigation [317].

Published in: Educational Psychology Review (2023).

Link to original: https://doi.org/10.1007/s10648-023-09730-8.

If you want to cite the papers, please refer to the original.

For technical reasons we could not place the page where the original quote is coming from.

Abstract-Summary

"In the USA, over 600,000 student-athletes participate in nationally organized intercollegiate sports and occupy socially prominent spaces on college campuses."

"Although their athletic accomplishments often garner much attention, there is growing interest in collegiate student-athletes' academic achievement and its precursors."

"Based on 73 studies using 79 unique samples of US collegiate student-athletes (N = 43,617), our meta-analytic findings indicated several factors were positive correlates (academic identity, academic motivation, role negotiation, and self-regulation) and negative correlates (athletic identity and athletic motivation) to grade point averages."

Literature Review

"In the subsequent sections, we review the literature on collegiate student-athletes using the following constructs: athletic identity and motivation, academic identity and motivation, role negotiation, and self-regulation."

"The majority of studies tend to support positive associations between student-athletes' academic identity and achievement (O'Quin [263])."

"Some studies pointed to negative correlations between academic and athletic identities among student-athletes, particularly Division I athletes (Feltz and others [264]; Yopyk & Prentice [265])."

"We reviewed the literature on six motivational, identity-based, and self-regulatory factors relevant to student-athletes' experiences in college."

"Meta-analyses on student success and motivational, identity-based, and self-regulatory factors have sparsely included studies specifically focused on student-athletes, if at all."

"In light of the accumulated research on student-athletes and the role of motivational, identity-based, and self-regulatory factors in predicting their academic achievement, a research synthesis on this topic is timely and important."

Method

"To be included, a study must have met the following criteria: (1) a sample of student-athletes who play for a US college-sponsored sports program through the NCAA, NJCAA, or the NAIA; (2) an academic achievement outcome (GPA); (3) a motivational, identity-based, and self-regulatory variable; and (4) data to derive a correlational effect size measure."

"When studies compared levels of motivational, identity-based, and self-regulatory factors between low- and high-achieving students, means and standard deviations were used to derive correlation effect sizes."

"The common metric used to combine findings from studies was Pearson's r. If a correlation was unavailable, but means and standard deviations were provided, a correlation was estimated based on established effect size conversion formulae (Cooper and others [266])."

"Effect sizes may vary even if they estimate the same underlying population value, as indicated by Cochran's Q, which distinguishes study heterogeneity from studies' sampling error."

Results

"The percentage of student-athletes receiving scholarships significantly moderated the relationship between athletic motivation and academic achievement ($ß = 0.79$, $SE = 0.23$, $p = 0.008$), while controlling for other athletic characteristics, study quality, and risk of bias measures."

"Simple slope analyses revealed that studies with less than two-thirds of the sample consisting of students on scholarship had more negative correlations between athletic motivation and academic achievement ($r = -0.06$, $SE = 0.23$, $p = 0.84$), compared to studies with larger proportion of student-athletes on scholarship (66–100% scholarship: $r = -0.02$, $SE = 0.04$, $p = 0.68$)."

"Simple slope analyses revealed that studies with less than one-third of the sample consisting of women had more negative correlations between athletic motivation and academic achievement ($r = -0.15$, $SE = 0.06$, $p = 0.03$), compared to studies with larger proportion of women (33–66% women: $r = 0.04$, $SE = 0.03$, $p = 0.11$; over 66%: $r = 0.16$, $SE = 0.03$, $p = 29$)."

Discussion

"Our study centers on collegiate student-athlete academic achievement and its motivational, identity-based, and self-regulatory correlates."

"Similar to role negotiation, self-regulated learning was positively correlated to academic achievement as well, suggesting that student-athletes who can effectively manage their time and resources and put their thoughts and intentions into action are more likely to have higher GPAs (Lombardi and others [267])."

"The percentage of women student-athletes positively moderated the correlation between athletic motivation and academic achievement."

"An emphasis on re-contextualizing student-athletes' academic and athletic identities and motivations may help them engage more in their studies."

"For a last direction for research, we encourage researchers to explore thoughtfully the ways student-athletes can be supported in their learning experiences and the motivational, identity-based, and self-regulatory mechanisms undergirding their academic achievement, building upon the decades of research conducted by sports psychologists and higher education scholars we highlighted in our synthesis."

Concluding Remarks

"Our synthesis highlights important motivational, identity-based, and self-regulatory factors associated with the academic achievement of student-athletes."

"Domain-specific motivations and identities are critical, especially for student-athletes who function in two domains and simultaneously endorse perceptions about both academic and athletic pursuits."

"Our study highlights the complexity of motivation and identity situated in student-athlete activities and the promising findings regarding role negotiation."

"Practical considerations include training students to self-regulate the learning process and the multiple demands on their time and cultivating students' identities and motivations for academics as a valuable pursuit in addition to non-academic interests."

[Section 6]

"Despite the lack of representation in mainstream educational psychology journals, work on the psychological antecedents of student-athletes' academic achievement has flowered in other areas using constructs from educational psychology, which we highlight in the present study."

"A multitude of factors has been proposed as antecedents to student-athletes' academic development and achievement."

"Many studies have explored student-athlete academic success and the role of institutional characteristics and college engagement activities such as collaborative learning or co-curricular activities (Carodine and others [268]; Gaston-Gayles & Hu [269]; Umbach and others [270])."

"The purpose of our study was to assess meta-analytically how motivational, identity-based, and self-regulatory factors in the contexts of academics and athletics are linked with the academic performance of student-athletes."

"A synthesis of this nature can resolve inconsistencies in the literature and also shed light on how motivational, identity, and self-regulatory factors are associated with academic achievement for students with multiple roles (e.g., student and athlete)."

Parent Perceptions of Elementary Classroom Management Systems and Their Children's Motivational and Emotional Responses [318]

This is a machine-generated summary of:

Kowalski, Monica J.; Froiland, John Mark: Parent perceptions of elementary classroom management systems and their children's motivational and emotional responses [318].

Published in: Social Psychology of Education (2020).

Link to original: https://doi.org/10.1007/s11218-020-09543-5.

If you want to cite the papers, please refer to the original.

For technical reasons we could not place the page where the original quote is coming from.

Abstract-Summary

"Classroom management systems emphasizing rewards and consequences have long existed in schools within the US, but parental perceptions of the effects of these systems have largely been ignored."

"Regression results showed that behavior chart systems were negatively associated with student autonomous motivation to learn, whereas parent–school relationships and teacher autonomy support were positively associated with autonomous motivation."

"In hierarchical regression models, the frequency of students receiving rewards was negatively related to parent–school relationships."

"Parents' open-ended responses revealed frustrations with systems that caused student anxiety, encouraged dependence on rewards, and oppressed students' personalities."

Introduction

"It is also important to consider whether teacher autonomy support and behavior management systems have an effect on the parent–school relationship because parents may trust teachers more when they support student autonomy and relatedness, whereas controlling teaching as manifested by an emphasis on rewards may threaten the parent–school relationship."

"In order to better understand the relationship between controlling behavior management systems and students' experiences in elementary schools, this study examines parents' perceptions of classroom management systems in schools, with particular emphasis on how external rewards and punishments are experienced."

Methods

"The PQCML short form involves five items from the longer version of the PQCML, which demonstrated strong internal consistency among parents of elementary school students in prior studies (Froiland [271, 272])."

"Parents responded to the following survey question: "In your child's classroom, is behavior monitored by a "behavior chart" system that includes students pulling a card, changing a color, or moving a clip along a chart based on their positive or negative choices?"."

"In order to see if the frequency of rewards was related to other variables, parents were asked, "How many times during this school year has a teacher or school administrator rewarded your child for positive behaviors at school?"."

"The survey also included open-ended questions asking parents to reflect on the types of behavior management systems they knew of in their children's schools."

Results

"The number of times a child was rewarded was significantly and negatively related with parent–school relationships, teacher autonomy support, and autonomous motivation to learn."

"In model 1, times rewarded was negatively and significantly related to parent–school relationships, suggesting that rewarding children for positive behavior may put parents at-risk for negative relationships with the school."

"Parents frustration, disgust, and disappointment with the reward systems complements the quantitative findings for hypothesis five, in that it is hard to imagine the parent–school relationship being characterized by trust, when they feel that the teacher is threatening a child's personality, administering rewards and consequences in a way that creates social injustices, is allowing the reward system to take center stage over learning, creates unnecessary grief, and wastes time and resources on ineffective behavioral management systems."

Discussion

"Teacher autonomy and relatedness support explained a great deal of the variance in parent–school relationships, suggesting that future longitudinal and intervention studies may wish to examine whether teacher autonomy support predicts the long-term health of parent–school relationships."

"Such studies may include working with a school to change the behavior management system to a more autonomy and relatedness supportive system and measuring changes in perceptions of teachers, parents, and students, as well as tracking frequency of infractions and even student achievement outcomes."

"Behavioral chart systems may jeopardize the parent–school relationship, whereas teacher autonomy and relatedness support in the classroom may strengthen the parent–school relationship."

"Future intervention studies could examine whether teacher autonomy support or teacher autonomy support plus behavioral chart systems, or behavioral chart systems alone have the best effects on motivation, relationships, and learning."

Emotion Research in Education: Theoretical and Methodological Perspectives on the Integration of Affect, Motivation, and Cognition [319]

This is a machine-generated summary of:

Linnenbrink, Elizabeth A.: Emotion Research in Education: Theoretical and Methodological Perspectives on the Integration of Affect, Motivation, and Cognition [319].

Published in: Educational Psychology Review (2006).

Link to original: https://doi.org/10.1007/s10648-006-9028-x.

If you want to cite the papers, please refer to the original.

For technical reasons we could not place the page where the original quote is coming from.

Abstract-Summary

"The authors also consider how to integrate affect, motivation, and cognition and discuss methodological challenges to investigating affect in educational settings."

"The authors consider the role of affect in education, making specific suggestions for practice as well as more general suggestions for theory development."

Introduction

"The contributors to this special issue agreed that it is important to pause to consider the theoretical similarities and disconnects among the various perspectives and to consider whether our current methodologies are in line with our understanding of affect in educational settings."

"This reflection began as part of a theoretical symposium that was presented at the annual meeting of the American Educational Research Association in April of 2005, entitled Reflections on Emotion Research: The Theoretical Integration of Affect, Motivation, and Cognition."

"Although there are differences among the theoretical perspectives and programs of research represented in these five articles, there is also considerable overlap in the way in which the authors define affect, moods, and emotions, their use of social psychological theories of affect, and their call for innovative methodologies to study the complex, dynamic, shifting, and sometimes elusive emotions that ebb and flow in educational settings."

How do You Define Affect? What Theories Inform Your Work on Affect?
"The articles by Schutz and others and Meyer and J. C. Turner use this distinction as their working definition of moods and emotions."

"Meyer and J. C. Turner also talk more generally about affective states, which they define as consisting of both moods and emotions."

"With respect to the theoretical basis for understanding affect, Pekrun, Schutz and others, Op 't Eynde and J. E. Turner, and Meyer and J. C. Turner primarily rely on social psychological theories of emotions."

"Rather than focusing on solely on discrete emotions, Ainley investigates affective states and draws upon circumplex models of affect (e.g., Feldman Barrett & Russell [273]; Tellegen, Watson, & Clark [274]), which highlight the important distinction between valence and activation."

"Readers especially interested in an in-depth discussion of the theoretical underpinnings of emotion should refer to the Pekrun, Op 't Eynde and J. E. Turner, and Schutz and others articles, which include elaborated discussions of how emotions can be defined in educational settings."

Level of Focus
"At the structure level, a number of the authors discuss the development of affective constructs specific to educational settings."

"All three of the aforementioned manuscripts also deal with process, as these authors seek to understand how the proposed constructs (e.g., academic emotions, affective state of interest, emotion regulation) develop in classrooms and how they relate to subsequent engagement and learning."

"Meyer and J. C. Turner and Op 't Eynde and J. E. Turner consider the process level, as they investigate students' and teachers' emotions that emerge in classroom settings."

"Schutz and his colleagues highlight the importance of this third level of analysis with respect to the personal histories of teachers entering education and teachers' emotions related to their profession."

"Pekrun recognizes the importance of the social–historical level in understanding gender and cultural differences in the types of emotions that students experience."

Links Between Affect, Motivation, and Cognitive Processing
"To the theoretical basis for affect, the contributing authors were asked to consider how to incorporate affect into existing models of motivation and cognitive processing or to propose alternative theoretical perspectives that integrate these three areas."

"This call for integrating affect is similar to earlier calls by motivational researchers to infuse motivation into existing theories of cognitive processing (e.g.,

Pintrich, Marx, & Boyle, 277 argued for the need for "hot" conceptual change that considered motivation as a critical variable in the conceptual change process)."

"Schutz and others, Pekrun, Meyer and J. C. Turner, and Op 't Eynde and J. E. Turner all draw upon appraisal theories in their work, suggesting that cognitive appraisals can alter affective experiences."

"This perspective calls for a dynamic, integrated model in which neither motivation, affect, nor cognition is given precedence—but rather all three are critical variables for understanding students' educational experiences."

Methodological Concerns

"Given the complexity of affect and the difficulty in assessing it, the contributing authors were also asked to discuss how they assess affect, to consider pressing methodological concerns, and to suggest potential ways that these concerns might be addressed."

"All of the authors have used self-report indicators, such as questionnaires, and many have developed their own scales to assess affect (see Pekrun's Achievement Emotions Questionnaire and Schutz and others's scales to assess emotion regulation during test taking)."

"Many of the authors have also employed other quantitative indicators, such as experience sampling methodology (ESM) and diaries, which allow them to tap into the dynamic, shifting aspects of affect in classrooms (see Meyer and J. C. Turner; Pekrun)."

"They also suggest that researchers take into account the possibility that display rules may interfere with the assessment of affect in that students and teachers may not display or report their true emotional states."

Informing Education

"The authors were asked to specifically consider the role of affect in educational settings."

"A number of the papers also make specific suggestions for how an understanding of affect in education can be beneficial for schooling."

"Using control-value theory as a guide, Pekrun makes a number of interesting suggestions about how educational settings can be altered to change the patterns of emotions that students experience."

"Schutz and his colleagues also make specific suggestions for educational settings, arguing that we need to consider emotions for teachers in pre-service and in-service teacher training and realize that there is a distinction between the emotions educators might experience and pedagogically desirable emotions."

Conclusion
"The five manuscripts presented in this special issue offer a critical overview of the current state of the field on emotions in education."

"What is clear from these manuscripts is that affect is critical to understanding students' and teachers' educational experiences, and that it is has been ignored for far too long."

"It is my hope that readers of this special issue will consider how affect might be infused into our understanding of both students' and teachers' experiences in educational settings, much in the same way that motivation has become infused into our understanding of education during the past two decades."

An Integrated Model of School Students' Academic Achievement and Life Satisfaction. Linking Soft Skills, Extracurricular Activities, Self-regulated Learning, Motivation, and Emotions [320]

This is a machine-generated summary of:

Feraco, Tommaso; Resnati, Dario; Fregonese, Davide; Spoto, Andrea; Meneghetti, Chiara: An integrated model of school students' academic achievement and life satisfaction. Linking soft skills, extracurricular activities, self-regulated learning, motivation, and emotions [320].

Published in: European Journal of Psychology of Education (2022).

Link to original: https://doi.org/10.1007/s10212-022-00601-4.

If you want to cite the papers, please refer to the original.

For technical reasons we could not place the page where the original quote is coming from.

Abstract-Summary

"The role of soft skills at school is still debated, but they have emerged as important factors for students' academic achievement and life satisfaction."

"This study focuses on the combined influence of soft skills (in terms of adaptability, curiosity, leadership, initiative, perseverance, and social awareness), extracurricular activities, achievement emotions, self-regulated learning, motivation, and cognitive abilities on academic achievement and life satisfaction."

"The results of a Bayesian path analysis based on meta-analytical priors show that soft skills were (i) directly positively associated with students' achievement emotions, self-regulated learning, motivation, and life satisfaction and (ii) indirectly related with academic achievement through the mediation of self-regulated learning and motivation."

"These results are the first to demonstrate the importance of soft skills and extracurricular activities when integrating all the above-mentioned factors in a model of students' academic achievement and life satisfaction."

Introduction

"Other studies highlighted the direct relation between students' single soft skills and emotional, behavioral, and cognitive components of learning, which could fully mediate the effect of soft skills on academic achievement."

"Taking this view, it was found that the six soft skills included in the WEF model compose a single factor, which is indirectly related to academic achievement through the mediation of SRL and motivation (Feraco and others [280])."

"The authors did not include emotional aspects or students' life satisfaction in their model, however, and studies considering soft skills together with other predictors of academic achievement are scarce, requesting further research in education."

"ECA could matter because they can promote school students' soft skills and cognitive abilities (Eccles [281]; Voyer & Jansen [282]), support their life satisfaction (Gilman & Huebner [283]), and influence factors such as academic achievement, SRL, motivation, and achievement emotions."

"Soft skills—we expect soft skills to directly relate to achievement emotions, motivation, SRL, and life satisfaction (Feraco and others [280]; Bruna and others [284]; Weber and others [285])."

Method

"It measures positive and negative emotions (10 items each) on a 5-point Likert scale indicating how much students experienced each of the 20 emotions at school during the previous weeks."

"The Satisfaction With Life Scale (SWLS, Diener and others [286]; Italian adaptation by Di Fabio & Gori [287]) measures overall life satisfaction with 5 items (e.g., "The condition of my life are excellent.") on a 7-point Likert scale."

"In the first model (m1), for each of these eight relations, we set a strongly informative prior (SD = 0.05) centered on the estimate calculated in the previous section (e.g., N[0.19, 0.05] on the association between soft skill and academic achievement)."

"In the second model (m2), the priors for the same relations were centered on 0, thereby representing a different hypothesis (e.g., N[0.00, 0.05] on the relation between soft skill and academic achievement)."

Results

"SRL, motivation, and cognitive abilities showed a positive direct relation with academic achievement, but PA, NA, soft skills, or ECA did not."

"PA and soft skills showed a positive direct relation with life satisfaction, and NA showed a negative direct relation with life satisfaction, while SRL, motivation, cognitive abilities, and ECA showed only a negligible association with it."

"PA and soft skills were positively related to SRL and motivation, while NA showed a small relation with motivation, but not with SRL."

"The analysis of the indirect effects showed that soft skills work as a mediator between ECA and PA, NA, motivation, and SRL."

"The effect of soft skills on academic achievement was also mediated by SRL and motivation, and the effect of soft skills on life satisfaction was not only direct, but also mediated by PA and NA."

Discussion

"With this study, we aimed to combine relevant soft skills within an exhaustive intraindividual model of learning that included ECA, achievement emotions, cognitive abilities, SRL, and motivation in relation to academic achievement and life satisfaction."

"SRL and academic motivation had previously been found positively related to students' life satisfaction (Antaramian [288]), but our study did not confirm this link suggesting that their role might be limited to learning outcomes, or that SRL and academic motivation are narrow factors that directly entail only academic variables, such as achievement or eventually school satisfaction, but that do not extend their influence on not-scholastic ones."

"Again, the role of soft skills seems to be important: they were the only other variable considered that directly related to life satisfaction."

"Our results support the importance of soft skills in favoring students' life satisfaction (both directly and indirectly), achievement emotions, and academic achievement (through the mediation of SRL and motivation)."

Conclusions
"Our findings support the importance of soft skills in school students."

"Soft skills thus support students' wellbeing, as well as their academic achievement, whereas SRL and motivation show no noteworthy association with life satisfaction (their importance is limited to school)."

"Our results indicate that it is important to consider an integrated intraindividual model of learning (Ben-Eliyahu [279]) in order to shed more light on the real role of each factor influencing school students' academic achievement and satisfaction."

Temporal Stability of Teachers' Judgment Accuracy of Students' Motivation, Emotion, and Achievement [321]

This is a machine-generated summary of:

Zhu, Chunjie; Urhahne, Detlef: Temporal stability of teachers' judgment accuracy of students' motivation, emotion, and achievement [321].

Published in: European Journal of Psychology of Education (2020).

Link to original: https://doi.org/10.1007/s10212-020-00480-7.

If you want to cite the papers, please refer to the original.

For technical reasons we could not place the page where the original quote is coming from.

Abstract-Summary
"The aim of this study was to investigate the temporal stability of teachers' judgment accuracy of students' motivation, emotion, and achievement."

"The correspondence between teacher judgments and student characteristics was measured twice within a 4-week period."

"Teachers judged each student's motivation, emotion, and achievement by single items."

"Results show that teachers were able to assess student achievement with high accuracy, student motivation with moderate to high accuracy, and student emotion mostly with low accuracy."

"It can be concluded that Chinese elementary school teachers are in a position to make fair judgments about student achievement and student motivation at different times."

"Student emotions, however, are hard to grasp for teachers."

Introduction
"The accuracy of teacher judgment is examined in terms of motivational, emotional, and cognitive student characteristics."

"The correspondence between teacher judgments and student characteristics is studied at two different points of time in order to determine the accuracy of teacher judgment over time."

"The accuracy of teacher judgment on student characteristics can be determined by three different components: rank, level, and differentiation component (Cronbach [289])."

"Longitudinal studies to measure the stability of teacher judgment accuracy are almost exclusively related to student achievement."

"Only the study by Givvin and others [290] provides some information on the temporal stability of teacher judgment accuracy on student motivation and emotion."

"This study examines the interplay of various motivational, emotional, and cognitive factors to holistically explore the accuracy of teacher judgments of student characteristics (Huber and Seidel [291])."

Method
"A standardized mathematics test for the sixth grade (DEMAT 6+, Götz and others [292]) was used to measure students' mathematics achievement."

"Mathematics teachers received a copy of the standardized test that was used to measure students' mathematics achievement."

"Expectancy for success and level of aspiration were measured differently: teachers had to predict the score that students would expect to get and would be satisfied with in the next mathematics test on the Chinese grading scale from 0 to 100."

"In the afternoon, the teachers themselves had the opportunity to assess test performance and motivational-affective characteristics of each student in class."

"The differences in the motivational-affective characteristics between classes varied between 0.5 (test anxiety) and 8.1% (expectancy for success) at the first point of measurement."

"At the first point of measurement up to 3.7% (effort) and at the second point of measurement up to 4.3% (test anxiety) of the student data had to be estimated."

Results

"Repeated measure analyses of variance were used to check whether student characteristics and teacher judgments changed over time."

"Teachers reduced their high expectation of student achievement from an average of 21 to 20 correctly solved items."

"Teachers assumed higher student self-concept, increased expectancy for success, and reduced test anxiety at the second point of measurement."

"Teacher judgments and student characteristics were related to each other."

"Teachers are able to judge student achievement with high accuracy, motivational student characteristics with moderate to high accuracy, and emotional student characteristics mostly with low accuracy."

"The heterogeneity of student achievement is judged more badly at the second point of measurement."

"It is striking that teacher judgment of students' motivation and emotion is a good predictor of teacher judgment of student achievement."

Discussion

"The aim of the study was to examine the temporal stability of teachers' judgment accuracy of students' motivation, emotion, and achievement."

"The results indicate that teachers can judge their students' achievement with high accuracy, motivation with moderate to high accuracy, and emotion mostly with low accuracy."

"Regarding students' motivation and emotion, teachers' judgment accuracy was consistent over the period of 4 weeks."

"The results slightly differ from prior research by Givvin and others [290] who found high stability of teachers' judgments about students' motivation over time, but argued that students' self-report motivation tended to be more differentiating and changeable."

"Teachers' judgments of students' motivation and emotion are robust predictors of their achievement judgments."

"The present study provided several strengths and found evidence for the temporal stability of teacher judgment accuracy on both student achievement and motivation."

Conclusions

"The best teacher ratings were found for students' academic achievement followed by students' motivation."

"All, the findings of the study confirm the temporal stability of teacher judgment accuracy for both cognitive and motivational-affective student characteristics."

"They should learn more about valid indicators of students' motivation and emotion if the same degree of judgment accuracy should be reached as in the cognitive domain."

"To gain more insight into teachers' judgment process, qualitative studies might be helpful that ask of teachers' strategies for assessing student traits."

Bibliography

1. Liljedahl, P. G. (2005). Mathematical discovery and affect: The effect of AHA! experiences on undergraduate mathematics students. International Journal of Mathematical Education in Science & Technology, 36(2/3), 219–235.
2. McLeod, D. B. (1988). Affective issues in mathematical problem solving: Some theoretical considerations. Journal for Research in Mathematics Education, 19(2), 134–141.
3. Linnenbrink, E. A., & Pintrich, P. R. (2002). Achievement goal theory and affect: An asymmetrical bidirectional model. Educational Psychologist, 37(2), 69–78.
4. Hannula, M. S. (2006). Motivation in mathematics: Goals reflected in emotions. Educational Studies in Mathematics, 63(2), 165–178.
5. Op 't Eynde, P., Corte, E., & Verschaffel, L. (2006). "Accepting emotional complexity": A socio-constructivist perspective on the role of emotions in the mathematics classroom. Educational Studies in Mathematics, 63(2), 193–207.
6. Meyer, D. K., & Turner, J. C. (2006). Re-conceptualizing emotion and motivation to learn in classroom contexts. Educational Psychology Review, 18(4), 377–390.
7. Op 't Eynde, P., & Turner, J. E. (2006). Focusing on the complexity of emotion issues in academic learning: A dynamical component systems approach. Educational Psychology Review, 18(4), 361–376.
8. Turner, J. C., & Patrick, H. (2008). How does motivation develop and why does it change? Reframing motivation research. Educational Psychologist, 43(3), 119–131.
9. Croft, A., & Ward, J. (2001). A modern and interactive approach to learning engineering mathematics. British Journal of Educational Technology, 32(2), 195–207.
10. Hodges, C. B. (2008). Self-efficacy, motivational email, and achievement in an asynchronous math course. Journal of Computers in Mathematics and Science Teaching, 27(3), 265–285.
11. Cohen, J. (1988). Statistical power analysis for the behavioral sciences (2nd ed.). Hillsdale, NJ: Lawrence Earlbaum.
12. Goetz, T., Frenzel, A. C., Hall, N. C., & Pekrun, R. (2008). Antecedents of academic emotions: Testing the internal/external frame of reference model for academic enjoyment. Contemporary Educational Psychology, 33(1), 9–33.
13. Pekrun, R., Goetz, T., Titz, W., & Perry, R. P. (2002). Academic emotions in students' self-regulated learning and achievement: A program of qualitative and quantitative research. Educational Psychologist, 37(2), 91–105.
14. Schallert, D. L., Reed, J. H., & Turner, J. E. (2004). The interplay of aspirations, enjoyment, and work habits in academic endeavors: Why is it so hard to keep long-term commitments? Teachers College Record, 106(9), 1715–1728.

15. Ford, M. E. (1992). Motivating humans: Goals, emotions, and personal agency beliefs. Newbury Park, CA: Sage.
16. Ainley, M. (2006). Connecting with learning: Motivation, affect and cognition in interest processes. Educational Psychology Review, 18(4), 391–405.
17. Goetz, T., Pekrun, R., Hall, N., & Haag, L. (2006). Academic emotions from a social-cognitive perspective: Antecedents and domain specificity of students' affect in the context of Latin instruction. British Journal of Educational Psychology, 76(2), 289–308.
18. Schutz, P. A., Hong, J. Y., Cross, D. I., & Osbon, J. N. (2006). Reflections on investigating emotion in educational activity settings. Educational Psychology Review, 18(4), 343–360.
19. Weiner, B. (1985). An attributional theory of achievement motivation and emotion. Psychological Review, 92(4), 548–573.
20. Meyer, D. K., & Turner, J. C. (2002). Discovering emotion in classroom motivation research. Educational Psychologist, 37(2), 107–114.
21. Ertmer, P., & Newby, T. J. (1993). Behaviorism, cognitivism, constructivism: Comparing critical features from an instructional design perspective. Performance Improvement Quarterly, 6(4), 50–71.
22. Astleitner, H. (2001). Designing emotionally sound instruction: An empirical validation of the FEASP approach. Journal of Instructional Psychology, 28(4), 209–219.
23. National Mathematics Advisory Panel (NMAP). (2008). Foundations for success: The final report of the National Mathematics Advisory Panel. Washington, DC: U.S. Department of Education.
24. National Center for Education Statistics (NCES). (2003). Remedial education at degree-granting post-secondary institutions in all 2000. Washington, DC: U.S. Department of Education.
25. Hagedorn, L. S., Siadat, M. V., Fogel, S. F., Nora, A., & Pascarella, E. T. (1999). Success in college mathematics: Comparisons between remedial and nonremedial first-year college students. Research in Higher Education, 40(3), 261–284.
26. Stage, F. K., & Kloosterman, P. (1995). Gender, beliefs, and achievement in remedial college level mathematics. Journal of Higher Education, 66(3), 294–311.
27. Belland, B. R., Kim, C., & Hannafin, M. J. (2013). A framework for designing scaffolds that improve motivation and cognition. Educational Psychologist, 48(4), 243–270. https://doi.org/10.1080/00461520.2013.838920
28. Davou, B. (2007). Interaction of emotion and cognition in the processing of textual material. Meta, 52(1), 37–47. https://doi.org/10.7202/014718ar
29. Greenspan, S. I., & Benderly, B. L. (1998). The growth of the mind: And the endangered origins of intelligence. Da Capo Press.
30. Kim, C., & Pekrun, R. (2014). Emotions and motivation in learning and performance. In J. M. Spector, M. D. Merrill, J. Elen, & M. J. Bishop (Eds.), Handbook of research on educational communications and technology (4th ed., pp. 65–75). Springer. https://doi.org/10.1007/978-1-4614-3185-5_6
31. Moreno, R., & Mayer, R. E. (2010). Techniques that increase generative processing In multimedia learning: Open questions for cognitive load research. In J. L. Plass, R. Moreno, & R. Brünken (Eds.), Cognitive load theory (pp. 153–177). Cambridge University Press.
32. Norman, D. A. (2002). Emotion & design: Attractive things work better. Interactions, 9(4), 36–42. https://doi.org/10.1145/543434.543435
33. Park, O., & Lee, J. (2004). Adaptive instructional systems. In D. H. Jonassen (Ed.), Handbook of research on educational communications and technology (2nd ed., pp. 651–684). Lawrence Erlbaum Associates Publishers.
34. Brom, C., Stárková, T., & D'Mello, S. K. (2018). How effective is emotional design? A meta-analysis on facial anthropomorphisms and pleasant colors during multimedia learning. Educational Research Review, 25, 100–119. https://doi.org/10.1016/j.edurev.2018.09.004
35. Feltman, R., & Elliot, A. J. (2012). Approach and avoidance motivation. In N.M. Seel (Eds.), Encyclopedia of the sciences of learning. Springer. https://doi.org/10.1007/978-1-4419-1428-6_1749

36. Elliot, A. J. (2008). Handbook of approach and avoidance motivation. Routledge. https://doi.org/10.4324/9780203888148
37. Frick T. W. (2018) The theory of totally integrated education (TIE). In M. Spector, B. Lockee, & M. Childress. (Eds.) Learning, design, and technology. Springer. https://doi.org/10.1007/978-3-319-17727-4_69-2
38. Maccia, G. S. (1987). Correspondence: Genetic epistemology of intelligent natural systems. Systems Research, 4(3), 213–218. https://doi.org/10.1002/sres.3850040308
39. Steiner, E. (1981). Educology of the free. Philosophical Library.
40. Barriball, K. L., & While, A. (1994). Collecting data using a semi-structured interview: A discussion paper. Journal of Advanced Nursing-Institutional Subscription, 19(2), 328–335. https://doi.org/10.1111/j.1365-2648.1994.tb01088.x
41. Spradley, J. P. (1979). The ethnographic interview. Harcourt Brace Jovanovich.
42. Pilgrim's Reception Office (2020). Informe estadístico Ano 2019. https://oficinadelperegrino.com/en/statistics/
43. Oviedo, L., de Courcier, S., & Farias, M. (2014). Rise of pilgrims on the Camino to Santiago: Sign of change or religious revival? Review of Religious Research, 56(3), 433–442.
44. Chemin, E. (2011). Pilgrimage in a secular age. University of Exeter. Doctoral dissertation.
45. Farias, M., Coleman, T. J., Bartlett, I. I. I., Oviedo, J. E., Soares, L., Santos, P., T., & Bas, M. D. C. (2019). Atheists on the Santiago Way: Examining motivations to go on pilgrimage. Sociology of Religion: A Quarterly Review, 80(1), 28–44.
46. Kim, H., Yilmaz, S., & Ahn, S. (2019). Motivational landscape and evolving identity of a route-based religious tourism space: A case of Camino de Santiago. Sustainability, 11(13), 1–20. https://doi.org/10.3390/su11133547
47. Kaufman, S. B. (2020). Transcend: The new science of self-actualization. A TeacherPerigee Book.
48. Chen, X. (2013). STEM attrition: College students' paths into and out of STEM Fields. Statistical analysis report. NCES 2014–001. National Center for Education Statistics.
49. Hunter, A.-B. (2019). Why undergraduates leave STEM majors: Changes over the last two decades. In E. Seymour & A.-B. Hunter (Eds.), Talking about leaving revisited: Persistence, relocation, and loss in undergraduate STEM education (pp. 87–114). Springer International Publishing. https://doi.org/10.1007/978-3-030-25304-2_3
50. Larcombe, W., Finch, S., Sore, R., Murray, C. M., Kentish, S., Mulder, R. A., Lee-Stecum, P., Baik, C., Tokatlidis, O., & Williams, D. A. (2016). Prevalence and socio-demographic correlates of psychological distress among students at an Australian university. Studies in Higher Education, 41(6), 1074–1091. https://doi.org/10.1080/03075079.2014.966072
51. Gasiewski, J. A., Eagan, M. K., Garcia, G. A., Hurtado, S., & Chang, M. J. (2012). From gatekeeping to engagement: A multicontextual, mixed method study of student academic engagement in introductory STEM courses. Research in Higher Education, 53(2), 229–261. https://doi.org/10.1007/s11162-011-9247-y
52. Weston, T. J., Seymour, E., Koch, A. K., & Drake, B. M. (2019). Weed-out classes and their consequences. In E. Seymour & A.-B. Hunter (Eds.), Talking about leaving revisited: Persistence, relocation, and loss in undergraduate STEM education (pp. 197–243). Springer. https://doi.org/10.1007/978-3-030-25304-2_7
53. Ben-Eliyahu, A., & Linnenbrink-Garcia, L. (2013). Extending self-regulated learning to include self-regulated emotion strategies. Motivation and Emotion, 37(3), 558–573. https://doi.org/10.1007/s11031-012-9332-3
54. Pelch, M. (2018). Gendered differences in academic emotions and their implications for student success in STEM. International Journal of STEM Education, 5(1), 33. https://doi.org/10.1186/s40594-018-0130-7
55. Simon, R. A., Aulls, M. W., Hubbard, K., Hall, N. C., & Dedic, H. (2015). Exploring student persistence in STEM programs: A motivational model. Canadian Journal of Education, 38(1), 1–27.

56. Szulecka, T. K., Springett, N. R., & De Pauw, K. W. (1987). General health, psychiatric vulnerability and withdrawal from university in first-year undergraduates. British Journal of Guidance & Counselling, 15(1), 82–91. https://doi.org/10.1080/03069888708251646
57. Jacobs, S. E., & Gross, J. J. (2014). Emotion regulation in education: Conceptual foundations, current applications, and future directions. In R. Pekrun & L. Linnenbrink-Garcia (Eds.), International handbook of emotions in education (pp. 183–201). Routledge.
58. Ahmed, W., van der Werf, G., Kuyper, H., & Minnaert, A. (2013). Emotions, self-regulated learning, and achievement in mathematics: A growth curve analysis. Journal of Educational Psychology, 105(1), 150–161. https://doi.org/10.1037/a0030160
59. Frenzel, A. C., Thrash, T. M., Pekrun, R., & Goetz, T. (2007). Achievement emotions in Germany and China: A cross-cultural validation of the academic emotions questionnaire—mathematics. Journal of Cross-Cultural Psychology, 38(3), 302–309.
60. Pekrun, R., Elliot, A. J., & Maier, M. A. (2009). Achievement goals and achievement emotions: Testing a model of their joint relations with academic performance. Journal of Educational Psychology, 101(1), 115–135. https://doi.org/10.1037/a0013383
61. Pekrun, R., Goetz, T., Frenzel, A. C., Barchfeld, P., & Perry, R. P. (2011). Measuring emotions in students' learning and performance: The achievement emotions questionnaire (AEQ). Contemporary Educational Psychology, 36(1), 36–48. https://doi.org/10.1016/j.cedpsych.2010.10.002
62. Artino, A. R., & Jones, K. D. (2012). Exploring the complex relations between achievement emotions and self-regulated learning behaviors in online learning. The Internet and Higher Education, 15(3), 170–175.
63. Naismith, L. M., & Lajoie, S. P. (2017). Motivation and emotion predict medical students attention to computer-based feedback. Advances in Health Sciences Education, 1–21.
64. Pintrich, P. R., Smith, D. A. F., Garcia, T., & McKeachie, W. J. (1991). A manual for the use of the motivated strategies for learning questionnaire (MSLQ). University of Michigan, National Center for Research to Improve Postsecondary Teaching and Learning.
65. Pekrun, R., & Perry, R. P. (2014). Control-value theory of achievement emotions. In R. Pekrun & L. Linnenbrink-Garcia (Eds.), International handbook of emotions in education (pp. 120–141). Routledge.
66. Dörnyei, Z. (1998). Motivation in second and foreign language learning. Language Teaching, 31(03), 117–135.
67. Dörnyei, Z., & Ushioda, E. (2013). Teaching and researching: Motivation. London: Routledge.
68. Dweck, C. S., & Leggett, E. L. (1988). A social-cognitive approach to motivation and personality. Psychological Review, 95(2), 256.
69. Ehrman, M. E., Leaver, B. L., & Oxford, R. L. (2003). A brief overview of individual differences in second language learning. System, 31(3), 313–330.
70. Dewaele, J. M., & Furnham, A. (1999). Extraversion: The unloved variable in applied linguistic research. Language Learning, 49(3), 509–544.
71. Flowers, S. (2015). Friendship and reciprocity as motivators in CSCL. JALT CALL Journal, 11(3), 191–212.
72. Benenson, J. F., & Dweck, C. S. (1986). The development of trait explanations and self-evaluations in the academic and social domains. Child Development, 57, 1179–1187.
73. Dweck, C. S. (1986). Motivational processes affecting learning. American Psychologist, 41, 1040–1048.
74. Tamir, M., Bigman, Y., Rhodes, E., Salerno, J., & Schreier, J. (2015). An expectancy-value model of emotion regulation: Implications for motivation, emotional experience, and decision-making. Emotion, 15, 90–103.
75. Gollwitzer, P. M. (1990). Action phases and mind-sets. In E. T. Higgins & R. M. Sorrentino (Eds.), Handbook of motivation and cognition: Foundations of social behavior (Vol. 2, pp. 53–92). Guilford Press.
76. Tamir, M. (2016). Why do people regulate their emotions? A taxonomy of motives in emotion regulation. Personality and Social Psychology Review, 20, 199–222.

77. Suri, G., Whittaker, K., & Gross, J. J. (2015). Launching reappraisal: It's less common than you might think. Emotion, 15(1), 73–77.
78. Liu, D. Y., & Thompson, R. J. (2017). Selection and implementation of emotion regulation strategies in major depressive disorder: An integrative review. Clinical Psychology Review, 57, 183–194.
79. Yoon, S., & Rottenberg, J. (2020). Why do people with depression use faulty emotion regulation strategies? Emotion Review, 12(2), 118–128.
80. Gross, J. J., Uusberg, H., & Uusberg, A. (2019). Mental illness and well-being: An affect regulation perspective. World Psychiatry, 18(2), 130–139.
81. Marchewka, A., Żurawski, Ł., Jednoróg, K., & Grabowska, A. (2014). The Nencki affective picture system (NAPS): Introduction to a novel, standardized, wide-range, high-quality, realistic picture database. Behavior Research Methods, 46, 596–610.
82. Shiota, M. N., Campos, B., Oveis, C., Hertenstein, M. J., Simon-Thomas, E., & Keltner, D. (2017). Beyond happiness: Building a science of discrete positive emotions. American Psychologist, 72, 617–634.
83. Carver, C. S., & Harmon-Jones, E. (2009). Anger is an approach-related affect: Evidence and implications. Psychological Bulletin, 135, 183–204.
84. Harmon-Jones, E. (2018). On motivational influences, moving beyond valence, and integrating dimensional and discrete views of emotion. Cognition and Emotion, 1–8.
85. Lench, H. C., Flores, S. A., & Bench, S. W. (2011). Discrete emotions predict changes in cognition, judgment, experience, behavior, and physiology: A meta-analysis of experimental emotion elicitations. Psychological Bulletin, 137, 834–855.
86. Harmon-Jones, E., Harmon-Jones, C., & Price, T. F. (2013b). What is approach motivation? Emotion Review, 5, 291–295.
87. Gable, P. A., & Harmon-Jones, E. (2008). Approach-motivated positive affect reduces breadth of attention. Psychological Science, 19, 476–482.
88. Reb, J., Allen, T., & Vogus, T. J. (2020). Mindfulness arrives at work: Deepening our understanding of mindfulness in organizations. Organizational Behavior and Human Decision Processes. https://doi.org/10.1016/j.obhdp.2020.04.001
89. Reb, J., & Atkins, P. (2015). Mindfulness in organizations: Foundations, research, and applications. Cambridge University Press.
90. Birtwell, K., Williams, K., van Marwijk, H., Armitage, C. J., & Sheffield, D. (2019). An exploration of formal and informal mindfulness practice and associations with wellbeing. Mindfulness, 10(1), 89–99. https://doi.org/10.1007/s12671-018-0951-y
91. Custers, R., & Aarts, H. (2005). Positive affect as implicit motivator: On the nonconscious operation of behavioral goals. Journal of Personality and Social Psychology, 89, 129–142. https://doi.org/10.1037/0022-3514.89.2.129 .
92. May, C. N., Nock, N. L., Bentley, D., & Demaree, H. A. (2016). Acute aerobic exercise increases implicit approach motivation for dessert images. Journal of Health Psychology, 1–11. https://doi.org/10.1177/1359105316657404 .
93. Orehek, E., Bessarabova, E., Chen, X., & Kruglanski, A. W. (2011). Positive affect as informational feedback in goal pursuit. Motivation and Emotion, 35(1), 44–51. https://doi.org/10.1007/s11031-010-9197-2 .
94. Cuddy, A. J. C., Schultz, S. J., & Fosse, N. E. (2018). P-curving a more comprehensive body of research on postural feedback reveals clear evidential value for power-posing effects: Reply to Simmons and Simonsohn (2017). Psychological Science, 29(4), 656–666. https://doi.org/10.1177/0956797617746749 .
95. Harmon-Jones, E., Price, T. F., & Harmon-Jones, C. (2015). Supine body posture decreases rationalizations: Testing the action-based model of dissonance. Journal of Experimental Social Psychology, 56, 228–234. https://doi.org/10.1016/j.jesp.2014.10.007 .
96. Saghaei, M. (2004). Random allocation software for parallel group randomized trials. BMC Medical Research Methodology, 4(1), 26. https://doi.org/10.1186/1471-2288-4-26 .
97. Bandura, A. (1997). Self-efficacy: The exercise of control. New York, NY: Freeman.

98. Rotter, J. B. (1966). Generalized expectancies for internal versus external control of reinforcement. Psychological Monographs: General and Applied, 80(1), 1–28.
99. Shell, D. F., & Husman, J. (2008). Control, motivation, affect, and strategic self-regulation in the college classroom: A multidimensional phenomenon. Journal of Educational Psychology, 100(2), 443–459.
100. Pekrun, R. (2000). A social cognitive, control-value theory of achievement emotions. In J. Heckhausen (Ed.), Motivational psychology of human development (pp. 143–163). Oxford: Elsevier.
101. Eccles, J. (1983). Expectancies, values, and academic behaviors. In J. T. Spence (Ed.), Achievement and achievement motives (pp. 75–146). San Francisco, CA: Freeman.
102. Pekrun, R. (1992). Facets of adolescents' academic motivation: A longitudinal expectancy-value approach. In M. Maehr & P. Pintrich (Eds.), Advances in motivation and achievement 8 (pp. 139–190). Greenwich, CT: JAI Press.
103. Pintrich, P. (1988). A process-oriented view of student motivation and cognitions. In J. S. Stark & L. A. Mets (Eds.), Improving teaching and learning through research (Vol. 57, pp. 65–80). San Francisco, CA: Jossey-Bass, Inc.
104. Pintrich, P. (1989). The dynamic interplay of student motivation and cognition in the college classroom. In M. Maehr & C. Ames (Eds.), Advances in motivation and achievement: Motivation enhancing environments (pp. 117–160). Connecticut: JAI Press.
105. Wigfield, A., & Eccles, J. (2000). Expectancy-value theory of achievement motivation. Contemporary Educational Psychology. Motivation and the Educational Process, 25(1), 68–81.
106. Berrenberg, J. L. (1987). The belief in Personal Control Scale: A measure of God-mediated and exaggerated control. Journal of Personality Assessment, 51, 194–206.
107. Ferguson, E., & Cox, T. (1996). An implicit theory of perceived control. European Psychologist, 1(4), 271–277.
108. Hall, N. C., Perry, R. P., Ruthig, J. C., Hladkyj, S., & Chipperfield, J. (2006). Primary and secondary control in achievement settings: A longitudinal field study of academic motivation, emotions, and performance. Journal of Applied Social Psychology, 36(6), 1430–1470.
109. Rothbaum, F., Weisz, J. R., & Snyder, S. S. (1982). Changing the world and changing the self: A two –process model of perceived control. Journal of Personality and Social Psychology, 42(1), 5–37.
110. McAdams, D. P., Josselson, R., & Lieblich, A. (2001a). Turns in the road: Narrative studies of lives in transition. Washington, D.C: American Psychological Association.
111. Duncan, T. G., & McKeachie, W. J. (2005). The making of the motivated strategies for learning questionnaire. Educational Psychologist, 40(2), 117–128. https://doi.org/10.1207/s15326985ep4002_6
112. Pintrich, P. R., & De Groot, E. V. (1990). Motivation and self-regulated learning components of classroom academic performance. Journal of Educational Psychology, 82(1), 33–40. https://doi.org/10.1037/0022-0663.82.1.33
113. Pintrich, P. R., Smith, D. A. F., Garcia, T., & Mckeachie, W. J. (1993). Reliability and predictive validity of the motivated strategies for learning questionnaire (MSLQ). Educational and Psychological Measurement, 53(3), 801–813. https://doi.org/10.1177/0013164493053003024
114. Wigfield, A., & Eccles, J. S. (2000). Expectancy-value theory of achievement motivation. Contemporary Educational Psychology, 25(1), 68–81. https://doi.org/10.1006/ceps.1999.1015
115. Pekrun, R., Hall, N. C., Goetz, T., & Perry, R. P. (2014). Boredom and academic achievement: Testing a model of reciprocal causation. Journal of Educational Psychology, 106(3), 696–710. https://doi.org/10.1037/a0036006
116. Holm, M. E., Hannula, M. S., & Björn, P. M. (2017). Mathematics-related emotions among Finnish adolescents across different performance levels. Educational Psychology, 37(2), 205–218. https://doi.org/10.1080/01443410.2016.1152354

117. Arroyo, I., Burleson, W., Tai, M., Muldner, K., & Woolf, B. P. (2013). Gender differences in the use and benefit of advanced learning technologies for mathematics. Journal of Educational Psychology, 105(4), 957–969. https://doi.org/10.1037/a0032748
118. Frenzel, A. C., Pekrun, R., & Goetz, T. (2007). Girls and mathematics—A "hopeless" issue? A control-value approach to gender differences in emotions towards mathematics. European Journal of Psychology of Education, 22(4), 497–514. https://doi.org/10.1007/BF03173468
119. Krinzinger, H., Kaufmann, L., & Willmes, K. (2009). Math anxiety and math ability in early primary school years. Journal of Psychoeducational Assessment, 27(3), 206–225. https://doi.org/10.1177/0734282908330583
120. Ahmed, W., van der Werf, G., Minnaert, A., & Kuyper, H. (2010). Students' daily emotions in the classroom: Intra-individual variability and appraisal correlates. British Journal of Educational Psychology, 80, 583–597. https://doi.org/10.1348/000709910X498544
121. Pekrun, R., Elliot, A. J., & Maier, M. A. (2006). Achievement goals and discrete achievement emotions: A theoretical model and prospective test. Journal of Educational Psychology, 98(3), 397–583.
122. Putwain, D. W., Larkin, D., & Sander, P. (2013). A reciprocal model of achievement goals and learning related emotions in the first year of undergraduate study. Contemporary Educational Psychology, 38, 361–374. https://doi.org/10.1016/j.cedpsych.2013.07.003
123. Zimmerman, B. J., & Schunk, D. (2011). Handbook of self-regulation of learning and performance. Routledge.
124. Pekrun, R. (2006). The control-value theory of achievement emotions: Assumptions, corollaries, and implications for educational research and practice. Educational Psychology Review, 18(4), 315–341.
125. Pekrun, R., Goetz, T., Daniels, L. M., Stupnisky, R. H., & Perry, R. P. (2010). Boredom in achievement settings: Control-value antecedents and performance consequences of a neglected emotion. Journal of Educational Psychology, 102(3), 531–549. https://doi.org/10.1037/a0019243
126. Webb, T. L., Miles, E., & Sheeran, P. (2012). Dealing with feeling: A meta-analysis of the effectiveness of strategies derived from the process model of emotion regulation. Psychological Bulletin, 138, 775–808.
127. Harley, J. M., Pekrun, R., Taxer, J. L., & Gross, J. J. (2019). Emotion regulation in achievement situations: An integrated model. Educational psychologist, 54(2), 106–126
128. Marsh, H. W. (1990). Self Description Questionnaire-I. APA PsycTests. https://doi.org/10.1037/t01843-000
129. Pekrun, R., Vogl, E., Muis, K. R., & Sinatra, G. M. (2017). Measuring emotions during epistemic activities: the Epistemically-Related Emotion Scales. Cognition and Emotion, 31(6), 1268–1276
130. Wild, K. P., & Schiefele, U. (1994). Lernstrategien im Studium: Ergebnisse zur Faktorenstruktur und Reliabilität eines neuen Fragebogens. Zeitschrift für Differentielle und Diagnostische Psychologie, 15(4), 185–200
131. Hudig, J., Scheepers, A. W. A., Schippers, M. C., & Smeets, G. (2021). Motives for studying and student wellbeing: Validation of the motivational mindset model. Frontiers in Psychology, 11, 753987. https://doi.org/10.3389/fpsyg.2021.753987
132. Morisano, D., Hirsh, J. B., Peterson, J. B., Pihl, R. O., & Shore, B. M. (2010). Setting, elaborating, and reflecting on personal goals improves academic performance. Journal of Applied Psychology, 95(2), 255–264. https://doi.org/10.1037/a0018478
133. Schippers, M. C., Morisano, D., Locke, E. A., Scheepers, A. W. A., Latham, G. P., & de Jong, E. M. (2020). Writing about personal goals and plans regardless of goal type boosts academic performance. Contemporary Educational Psychology, 60, 101823. https://doi.org/10.1016/J.CEDPSYCH.2019.101823
134. Schippers, M. C., & Ziegler, N. (2019). Life crafting as a way to find purpose and meaning in life. Frontiers in Psychology, 10. https://doi.org/10.3389/fpsyg.2019.02778

135. Bronk, K. C., Leontopoulou, S., & McConchie, J. (2019b). Youth purpose during the great recession: A mixed-methods study. Journal of Positive Psychology, 14(4), 405–416. https://doi.org/10.1080/17439760.2018.1484942
136. Dekker, I, Schippers, M. C. & Van Schooten, E. (2021). Reflective goal-setting improves academic performance in teacher and business education: A large-scale field experiment. SSRN [Preprint]. https://doi.org/10.2139/ssrn.3778544 .
137. Cerasoli, C. P., Nicklin, J. M., & Ford, M. T. (2014). Intrinsic motivation and extrinsic incentives jointly predict performance: A 40-year Meta-analysis. Psychological Bulletin. https://doi.org/10.1037/a0035661
138. D'Lima, G. M., Winsler, A., & Kitsantas, A. (2014). Ethnic and gender differences in first-year college students' goal orientation, self-efficacy, and extrinsic and intrinsic motivation. Journal of Educational Research, 107(5), 341–356. https://doi.org/10.1080/00220671.2013.823366
139. Kaufman, J. C., Agars, M. D., & Lopez-Wagner, M. C. (2008). The role of personality and motivation in predicting early college academic success in non-traditional students at a Hispanic-serving institution. Learning and Individual Differences, 18, 492–496. https://doi.org/10.1016/j.lindif.2007.11.004
140. Richardson, M., Abraham, C., & Bond, R. (2012). Psychological correlates of university students' academic performance: A systematic review and meta-analysis. Psychological Bulletin, 138(2), 353–387. https://doi.org/10.1037/a0026838
141. Christie, H., Munro, M., & Fisher, T. (2004). Leaving university early: Exploring the differences between continuing and non-continuing students. Studies in Higher Education. https://doi.org/10.1080/0307507042000261580
142. Van Rooij, E., Brouwer, J., Fokkens-Bruinsma, M., Jansen, E., Donche, V., & Noyens, D. (2017). A systematic review of factors related to first-year students' success in Dutch and flemish higher education. Pedagogische Studien, 94(5), 360–405.
143. Willcoxson, L., Cotter, J., & Joy, S. (2011). Beyond the first-year experience: The impact on attrition of student experiences throughout undergraduate degree studies in six diverse universities. Studies in Higher Education, 36(3), 331–352. https://doi.org/10.1080/03075070903581533
144. Inspectie van het Onderwijs [Inspection of Education]. (2021). De staat van het onderwijs. Onderwijsverslag 2020/2021 [the current state of education. Educational report 2020/2021]. Utrecht, The Netherlands. Available at: https://www.onderwijsinspectie.nl/binaries/onderwijsinspectie/documenten/rapporten/2021/04/14/de-staat-van-het-onderwijs-2021/Staat-van-het-Onderwijs-2021.pdf . Accessed 2 July 2021
145. Van den Broek, A., Warps, J., de Korte, K., Cuppen, J., Termorshuizen, T., & Lodewick, J. (2019). Monitor Beleidsmaatregelen 2019–2020: Studiekeuze, studiegedrag & leengedrag in relatie tot beleidsmaatregelen in het hoger onderwijs. ResearchNed.
146. Côté, J. E., & Levine, C. (1997). Student motivations, learning environments, and human capital acquisition: Toward an integrated paradigm of student development. Journal of College Student Development, 38(3), 229–243.
147. Henderson-King, D., & Smith, M. N. (2006). Meanings of education for university students: Academic motivation and personal values as predictors. Social Psychology of Education, 9(2), 195–221. https://doi.org/10.1007/s11218-006-0006-4
148. Miele DB, Scholer AA, Fujita K. Metamotivation: Emerging research on the regulation of motivational states. In: Elliot AJ, editor. Advances in Motivation Science. Elsevier; 2020. p. 1–42. Available from: http://www.sciencedirect.com/science/article/pii/S221509191930015X
149. Norouzi A, Parmelee D, Shariati M, Norouzi S, Nedjat S, Alizadeh M. The development and validation of metamotivational strategies in medical students questionnaire. Med Teach. 2021;43(10):1186–95. Available from: https://www.tandfonline.com/doi/full/https://doi.org/10.1080/0142159X.2021.1922656 .

150. Miele DB, Scholer AA. The role of metamotivational monitoring in motivation regulation. Educ Psychol. 2018;53(1):1–21. Available from: https://www.tandfonline.com/doi/full/https://doi.org/10.1080/00461520.2017.1371601 .
151. Wolters CA. Regulation of motivation: evaluating an underemphasized aspect of self-regulated learning. Educ Psychol. 2003;38(4):189–205. Available from: http://www.tandfonline.com/doi/abs/https://doi.org/10.1207/S15326985EP3804_1 .
152. Sheppes, G., & Gross, J. J. (2011). Is timing everything? Temporal considerations in emotion regulation. Personality & Social Psychology Review, 15, 319–331. https://doi.org/10.1177/1088868310395778 .
153. Yoon, K. L., & Joormann, J. (2012). Is timing everything? Sequential effects of rumination and distraction on interpersonal problem solving. Cognitive Therapy & Research, 36, 165–172. https://doi.org/10.1007/s10608-010-9330-2 .
154. Urry, H. L. (2009). Using reappraisal to regulate unpleasant emotional episodes: Goals and timing matter. Emotion, 9, 782–797. https://doi.org/10.1037/a0017109 .
155. Aupperle, R. L., & Paulus, M. P. (2010). Neural systems underlying approach and avoidance in anxiety disorders. Dialogues in Clinical Neuroscience, 12, 517–531. Retrieved from: http://www.ncbi.nlm.nih.gov/pmc/articles/PMC3181993/
156. Fauerbach, J. A., Lawerence, J. W., Fogel, J., Richter, L., Magyar-Russell, G., McKibben, J. B. A., & McCann, U. (2009). Approach–avoidance coping conflict in a sample of burn patients at risk for posttraumatic stress disorder. Depression and Anxiety, 26, 838–850. https://doi.org/10.1002/da.20439 .
157. Gray, J. A., & McNaughton, N. (2000). The neuropsychology of anxiety (2nd ed.). New York: Oxford University Press.
158. Mennin, D. S., & Fresco, D. M. (2014). Emotion regulation therapy. In J. J. Gross (Ed.), Handbook of emotion regulation (2nd ed., pp. 469–490). New York: Guilford Press.
159. Aldao, A., & Nolen-Hoeksema, S. (2013). One versus many: Capturing the use of multiple emotion regulation strategies in response to an emotion-eliciting stimulus. Cognition and Emotion, 27, 753–760. https://doi.org/10.1080/02699931.2012.739998 .
160. Corr, P. J. (2004). Reinforcement sensitivity theory and personality. Neuroscience and Biobehavioral Reviews, 28, 317–332. https://doi.org/10.1016/j.neubiorev.2004.01.005 .
161. Robinson, M. D., Wilkowski, B. M., & Meier, B. P. (2008). Approach, avoidance, and self-regulatory conflict: An individual differences perspective. Journal of Experimental & Social Psychology, 44, 65–79. https://doi.org/10.1016/j.jesp.2007.02.008 .
162. Ames, C. (1992a). Classrooms: Goals, structures, and student motivation. Journal of Educational Psychology, 84, 261–271.
163. Maehr, M. L. (1984). Meaning and motivation: Toward a theory of personal investment. In C. Ames & R. Ames (Eds.), Research on motivation in education, vol. 1, ch. 4 (pp. 115–144). New York: Academic.
164. Nicholls, J. G. (1984). Achievement motivation: Conceptions of ability, subjective experience, task choice, and performance. Psychological Review, 91, 328–346.
165. Anderman, E. M., & Wolters, C. (2006). Goals, values and affects: Influences on student motivation. In P. Alexander & P. Winne (Eds.), Handbook of educational psychology (2nd ed.). New York: Simon & Schuster.
166. Elliot, A. J. (2005). A conceptual history of the achievement goal construct. In A. J. Elliot & C. S. Dweck (Eds.), Handbook of competence and motivation (pp. 52–72). New York: Guildford.
167. Meece, J. L., Anderman, E. M., & Anderman, L. H. (2006). Classroom goal structure, student motivation and academic achievement. Annual Review of Psychology, 57, 487–503.
168. Pintrich, P. R., & Schunk, D. (2002). Motivation in education: theory, research, and applications (2nd ed.). Upper Saddle River, New Jersey: Prentice-Hall.
169. Midgley, C. (Ed.) (2002). Goals, goal structures, and patterns of adaptive learning. Mahwah, New Jersey: Erlbaum.

170. Brookhart, S. M., Walsh, J. M., & Zientarski, W. A. (2006). The dynamics of motivation and effort for classroom assessments in middle school science and social studies. Applied Measurement in Education, 19, 151–184.
171. Kaplan, A., & Maehr, M. L. (1999). Achievement goals and student well-being. Contemporary Educational Psychology, 24, 330–358.
172. Cury, F., Elliot, A. J., Da Fonseca, D., & Moller, A. C. (2006). The social-cognitive model of achievement motivation and the 2 × 2 achievement goal framework. Journal of Personality and Social Psychology, 90, 666–679.
173. Grant, H., & Dweck, C. S. (2003). Clarifying achievement goals and their impact. Journal of Personality and Social Psychology, 85, 541–553.
174. Harackiewicz, J. M., Barron, K. E., Carter, S. M., Lehto, A. R., & Elliot, A. J. (1997). Predictors and consequences of achievement goals in the college classroom: Maintaining interest and making the grade. Journal of Personality and Social Psychology, 73, 1284–1295.
175. Harackiewicz, J. M., & Sansone, C. (2000). Rewarding competence: The importance of goals in the study of intrinsic motivation. In C. Sansone & J. M. Harackiewicz (Eds.), Intrinsic and extrinsic motivation: The search for optimal motivation and performance (pp. 79–103). New York: Academic.
176. Elliot, A. J. (1997). Integrating the 'classic and 'contemporary' approaches to achievement motivation: A hierarchical model of approach and avoidance achievement motivation. In M. L. Maehr & P. R. Pintrich (Eds.), Advances in motivation and achievement, vol. 10 (pp. 143–179). Greenwich, Connecticut: JAI.
177. Maehr, M. L., & Braskamp, L. A. (1986). The motivation factor: A theory of personal investment. Lexington, Massachusetts: Lexington Books.
178. Maehr, M. L., & Nicholls, J. G. (1980). Culture and achievement motivation: A second look. In N. Warren (Ed.), Studies on cross-cultural psychology, vol. 2 (pp. 221–267). New York: Academic.
179. Bempechat, J., & Boulay, B. A. (2000). Beyond dichotomous characterizations of student learning: New directions in achievement motivation research. In D. M. McInerney & S. Van Ettern (Eds.), Research on sociocultural influences on motivation and learning, vol. 1 (pp. 15–36). Greenwich, Connecticut: Information Age.
180. Dowson, M., & McInerney, D. M. (2001). Psychological parameters of students' social and work avoidance goals: A qualitative investigation. Journal of Educational Psychology, 93, 35–42.
181. MacCallum, J. (2000). The contexts of individual motivational change. In D. M. McInerney & S. Van Ettern (Eds.), Research on sociocultural influences on motivation and learning, vol. 1 (pp. 61–97). Greenwich, Connecticut: Information Age.
182. Nicholls, J. G., Patashnick, M., & Nolen, S. B. (1985). Adolescents' theories of education. Journal of Educational Psychology, 77, 683–692.
183. Lemos, M. S. (1996). Students' and teachers' goals in the classroom. Learning and Instruction, 6, 151–171.
184. Nicholls, J. G. (1989). The competitive ethos and democratic education. Cambridge, Massachusetts: Harvard University Press.
185. Baumeister, R. F. (1997). Esteem threat, self-regulatory breakdown, and emotional distress as factors in self-defeating behavior. Review of General Psychology, 1, 145–174.
186. Higgins, E. T. (1987). Self-discrepancy: A theory relating self and affect. Psychological Review, 94, 319–340.
187. Flum, H., & Kaplan, A. (2006). Exploratory orientation as an educational goal. Educational Psychologist, 41, 99–100.
188. Kaplan, A. & Flum, H. (2005, August). Achievement goal orientations and identity formation. Paper presented at the annual meeting of the American Psychological Association, Washington, District of Columbia, USA.
189. Kashima, Y., Foddy, M., & Platow, M. (Eds.) (2002). Self and identity: Personal, social, and symbolic. Mahwah, New Jersey: Erlbaum.

190. Kluger, A. N., & DeNisi, A. (1996). The effects of feedback interventions on performance: A historical review, a meta-analysis, and a preliminary feedback intervention theory. Psychological Bulletin, 119, 254–284.
191. Schlenker, B. R., Britt, T. W., & Pennington, J. (1996). Impression regulation and management: Highlights of a theory of self-identification. In R. M. Sorrentino & E. T. Higgins (Eds.), Handbook of motivation and cognition: Vol. 3: The interpersonal context. New York: Guilford.
192. Maehr, M. L. (1989). Thoughts about motivation. In C. Ames, & R. Ames (Eds.), Research on motivation in education: (Vol. 3) Goals and cognitions (pp. 299–315). San Diego, California: Academic.
193. Fasihuddin, H., Skinner, G., & Athauda, R.(2016). A comprehensive review of adaptive, open, and online learning literature. International Conference on Computer Science Education Innovation & Technology (pp. 1–10).
194. Brusilovsky, P. (2001). Adaptive educational hypermedia. Proceedings of Tenth International PEG Conference, 8–12.
195. Bawa, P. (2016). Retention in online courses: exploring issues and solutions. SAGE Open, 6(1), 1–11.
196. Sunar, A. S., Abdullah, N. A., White, S., & Davis, H. (2015). Personalisation of MOOCs: The state of the art. Proceedings of the 7th International Conference on Computer Supported Education, Lisbon, Portugal (pp. 88–97).
197. Aleven, V., McLaughlin, E. A., Glenn, R. A., & Koedinger, K. R. (2017). Instruction based on adaptive learning technologies. In R. E. Mayer & P. Alexander (Eds.), Handbook of research on learning and instruction (2nd ed., pp. 522–560). New York: Routledge.
198. Nakic, J., Granic, A., & Glavinic, V. (2015). Anatomy of student models in adaptive learning systems: A systematic literature review of individual differences from 2001 to 2013. Journal of Educational Computing Research, 51(4), 459–489.
199. Mory, E. H. (1994). Adaptive feedback in computer-based instruction: Effects of response certitude on performance, feedback-study time, and efficiency. Educational Computing Research, 11(3), 263–290.
200. Privitera, G. J. (2016). Research methods for the behavioral sciences. Sage Publications.
201. Martinez, M. A. (2001). Key design considerations for personalized learning on the web. Educational Technology & Society, 4(1), 26–40.
202. Pintrich, P. R. (1999). The role of motivation in promoting and sustaining self-regulated learning. International Journal of Educational Research, 31(6), 459–470.
203. Doppelt, Y., & Schunn, C. D. (2008). Identifying students' perceptions of the important classroom features affecting learning aspects of a design-based learning environment. Learning Environments Research, 11(3), 195–209.
204. Brown, T. (2008). Design thinking. Harvard Business Review, 86(6), 84–92.
205. Mehalik, M. M., Doppelt, Y., & Schuun, C. D. (2008). Middle-school science through design-based learning versus scripted inquiry: better overall science concept learning and equity gap reduction. Journal of Engineering Education, 97(1), 71–85.
206. Gerber, E. M., Marie Olson, J., & Komarek, R. L. (2012). Extracurricular design-based learning: preparing students for careers in innovation. International Journal of Engineering Education, 28(2), 317–324.
207. Kolodner, J. L., Camp, P. J., Crismond, D., Fasse, B., Gray, J., Holbrook, J., et al. (2003). Problem-based learning meets case-based reasoning in the middle-school science classroom: putting learning by design (tm) into practice. The Journal of the Learning Sciences, 12(4), 495–547.
208. Doppelt, Y. (2009). Assessing creative thinking in design-based learning. International Journal of Technology and Design Education, 19(1), 55–65.
209. Fontaine, L. (2014). Learning design thinking by designing learning experiences: a case study in the development of strategic thinking skills through the design of interactive museum exhibitions. Visible Language, 48(2), 48–69.

210. Scherer, K. R., Schorr, A., & Johnstone, T. (Eds.). (2001). Appraisal processes in emotion. Oxford: Oxford University Press.
211. Goetz, T., Cronjaeger, H., Frenzel, A. C., Lüdtke, O., & Hall, N. C. (2010). Academic self-concept and emotion relations: Domain specificity and age effects. Contemporary Educational Psychology, 35, 44–58. https://doi.org/10.1016/j.cedpsych.2009.10.001
212. Gumora, G., & Arsenio, W. F. (2002). Emotionality, emotion regulation, and school performance in middle school children. Journal of School Psychology, 40(5), 395–413.
213. Bandura A, Schunk DH. Cultivating competence, self-efficacy, and intrinsic interest through proximal self-motivation. J Pers Soc Psychol. 1981;41(3):586–98.
214. Miles, M. B., Huberman, A. M., & Saldaña, J. (2013). Qualitative data analysis: A methods sourcebook. Thousand Oaks: Sage.
215. NRC. (2010). Exploring the intersection of science education and twenty-first century skills: A workshop summary. Washington, DC: The National Academies Press.
216. Sadler, P. M., Sonnert, G., Hazari, Z., & Tai, R. (2012). Stability and volatility of STEM career interest in high school: a gender study. Science Education, 96(3), 411–427.
217. Zimmerman, B. J. (2006). Enhancing students' academic responsibility and achievement: A Social-cognitive self-regulatory account. In R. J. Sternberg & R. Subotnik (Eds.), Optimizing student success in school with the other three Rs: Reasoning, resilience, and responsibility (pp. 179–197). Greenwich, CT: Information Age.
218. Zimmerman, B. J., & Kitsantas, A. (2005). Homework practices and academic achievement: The mediating role of self-efficacy and perceived responsibility beliefs. Contemporary Educational Psychology, 30(4), 397–417. https://doi.org/10.1016/j.cedpsych.2005.05.003 .
219. Ames, C., & Archer, J. (1988). Achievement goals in the classroom: Students' learning strategies and motivation processes. Journal of Educational Psychology, 80, 260–267.
220. Su, Y., & Reeve, J. (2011). A meta-analysis of the effectiveness of intervention programs designed to support autonomy. Educational Psychology Review, 23, 159–188. https://doi.org/10.1007/s10648-010-9142-7
221. Assor, A., Kaplan, H., Kanat-Maymon, Y., & Roth, G. (2005). Directly controlling teaching behaviors as predictors of poor motivation and engagement in girls and boys. The role of anger and anxiety. Learning and Instruction, 15, 397–413. https://doi.org/10.1016/j.learninstruc.2005.07.008
222. Hofferber, N., Basten, M., Großmann, N., & Wilde, M. (2016). The effects of autonomy-supportive and controlling teaching behaviour in biology lessons with primary and secondary experiences on students' intrinsic motivation and flow-experience. International Journal of Science Education, 38(13), 2114–2132. https://doi.org/10.1080/09500693.2016.1229074 .
223. Reeve, J. (2015). Understanding motivation and emotion (6th ed.). Wiley.
224. Ryan, R. M., & Deci, E. L. (2017). Self-determination theory - Basic psychological needs in motivation, development, and wellness. Guilford Press.
225. Budde, J. (2008). Jungen, Männlichkeit und Schule [Boys, masculinity, and school]. In Bildungsministerium für Bildung und Forschung (Ed.), Bildungs(miss)erfolge von Jungen und Berufswahlverhalten bei Jungen/männlichen Jugendlichen (pp. 39–45). Bonn, Berlin, Germany. https://doi.org/10.1007/978-3-531-91396-4_25
226. Dietze, J., Gehlhaar, K.-H., & Klepel, G. (2005). Untersuchungen zum Entwicklungsstand von Biologieinteressen bei Schülerinnen und Schülern der Sekundarstufe II [Investigations on the state of development of student interest in biology at the secondary level]. In R. Klee, A. Sandmann, & H. Vogt (Eds.), Lehr- und Lernforschung in der Biologiedidaktik (pp. 133–145). Studien Verlag.
227. Hannover, B., & Kessels, U. (2002). In J. Doll & M. Prenzel (Eds.), Bildungsqualität von Schule: Schulische und außerschulische Bedingungen mathematischer, naturwissenschaftlicher und überfachlicher Kompetenzen (pp. 341–358). Beltz.
228. Kahle, J. B., Parker, L. H., Rennie, L. J., & Riley, D. (1993). Gender differences in science education. Building a Model. Educational Psychologist, 28(4), 379–404. https://doi.org/10.1207/s15326985ep2804_6

229. Quaiser-Pohl, C. (2012). Mädchen und Frauen in MINT: Ein Überblick [Girls and women in STEM: An overview]. In H. Stöger, A. Ziegler, & M. Heilemann (Eds.), Mädchen und Frauen in MINT: Bedingungen von Geschlechtsunterschieden und Interventionsmöglichkeiten (pp. 13–40). Berlin, Germany: LIT.
230. Desch, I., Basten, M., Großmann, N., & Wilde, M. (2017). Geschlechterdifferenzen in der wahrgenommenen Erfüllung der Prozessmerkmale gemäßigt konstruktivistischer Lernumgebungen – Die Effekte von Autonomieförderung durch Schülerwahl [Gender differences in the perception of moderate constructivist learning environments - The effects of autonomy support by means of students' choice]. Journal for Educational Research Online, 9(2), 156–182. https://doi.org/10.25656/01:14933 .
231. Lietaert, S., Roorda, D., Laevers, F., Verschueren, K., & De Fraine, B. (2015). The gender gap in student engagement: The role of teachers' autonomy support, structure, and involvement. British Journal of Educational Psychology, 85, 498–518. https://doi.org/10.1111/bjep.12095
232. Patall, E. A., Steingut, R. R., Freeman, J. L., Pituch, K. A., & Vasquez, A. C. (2018). Gender disparities in students' motivational experiences in high school science classroom. Science Education, 102, 951–957. https://doi.org/10.1002/sce.21461
233. Lüftenegger, M., Schober, B., Finsterwald, M., Wagner, P., & Spiel, C. (2011). Wahrgenommene Autonomie und Feedback im Unterricht als Determinanten von Motivation und Leistung in der Schule [Perceived autonomy and feedback as determinants of motivation and performance at school]. In M. Dresel & L. Lämmle (Eds.), Motivation, Selbstregulation und Leistungsexzellenz (pp. 199–217). LIT.
234. Schweder, S., & Raufelder, D. (2021). Needs satisfaction and motivation among adolescent boys and girls during self-directed learning intervention. Journal of Adolescence, 88, 1–13. https://doi.org/10.1016/j.adolescence.2021.01.007
235. Vallerand, R. J., & Ratelle, C. F. (2002). Intrinsic and extrinsic motivation. A hierarchical model. In E. L. Deci, & R. M. Ryan (Eds.), Handbook of self-determination research (pp. 37–63). Rochester, NY: University of Rochester Press.
236. Taylor, C. M., Schepers, J., & Crous, F. (2006). Locus of control in relation to flow. Journal of Institutional Psychology, 32(3), 63–71. https://doi.org/10.4102/sajip.v32i3.438
237. Froiland, J. M., Davison, M. L., & Worrell, F. C. (2017). Aloha teachers: Teacher autonomy support promotes Native Hawaiian and Pacific Islander students' motivation, school belonging, course-taking and math achievement. Social Psychology of Education, 19(4), 879–894. https://doi.org/10.1007/s11218-016-9355-9
238. Haerens, L., Aelterman, N., Vansteenkiste, M., Soenens, B., & Van Petegem, S. (2015). Do perceived autonomy-supportive and controlling teaching relate to physical education students' motivational experiences through unique pathways? Distinguishing between the bright and dark side of motivation. Psychology of Sport and Exercise, 16(3), 26–36. https://doi.org/10.1016/j.psychsport.2014.08.013
239. Tessier, D., Sarrazin, P., & Ntoumanis, N. (2010). The effect of an intervention to improve newly qualified teachers' interpersonal style, students' motivation and psychological need satisfaction in sport-based physical education. Contemporary Educational Psychology, 35, 242–253. https://doi.org/10.1016/j.cedpsych.2010.05.005
240. De Meyer, J., Tallir, I. B., Soenens, B., Vansteenkiste, M., Aelterman, N., Van den Berghe, L., et al. (2014). Does observed controlling teaching behavior relate to students' motivation in physical education? Journal of Educational Psychology, 106(2), 541–554. https://doi.org/10.1037/a0034399
241. Guggenbühl, A. (2008). Die Schule - ein weibliches Biotop? [School - a feminine biotope?]. In M. Matzner & W. Tischner (Eds.), Handbuch Jungenpädagogik (pp. 150–167). Beltz.
242. Lindsey, E. W., Mize, J., & Pettit, G. S. (1997). Differential play patterns of mothers and fathers of sons and daughters. Implications for children's gender role development. Sex Roles, 37(9–10), 643–661. https://doi.org/10.1007/BF02936333
243. Maccoby, E. E. (1998). The two sexes: Growing up apart, coming together. Harvard University Press.

244. Reeve, J., Nix, G., & Hamm, D. (2003). Testing models of the experience of self-determination in intrinsic motivation and the conundrum of choice. Journal of Educational Psychology, 95(2), 375–392. https://doi.org/10.1037/0022-0663.95.2.375
245. Pintrich PR, De Groot EV. Motivational and self-regulated learning component of classroom academic performance. J Educ Psychol. 1990;82:33–40.
246. Barbuto, J. E., & Scholl, R. W. (1998). Motivation sources inventory: development and validation of new scales to measure an integrative taxonomy of motivation. Psychological Reports, 82, 1011–1022.
247. Nicholls, J. G. (1990). What is ability and why are we mindful of it? A developmental perspective. In R. Sternberg & J. Kolligian (Eds.), Competence considered (pp. 11–40). New Haven: Yale University Press.
248. Sideridis, G. D., & Mouratidis, A. (2008). Forced choice versus open-ended assessments of goal orientations: a descriptive study. International Review of Social Psychology, 1, 217–246.
249. Wolters, C. A., Pintrich, P. R., & Karabenick, S. A. (2005). Assessing academic self-regulated learning. In L. H. Lippman, K. A. Moore, L. Guzman, G., Matthews, G., & J. Hamilton (2005). What Do Children Need to Flourish? (pp. 251–270). Springer.
250. Butler, D. L., & Cartier, S. C. (2004). Promoting effective task interpretation as an important work habit: a key to successful teaching and learning. Teachers College Record, 106, 1729–1758.
251. Kaplan, A., Lichtinger, E., & Margulis, M. (2011). The situated dynamics of purposes of engagement and self-regulation strategies: A mixed-methods case study of writing. Teachers College Record, 113, 284–324.
252. Gureasko-Moore, S., DuPaul, G., & White, G. (2007). Self management of classroom preparedness and homework: effects on school functioning of adolescents with attention deficit hyperactivity disorder. School Psychology Review, 36, 647–664.
253. Sideridis, G. D. (2005b). Classroom goal structures and hopelessness as predictors of day-to-day experience at school: differences between students with and without learning disabilities. International Journal of Educational Research, 43, 308–328.
254. Zimmerman, B. J. (2008). Investigating self-regulation and motivation: historical background, methodological developments, and future prospects. American Educational Research Journal, 45(1), 166–183.
255. Switzky, H. N. (2006). The importance of cognitive–motivational variables in understanding the outcome performance of persons with mental retardation: a personal view from the early twenty first century. International Review of Research in Mental Retardation, 31, 1–29.
256. Fulk, B. M., Brigham, F. J., & Lohman, D. A. (1998). Motivation and self-regulation: a comparison of students with learning and behavior problems. Remedial and Special Education, 19, 300–309.
257. Garcıa-Sanchez, J. N., & de Caso-Fuertes, A. M. (2005). Comparison of the effects on writing attitudes and writing self-efficacy of three different training programs in students with learning disabilities. International Journal of Educational Research, 43, 272–289.
258. Graham, S., Harris, K. R., & McKeown, D. (2013). The writing of students with learning disabilities: Meta-analysis of self-regulated strategy development writing intervention studies, and future directions. In H. L. Swanson, K. R. Harris & S. Graham (Eds.). Handbook of learning disabilities (pp. 405–438). Guilford Press.
259. Butler, D. L., Elaschuk, C. L., & Poole, S. (2000). Promoting strategic writing by postsecondary students with learning disabilities: a report of three case studies. Learning Disability Quarterly, 23(3), 196–213.
260. Winne, P. H., & Hadwin, A. F. (1998). Studying as self-regulated learning. In D. J. Hacker, J. Dunlosky, & A. C. Graesser (Eds.), Metacognition in educational theory and practice (pp. 277–304). Mahwah: Erlbaum.
261. Johnson, A., Azevedo, R., & D'Mello, S. K. (2011). The temporal and dynamic nature of self-regulatory processes during independent and externally assisted hypermedia learning. Cognition and Instruction, 29, 471–504.

262. Lichtinger, E., & Kaplan, A. (2011). The role of purpose of engagement in self-regulation. New Directions for Teaching and Learning, 126, 9–19.
263. O'Quin, M. (2020). Understanding the similarities and differences of academic aspirations between regular and specially admitted student-athletes (Publication No. 28000670) [Doctoral Dissertation, California State University, Fresno]. ProQuest Dissertations & Theses Global.
264. Feltz, D. L., Schneider, R., Hwang, S., & Skogsberg, N. J. (2013). Predictors of collegiate student-athletes' susceptibility to stereotype threat. Journal of College Student Development, 54(2), 184–201. https://doi.org/10.1353/csd.2013.0014
265. Yopyk, D. J. A., & Prentice, D. A. (2005). Am I an athlete or a student? Identity salience and stereotype threat in student-athletes. Basic and Applied Social Psychology, 27(4), 329–336. https://doi.org/10.1207/s15324834basp2704_5
266. Cooper, H., Hedges, L. V., & Valentine, J. C. (Eds.). (2019). The handbook of research synthesis and meta-analysis (3rd ed.). Russell Sage Foundation.
267. Lombardi, A., Conley Downs, R., Downs, A., & Conley, D. (2012). Development of a college readiness screening measure for student athlete recruits. Journal of Issues in Intercollegiate Athletics, 5, 64–78. https://bit.ly/3lYEe10
268. Carodine, K., Almond, K. F., & Grotto, K. K. (2001). College student athlete success both in and out of the classroom. New Direction for Student Services, 93, 19–33. https://doi.org/10.1002/ss.2
269. Gaston Gayles, J., & Hu, S. (2009). The influence of student engagement and sport participation on college outcomes among division I student athletes. The Journal of Higher Education, 80(3), 315–333. https://doi.org/10.1080/00221546.2009.11779015
270. Umbach, P. D., Palmer, M. M., Kuh, G. D., & Hannah, S. J. (2006). Intercollegiate athletes and effective educational practices: Winning combination or losing effort? Research in Higher Education, 47(6), 709–733. https://doi.org/10.1007/s11162-006-9012-9
271. Froiland, J. M. (2011). Parental autonomy support and student learning goals: A preliminary examination of an intrinsic motivation intervention. Child & Youth Care Forum,40(2), 135–149.
272. Froiland, J. M. (2014). Inspired childhood: Parents raising motivated, happy, and successful students from preschool to college. Seattle, WA: Amazon.
273. Barrett, L. F., & Russell, J. A. (1999). The structure of current affect: Controversies and emerging consensus. Current Directions in Psychological Science, 8(1), 10–14.
274. Tellegen, A., Watson, D., & Clark, L. A. (1999). On the dimensional and hierarchical structure of affect. Psychological Science, 10, 297–309.
275. Pintrich, P. R., Marx, R., & Boyle, R. (1993). Beyond cold conceptual change: The role of motivational beliefs and classroom contextual factors in the process of conceptual change. Review of Educational Research, 63, 167–199.
276. European Commission. (2016). A new skills agenda for Europe. Brussels, Belgium
277. Ministry of Education, University and Research, MIUR. (2018). Percorsi per le competenze trasversali e l'orientamento. Rome, Italy.
278. Pellegrino, J. W., & Hilton, M. L. (2012). Education for life and work: Developing transferable knowledge and skills in the 21st century. National Academy of Sciences.
279. Ben-Eliyahu, A. (2019). Academic emotional learning: A critical component of self-regulated learning in the emotional learning cycle. Educational Psychologist, 54, 84–105. https://doi.org/10.1080/00461520.2019.1582345
280. Feraco, T., Resnati, D., Fregonese, D., Spoto, A., & Meneghetti, C. (2021). Soft skills and extracurricular activities sustain motivation and self-regulated learning at school. The Journal of Experimental Education, 1–20. https://doi.org/10.1080/00220973.2021.1873090
281. Eccles, J. S. (1999). The development of children ages 6 to 14. The future of children, 9, 30–44. https://doi.org/10.2307/1602703
282. Voyer, D., & Jansen, P. (2017). Motor expertise and performance in spatial tasks: A meta-analysis. Human Movement Science, 54, 110–124. https://doi.org/10.1016/j.humov.2017.04.004

283. Gilman, R., & Huebner, E. S. (2006). Characteristics of adolescents who report very high life satisfaction. Journal of Youth and Adolescence, 35, 293–301. https://doi.org/10.1007/s10964-006-9036-7
284. Bruna, M. O., Brabete, A. C., & Izquierdo, J. M. A. (2019). Reliability generalization as a seal of quality of substantive meta-analyses: The case of the VIA Inventory of Strengths (VIA-IS) and their relationships to life satisfaction. Psychological Reports, 122, 1167–1188. https://doi.org/10.1177/0033294118779198
285. Weber, M., Wagner, L., & Ruch, W. (2016). Positive feelings at school: On the relationships between students' character strengths, school-related affect, and school functioning. Journal of Happiness Studies, 17, 341–355. https://doi.org/10.1007/s10902-014-9597-1
286. Diener, E., Emmons, R. A., Larsen, R. J., & Griffin, S. (1985). The satisfaction with life scale. Journal of Personality Assessment, 49, 71–75. https://doi.org/10.1207/s15327752jpa4901_13
287. Di Fabio, A., & Gori, A. (2016). Measuring adolescent life satisfaction: Psychometric properties of the Satisfaction With Life Scale in a sample of Italian adolescents and young adults. Journal of Psychoeducational Assessment, 34, 501–506. https://doi.org/10.1177/0734282915621223
288. Antaramian, S. (2017). The importance of very high life satisfaction for students' academic success. Cogent Education, 4, 1307622. https://doi.org/10.1080/2331186X.2017.1307622
289. Cronbach, L. J. (1955). Processes affecting scores on "understanding of others" and "assumed similarity." Psychological Bulletin, 52, 177–193. https://doi.org/10.1037/h0044919
290. Givvin, K. B., Stipek, D. J., Salmon, J. M., & MacGyvers, V. L. (2001). In the eyes of the beholder: Students' and teachers' judgments of students' motivation. Teaching and Teacher Education, 17(3), 321–331. https://doi.org/10.1016/S0742-051X(00)00060-3 .
291. Huber, S. A., & Seidel, T. (2018). Comparing teacher and student perspectives on the interplay of cognitive and motivational-affective student characteristics. PLoS One, 13(8), e0200609. https://doi.org/10.1371/journal.pone.02006009 .
292. Götz, L., Lingel, K., & Schneider, W. (2013). DEMAT 6+. Deutscher Mathematiktest für sechste Klassen [German mathematics test for the sixth grade]. Göttingen: Hogrefe.
293. Kim, ChanMin; Hodges, Charles B. Effects of an emotion control treatment on academic emotions, motivation and achievement in an online mathematics course. Instructional Science (2011). https://doi.org/10.1007/s11251-011-9165-6
294. Tomita, Kei Visual design as a holistic experience: how students' emotional responses to the visual design of instructional materials are formed. Educational technology research and development (2022). https://doi.org/10.1007/s11423-022-10088-x
295. Brumec, Snežana; Lavrič, Miran; Naterer, Andrej Examining Motivations to Walk the Camino de Santiago: A Typology of Pilgrims. Pastoral Psychology (2023). https://doi.org/10.1007/s11089-023-01071-1
296. Jarrell, Amanda; Lajoie, Susanne P.; Hall, Nathan C.; Horrocks, Phoenix T. M. Antecedents and Consequences of Emotion Regulation in STEM Degree Programs. Innovative Higher Education (2022). https://doi.org/10.1007/s10755-021-09587-1
297. Liang, Hsin-Yi; Kelsen, Brent Influence of Personality and Motivation on Oral Presentation Performance. Journal of Psycholinguistic Research (2018). https://doi.org/10.1007/s10936-017-9551-6
298. Vogler, Jane S.; Bakken, Linda Motivation across domains: Do goals and attributions change with subject matter for Grades 4 and 5 students?. Learning Environments Research (2007). https://doi.org/10.1007/s10984-007-9021-4
299. Gutentag, Tony; Tamir, Maya Putting Effort into Emotion Regulation: Manipulating Desirability and Motivational Strength. Affective Science (2022). https://doi.org/10.1007/s42761-022-00155-0
300. Kaczmarek, Lukasz D.; Behnke, Maciej; Enko, Jolanta; Kosakowski, Michał; Guzik, Przemysław; Hughes, Brian M. Splitting the affective atom: Divergence of valence and approach-avoidance motivation during a dynamic emotional experience. Current Psychology (2019). https://doi.org/10.1007/s12144-019-00264-3

301. Masters-Waage, Theodore C.; Reb, Jochen; Tov, William; Bandara, Udana An Initial Examination of State and Longitudinal Effects of Loving-Kindness Practice on Affective and Motivational States at Work. Mindfulness (2021). https://doi.org/10.1007/s12671-021-01781-8
302. Miragall, Marta; Vara, M. Dolores; Cebolla, Ausiàs; Etchemendy, Ernestina; Baños, Rosa M. Leaning forward to increase approach motivation! The role of joy, exercise, and posture in achieving goals. Current Psychology (2019). https://doi.org/10.1007/s12144-019-00175-3
303. Turner, Jeannine E.; Goodin, Joel B.; Lokey, Cathryn Exploring the Roles of Emotions, Motivations, Self-Efficacy, and Secondary Control Following Critical Unexpected Life Events. Journal of Adult Development (2012). https://doi.org/10.1007/s10804-012-9148-0
304. Park, Seung Won; Cho, Moon-Heum; Lim, Seongmi; Hwang, Seokwon Predicting students' negative emotions in college remedial mathematics courses. European Journal of Psychology of Education (2023). https://doi.org/10.1007/s10212-023-00697-2
305. Schubert, Sandra; Pekrun, Reinhard; Ufer, Stefan The role of epistemic emotions in undergraduate students' proof construction. ZDM – Mathematics Education (2022). https://doi.org/10.1007/s11858-022-01413-y
306. Litmanen, Topi; Lonka, Kirsti; Inkinen, Mikko; Lipponen, Lasse; Hakkarainen, Kai Capturing teacher students' emotional experiences in context: does inquiry-based learning make a difference?. Instructional Science (2012). https://doi.org/10.1007/s11251-011-9203-4
307. Hudig, Job; Scheepers, Ad W. A.; Schippers, Michaéla C.; Smeets, Guus Motivational mindsets, mindset churn and academic performance: The role of a goal-setting intervention and purpose in life. Current Psychology (2022). https://doi.org/10.1007/s12144-022-03462-8
308. Norouzi, Ali; Parmelee, Dean; Shariati, Mohammad; Norouzi, Saiideh; Alizadeh, Maryam The role of motivational components in metamotivational monitoring in medical students: a mixed method study. BMC Medical Education (2023). https://doi.org/10.1186/s12909-023-04081-y
309. Aldao, Amelia; Wisco, Blair E. Motivational conflict influences the timing of emotions and their regulation. Motivation and Emotion (2015). https://doi.org/10.1007/s11031-015-9496-8
310. Kaplan, Avi; Maehr, Martin L. The Contributions and Prospects of Goal Orientation Theory. Educational Psychology Review (2006). https://doi.org/10.1007/s10648-006-9012-5
311. Wang, Huanhuan; Lehman, James D. Using achievement goal-based personalized motivational feedback to enhance online learning. Educational technology research and development (2021). https://doi.org/10.1007/s11423-021-09940-3
312. Vongkulluksn, Vanessa W.; Matewos, Ananya M.; Sinatra, Gale M.; Marsh, Julie A. Motivational factors in makerspaces: a mixed methods study of elementary school students' situational interest, self-efficacy, and achievement emotions. International Journal of STEM Education (2018). https://doi.org/10.1186/s40594-018-0129-0
313. Los, Ryan; Schweinle, Amy The interaction between student motivation and the instructional environment on academic outcome: a hierarchical linear model. Social Psychology of Education (2019). https://doi.org/10.1007/s11218-019-09487-5
314. Großmann, Nadine; Hofferber, Natalia; Wilde, Matthias; Basten, Melanie Students' motivation in biology lessons—can student autonomy reduce the gender gap?. European Journal of Psychology of Education (2022). https://doi.org/10.1007/s10212-022-00604-1
315. Talley, Kimberly Grau; Martinez Ortiz, Araceli Women's interest development and motivations to persist as college students in STEM: a mixed methods analysis of views and voices from a Hispanic-Serving Institution. International Journal of STEM Education (2017). https://doi.org/10.1186/s40594-017-0059-2
316. Lichtinger, Einat; Kaplan, Avi Employing a case study approach to capture motivation and self-regulation of young students with learning disabilities in authentic educational contexts. Metacognition and Learning (2015). https://doi.org/10.1007/s11409-014-9131-1
317. García, Agustín J.; Fong, Carlton J.; Regalado, Yvette M. Motivational, Identity-Based, and Self-Regulatory Factors Associated with Academic Achievement of US Collegiate Student-Athletes: A Meta-Analytic Investigation. Educational Psychology Review (2023). https://doi.org/10.1007/s10648-023-09730-8

318. Kowalski, Monica J.; Froiland, John Mark Parent perceptions of elementary classroom management systems and their children's motivational and emotional responses. Social Psychology of Education (2020). https://doi.org/10.1007/s11218-020-09543-5
319. Linnenbrink, Elizabeth A. Emotion Research in Education: Theoretical and Methodological Perspectives on the Integration of Affect, Motivation, and Cognition. Educational Psychology Review (2006). https://doi.org/10.1007/s10648-006-9028-x
320. Feraco, Tommaso; Resnati, Dario; Fregonese, Davide; Spoto, Andrea; Meneghetti, Chiara An integrated model of school students' academic achievement and life satisfaction. Linking soft skills, extracurricular activities, self-regulated learning, motivation, and emotions. European Journal of Psychology of Education (2022). https://doi.org/10.1007/s10212-022-00601-4
321. Zhu, Chunjie; Urhahne, Detlef Temporal stability of teachers' judgment accuracy of students' motivation, emotion, and achievement. European Journal of Psychology of Education (2020). https://doi.org/10.1007/s10212-020-00480-7

Educational Motivation 3

Introduction by the Editor

Motivation has long been a subject of intense study in the field of education, as it is widely recognised as a critical factor in shaping students' academic achievement, engagement and overall learning experience. The field of motivation science, which draws on theories and research from a range of disciplines including psychology, neuroscience and education, has made significant impact on understanding the complexities of human motivation and its role in the educational context. At the core of motivation research in education is the fundamental question: What motivates individuals to engage in learning activities, persevere in the face of challenges, and ultimately achieve their academic goals? Researchers have proposed numerous theoretical frameworks and models to elucidate the intricate interplay of cognitive, affective and environmental factors that influence motivation.

Educational motivation is a critical factor influencing students' learning processes, academic achievement, and overall development. It encompasses the internal and external forces that drive students to engage in educational activities, persist in the face of challenges, and strive for academic success. This essay explores the various dimensions of educational motivation, examining theoretical frameworks, empirical evidence, and practical applications. By understanding the factors that influence motivation, educators, policymakers, and researchers can develop strategies to enhance student engagement and academic outcomes.

Educational Motivation

One of the most influential theories in this area is Self-Determination Theory (SDT), developed by psychologists Ryan and Deci (2017). SDT posits that individuals have three innate psychological needs: autonomy (the need for self-direction and volition), competence (the need to feel capable and effective), and relatedness (the need for belonging and meaningful relationships). According to SDT, when

M. S. Khine (ed.), *Motivation Science*, https://doi.org/10.1007/978-981-97-9247-4_3

these needs are met in the educational environment, students are more likely to experience intrinsic motivation, which refers to the inherent desire to engage in an activity for its own sake, driven by curiosity, enjoyment and personal fulfilment. Building on SDT, researchers have explored various teaching strategies and classroom practices that can foster intrinsic motivation in students. For example, providing students with a learning environment that supports choice and autonomy, fostering a mastery-oriented classroom climate, and promoting positive teacher-student relationships have all been shown to increase intrinsic motivation and engagement.

Intrinsic motivation is widely regarded as the most desirable form of motivation, as it is associated with deeper learning (Alamri et al., 2020). Intrinsically motivated students approach their studies with curiosity, embrace challenges as opportunities for growth, and derive satisfaction from the process of learning itself. Fostering intrinsic motivation requires creating learning environments that support student autonomy, offer optimal levels of challenge, and provide meaningful feedback (Ryan & Vansteenkist, 2023). Extrinsic motivation, on the other hand, encompasses external factors such as grades, rewards, or parental pressure. While extrinsic motivation can serve as a valuable motivator, particularly in the initial stages of learning, an overreliance on it can undermine intrinsic motivation and lead to diminished engagement, decreased creativity, and a narrow focus on performance rather than mastery (Howard et al., 2020). Extrinsic motivation refers to the pursuit of an activity due to external factors, such as rewards, punishments, or social pressures. Motivation science has delved into the complexities of extrinsic motivation, examining how different types of external motivators (e.g., grades, praise, rewards) can either undermine or support students' intrinsic motivation and overall engagement.

Apart from self-determination theory, several other frameworks have been proposed to understand the complexities of educational motivation. Eccles and Wigfield's (2020) expectancy-value theory suggests that students' motivation is influenced by their beliefs about their ability to succeed (expectancy) and the value they place on the task or subject matter. Dweck and Yeager's (2020) mindset theory suggests that students with a growth mindset, who believe that their abilities can be developed through effort, are more likely to be motivated and persevere in the face of challenges than those with a fixed mindset. In addition, social cognitive theories, such as Bandura's (1997) theory of self-efficacy, highlight the role of self-belief and self-regulation in shaping motivation. Students with high self-efficacy beliefs, or confidence in their abilities, are more likely to set challenging goals, persevere in the face of obstacles, and demonstrate greater motivation and academic achievement (Schunk & DiBenedetto, 2020).

To foster motivation, educational practices should consider various theoretical perspectives and create supportive learning environments. This can be achieved by promoting autonomy through choices and self-directed learning, encouraging a growth mindset with challenging yet achievable tasks, building self-efficacy with constructive feedback, and adopting a mastery-oriented approach to learning.

Additionally, it's essential to acknowledge the influence of socio-cultural and contextual factors on educational motivation. Students from diverse backgrounds may encounter unique motivational obstacles, such as stereotype threat, limited role models, or restricted access to resources. To ensure equal opportunities for all students to cultivate and maintain motivation, it is crucial to address these challenges through culturally responsive teaching methods, inclusive curricula, and targeted interventions. Fostering motivation in education has far-reaching consequences that extend beyond academic success. Motivated learners develop resilience, embrace lifelong learning, and cultivate a growth mindset that transcends formal education. This enables them to navigate the complexities of a rapidly changing world, adapt to new challenges, and pursue personal and professional fulfillment throughout their lives.

Some of the examples of the articles included in this section are the psychometric evaluation of a new internalization of learning motivation (ILM) scale, motivational predictors of successful transition from grades 11 to 12 in South Africa, a study of motivational influences on academic achievement, feedback-seeking behavior in second language writing: motivational mechanisms, and motivational predictors of struggling readers' reading comprehension: the effects of mindset, achievement goals, and engagement. This section also includes research reports related to the role of students' beliefs, parents' educational level, and the mediating role of attitude and motivation in students' mathematics achievement, study of the relationship between familial factors and academic motivation and achievement in high school students, and motivation and parent and teacher effects on academic coping and classroom engagement: testing a motivational model.

References

Alamri, H., Lowell, V., Watson, W., & Watson, S. L. (2020). Using personalized learning as an instructional approach to motivate learners in online higher education: Learner self-determination and intrinsic motivation. *Journal of Research on Technology in Education, 52*(3), 322–352.

Bandura, A. (1997). Self-efficacy: The exercise of control. W.H. Freeman and Company.

Dweck, C., & Yeager, D. (2020). A growth mindset about intelligence. *Handbook of wise interventions: How social psychology can help people change*, 9–35.

Eccles, J. S., & Wigfield, A. (2020). From expectancy-value theory to situated expectancy-value theory: A developmental, social cognitive, and sociocultural perspective on motivation. Contemporary Educational Psychology, 61, 101859.

Howard, J. L., Chong, J. X., & Bureau, J. S. (2020). The tripartite model of intrinsic motivation in education: A 30-year retrospective and meta-analysis. *Journal of Personality, 88*(6), 1268–1285.

Ryan, R. M., & Deci, E. L. (2017). *Self-determination theory: Basic psychological needs in motivation, development, and wellness.* New York: Guilford Publishing.

Ryan, R. M., & Vansteenkiste, M. (2023). Self-determination theory. In *The Oxford Handbook of Self-Determination Theory* (pp. 3–30). Oxford University Press.

Schunk, D. H., & DiBenedetto, M. K. (2020). Motivation and social cognitive theory. Contemporary Educational Psychology, 60, 101832.

Machine Generated Summaries

Disclaimer: The summaries in this chapter were generated from Springer Nature publications using extractive AI auto-summarization: An extraction-based summarizer aims to identify the most important sentences of a text using an algorithm and uses those original sentences to create the auto-summary (unlike generative AI). As the constituted sentences are machine selected, they may not fully reflect the body of the work, so we strongly advise that the original content is read and cited. The auto generated summaries were curated by the editor to meet Springer Nature publication standards. To cite this content, please refer to the original papers.

Machine generated keywords: student, language, goal, academic, mastery, achievement, education, achievement goal, science, teacher, learner, selfdetermination, intrinsic, dweck, school.

Psychometric Evaluation of a New Internalization of Learning Motivation (ILM) Scale [267]

This is a machine-generated summary of:

Wang, Yurou; Wind, Stefanie A.: Psychometric evaluation of a new internalization of learning motivation (ILM) scale [267]

Published in: Current Psychology (2020).

Link to original: https://doi.org/10.1007/s12144-020-00909-8.

If you want to cite the papers, please refer to the original.

For technical reasons we could not place the page where the original quote is coming from.

Abstract-Summary
"Internalization is an important component of motivation and self-determination."

"This article presents a new Internalization of Learning Motivation (ILM) Scale based on multiple motivational forces."

"Of four dimensions according to the self-determination theory, five dimensions were identified: lack of internalization (INL), partial ego internalization (INPE), partial guilt internalization (INPG), identified internalization (INID), and integrated internalization (ININ)."

"The results indicated that ILM items worked together to form a continuum that represents internalization."

"The ILM Scale can distinguish between participants with different levels of internalization and provide a deeper understanding of how specific motivation forces facilitate internalization of learning motivation."

"Educators and researchers can use this scale to identify students' level of internalization in order to inform appropriate interventions."

Conceptual Framework: Stages of Internalization

"The conceptual framework for this study is based on self-determination theory, with a specific focus on motivation towards learning tasks."

"Researchers have frequently used self-determination theory as a framework for examining students' learning motivation."

"We use self-determination as a framework through which to consider the development of internalization."

"Self-determination theory divides motivation into two main categories: (1) intrinsic motivation, and (2) extrinsic motivation."

"We present four hypothesized stages in the development of internalization toward intrinsic motivation; from this perspective, all of the internalization stages reflect extrinsic motivation."

Lack of Internalization

"On the non-self-determined end of the continuum, people are motivated to engage in a task almost entirely for the purpose of gaining rewards or avoiding punishment; enjoyment or interest in the task is absent in this level of regulation."

"Even though external forces are the primary motivating factors at this stage, learning action or behavior is starting to happen."

"In this stage, people rely on commands, rewards, or punishment to motivate them to conduct certain behaviors."

Partial Internalization

"In this stage, individuals begin to associate environmental information (e.g., cultural value, social expectation, family legacy) with some internal meaning or value."

"People partially accept the value of the activity; they perform a certain behavior to avoid guilt, shame or the probability of being looked down upon by others."

"Individuals at the partial internalization stage agree that they should do certain things when other people ask them to do so, but they do not understand why."

Identified Internalization

"Identified internalization is the third stage on the internalization continuum."

"In this stage, using environmental information, individuals start to perceive learning tasks as self-relevant and having personal importance, so they are internally motivated to finish tasks."

"Individuals in this stage agree that what they do is important and meaningful, but they do not necessarily find completing tasks to be joyful or exciting."

"They understand that what the external regulations (e.g., teachers' requirement or parents' expectation) are asking for is important and valuable."

Integrated Internalization

"The fullest stage of internalization is integrated internalization."

"Internalization researchers are convinced that if children are learning for their own sake—that is, out of pure interest and fascination for the learning content itself—they will process the learning material more deeply, be more persistent when facing obstacles, and eventually obtain better grades (Ryan and Deci [1])."

"We considered this stage to reflect the fullest form of internalization."

"According to self-determination theory, in order to form an ideal type of internalization, or move students from lower stage of internalization (i.e. lack of internalization) to a higher stage (i.e. identified internalization), students' needs for autonomy, competence, and relatedness should be satisfied (Ryan and Deci [2, 3])."

Measuring Internalization

"No studies have been published in which researchers have measured internalization using indicators of different motivational forces."

"Motivation measurements include unidimensional scales, such as the Perception of Ability Scale for Students (Boersma and Chapman [4]), and multi-dimensional approaches, such as the School Motivation Questionnaire (Marsh and others [5]), which includes integrated achievement goals and intrinsic motivation."

"Researchers have developed numerous tools to measure motivation in specific academic domains, such as reading (e.g., the Reading Motivation Questionnaire; Peklaj and Bucik [6]), math (e.g., the Motivation Orientation Scales; Nicholls and others [7]), and sport (Duda and Nicholls [8])."

"To measuring motivation is to assess the level of autonomy associated with different kinds of motivation."

"From a review of the assessment of motivation, there is strong evidence for a continuum of autonomy characterized by different levels of regulation."

"The currently available assessments primarily reflect autonomy, and do not specifically focus on measuring the motivational force of internalization."

"But not the least, the current motivation measurements reflect an emphasis on intrinsic motivation."

Purpose

"Motivation can be formed and developed over time when individuals are provided with the right kind of motivational force."

"We operate under the assumption that with the right guidance, individuals can develop proximal motivation."

"The purpose of this study is to present the Internalization of Learning Motivation (ILM) scale for measuring students' internalization as a continuum consisting of four components: (1) lack of internalization (INL); (2) partial internalization (INP); (3) identified internalization (INID); and (4) integrated internalization (ININ)."

"To what extent can internalization be measured as a single continuum consisting of INL, INP, INID, and ININ?"

Method

"Our main analytic technique involved applying a modern measurement theory model to the item responses in order to evaluate the degree to which they provide measures of participant locations on a linear continuum that reflects the internalization construct."

"Our analyses focused mainly on estimates of participant, item, and component locations on the logit scale that represented internalization, along with indicators of fit specific to each of these facets."

"In keeping with the traditional approach to evaluating psychometric properties within the framework of Rasch measurement theory, we used two popular residual fit statistics to evaluate the degree to which the responses associated with each participant, item, and component matched what would be expected if data fit the PC-MFR model perfectly: Infit mean square error (MSE) and outfit MSE."

"Model-expected responses are calculated for each participant-item-component combination using the location estimates."

Results

"The five-factor model was tested in which INL has 7 items, INPE has 6 items, INPG has 4 items, INID has 6 items, and ININ has 10 items, so 32 items in total after removing the items have large modification indexes."

"We observed relatively high reliability of separation statistics for participants: 0.81, items: 0.99, and components > 0.99; these statistics summarize the degree to which individual elements within facets (i.e., individual participants, items, and components) have distinct locations on the linear continuum that represents the internalization construct, and values that approach 1.00 provide evidence of distinct estimates."

"The first column shows the linear continuum on which we estimated locations for each participant, item, and component; this scale represents internalization, where higher numbers represent more internalization, and lower numbers represent less internalization."

Discussion

"Results from analysis of the ILM items with a Rasch measurement theory model supported the presence of a continuum that reflects ordered components of internalization of learning motivation."

"Studies, researchers should examine the psychometric properties of the ILM, including the ordering of the internalization components, with samples that reflect other populations."

"Better understanding of the underlying forces that drive students' motivation and the stage of internalization could help parents, educations, and researchers to study the association between motivation and academic performance more thoroughly."

"Practice, the ILM scale can help school and teachers to identify characteristics of learning environments that can contribute to students' internalization of learning motivation (e.g., identified internalization or integrated internalization) and provide insight into characteristics of family structures that can promote an integrated view of learning."

[Section 11]

"Self-determination is one of the dominant theories in motivation studies, and internalization is a crucial component of self-determination theory that represents the process of transforming external learning regulation into internal regulation (Deci and others [9])."

"Current measures of internalization either fail to capture one or more of the four stages or focus on other motivational dimensions such as level of autonomy (Ryan and Deci [2, 3])."

"We develop and test a new measure of internalization of learning motivation designed to empirically examine this hypothesized continuum of internalization."

Motivational Predictors of Successful Transition from Grades 11 to 12 in South Africa [268]

This is a machine-generated summary of:

Grobler, Adelene A.; Knight, Mirriam R.; Lens, Willy; Lacante, Marlies: Motivational predictors of successful transition from grades 11 to 12 in South Africa [268].

Published in: European Journal of Psychology of Education (2014).

Link to original: https://doi.org/10.1007/s10212-014-0223-8.

If you want to cite the papers, please refer to the original.

For technical reasons we could not place the page where the original quote is coming from.

Abstract-Summary
"This study was carried out to explore whether, aside from gender and school environment, learners' motivation can serve as a predictor of the (un)successful transition of learners from grades 11 to 12 in South Africa."

"A total of 1,355 grade 11 learners from previously advantaged, transition, and previously disadvantaged school environments in the Free State's Motheo district participated in the study."

"Results showed that the number of unsuccessful learners was much higher among boys and in previously disadvantaged schools."

"We did, however, find that successful transition is associated with higher levels of learner motivation, a more positive attitude towards learning and lower levels of educational anxiety."

"In comparison with their unsuccessful peers, learners who were successful in the transition to grade 12 were significantly more autonomously motivated and showed significantly less controlled motivation."

Introduction
"Numerous authors have indicated that academic achievement, student motivation, and school environment play a significant role in learners continuing their education or dropping out (Battin-Pearson and others [10]; Harde and Reeve [11]; Thornburgh [12]; Vallerand and others [13]; Wasonga and others [14])."

"In cases where academic achievement was not the only significant predictor of high school dropout, it remained an important mediator of other independent factors such as perceived incompetence, lack of motivation, negative attitude towards education, educational anxiety, lack of future goals, etc. (Battin-Pearson and others [10])."

"The learners' level of motivation has significant implications with regard to the learners' dropping out or persisting in high school (Vallerand and others [13])."

"SDT attaches more importance to the quality than to the quantity of motivation: intrinsic goals and autonomous motivation or behavioural regulation create a high quality of motivation whereas extrinsic goals and controlled motivation are of low quality."

The Present Study

"Although various studies on high school dropout have been carried out in various parts of the world, little has been done to investigate the transition from grades 11 to 12 with a specific focus on reasons for drop out at this particularly late stage (i.e. when the learner is very close to completing secondary education)."

"Regarding the quality of students' motivation, it can be hypothesised that the type of academic goals or aspirations (i.e. intrinsic versus extrinsic goals) learners pursue and the autonomous versus controlled nature of their motivation to study will be related to persisting versus dropping out of high school (Bukatko and Kaehler [15]; Vansteenkiste and others [16])."

"The final school environment described in this study refers to previously advantaged schools, which continue to serve a predominantly advantaged group of learners (Msila [17])."

Method

"For the purposes of the present study, only the motivation-related subscales (i.e. motivation, attitude, and anxiety) were explored as predictors of students' successful or unsuccessful transition from grades 11 to 12."

"The quality of students' motivation/behavioural regulation was measured with a questionnaire listing 16 reasons for continuing their education based on research by Ryan and Connell [18]."

"The questionnaire measures the importance of four different types of motivation or underlying reasons for behaviour—external reasons (e.g. because others force me to do so, $\alpha = 0.75$); introjected reasons (e.g. because I would feel guilty if I did not do so, $\alpha = 0.75$); identified reasons (e.g. because I find it personally valuable, $\alpha = 0.78$); intrinsic motivation (e.g. because studying is an exciting activity, $\alpha = 0.86$)."

Results

"The disadvantaged mixed ethnicity schools did not differ from the advantaged schools but scored significantly lower (this implies a higher level of anxiety) than the transition schools and the disadvantaged black schools."

"The two types of disadvantaged environments did not differ from each other and scored significantly higher than the advantaged schools and the transition schools, which did not differ from each other."

"For intrinsic motivation ($F(3, 1{,}312) = 78.45$; $p < 0.000$; $\eta_p^2 = 0.152$), the advantaged schools scored significantly and unexpectedly lower and the disadvantaged black schools significantly higher than the other three school types."

"When external and introjected motivation were combined in a score for controlled motivation, a significant main effect of school environment ($F(3, 1{,}257) = 44.68$; $p < 0.000$; $\eta_p^2 = 0.096$) was found."

Discussion and Conclusion

"The aim of this study was to explore the predictive value of motivational variables for the successfulness of South African learners' transition from grades 11 to 12 within different school environments."

"Schools and family environments can have strong positive effects on students' motivation by promoting intrinsic goals and by creating an autonomy-supportive, non-controlling environment."

"Evidence for the importance of students' motivation as a positive predictor (hypothetical "determinant") of successful transition was also found in the LASSI results."

"The above results are in agreement with other studies, suggesting that also in poorer and less advantaged communities, motivational aspirations, and the motivation to study contribute significantly to learners' decisions to persist with educational pursuits in the latter stages of high school learning (Bridgeland and others [19])."

"SDT advocates the fostering of interest in learning and the value of education by inducing intrinsic rather than extrinsic goals and by supporting students' autonomous motivation and reducing their controlled motivation."

Limitations and Suggestions for Future Research

"Although the sample included in this study was representative of the general learner population in the Motheo District, the results cannot be generalised to the entire learner population in South Africa, particularly due to the tremendous cultural and socioeconomic diversities applicable to the nation as a whole."

"In previous studies (Jimerson and others [20]), poor academic achievement might be an early indicator of unsuccessful transition and dropout rather than the root cause itself."

"The results of this study are supportive of the results obtained by Suh and others [21], who indicate that this has implications for counselling in the following manner."

"Continuous support for learners should be provided throughout the school year, so that indications of possible drop out might be recognised earlier on and addressed in a preventative manner."

A Study of Motivational Influences on Academic Achievement [269]

This is a machine-generated summary of:

Dishon-Berkovits, Miriam: A study of motivational influences on academic achievement [269].

Published in: Social Psychology of Education (2014).

Link to original: https://doi.org/10.1007/s11218-014-9257-7.

If you want to cite the papers, please refer to the original.

For technical reasons we could not place the page where the original quote is coming from.

Abstract-Summary
"The theoretical rationale for the study is grounded in bridging two influential yet isolated literatures of motivation: goal setting theory and achievement goal theory."

"Although it is clear that academic performance should be influenced by assigned learning goals, surprisingly, goal setting theory has rarely been used as a theoretical framework to further our understanding of academic performance."

"Internal achievement goals were assessed and academic performance was measured."

"The combination of internal mastery achievement goals and performance-approve achievement goals predicted academic performance, thereby supporting a multiple achievement goals perspective."

Introduction
"Several studies revealed that the highest performance outcomes on complex tasks are attained by individuals who were provided with high assigned learning goals."

"The first research question explored in the current study is whether providing students with assigned learning goals would result in higher academic achievement than providing students with assigned performance goals."

"Based on the dozen studies published in peer-reviewed journals which describe the differential effects of assigned learning compared to assigned performance goals in complex laboratory tasks, and based on Brown and Latham [22], it is predicted that students who are provided with high and specific assigned learning goals will perform academically better, i.e., score a higher grade on the final comprehensive exam in the course, than students who are provided with high and specific assigned performance goals."

Methods

"In the first weeks of classes, 49 students in one section of the course were provided with distal assigned performance goals."

"The scale assessing performance-approach achievement goals consisted of 6 items (example: "it is important to me to do better than other students")."

"The scale assessing performance-avoidance achievement goals consisted of 6 items (example: "I often think to myself, 'What if I do badly in this class'?")."

"The scale assessing mastery achievement goals consisted of 6 items (example: "I want to learn as much as possible from this class")."

"Academic achievement in the course was assessed via the grade on the final examination in the course. (In social studies in colleges and universities in Israel, there is typically only one examination in a course which is comprehensive and takes place at the end of the semester)."

Results

"Providing students with assigned learning goals led to higher academic achievement."

"The second research question explored what are the best possible internal achievement goals with which students come to the classroom that would positively influence academic performance."

"The final research question explored whether to enhance academic achievement, the optimal goal a teacher should assign in a classroom depends on students' internal achievement goals."

"The findings reveal that a type of assigned goal did not interact with any of the internal achievement goals to reveal significant effects on academic performance."

"There was no benefit for students' grades for approaching a situation with mastery achievement goals when assigned learning goals were provided, nor was an advantage found for endorsing performance-approve achievement goals when assigned performance goals were given."

Discussion
"The finding that a specific, challenging assigned learning goal led to higher academic achievement than a specific challenging assigned performance goal is in line with the results of the few recent studies in the goal setting literature that compare the effects of assigned learning and assigned performance goals on performance outcomes (e.g., Brown and Latham [22]; Seijts and others [23])."

"The findings of the current study also show that endorsing mastery achievement goals along with performance-approve achievement goals results in high academic achievement."

"In light of current findings, one is led to conclude that higher academic achievement would be better facilitated by attempts by teachers, parents and educators to provide students with assigned learning rather than assigned performance goals."

"Results of the current study indicate that academic performance is situationally malleable and teachers can actively influence their students' academic achievement by providing them with assigned learning goals to ensure their academic success."

Feedback-Seeking Behavior in Second Language Writing: Motivational Mechanisms [270]

This is a machine-generated summary of:

Papi, Mostafa; Bondarenko, Anna Vitalyevna; Wawire, Brenda; Jiang, Chen; Zhou, Shiyao: Feedback-seeking behavior in second language writing: motivational mechanisms [270].

Published in: Reading and Writing (2019).

Link to original: https://doi.org/10.1007/s11145-019-09971-6.

If you want to cite the papers, please refer to the original.

For technical reasons we could not place the page where the original quote is coming from.

Abstract-Summary
"Focused on the effects of different type of feedback on learners' written products, research on written corrective feedback (WCF) has cast second language writers as passive recipients rather than proactive agents in the feedback process."

"Multiple regression and mediation analyses showed that a growth language mindset predicted the value of feedback, which, in turn, was a strong predictor of both feedback monitoring and feedback inquiry."

"A fixed language mindset, on the other hand, predicted the cost of feedback seeking, which, in turn, negatively predicted feedback monitoring."

Introduction

"These motivational characteristics have been argued to exert their effects on learners' choice of feedback-seeking strategies through different meaning systems that influence individuals' perceptions of the costs and values associated with various feedback-seeking strategies."

"According to this framework, learners' mindsets exert their effects on learners' choice of feedback-seeking strategies through different meaning systems that influence individuals' internal analysis of the cost and value associated with various feedback-seeking strategies."

"Since the introduction of the model of FSB, many studies have been conducted to examine the effects of different motivational and contextual factors on the perceived value and cost of feedback-seeking strategies, which, in turn, affect the individual's method, frequency, timing, source, and topic of FSB (see Anseel and others [24]; Ashford [25]; Ashford, De Stobbeleir, & Nujella [26])."

"To mindsets and FSB, in the present study the perceived self-presentation cost and value of FSB are measured to examine whether they mediate the relationships between mindsets and different feedback-seeking strategies."

Methods

"The first part included items measuring the participants' language mindsets, feedback-seeking strategies, and value and self-presentation cost of feedback seeking."

"To examine participants' perceptions of the value and self-presentation cost of feedback seeking, Ashford's [25] risk and value scales were adapted to the L2 writing context:Value: eight items measuring the amount of value learners associated with the feedback they receive from their teachers;Self-Presentation Cost: 10 items examining learners' perception of the risk of face-loss and embarrassment associated with seeking WCF."

"The first factor explained 53.7% of the variance (eigenvalue = 7.36) and included seven items measuring the participants' use of Feedback Monitoring (Cronbach's alpha = 0.92); the second factor, labeled Feedback Inquiry (Cronbach's alpha = 0.89), explained 7.9% of the variance (eigenvalue = 1.40) and included six items examining the participants' frequency of directly asking their teachers for WCF."

Results and Discussion

"When Cost was added to the base model, it emerged as a strong and negative predictor of Feedback Monitoring ($\beta = -0.43$, $p < 0.001$) and the beta value for Fixed Mindset approached statistical significance ($\beta = 0.17$, $p = 0.09$), suggesting mediation."

"These results confirm that learners with a growth mindset seek WCF by method of monitoring because they value feedback as a source to develop their L2 writing competence."

"Learners with a fixed mindset do not monitor feedback on their writing because of the low value and high cost they associate with receiving WCF."

"Growth Mindset positively predicted Feedback Monitoring through Value, and Fixed Mindset did not predict Feedback Monitoring because of the Self-Presentation Cost of feedback seeking."

Conclusions

"The present study bridged these two gaps by introducing the notion of feedback-seeking behavior from organization psychology and examining it in relation to learners' mindsets from educational psychology."

"The results of this study confirmed that depending on their implicit beliefs about the malleability of their intelligence, L2 learners associate different costs and values with feedback-seeking, which, in turn, influence their decision whether or not to seek feedback and by what method."

"These findings confirm the basic premise of this study that learners are proactive agents of learning in the feedback process and their strategic and agentic involvement in this process is influenced by both dispositional and contextual motivational mechanisms."

"Perceived as a learning resource, feedback is personally-relevant information that students seek in any information environment, inside or outside the instructional settings, to meet their valued L2 writing goals."

Limitations and Future Research Directions

"The mindsets were examined as antecedents of the self-presentation cost and value of FSB."

"These can include students' previous experiences with feedback, and performance level, teacher–student relationships, feedback environment and attitudes, how errors are perceived in the class, classroom goals, objectives, and assessment standards, and basically any factors that might affect learners' perceptions of the cost and value associated with FSB."

"Only the learning value and self-presentation cost of feedback seeking were examined."

"Future studies can explore other costs and values such as ego and effort costs (Ashford [25]), and image/ego-enhancement and appraisal values (Park and others [27]) to get a more comprehensive picture of the motivational mechanisms underlying FSB."

Pedagogical Implications

"Promoting a growth L2 mindset and learning goals can increase the value and decrease the cost of feedback seeking, thereby contributing to learners' FSB."

"To mindset interventions, teachers can promote learning goals in their classes through setting learning rather than performance standards of progress, make the process of writing development rather than product of writing the focal point of their teaching, treat errors as signs of development rather than symptoms of weakness, establish an atmosphere of collaborative learning, minimize the sense of competition and social comparison, and finally, assess learners based on their intra-individual rather than normative progress."

"Increasing FSB's value and reducing its cost can be done through different means including but not limited to establishing feedback-seeking norms and role modeling (e.g., Williams, Miller, Steelman, & Levy [28]), improving the feedback seeker–source relationships (e.g., Levy, Cober, & Miller [29]; VandeWalle and others [30]), and creating a FSB-friendly environment through promoting intellectual stimulation, critical thinking, and problem-solving skills among students (Anseel and others [24])."

Motivational Predictors of Struggling Readers' Reading Comprehension: The Effects of Mindset, Achievement Goals, and Engagement [271]

This is a machine-generated summary of:

Cho, Eunsoo; Toste, Jessica R.; Lee, Minhye; Ju, Unhee: Motivational predictors of struggling readers' reading comprehension: the effects of mindset, achievement goals, and engagement [271].

Published in: Reading and Writing (2018).

Link to original: https://doi.org/10.1007/s11145-018-9908-8.

If you want to cite the papers, please refer to the original.

For technical reasons we could not place the page where the original quote is coming from.

Abstract-Summary

"Mindset and achievement goals provide a useful explanatory framework for understanding struggling readers' engagement and reading comprehension."

"This study examined the extent to which mindset and achievement goals are associated with reading comprehension; as well as the mechanism through which engagement mediates this relation, controlling for the effects of vocabulary, word recognition, and limited English proficiency status."

"Structural equation modeling of four and fifth grade struggling readers (N = 107) demonstrated the importance of achievement goals in predicting reading comprehension."

"Effects of mindset on engagement and reading comprehension were completely mediated by mastery and performance-avoidance goals; further, these goals had indirect relations to reading comprehension through emotional engagement."

Introduction

"Achievement goals can provide critical links between students' psychological dispositions, such as mindset, to engagement and, possibly, to reading comprehension."

"Empirical investigations of mindset and achievement goals as they relate to reading comprehension are scarce."

"Do various dimensions of engagement mediate the relations of fixed mindset and achievement goals to reading comprehension?"

"The first partial mediation model specifies the direct effect of fixed mindset and the indirect effects via achievement goals to reading comprehension (Model 2a)."

"In the second model, the direct effects of fixed mindset is removed, assuming that achievement goals and engagement completely mediate the effects of fixed mindset to reading comprehension (Model 2b)."

"This model hypothesize that only the engagement variables have direct relations to reading comprehension, completely mediating the relation of fixed mindset and each achievement goal to reading comprehension (Model 2c.1–2c.3)."

Methods

"Mastery goals were assessed with six items that asks students' orientation to learning and understanding (e.g., "An important reason why I do my work in class is because I want to get better at it.")."

"Performance-approach goals were assessed with five items that asks students' orientation to demonstrating their ability (e.g., "Doing better than other students in class is important to me.")."

"Performance-avoidance goals were assessed using six items that asks students' orientation to hiding lack of ability (e.g., "One of my main goals is to avoid looking like I can't do my work.")."

"Three items assessing students' on-task attention and persistence were used for behavioral engagement (e.g., This student stays on task.) and four items measuring students' cognitive and metacognitive monitoring of one's work were used for cognitive engagement (e.g., This student critically evaluates own work.)."

Results

"Reading comprehension was negatively correlated with fixed mindset ($r = -0.29$, $p < 0.05$) and both performance goals ($-0.26 \leq rs \leq -0.25$, $ps < 0.05$)."

"In Model 1c, fixed mindset predicted performance-avoidance goals ($B = 0.54$, $SE = 0.21$, $p < 0.05$) and mastery goals ($B = -0.35$, $SE = 0.18$, $p = 0.05$)."

"Performance-approach goals did not predict any of the engagement dimensions ($-0.22 \leq Bs \leq 0.15$, $0.09 \leq SEs \leq 0.14$, $ps < 0.05$) and performance-avoidance goals predicted only emotional engagement ($B = -0.25$, $SE = 0.09$, $p < 0.05$)."

"Eliminating the direct path to reading comprehension from mastery goals (Model 2c.1), followed by the path from performance-avoidance goals (Model 2c.2) did not significantly lower the fit ($p = 0.41$ and $p = 0.53$, respectively)."

"Further removing the direct path from performance-approach goals to reading comprehension (Model 2c) significantly lower the fit ($p < 0.05$)."

Discussion

"Mindset and achievement goal theory provide a useful explanatory framework for understanding struggling readers' engagement and reading comprehension."

"We did not find a direct effect of mindset on reading comprehension, but this relation was found to be mediated by achievement goals and engagement."

"Mastery goals and performance-avoidance goals both served as important mediators, linking fixed mindset and emotional engagement, which in turn, predicted reading comprehension."

"Coupled with the significant indirect relation of performance-avoidance goals to reading comprehension, our findings point to the negative aspects of struggling readers pursuing either of the performance goals."

"Our findings suggest that the effects of mindset is completely mediated by achievement goals (mastery and performance-avoidance goals) and emotional engagement."

"Our findings suggest that future attempts to remediate reading difficulties need to attend to students' motivational needs and couple instructional practices that promote growth mindset and mastery goals in struggling readers with reading intervention."

The Role of Students' Beliefs, Parents' Educational Level, and the Mediating Role of Attitude and Motivation in Students' Mathematics Achievement [272]

This is a machine-generated summary of:

Hidayatullah, Achmad; Csíkos, Csaba: The Role of Students' Beliefs, Parents' Educational Level, and The Mediating Role of Attitude and Motivation in Students' Mathematics Achievement [272].

Published in: The Asia–Pacific Education Researcher (2023).

Link to original: https://doi.org/10.1007/s40299-023-00724-2.

If you want to cite the papers, please refer to the original.

For technical reasons we could not place the page where the original quote is coming from.

Abstract-Summary
"Beliefs, parents' educational background, motivation, and attitudes have been proven significantly influence achievement."

"To contribute to this gap in knowledge, the purpose of this study is to examine the structural relationships among beliefs, parents' educational level, attitude, motivation, and achievement in mathematics learning."

"Beliefs are positively associated with students' achievement ($\beta = 0.20$, $p < 0.001$), attitude ($\beta = 0.82$, $p < 0.001$), and motivation ($\beta = 0.68$, $p < 0.001$)."

"Parents' educational level is positively associated with achievements ($\beta = 0.17$, $p < 0.001$) and motivation ($\beta = 0.07$, $p = 0.04$)."

"Beliefs were indirectly associated with achievements through attitude ($\beta = 0.31$, $p < 0.001$) and motivation ($\beta = 0.08$, $p = 0.01$)."

"The indirect association between parents' educational level and achievement through motivation was insignificant."

"This study is valuable because it helps unpack the relationship between beliefs, parents' educational level, attitudes, motivation, and achievement."

Introduction

"Research also indicated that beliefs are not only associated with achievements but also it is associated with attitudes and motivation (Hofer [32]; Rarujanai and others [33])."

"Although there was evidence indicating achievements are associated with beliefs and PED, little is known whether beliefs and PED can also predict and control attitudes and motivation, especially in mathematics learning."

"Less attention has been directed to the mediation of attitudes and motivation for the relationship between beliefs, PED, and achievements in mathematics learning."

"To respond to the gap in the previous studies, the aim of this study is to examine how beliefs and PED are associated with attitudes and motivation and predict students' achievements in mathematics learning."

"We develop a model that examines the structural association among beliefs, PED, attitudes, motivation, and achievement in mathematics learning."

"We study whether attitude and motivation mediate beliefs and achievement and whether motivation mediates the relationship between PED and achievement."

Theoretical Framework

"Kim and Keller [34] empirically identified a significant influence of students' beliefs on their motivation toward mathematics."

"Habók and others [35] suggested that when students hold strong beliefs about their capability, they have good motivation to achieve the best performance."

"The influence of beliefs on achievement, attitude, and motivation is investigated in the present study."

"Voica and others [36] suggested that the level of students' motivation in mathematics is determined by their individual beliefs."

"Students' beliefs strengthen and stimulate their motivation in mathematics."

"Previous studies suggested that beliefs, PED, attitudes, and motivation play a key role in mathematics achievements."

"We examine the mediation of attitudes and motivation for the relationship between beliefs, PED, and achievements."

"Our investigation follows the hypothesis: H1: Beliefs are expected to be positively associated with attitudes, motivation, and mathematics achievements."

"H4: Motivation mediated the association between beliefs and mathematics achievements."

Method

"We selected 9 items from the mathematics-related beliefs system questionnaire developed by Eynde and De Corte [37] to measure students' beliefs about mathematics, such as "I can understand the difficult topic presented in mathematic" and "My teacher tries to make learning mathematics interesting"."

"The result showed that the model of beliefs about mathematics education instrument is fit, CFI = 0.94, TLI = 0.91, RMSEA = 0.06, and SRMR = 0.04."

"We measured students' motivation with five items, such as "I want top grades in mathematics learning" and "Whatever I do, I want to be the best in mathematics learning"."

"The model of motivation toward mathematics had a good fit model (CFI = 0.99, TLI = 0.97, RMSEA = 0.06, SRMR = 0.02) and a reliability value of 0.082 (α = 0.82)."

Procedure

"Since schools in Indonesia have reopened and started conducting offline classes, we collected our data using paper–pencil tests to obtain good responses from the students."

"We examined the construct validity and reliability of each questionnaire."

"Confirmatory factor analysis (CFA) was used to examine construct validity."

"The Tucker–Lewis index (TLI), comparative fit index (CFI), and root mean square error of approximation (RMSEA) were used to assess the model fit."

"In the third step, we performed descriptive statistical analysis to describe our data (mean and standard deviation [SD]) and the correlation between factors."

"We examined our hypothesis model via full structural equation modeling (SEM) in the third step."

"The model fit criteria were the same as those for the CFA (TLI and CFI ~ 0.95, RMSEA ≤ 0.08)."

Result

"Beliefs about mathematics were positively associated with students' achievement ($\beta = 0.20$, $p < 0.001$), attitude ($\beta = 0.82$, $p < 0.001$), and motivation ($\beta = 0.68$, $p < 0.001$)."

"Beliefs were positively and indirectly related to students' achievement ($\beta = 0.31$, $p < 0.001$) through attitude toward mathematics."

"Beliefs were also indirectly associated with achievement ($\beta = 0.08$, $p = 0.01$) through motivation."

"The direct association between PED and motivation was significant but weak ($\beta = 0.07$, $p = 0.04$)."

"The indirect association between PED and achievement through motivation was insignificant ($\beta = 0.01$, $p = 0.11$)."

"Beliefs were positively associated with motivation and attitude in both grades."

"While the mediation motivation for the relations of PED and achievements in both grades was not significant."

Discussion

"The novelty of this research lies in the proposed structural model consisting of students' beliefs, PED, attitudes, motivation, and achievement, which have been rarely investigated."

"The simultaneous contribution of all the variables (beliefs, attitude, motivation, and PED) explains mathematics achievements by 49%."

"We find the mediation of attitudes and motivation for the association of beliefs, and achievements."

"Our research confirmed that beliefs about mathematics are associated with mathematics achievement, attitude, and motivation."

"This finding supports Hypothesis 1, which holds that beliefs are associated with achievements, attitudes, and motivation."

"The direct association between beliefs and attitudes is the strongest relation among other variables (motivation and achievements)."

"Attitudes are the strongest predictor for students' achievements among other variables latent (beliefs, motivation, and PED)."

"Our study told us that attitudes and motivation can mediate the relationship between beliefs and achievement."

Limitations and Suggestions for Future Research

"A longitudinal study is required for future research to prove the predictive power of the aforementioned variables for achievement."

"We only used the PED variable as an external factor to explain achievements."

"Parents' education is not the sole factor that influences students' achievement and motivation."

"Other variables, such as self-regulated learning and students' attitude toward digital technology in mathematics, should likewise be investigated in future research."

Implications

"For practice, considering the mediation of attitudes and motivation for the relationship between beliefs and achievements, mathematics educators should consider how to shape students' beliefs, because these beliefs would control students' attitudes as well as motivation, in that way their achievements will increase."

"Mathematics educators should train students on how to regulate their motivation as well as their attitudes toward mathematics."

"Additional support services can be provided by mathematics educators to improve students' achievements by communicating with parents on how to encourage parents to increase their involvement."

A Study of the Relationship Between Familial Factors and Academic Motivation and Achievement in High School Students [273]

This is a machine-generated summary of:

Islam, Md. Nurul; Chakrabarty, Sumi: A Study of the Relationship between Familial Factors and Academic Motivation and Achievement in High School Students [273].

Published in: Psychological Studies (2019).

Link to original: https://doi.org/10.1007/s12646-019-00500-7.

If you want to cite the papers, please refer to the original.

For technical reasons we could not place the page where the original quote is coming from.

Abstract-Summary

"Educational psychologists, nowadays, are more concerned about the students who are motivated in learning and interested in achieving their academic goals."

"This study aimed to know the relationship between the most important familial factors (e.g., parental education, parental occupation, family type, monthly household income, sibling's achievement) with students' academic motivation and achievement."

"The result showed that all of the familial factors except the family type had significant effects on both students' academic motivation and achievement scores."

"A significant positive correlation was found between students' academic achievement and academic motivation scores ($r = 0.339$, $p < 0.01$)."

"All of the familial factors were significant in explaining students' academic achievement as well as their academic motivation."

"These factors combinedly explained 57.5% of the variance in students' academic achievement ($R^2 = 0.575$, $F(13, 166) = 17.281$, $p < 0.01$) and explained 30.5% of the variance in students' academic motivation ($R^2 = 0.305$, $F(13, 166) = 5.614$, $p < 0.01$)."

Introduction

"Most of these factors are relevant to students' family backgrounds (e.g., the family's monthly income, family types, parents' education, parents' occupation, siblings' achievement, etc.), highly interconnected with each other, and have a joint or a single effect on students' academic motivation as well as their academic performance (Duncan & Magnuson [38]; Grolnick, Friendly, & Bellas [39]; Usher & Kober [40])."

"The author found a significant relationship between familial factors (e.g., father's education and occupation, mother's education and occupation, family size and learning facilities in family) and students' academic achievement motivation."

"The present study was designed to investigate the influence of family factors (e.g., parents' education, parents' occupation, family's monthly household income, family types, and siblings' achievement) on students' academic motivation and achievement."

"The findings of the study will help us to sort out the important significant family factors affecting students' academic motivation and achievement."

Method

"180 high school students in eighth to tenth grade, age ranging from 13 to 16 years (M = 14.20, SD = 0.960), were selected through purposive sampling from six high schools, in Chittagong district, Bangladesh."

"Two instruments were used in the data collection for the study: (a) Academic Motivation Scale (AMS-28) and (b) personal information form (PIF)."

"The Bangla version of AMS-28 had a satisfactory level of internal consistency reliability in the present study ($\alpha = 0.803$), it had no adequate factor structure replicated with the original scale."

"A PIF was provided along with the Bangla version of AMS-28 to collect data on some demographic and familial factors such as age, gender, class, residence, parent's education, parent's occupation, family type, monthly household income, student's academic achievement, and sibling's academic achievement."

Results

"A Kruskal–Wallis test was performed to know whether there is an effect of parental education on student's academic motivation and achievement."

"To know the effect of sibling's achievement on student's academic motivation and achievement, an independent sample t test was performed."

"A multiple regression analysis was then performed to know whether there are significant predictors of student's academic motivation and achievement."

"Except the 'family type,' the other six familial factors were found to be the most significant predictors in explaining academic motivation of students."

"Two predictors, namely the 'father's occupation' and the 'family type,' did not significantly explain student's academic achievement."

"The other five predictors were the most significant in explaining the academic achievement of students."

Discussion

"The present study was aimed to find out the important familial factors affecting high school students' academic motivation and achievement."

"This study was to investigate the significant influence of each familial factor on students' academic motivation and achievement."

"In respect of the current study, it can be said that the students who came from the single or nuclear families probably had greater academic facilities and support, had no shared educational necessities, got more attention and more access to resources from their parents, etc. Students who were from the families having more amount of monthly household income had better academic motivation and achievement than the students who were from the families having low amount of monthly household income."

"The result of the study revealed that both the mother's and father's education had significant main effects on student's academic motivation and achievement."

Motivation and Engagement with Physics: A Comparative Study of Females in Single-Sex and Co-educational Classrooms [274]

This is a machine-generated summary of:

Abraham, Jessy; Barker, Katrina: Motivation and Engagement with Physics: a Comparative Study of Females in Single-Sex and Co-educational Classrooms [274].

Published in: Research in Science Education (2018).

Link to original: https://doi.org/10.1007/s11165-018-9770-3.

If you want to cite the papers, please refer to the original.

For technical reasons we could not place the page where the original quote is coming from.

Abstract-Summary

"The aim of this study was to identify the differences in motivation, engagement and sustained enrolment plans in relation to physics between female students in single-sex from those in co-educational settings."

"Results revealed that both cohorts have high levels of achievement motivation in relation to physics irrespective of their classroom composition; however, females from single-sex schools showed higher values for some constructs across various physics topics."

"This paper highlights the scarcity of research into females' sustained participation in physics, once they commence studying physics, and the difference class composition makes to females' achievement motivation and future enrolment intentions."

Materials and Procedures

"Performance perceptions (perfperc) represents the task-specific beliefs are labelled as students' performance perceptions in that particular physics module."

"This construct subsumes two highly related constructs, namely students' self-concept of ability in the particular module and their perceptions of task difficulty of the physics module."

"Sustained engagement (engage) was measured as a broad construct that is subsumed by interrelated aspects of behaviour, emotion and cognition in the year 11 physics module (Willms [41])."

"Sustained enrolment intentions in physics (choicein) represented students' sustained intention to choose further physics at the completion of a specific physics module (Abraham and Barker [42])."

"Using the SEMP constructs described above, this study sought to examine whether there is a degree of difference (Martin [43]) in achievement motivation, sustained engagement and retention intentions of females in relation to physics in female-only and co-ed classroom settings."

"Data was collected using the Physics Motivation Questionnaire (PMQ) (Abraham and Barker [44])."

Sample Size and Characteristics

"The participants in this study were year 11 female physics students from eight New South Wales (NSW) high schools (five Government and three Catholic schools) located in Western and Northern Sydney."

"The total sample size across the eight schools varied across the modules (92, 100, 92 and 82, respectively) reflecting the significantly lower female participation in Australian physics classrooms (Lyons [45])."

"Disentangling the effect of the students' home and background factors as well as school factors is one of the challenges faced in comparative studies."

"Purposive sampling secured SS and co-ed schools with similar socio-educational advantage and this equivalence reduced sampling differences."

"The participants' socio-economics status (SES) background data were collected using the Australian National University SES (ANU4) scale (Jones and McMillan [46]), and this showed equivalence across the SS and co-ed sample for SES occupations."

Results

"It shows that participants reported higher or near-average values (mean of the constructs = 3.5) for all variables except for sexstereo, across most physics modules."

"To verify whether the difference across the two groups were significant or not, the standardised factor correlations of the constructs with type of school (TOS) were examined (Bodkin-Andrews and others [47]; Kelloway [48])."

"The results revealed significant differences in the interest (for waves and cosmic engine modules) and utility (cosmic engine module) value variables, where SS females tended to hold higher values."

"For other constructs, no statistically significant differences were noted across the modules."

"At the completion of the year 11, when the students were making the crucial decision about continuing physics to their final year of senior secondary (year 12) or exiting from physics (discontinuing the subject at year 11), their motivation levels were found the same for both cohorts."

"The two value constructs (interest and utility), and performance perceptions were the recurring themes in students' responses from both cohorts for continuing physics."

Discussion

"The results provided strong evidence of similarities in physics achievement motivation between females in SS schools and those in co-ed schools."

"Both cohorts were highly motivated on average; however, there were indications that SS females had higher values for some modules."

"Although there was evidence for some degree of difference between these cohorts in favour with SS students, they do not necessarily support the belief that females overall would be better motivated in SS classes since females in co-ed settings also held high values for the same constructs."

"The lack of uniformity of the results across the modules (e.g. SS females scored higher for one module in their perfperc than co-ed females but for other modules they were not different from their counterparts) suggests that there could be other factors that are more influential than the physical absence of males in the classroom that affect their self-concept and interest levels in the subject."

Limitations of the Study and Future Directions

"The present study provided a novel comparison of physics motivation of a self-selected group of females at a topic-specific level in single-sex and co-ed high school settings."

"In order to keep the situation as normal as possible, PMQ was cautious not to include more direct measures on the presence of males in classrooms fearing it might induce ST among females in co-ed classrooms and distort the actual data."

"Future research should include a larger sample size to conduct invariance testing to provide even more rigour in developing their conclusions."

"Future research into understanding gendered patterns of student motivation in physics should include more than two categories of sex given that not all students identify with being male or female."

Conclusion
"The assumption behind the arguments for same-sex schooling, that is, an all-female environment may diminish the impact of masculine stereotypes and improve females' engagement and sustain enrolment motivation, was not held in this study."

"Once females commence physics education at a senior high school level, factors such as the utility value and interest value drive their achievement motivation and consequently the class composition itself is less important."

"During an extensive period of physics study, females in senior high school are not found to hold stereotypical views about physics being a "masculine" domain or that they are less capable in physics than males irrespective of classroom setting."

[Section 7]
"Several studies examining physics in high school report that compared with males, females have markedly lower levels of extrinsic and intrinsic motivation, poorer engagement and, subsequently, reduced intentions to select physics as a subject to study in their senior years of school and are less likely to pursue a physics career path (e.g. Ainley and others [49]; Hazari and others [50]; Kennedy and others [51]; Stokking [52]; Taasoobshirazi and Carr [53]; Whitelegg and others [54])."

"Haussler and Hoffman's [55] intervention study found that teaching females in a SS setting catered for their differing learning preferences and this adaption has led to improvements to female students' confidence in physics as well as bolstered their physics self-concept."

"A limitation of the research conducted to date on the differences between SS and co-ed schools and the uptake of physics is that they map the physics motivational profile of females in both educational settings at a domain specific level."

"The sampling for this study allows for the examination of whether females' physics motivation across topics varies as a function of being at a SS or co-ed classroom."

Parent and Teacher Effects on Academic Coping and Classroom Engagement: Testing a Motivational Model [275]

This is a machine-generated summary of:

Raftery-Helmer, Jacquelyn N.; Grolnick, Wendy S.: Parent and teacher effects on academic coping and classroom engagement: Testing a motivational model [275].

Published in: Motivation and Emotion (2018).

Link to original: https://doi.org/10.1007/s11031-018-9676-4.

If you want to cite the papers, please refer to the original.

For technical reasons we could not place the page where the original quote is coming from.

Abstract-Summary

"One hundred and eighty-three-sixth grade students participated in a study that examined a process model whereby children's perceptions of parent and teacher behaviors were associated with children's coping strategies through children's appraisals of failure."

"Results revealed that parent autonomy support was indirectly associated with defensive coping through its effects on children's appraisals of failure."

"There was also some support that cognitive appraisals, particularly threat appraisals were indirectly associated with child reported engagement through coping, suggesting that coping appraisals impact children's engagement and participation in the classroom through their effects on coping behavior/strategy use."

Introduction

"Supporting this, studies have shown that contextual factors, such as parent and teacher behavior (e.g., Hamre and Pianta [56]; Jang and others [57, 58]; Raftery-Helmer and Grolnick [59]; Swanson and others [60]; Zimmer-Gembeck and Locke [61]) are associated with children's coping."

"While it is clear that parent and teacher behavior can predict children's coping strategy use, it is not known whether this translates into more distal academic outcomes, such as classroom engagement."

"Further, it is not clear how or why parent and teacher behaviors predict academic coping."

"This study examines relations between parent and teacher behavior and children's appraisals of academic failure, coping, and classroom engagement."

"A better understanding of the antecedents (parent/teacher behavior and cognitive appraisals) and consequences (classroom engagement) of academic coping will represent a significant theoretical contribution to academic coping as an emerging field."

Engagement
"Cognitive engagement within this model also includes student participation, focus, and attention at school or during learning activities."

"Disengaged students have often had a host of negative school-related experiences, particularly those involving failure."

"Students who drop out of school are more likely to have had poor school performance, limited attendance, and early academic failure, such as having had to repeat a grade (Barrington and Hendricks [62]; Cairns and others [63]; Ekstrom and others [64]; Grissom and Shepard [65]; Wehlage and Rutter [66])."

"These factors may be especially relevant in the context of perceived or actual failure experiences that may undermine students' active participation in academic activities."

"We employ a theoretical view, Self-Determination Theory (Deci and Ryan [67]), which delineates what children need to stay engaged when faced with academic challenges."

"Self-Determination Theory also delineates what features of a child's context might affect academic coping and engagement and why."

Self-Determination Theory: An Approach to Understanding Academic Challenges
"The extent to which children experience themselves as autonomous, competent, and connected will predict appraisals of setbacks and failure as either a threat to well-being or a challenge to overcome (Skinner and Wellborn [68])."

"If students experience themselves as being controlled, incompetent, and disconnected, failure is likely to be experienced as threatening, as a danger to their well-being or esteem in the context of minimal support, resources, or capacity to manage the threat."

"When students don't feel connected to important adults in their life, they will perceive others as unavailable for support and thus failure may be experienced as an overwhelming threat that must be managed in isolation."

"Students who feel connected to important adults in their lives, such as parents or teachers, may see failure as a challenge because they believe they can leverage supportive resources when things become difficult."

Context
"If students experience adults in their life as autonomy supportive, as providing structure, and as involved, their behavior will be less defensive and they will instead channel their efforts into mastering the environment (Skinner and Wellborn [69]; Skinner and others [70]) by actively problem solving and seeking information or support that can help them understand do better next time."

"A recent empirical study provided support for this hypothesis, showing that parental autonomy support, involvement, and structure demonstrate independent effects on children's coping with failure (Raftery-Helmer and Grolnick [59])."

"Zimmer-Gembeck and Locke [61] found that a composite measure of teacher behavior (including teacher autonomy support, involvement, and structure) predicted active coping (including strategies such as cognitive decision-making, direct problem solving, and seeking understanding), using a measure of coping that allowed children to self-identify their school stressor."

Current Project

"Others have found that positive emotions, but not negative emotions, are associated with mastery coping strategies (i.e., problem solving, social support seeking; Reschly and others [71]) we predict that challenge appraisals will be uniquely associated with mastery coping strategies, while threat appraisals will be uniquely associated with defensive strategies. (3) Greater endorsement of mastery coping and lower endorsement of defensive coping will be associated with higher academic engagement. (4) Relations between parent and teacher autonomy support, involvement, and structure, and coping (i.e., mastery and defensive) will be mediated by cognitive appraisals."

"Higher parent and teacher autonomy support, involvement, and structure will be associated with greater endorsement of challenge appraisals and lower endorsement of threat appraisals, which in turn will be associated with more mastery coping and less defensive coping. (5) Mastery and defensive coping will mediate relations between challenge and threat appraisals and academic engagement."

Method

"Caregivers varied in their levels of education: mothers: 17% did not complete high school, 28% completed high school or received their GED, 20% completed some college, received their associates degree, or reported some vocational training, 23% completed college, and 11% reported schooling beyond college; fathers: 14% did not complete high school, 37% completed high school or received their GED, 13% completed some college, received their associates degree, or reported some vocational training, 26% completed college, and 10% reported schooling beyond college."

"In terms of race and ethnicity: 25% of the children were European-American, 34% were Hispanic, 15% were African American, 7% were Asian, 2% were African, and 17% were multi-ethnic."

"The student participants were representative of students enrolled in the district, with district enrollment data showing that approximately a third of students were European-American and that approximately 40% of students were Hispanic."

"Similar to our sample, district enrollment data also showed that boys and girls were equally represented."

Procedure
"Students were informed about the study in their classrooms and asked to bring home a letter describing the study and inviting child participation."

"Of the letters returned, 94% of parents provided consent and 6% declined participation."

"Children whose parents provided consent completed questionnaires at school."

"All study procedures were approved by the Research and Accountability Office of the participating school district and University Institutional Review Board."

Measures
"Seven items measured children's perceptions of their parents' autonomy support versus control (e.g., "My parents allow me to solve school problems my own way"), six items measured involvement (e.g., "My parents want to know about my school day"), and 15 items measured structure (e.g., "I don't know what my parents expect of me in school"—reverse coded) in relation to school."

"Children indicated on a 4-point Likert scale ranging from not true at all to very true how much they agreed with each statement."

"The negative evaluation by others and negative self-evaluation appraisals subscales were highly correlated ($r = 0.59$, $p < 0.01$) and so were combined to form a negative evaluation composite (Cronbach's $\alpha = 0.86$)."

"The challenge and resource subscales were highly correlated ($r = 0.67$, $p < 0.01$) and when combined formed a highly reliable scale (Cronbach's $\alpha = 0.83$)."

Results
"Model 1 tested whether parenting dimensions of autonomy support, involvement and structure would be linked to coping through children's cognitive appraisals and whether coping would mediate the relations between appraisals and children's engagement."

"A model including direct and indirect effects, which tested direct paths from parenting to coping strategies and direct paths from cognitive appraisal to classroom engagement, was tested."

"We examined the extent to which teacher dimensions of autonomy support, involvement, and structure were associated with classroom engagement through their effects on children's cognitive appraisals and coping strategy use utilizing the same model testing strategy."

"All parent and teacher dimensions, both cognitive appraisals, and mastery and defensive coping strategies were associated with classroom engagement."

"Because maternal education was associated with classroom engagement and child gender was associated with several of the teacher dimensions and coping variables, both were included as covariates in subsequent analyses."

Model 1: Parents

"The indirect effect of perceived parental autonomy support on defensive coping through threat appraisals was significant [$\beta = -0.36$, SE(β) = 0.08 $p = 0.01$; bootstrap CI − 0.503, − 0.234]."

"The indirect effect of parental autonomy support on mastery coping through either challenge or through threat appraisals was not significant."

"As hypothesized, challenge appraisals related to mastery [$\beta = 0.43$, SE(β) = 0.12, $p < 0.01$] but not to defensive coping strategies and threat appraisals related to defensive [$\beta = 0.62$, SE(β) = 0.12, $p < 0.01$] but not to mastery coping strategies The indirect relation of threat appraisals to academic engagement (through defensive coping) was significant $\beta = -0.41$, SE(β) = 0.34, $p = 0.010$, bootstrap CI [-1.070, − 0.162] while the indirect effect of challenge appraisals on engagement did not reach statistical significance."

Model 2: Teachers

"While teacher involvement was not directly associated with defensive coping, there was a direct effect of teacher involvement on mastery coping [$\beta = 0.57$, SE(β) = 0.65 $p = 0.016$]."

"For teacher structure, there were no significant relations with threat and challenge appraisals and no direct effects of mastery or defensive coping, controlling for the other dimensions."

"As hypothesized, challenge appraisals related to mastery [$\beta = 0.57$, SE(β) = 0.21, $p = 0.015$] but not to defensive coping strategies and threat appraisals related to defensive [$\beta = 0.65$ SE(β) = 0.17, $p = 0.010$] but not to mastery coping strategies The indirect relation of threat appraisals to academic engagement (through defensive coping) was significant $\beta = -0.43$, SE(β) = 0.18, $p = 0.010$, bootstrap CI [− 0.785, − 0.180] as was the indirect effect of challenge appraisals on engagement through mastery coping; $\beta = 0.41$, SE(β) = 0.21 $p = 0.010$, bootstrap CI [0.194, 0.913]."

Discussion

"This project fills a gap in the literature by exploring whether parent and teacher behavior is associated with children's coping strategies by influencing how children appraise failure as a stressor; and whether these coping strategies are associated with students' classroom engagement."

"And contrary to findings from earlier studies (e.g., Raftery-Helmer and Grolnick [59]), neither parent nor teacher structure were uniquely associated with children's cognitive appraisals, coping, or engagement, over and above the effects of involvement and autonomy support."

"The findings from this study that both parent and teacher structure were not exerting unique influences on cognitive appraisals, coping, and engagement may not generalize to younger students, less comfortable with navigating academics, for whom structure may be more important."

"Results point to the need for interventions aimed at helping parents and teachers support autonomy, given its particularly salient contributions to children's appraisals, coping, and engagement, and its notable absence from most evidence-based family or school-based interventions (e.g., Sugai and others [72])."

Humanistic Education and Students' Educational Motivation in Tehran Primary Schools [276]

This is a machine-generated summary of:

PourAli, PariNaz; SeifNaraghi, Maryam; Naderi, Ezatollah: Humanistic Education and students' Educational Motivation in Tehran Primary Schools [276].

Published in: International Journal of Mental Health and Addiction (2016).

Link to original: https://doi.org/10.1007/s11469-016-9703-1.

If you want to cite the papers, please refer to the original.

For technical reasons we could not place the page where the original quote is coming from.

Abstract-Summary

"The present article surveys teachers and principals on perceived elementary students' educational motivation and achievement in the 2nd education district in Tehran."

"Two groups from the 2nd district in Tehran were examined: Principals (managers & assistants) in public and private primary schools (boys and girls schools), and teachers in public and private primary schools (boys and girls schools)."

"The questionnaire examined 4 primary components of humanistic education: the teaching–learning environment, the role of the teacher, the role of student and methods of evaluation."

Materials and Methods

"The sample consisted of two groups: principals and teachers of public and private primary schools (boy and girl schools) in Tehran's 2nd district."

"A stratified random sampling method was used for recruitment."

"A reliable questionnaire by the researchers was used to gather data, along with the under-evaluation elements of the related process."

Discussion

"In terms of social skills, a meta-analysis of social skills interventions in young students with emotional and behavioural disorders (EBD) found that current theoretical frameworks are ineffective at helping these students develop the social skills necessary to perform well in school (Mathur and others [73])."

"An analysis of 150 studies on open education found that students taught in this framework showed greater independence, creativity, emotional adjustment and improved attitudes towards learning, school and their teacher compared to students taught in traditional frameworks."

"It can be said that the humanistic approach uses a higher level of learning, one where the self-motivated individual may utilize behavioural approaches like memorization or cognitive processes like critical thinking in their educational pursuits."

"It is proposed that humanistic educational approaches may be helpful in improving social, mental and academic outcomes in students."

[Section 4]

"It encourages competition among students, fails to recognize differences in learning styles and abilities, and may actually encourage maladaptive behaviours if they are reinforced with rewards (Cannella [74]; Weiner [75]; Wolpe and Rowan [76])."

"While this process is definitely more reflexive and produces better outcomes than behaviourism, it still utilizes a one-size-fits-all approach to learning, limits self-direction to the confines of a structured curriculum and fails to focus on motivating students to take ownership of their education."

"Education systems should foster environments where students are self-motivated, take control of their learning and learn to self-evaluate."

"3) Teacher as facilitator: teachers create positive learning environments, supply resources as the student gains interest/aptitude in certain areas, develops emotional stability and self-esteem, promotes tolerance and understands differences in learning styles and abilities."

"It may be possible that many of the problems that students are experiencing in their psycho-social development could be resolved if education systems utilized humanistic theoretical perspectives in their practical frameworks."

Self-Focused Attention, Performance Expectancies, and the Intensity of Effort: Do People Try Harder for Harder Goals? [277]

This is a machine-generated summary of:

Silvia, Paul J.; McCord, Daniel M.; Gendolla, Guido H. E.: Self-focused attention, performance expectancies, and the intensity of effort: Do people try harder for harder goals? [277].

Published in: Motivation and Emotion (2010).

Link to original: https://doi.org/10.1007/s11031-010-9192-7.

If you want to cite the papers, please refer to the original.

For technical reasons we could not place the page where the original quote is coming from.

Abstract-Summary
"Brehm's motivational intensity theory, however, predicts that the intensity of motivation is only as high as necessary, so people will try harder for difficult tasks than for easy tasks, all else equal."

"Neither high self-focus nor an easy task per se caused increased effort; instead, high self-focus significantly increased systolic reactivity when the task was difficult."

"Effort was thus higher despite less optimistic goal expectancies, a finding that is predicted by Brehm's motivational intensity theory but not by traditional self-regulation models."

Introduction
"We consider the dynamics of effort involved in self-focused attention."

"Many researchers have studied how self-focused attention promotes trying harder versus withdrawing, particularly with regard to how optimistic and pessimistic performance expectancies influence motivation (Carver and Scheier [77]; Duval and others [78])."

"We consider the implications of Brehm's motivational intensity theory (Brehm and Self [79]; Brehm and others [80]) for the classic problem of how self-focus and performance expectancies influence motivation."

"Applying the principles of Brehm's theory to self-regulation makes novel and non-intuitive predictions about the role of self-focus in effort and integrates disparate findings."

Self-focused Attention, Expectancies, and Motivated Action

"In the initial studies, researchers showed several main effects of self-awareness on measures of persistence and achievement, such as how long people spent on a task, how well people performed, and how many responses people generated for open-ended tasks (e.g., Duval and Wicklund [81]; Liebling and Shaver [82]; McDonald [83]; Wicklund and Duval [84])."

"Several studies have found that self-focus amplifies the effects of expectancies on measures of persistence, behavioral approach and avoidance, and performance quality (Burgio and others [85]; Carver and others [86, 87])."

"In an extension of this work, Duval and others [78] found that the expected rate of progress influenced how self-focus affected time spent persisting on a task."

"The effects of expectancies go beyond measures of performance; for example, people who feel unable to meet a standard are more likely to blame external factors (Duval and Silvia [88]) and other people (Silvia and Duval [89]) for their failure."

Brehm's Motivational Intensity Theory

"Importance determines potential motivation, the maximum level of effort people would mobilize; difficulty determines actual motivation, the level of effort actually mobilized."

"Because unfixed tasks lack a fixed difficulty parameter, Brehm's model predicts that effort during unfixed tasks is determined by potential motivation: people will put forth effort as a function of the goal's importance (Wright and others [90])."

"Self-focus increased effort for an unfixed task (Gendolla and others [91], Study 1)."

"Because effort for unfixed tasks is a function of importance (Wright and others [90]), this effect implies that the task was more important for self-focused people."

"People low in self-focus put forth minimal effort in both cases, suggesting that neither task was worth the high level of effort needed."

"People high in self-focus, however, tried hard for the difficult task, which suggests that self-focus made the task important enough to merit the high effort (Gendolla and others [91], Study 2)."

The Present Research

"Our past work, however, was unable to evaluate the major point of contention between past models of expectancies and Brehm's model of effort: whether people try harder when expectancies are high."

"Although a manipulation check showed that self-awareness was increased by this manipulation, the induction could have been confounded with evaluation apprehension, which has the same effect on effort mobilization as self-focus (e.g., Gendolla and Richter [92]; Wright and others [90])."

"We created two levels of self-focus and had people work on a cognitive classification task that was either fairly easy or pretty hard."

"Based on Brehm's model, we predicted that effort would be highest when self-focused people confronted a difficult task."

"High effort is needed only when the task is hard, however, so self-focus will boost effort only in the difficult condition."

Method

"After completing an informed consent form, the participants were told that the study was concerned with how the body responds during mental tasks."

"The blood pressure cuff was applied and the experiment proceeded with an assessment of cardiovascular baseline values while participants completed questionnaires containing demographic information and filler scales."

"After the baseline measurement period, participants received instructions explaining the task used in the performance period."

"In the high self-awareness condition, a $24'' \times 36''$ mirror was placed behind the monitor on the participant's table."

"Participants were then informed that the standard for performance was to be correct on 90% of the trials."

"After the practice block, participants were given a short "pre-task questionnaire" assessing their expectations of success."

"Such manipulations have value, but for the present purposes we wanted to manipulate expectancies directly (experience performing an objectively easier or harder task) rather than obliquely (persuasion, false information, or social comparison)."

Results

"A 2 (self-focus: no mirror, mirror) by 2 (task difficulty: easy, hard) between persons ANOVA on pre-task performance expectancies revealed a significant main effect for task difficulty, $F(1, 66) = 37.50$, $p < 0.001$, no effect for self-focus, $F < 1$, and no interaction, $F(1, 66) = 1.12$, $p = 0.295$."

"For response times, a 2 (self-focus: no mirror, mirror) by 2 (task difficulty: easy, hard) ANOVA revealed a significant main effect for task difficulty, $F(1, 64) = 144.3$, $p < 0.001$, a main effect for self-focus, $F(1, 64) = 4.68$, $p = 0.034$, and a significant interaction, $F(1, 64) = 4.41$, $p = 0.040$."

"For error percentages, a similar ANOVA found only a main effect of task difficulty, $F(1, 64) = 97.9$, $p < 0.001$; the main effect of self-focus and the interaction were not significant, both $Fs < 1$."

Discussion

"Effort—assessed as cardiovascular reactivity, particularly SBP change—was a function of both self-focus and task difficulty, but the pattern was more consistent with Brehm's motivational intensity model (e.g., Brehm and Self [79]) than with self-awareness theories (e.g., Carver and Scheier [77])."

"When self-focus was low, effort was low regardless of task difficulty (and hence of performance expectancies)."

"When self-focus was high and success was thus more important, effort was higher for the hard task than for the easy task."

"If the difficult task was threatening to the participants, and if self-focused attention magnified the feelings of threat (e.g., Scheier and others [93]), then one might expect higher cardiovascular reactivity."

"Someone can show high task persistence but low effort, such as when people listlessly perform boring tasks."

"For an easy task self-focus doesn't boost effort, but it could nevertheless boost performance and persistence."

Motivating the Academic Mind: High-Level Construal of Academic Goals Enhances Goal Meaningfulness, Motivation, and Self-Concordance [278]

This is a machine-generated summary of:

Davis, William E.; Kelley, Nicholas J.; Kim, Jinhyung; Tang, David; Hicks, Joshua A.: Motivating the academic mind: High-level construal of academic goals enhances goal meaningfulness, motivation, and self-concordance [278].

Published in: Motivation and Emotion (2015).

Link to original: https://doi.org/10.1007/s11031-015-9522-x.

If you want to cite the papers, please refer to the original.

For technical reasons we could not place the page where the original quote is coming from.

Abstract-Summary

"Two experiments tested the hypothesis, inspired by work on meaning in life, action identification theory, and expectancy-value theory, that high-level construal of an academic goal should enhance motivation to pursue that goal."

"In each experiment, we manipulated high-level versus low-level construal of an academic goal and assessed several variables related to the goal: the perceived meaningfulness of the goal, motivation to pursue the goal, and goal self-concordance."

"Supporting the hypothesis, individuals who thought about their academic goal in a high-level manner viewed their goal as more meaningful, reported being more motivated to pursue the goal, and reported the goal to be more self-concordant."

Introduction

"Conceptualizing a goal in a high-level manner (i.e., high levels of meaning) not only enhances the perceived meaningfulness of a goal, but can enhance one's motivation to pursue a goal by placing it in a broader context of integrated relationships and connections."

"As in high levels of meaning, identifying an action in terms of intention rather than method may enhance motivation for goal pursuit by making the goal more purposeful and meaningful."

"Given on the established relationship between intrinsic motivation and meaning, we explored the possibility that not only do self-relevant goals feel more meaningful, but increasing the perceived meaningfulness of goals (via high-level construal) can make them feel more self-concordant."

"The current research tested the hypothesis, inspired by theoretical perspectives on meaning and action identification theory, that high-level goal conceptualizations should increase goal motivation and perceived meaningfulness in an academic context."

Experiment 1

"We randomly assigned participants to think about a specific academic goal ("get a good grade in one of my classes") in either a high-level or low-level manner and assessed the extent to which the goal provides their life with meaning and purpose, their motivation to pursue the goal, and goal self-concordance."

"We first conducted a one-way multivariate analysis of variance (MANOVA) with construal (high-level vs. low-level), goal domain (academic vs. non-specific), and their interaction predicting the dependent variables of goal meaningfulness, motivation, and self-concordance."

"There was also a marginal main effect of goal domain such that participants reported greater self-concordance for academic goals ($M = 2.33$, $SD = 3.64$) than a non-specific weekly goal ($M = 1.34$, $SD = 3.90$), $F(1,177) = 3.30$, $p = 0.07$, $\eta = 0.018$."

Experiment 2

"Results revealed a significant effect of the construal manipulation across the dependent variables of goal meaningfulness, motivation, self-concordance, and number of tips viewed, $F(4,179) = 8.25$, $p < 0.001$, Wilks' $\Lambda = 0.79$, $\eta = 0.16$."

"Participants who wrote about why they pursue an academic goal ($M = 6.95$, $SD = 6.08$) did not view significantly more tips than those who wrote about how they pursue an academic goal ($M = 6.41$, $SD = 5.97$), $F(1,182) = 0.36$, $p = 0.55$, $\eta = 0.002$."

"Correlational analyses further revealed that the number of tips participants elected to view was not strongly associated with the reported meaningfulness of the academic goal, $r(184) = 0.13$, $p = 0.08$; motivation to pursue the goal, $r(184) = 0.02$, $p = 0.81$; or self-concordance, $r(184) = 0.15$, $p = 0.05$."

General Discussion

"In the current experiments, we tested the effects of high-level versus low-level construal on goal meaningfulness, motivation, and self-concordance in an academic context."

"Both experiments provide consistent evidence that high-level goal construal led participants to view their academic goals as more meaningful, report more motivation to pursue those goals, and perceive those goals to be more self-concordant."

"Although high-level construals consistently led to enhanced goal meaningfulness, motivation, and self-concordance for the academic goal in the current experiments, it is important to consider that high-level construals may not always be more beneficial than low-level construals."

"Whereas high-level construals enhanced meaningfulness, motivation, and self-concordance for the academic goal, only meaningfulness was enhanced when participants were asked to consider a non-specific goal."

"The current research demonstrated that high-level construal of an academic goal made the goal feel more meaningful, enhanced motivation to pursue the goal, and made the goal feel more self-concordant."

Testing a Motivational Model of Achievement: How Students' Mathematical Beliefs and Interests Are Related to Their Achievement [279]

This is a machine-generated summary of:

Jones, Brett D.; Wilkins, Jesse L. M.; Long, Margaret H.; Wang, Feihong: Testing a motivational model of achievement: How students' mathematical beliefs and interests are related to their achievement [279].

Published in: European Journal of Psychology of Education (2011).

Link to original: https://doi.org/10.1007/s10212-011-0062-9.

If you want to cite the papers, please refer to the original.

For technical reasons we could not place the page where the original quote is coming from.

Abstract-Summary
"Blackwell and others (Child Development 78(1):246–263, 94) tested a motivational model of achievement in which an incremental theory of intelligence leads to learning goals and positive effort beliefs, which leads to fewer ability-based, helpless attributions, and more positive strategies, which leads to improved grades."

"We tested this model by using a different population, using domain-specific items, comparing the results across gender, including "interest" as another variable in the model, and assessing the impact of achievement as an antecedent and outcome variable in the model."

"We established that the model had a relatively good fit to the data and all paths were statistically significant when achievement was used as both an antecedent and outcome variable."

"We documented that interest could be included as another mediating variable in the model."

Introduction
"To address this problem, Blackwell and others [94] conducted the first simultaneous test of a motivational model of achievement that specifies the process through

which students' beliefs about the changeability of intelligence and achievement goals are related to their school grades."

"To theories of intelligence and achievement goals, their model included beliefs about strategies (i.e., after doing poorly on a task, would they work harder or use effort-avoidant strategies), effort beliefs (i.e., the extent to which students believed that effort leads to positive outcomes), and helpless attributions."

"Blackwell and others tested this model and documented that students with malleable views of intelligence were more committed to mastery goals and were more likely to believe that their academic effort was necessary and effective in achievement, than students with fixed views of intelligence."

"Students' grades in a domain could affect their theories of intelligence in that domain by impacting their interest, achievement goals, strategies, helpless attributions, and effort beliefs."

Method

"Because one of the purposes of this study was to use domain-specific questionnaire items, each of the following scales was adapted to measure students' beliefs specifically in the domain of math."

"This five-item scale measures the level of students' mastery goals (Midgley and others [95])."

"Because the Midgley and others [95] scale measures students' general mastery goals for school and we wanted to measure mathematical mastery goals, we modified the original scale by adding the word "math" in each item."

"This nine-item scale measures the extent to which students believe that effort leads to positive outcomes (Blackwell [96])."

"We substituted the word "math" into each item in place of more general words such as "schoolwork," "a subject," and "something" because we wanted to measure students' effort beliefs specifically in math."

"Four positive items measured students' belief that effort leads to positive outcomes (e.g., "The harder you work at math, the better you will be at it.")."

Results

"Regressing positive strategies on mastery goals and controlling for interest, we established a positive and statistically significant relationship ($\beta = 0.52$, $p < 0.01$)."

"We tested the hypothesis that interest mediates the relationship between mastery goals and positive strategies."

"The correlation between mastery goals and positive strategies was statistically significant ($r = 0.57$, $p < 0.001$) and the correlation between mastery goals and interest (the potential mediator) was statistically significant ($r = 0.46$, $p < 0.001$)."

"Although we found no direct effect of interest on positive strategies, we documented a small but statistically significant indirect relationship between interest and positive strategies ($\beta = 0.06$, $p < 0.01$), mediated by mastery/learning goals."

"We also documented a statistically significant indirect relationship between interest and positive strategies ($\beta = 0.10$, $p < 0.05$), mediated by mastery/learning goals."

Discussion and Conclusion

"The purpose of the study was to test the Blackwell and others [94] model of achievement motivation with a different population, using domain-specific items, comparing the results across gender, including "interest" as another variable in the model, and assessing the impact of achievement on the model."

"The fact that the findings from these two studies are similar is difficult to interpret because it could lead to a couple of different conclusions: (a) that students' beliefs related to these motivation variables are invariant across subject areas (because it did not matter whether they were asked about their general academic motivation beliefs in the Blackwell and others study or more specifically about their beliefs in the domain of math as in the present study), or (b) that students' motivation beliefs in the domain of math are similar to their general academic motivation beliefs, yet may be very different from their beliefs in other subject areas."

Motivational Component Profiles in University Students Learning Histology: A Comparative Study Between Genders and Different Health Science Curricula [280]

This is a machine-generated summary of:

Campos-Sánchez, Antonio; López-Núñez, Juan Antonio; Carriel, Víctor; Martín-Piedra, Miguel-Ángel; Sola, Tomás; Alaminos, Miguel: Motivational component profiles in university students learning histology: a comparative study between genders and different health science curricula [280].

Published in: BMC Medical Education (2014).

Link to original: https://doi.org/10.1186/1472-6920-14-46.

If you want to cite the papers, please refer to the original.

For technical reasons we could not place the page where the original quote is coming from.

Abstract-Summary

"The purpose of this study was to investigate the influence of different components of motivation (intrinsic motivation, self-determination, self-efficacy and extrinsic -career and grade- motivation) on learning human histology in health science curricula and their relationship with the final performance of the students in histology."

"Glynn Science Motivation Questionnaire II was used to compare students' motivation components to learn histology in 367 first-year male and female undergraduate students enrolled in medical, dentistry and pharmacy degree programs."

"For intrinsic motivation, career motivation and self-efficacy, the highest values corresponded to medical students, whereas dentistry students showed the highest values for self-determination and grade motivation."

"Career motivation and self-efficacy components correlated with final performance in histology of the students corresponding to the three curricula."

"Our results show that the overall motivational profile for learning histology differs among medical, dentistry and pharmacy students."

Background

"The influence of different components of motivation (e.g. intrinsic motivation, self-determination, self-efficacy and extrinsic motivation) on human histology learning in health science curricula is poorly understood."

"A question we want to investigate in the present study was the interrelation among different components in students of human histology enrolled in three different undergraduate health science programs, and to determine if the components of motivation are related to histology course performance."

"Because different components of motivation can influence students' science-learning behavior, knowledge about the influence of each component will be useful to more effectively foster students' motivation to learn histology in health sciences [97, 98]."

"Another question of the present study was to investigate male and female students' motivations to learn human histology in medical, dentistry and pharmacy curricula and to determine the incidence of gender on the participation of the different components of motivation."

Methods

"This study was carried out to evaluate the motivational component profiles in university students learning histology using a validated questionnaire described by Glynn and cols [99]."

"To compare the results among different health science curricula, the study was carried out during the same course, in the same period of the year and with instructors belonging to the same Department -Department of Histology-, which is responsible for teaching histology in Medicine, Dentistry and Pharmacy Schools."

"To evaluate the different components of the motivation to learn histology in students at the medical, dentistry and pharmacy schools, all participants responded to the Science Motivation Questionnaire II (SMQII) developed by Glynnn and cols. [99], which uses five items to assess each of the five components of motivation: intrinsic motivation, self-determination, self-efficacy, career motivation and grade motivation."

"The items in the SMQII were designed to serve as empirical indicators of components of students' motivation not only to learn science in university courses but also to learn a specific science discipline."

Results

"Statistical analysis showed that there were significant differences ($p < 0.05$) for the three comparisons between groups of students in the overall mean scores for components 2 and 5 (medicine vs. dentistry, medicine vs. pharmacy and dentistry vs. pharmacy students)."

"We found significant differences in component 1 scores between dentistry and pharmacy students, and between dentistry and medical students."

"When the overall mean scores for the different components were correlated with the final performance in histology of the students corresponding to the three curricula, we found a significant positive correlation ($p < 0.05$ and $r = 0.666$) for the components 2 and 4."

Discussion

"In the career motivation component there were significant differences when the overall mean scores for the three groups were compared (medicine vs. dentistry, medicine vs. pharmacy and dentistry vs. pharmacy students) whereas differences were not significant in self-efficacy."

"The similarities in the response profiles for these components among the groups shows that a relationship exists between self-determination, and hence the control that students believe they have over their learning process, and the grade motivation, that is, the short-term primary goals of students enrolled in different degree programs."

"Our results suggest, as Boiche and cols. [100] showed for physical education, an adaptive role for the self-determination component of motivation towards histology among dentistry and medicine students, with high levels of self-regulation thanks to which individuals do not act without feelings of control or competence to achieve success during their undergraduate degree program."

Conclusions

"Our results show that our histology students enrolled in each of three undergraduate health science programs differed in their personal motivational profiles."

"This finding is potentially useful to foster their learning process, because if they are metacognitively aware of their motivation they will be better equipped to self-regulate their science-learning behavior in histology."

"Our results also show that the overall motivational profile for learning histology differs among medical, dentistry and pharmacy students, and that instructors as well as education policy makers can use their awareness of these differences to enhance curricula so that they focus not only on fostering the cognitive component of learning but also integrate students' levels and types of motivation into the processes of planning, delivery and evaluation of medical education."

Authors' Information

"JALN is a professor in the Department of Didactics and School Organization at the Faculty of Education Sciences, University of Granada."

"VD is a Master Degree and PhD in Tissue Engineering and a research fellow in the Department of Histology at the University of Granada, Pharmacy School."

"TS is a full professor in the Department of Didactics and School Organization at the Faculty of Education Sciences, University of Granada."

"MA is a full professor in the Department of Histology at the Medical School, University of Granada."

The Entrepreneurial Motivation in Academia: A Multidimensional Construct [281]

This is a machine-generated summary of:

Morales-Gualdrón, Silvia T.; Gutiérrez-Gracia, Antonio; Roig Dobón, Salvador: The entrepreneurial motivation in academia: a multidimensional construct [281].

Published in: International Entrepreneurship and Management Journal (2009).

Link to original: https://doi.org/10.1007/s11365-009-0113-8.

If you want to cite the papers, please refer to the original.

For technical reasons we could not place the page where the original quote is coming from.

Abstract-Summary

"We propose a model to analyse the motivation of academic entrepreneurs that comprises six dimensions: personal, relating to the entrepreneurial opportunity, to scientific knowledge, to the availability of resources, to the incubator organization, and to the social environment."

"Our results show that entrepreneurial opportunity is not part of the entrepreneurial motivation, but is of the utmost importance to academic entrepreneurs."

Introduction

"Interest in academic spin-offs has increased (Landry and others [101]; Meyer [102])."

"Academic spin-offs, therefore, are tangible evidence of the implementation of entrepreneurialism in universities (Chiesa and Piccaluga [103]; O'Shea and others [104]; Searle [105])."

"In the European context, promoting the establishment of academic spin-offs is a complex task."

"Why would we expect researchers to want to create spin-offs?"

"These works tend to focus on the analysis of specific cases, such as the US and Canada, where the phenomenon of academic spin-offs is fully consolidated (Doutriaux and Peterman [106]; Louis and others [107]; Samsom and Gurdon [108]; Doutriaux [109]; Doutriaux and Dew [110]; Chrisman and others [111]; Henrekson and Rosenberg [112]; Shane and Khurana [113]; Shane [114]; Lehrer and Asakawa [115]; Searle [105]; Ding and Stuart [116]; Zhang [117]; Landry and others [101]), while studies on Europe are less common (Jones-Evans [118]; Klofsten and Jones-Evans [119]; Henrekson and Rosenberg [112]; Laukannen [120]; Vohora and others [121]; Mosey and Wright [122])."

"Our paper contributes to filling this gap in the literature, exploring the reasons why researchers create academic spin-offs."

Literature Review

"Samsom and Gurdon [108], in a study of 22 US and Canadian academic entrepreneurs, found that their main motivations for establishing firms were the advancement of science and its applications, the personal opportunity to become an entrepreneur and the opportunity to make money."

"Doutriaux and Dew [110] examined in depth the motivations and the predisposition towards the creation of new businesses as a tool for technology transfer, in a set of 26 Canadian researchers, and found there were three types of academic entrepreneurs: genuine, casual and reluctant."

"Shane [114], in his study of academic entrepreneurs from the Massachusetts Institute of Technology (MIT), presents qualitative evidence on a diverse range of motivations for creating spin-off, which reveals a more complete picture."

"We try to present a fuller picture of entrepreneurial motivation in academia through analysis of a broad set of motivations and by exploring the creation of academic spin-offs in a country where the phenomenon is just emerging."

A model for the Study of Entrepreneurial Motivation in Academia

"The results of studies concerning the entrepreneurial motivation in non-academic settings suggest that entrepreneurs have a variety of reasons for deciding to create a company (Shane and others [123]; Birley and Westhead [124]; Dubini [125]; Carter and others [126]; Shane and others [127]; Segal and others [128]), and academic entrepreneurs are no exception."

"The analysis of qualitative and quantitative works related to academic entrepreneurs and their motivations allows us to identify several issues relating to the decision to found a spin-off."

"Our hypothesis is that the entrepreneurial motivation is a construct which is comprised of six major groups or motivational dimensions: personal, related to scientific knowledge, to the entrepreneurial opportunity, to the availability of resources to create business, to the "incubator organization" and to the social environment."

"Weatherston [129], Doutriaux and Peterman [106], Chiesa and Piccaluga [103] and Shane [114] provide quantitative and qualitative evidence that suggests that the desire for independence is an important motivation for academic entrepreneurs to set up businesses despite the high degree of autonomy and the relative freedom they have in their university jobs."

Methodology and Data Collection

"Is the need for achievement (F1), which is measured by three items: V1-Desire to prove own ability to establish a new firm; V2-Personal achievement motivation; V3-Desire to develop one's own ideas."

"The dimension of entrepreneurial opportunity (F4) was measured by three items: V8-Analysis of business opportunity developed by the OTT, V9-A new idea for a product/service; V10-Perceived customer's needs or deficiencies in existing products."

"The subdimension related to social networks (F8) contains three items: V19-Availability of person suitable to be manager of the firm; V20-Good contacts for establishing the company (e.g., knowledge about potential partners); and V21-Good contacts in the potential market."

"The motivations related to incubator organization (F17) are measured by six items."

"The second subdimension, attitudes towards entrepreneurship (F13), also includes three items: V34-Advice received from external organizations; V35-Campaigns aimed at encouraging entrepreneurship and V36-Society's attitudes towards entrepreneurship."

Results and Discussion

"The process of assessing the multidimensional nature of the entrepreneurial motivation consists of two steps: assessment and adjustment of the measure model, and verification of the multidimensionality of the concept."

"We assessed and adjusted the measurement model proposed for analysis of the entrepreneurial motivation."

"Validity of the entrepreneurial motivation scale was measured in two ways: content validity and construct validity."

"Discriminant validity for the entrepreneurial motivation scale is supported since no range includes the value 1.0."

"The second step in the analysis was to verify the multidimensionality of the entrepreneurial motivation; thus we needed to very whether the dimensions proposed converged in a single factor."

"Items related to the entrepreneurial opportunity behave differently from the other dimensions and the situation is similar for the indicators of the subdimension desire for wealth and supporting infrastructure, which were eliminated in our adjustment to the scale."

Conclusions

"We proposed and evaluated a model for entrepreneurial motivation comprising six dimensions that represent different types of motivations: personal, related to scientific knowledge, to the entrepreneurial opportunity, to the availability of resources to create business, to the incubator organization and to the social environment."

"The motivations related to scientific knowledge are important in the decision to create an academic spin-off."

"Incubator organizations are shown not to be important motivating factors for the creation of academic spin-offs in Spain."

"There are some areas where the incubator organizations play a more positive role in the generation of these companies, for example, training in business management, reducing the academic loads of researchers engaged in creating spin-offs, and consultancy for various aspects of their business plans."

"Studying the impact of the motivation dimensions for the survival and growth of academic spin-offs would make an interesting research topic."

Evaluating Motivational Characteristics in Vocational Music Education Within the Perspective of Self-Determination Theory [282]

This is a machine-generated summary of:

Virkkula, Esa: Evaluating motivational characteristics in vocational music education within the perspective of self-determination theory [282].

Published in: Empirical Research in Vocational Education and Training (2020).

Link to original: https://doi.org/10.1186/s40461-020-00098-5.

If you want to cite the papers, please refer to the original.

For technical reasons we could not place the page where the original quote is coming from.

Abstract-Summary
"This article examines the influence of vocational music education workshops on the learning motivation of popular and jazz music students in a self-determination theory's (SDT) framework."

"Psychological needs will become satisfied when the learner autonomously influences the choices and decisions of his or her learning."

"The present article is based on a qualitative research project that examined 62 Finnish conservatory popular and jazz music students who participated in 11 music performance workshops with professional musicians in 2003–2011."

"The research data consists of (a) workshop plans and learning goals written by the students before the workshops and (b) reflective essays of experiences after the workshops."

"The students were responsible for the planning and implementation of music performances with professional musicians."

Introduction

"It is important for a music teacher to consider what kind of learning environments support student motivation and contribute to the development of competence (Barrett and Bond [130]; Cogdill [131])."

"The present study will clarify the learning motivation factors of popular and jazz music students in conservatories according to the self-determination theory (SDT) (Deci and Ryan [132–135])."

"A research question was set to focus on how the cooperation of a professional musician and conservatory students is in connection to the students' learning motivation evaluated in the perspective of self-determination theory?"

"Project, the musicians' vocational learning at work was arranged in a new type of workshop in which the students planned and implemented a performance in cooperation with professional musicians according to popular and jazz music tradition (see Berliner [136], p. 44–45, also Gatien [137]; Gridley [138])."

Research Method, Data and Analysis

"After the workshop (phase 2), the students (N = 62) wrote about their experiences of the workshops guided by the following questions: "Describe your cooperation with the professional musician", "What did you learn in the workshop?"."

"The researcher analysed the study data produced by the students (phase 3) using a theory-oriented content analysis method, which is widely used for systematic analyses of diverse data in qualitative research (Schreier [139], p. 2–5)."

"The data was transcribed, after which the grouped textual expressions were tabulated into the categories describing SDT."

"Aided by tabulation, the researcher could survey the data in student-specific, workshop-specific and SDT-specific ways and also collect quantitative data out of the data."

"The researcher's personal experiences and their analyses concerning the work of a musician widened the understanding of learning in workshops (Virkkula and Nissilä [140])."

Results: Self-determination Theory in Workshops

"By training with the repertoire and improvisation. (A16) (The musical) material was really challenging and you had to work hard on it to survive (…). (A19) The professional musicians who played with the students in the workshops appeared to be inspiring co-partners in ensembles."

"It would have been nice to be able to participate in choosing the pieces. (A5) In the written expressions that refer to supporting the competence of acting as a musician (f = 110, f % = 32.3%), the students described the connections of the workshop in terms of learning to play, clarifying the structures of music, the development of self-confidence and the mutual communication of the players."

Discussion

"The task of this case study was to explore how the cooperation between popular and jazz music students and professional musicians in the music workshops in the Conservatory of Oulu in Finland was connected to the students' learning motivation from the perspective of SDT."

"The students' (N = 62) written expressions (F = 341) revealed that working with a professional musician supported autonomy, competence and relatedness according to SDT in 88.6% (f = 302) of responses."

"The professional musician's encouraging scaffolding in the workshop had a significant role for students' development (e.g. Küpers and others [141])."

"Although controlling learning has been found to be connected to a decrease in student motivation, it does not necessarily come about if the external control does not thwart learner autonomy (Radel and others [142], p. 101)."

Conclusion—Implications for Music Education

"The difference between a traditional master class and the workshop method employed in this research is in the activation of students, their interaction and their reflection, as well as in the role of the professional musician (Green [143], 144)."

"According to the student responses, they regarded it as very meaningful to be allowed to participate in the workshop planning, to contribute to its implementation and apply to their own learning."

"The students' tasks only involved being present at given times and being prepared to play, which did not align with the original concept of the workshop method."

"The present research (also Burwell [145], p. 284; Evans [146], p. 77–79; Gatien [137], 96; Lebler [147], p. 207–208) suggests that, in music education institutes, learning environments should be developed in which the musician's work can be confronted and experienced: the students should (1) plan and implement performances, (2) critically evaluate the learning process and the results attained from it, and (3) work in cooperation with professional musicians."

Motivational Clusters and Performance in a Real-Life Setting [283]

This is a machine-generated summary of:

Gillet, Nicolas; Vallerand, Robert J.; Rosnet, Elisabeth: Motivational clusters and performance in a real-life setting [283].

Published in: Motivation and Emotion (2009).

Link to original: https://doi.org/10.1007/s11031-008-9115-z.

If you want to cite the papers, please refer to the original.

For technical reasons we could not place the page where the original quote is coming from.

Abstract-Summary
"The present research investigated whether assessing adolescent elite athletes' motivational profiles at the beginning of the season would allow us to predict their subsequent performance over the course of a competitive season."

"In two studies, athletes completed the French version of the Sport Motivation Scale (Brière and others, Int J Sport Psychol 26:465–489 [148]) at the beginning of the season."

"Results revealed the presence of a four-cluster solution, reflecting different levels of autonomous and controlled motivations."

"Results revealed that controlling for past performance, athletes with the least self-determined motivational profile obtained lower levels of subsequent tennis performance than those in the three other (more self-determined motivational) clusters."

"Results revealed a three-cluster solution very similar to that of Study 1."

"In line with Study 1, results revealed that the least self-determined motivational profile led to the lowest level of performance."

Introduction
"The more autonomous or self-determined forms of motivation (intrinsic motivation as well as identified and integrated regulations) have been found to lead to a number of positive cognitive (e.g., concentration; e.g., Brière and others [148]), affective (e.g., positive affect, flow; e.g., Kowal and Fortier [150]; Pelletier and

others [151]), and behavioral (e.g., persistence; e.g., Pelletier and others [152]; Sarrazin and others [153]) outcomes in sport settings."

"As Vallerand [154] suggested, a number of motivational profiles may exist in actual life domains such as sport (e.g., high levels of both autonomous and controlled motivation) and it is possible that in certain contexts, profiles that deviate from the one proposed by SDT, lead to equivalent or even more positive outcomes."

"While increasing research has looked at the motivational clusters that emerge in sport (e.g., Harwood and others [155]; Hodge and others [156]; McNeill and Wang [157]), exercise (Cumming and Hall [158]; Matsumoto and Takenaka [159]; Wang and Biddle [160]), and physical education (Ntoumanis [161]; Wang and others [162]) settings, researchers have typically used a mix of theoretical approaches in determining the clusters within the purview of the same study."

Study 1

"In line with past research using cluster analysis in education and sport (e.g., Ratelle and others [149]; Vlachopoulos and others [163]), it was expected that at least three clusters would be uncovered: (1) a high autonomous–low controlled motivation cluster; (2) a high autonomous–high controlled motivation cluster; and (3) a low autonomous–high controlled motivation cluster."

"In line with SDT and past studies using cluster analysis in education (e.g., Ratelle and others [149]), it was hypothesized that the least self-determined cluster would display the lowest level of performance."

"This is because research with athletes from a variety of sports (e.g., Brière and others [148]; Pelletier and others [151]) has shown positive relationships between autonomous forms of motivation and concentration which may represent one the most important predictors of performance (Vallerand [164])."

Method

"The ratio between the number of victories and the number of matches played during the tennis season following questionnaire completion was used as a performance measure (i.e., Performance 1)."

"We used the same method to calculate a total performance score for the two seasons following questionnaire completion (i.e., Performance 2; $M = 117.5$ matches; $SD = 41.1$)."

"We also obtained two other measures of performance with respect to the scores as determined by the French Tennis Federation for the tennis season following data collection (i.e., Performance 3), and for the next two seasons (i.e., Performance 4)."

"The ratio between the number of victories and the number of matches played (M = 63.1 matches; SD = 18.4) during the tennis season prior to data collection was used as a performance score for the previous season."

Results

"Participants in the first cluster represented 18% of the sample (n = 30) and included athletes who displayed high levels of intrinsic motivation, identified regulation, introjected regulation, and external regulation, but low levels of amotivation."

"The third cluster represented 28% of the sample (n = 47) and included athletes who displayed a high level of autonomous motivation, and low to moderate levels of controlled motivation and amotivation."

"Results from an ANOVA, $F(3, 166) = 15.25$, $P < 0.001$, $\eta 2 = 0.22$, followed by post hoc tests using Fisher's LSD revealed that participants in the mod AU–high C group (M = 4.73, SD = 3.72) exhibited significantly lower scores on the self-determination index compared to those in the high AU–high C group (M = 9.50, SD = 2.54), the high AU–mod C group (M = 8.82, SD = 2.03), and the mod AU–low C group (M = 7.84, SD = 2.89)."

"Post hoc tests indicated that the least self-determined cluster, the mod AU–high C group, obtained significantly lower levels of performance (M = 0.46) than all other clusters that did not differ among them."

Discussion

"Further, the present findings showed that athletes with the least self-determined motivational profile at the beginning of the tennis season (i.e., mod AU–high C profile), displayed the lowest level of subsequent performance."

"This result on the role of motivation in performance is particularly striking when one considers that these athletes were among the best of their respective age group in a country recognized for its high level of tennis development (for instance, the International Tennis Federation world men junior rankings for the year 2006 showed that France was the only country to have two athletes in the top 10)."

"Future research is needed to replicate the present findings with athletes from different sport activities because as suggested by Ratelle and others (149), athletes' motivational profiles might be context sensitive."

Study 2

"While our first prospective investigation has provided us with important information regarding athletes' motivational profiles in elite sport, it seems imperative to conduct an additional study with elite athletes practicing another competitive activity."

"The purpose of this second study was twofold."

"We examined the relationships between these motivational profiles and sport performance."

Method

"Each participant volunteered to complete a questionnaire before the beginning of the fencing season."

"At the end of each season, a measure of sport performance was obtained via the French Fencing Federation."

"The first performance score reflects the national ranking in fencing of each athlete at the end of the season following data collection (i.e., Performance 1)."

"The national ranking determined by the French Fencing Federation was also used as a performance measure for the previous season."

"The lower the national ranking (e.g., first), the worse the athletes' performance was during the fencing season."

Results

"The second group (mod AU–mod C), which constituted 33% of the sample ($n = 81$), included athletes whose motivational profile was characterized by moderate levels of both autonomous and controlled motivation and low levels of amotivation."

"Results from an ANOVA on the self-determination index revealed a significant effect for clusters, $F(2, 243) = 14.98$, $P < 0.001$, $\eta 2 = 0.11$."

"A first ANOVA and Fisher's LSD tests were conducted to determine whether the motivational profile groups differed significantly with respect to the fencing performance during the season prior to data collection (previous performance)."

"The results of a second ANOVA on the subsequent performance (i.e., Performance 1) revealed a significant effect for clusters, $F(2, 236) = 3.20$, $P < 0.05$, $\eta 2 = 0.03$."

Discussion

"The first purpose of the present study was to identify athletes' motivational profiles in a sample of junior elite fencers."

"Contrary to the first study in tennis, the present results revealed the existence of three readily interpretable clusters."

"The present sample was composed of junior elite fencers, while the first study included junior elite tennis players."

"In line with recent investigations (e.g., Ratelle and others 149, Study 3), the present findings suggest that having a motivational profile characterized by low to moderate levels of autonomous and controlled motivations toward an activity is counterproductive with respect to performance in this activity."

General Discussion

"By finding that the least self-determined profile was associated with the worst sport performance, the results of both studies provide support for SDT and past investigations that showed that non-self determined motivation was associated with negative sport outcomes such as distraction (e.g., Brière and others [148]; Pelletier and others [151]), dropout (e.g., Pelletier and others [152]) and burnout (e.g., Cresswell and Eklund [165, 166]; Lemyre and others [167])."

"Although these results were obtained in a specific cultural context (Bulgaria was under the controlling communist regime when the Chantal and others study was conducted in 1989), these findings suggest that sport performance may be related to high levels of both autonomous and controlled motivations."

"What is striking in the present findings, is that a truly self-determined motivation cluster (i.e., a high autonomous–low controlled motivation group) was not obtained."

Motivational Assistance System Design for Industrial Production: From Motivation Theories to Design Strategies [284]

This is a machine-generated summary of:

Dostert, Janik; Müller, Romy: Motivational assistance system design for industrial production: from motivation theories to design strategies [284].

Published in: Cognition, Technology & Work (2020).

Link to original: https://doi.org/10.1007/s10111-020-00643-y.

If you want to cite the papers, please refer to the original.

For technical reasons we could not place the page where the original quote is coming from.

Abstract-Summary

"The design of human–machine interaction often does not foster the motivation to learn more about their machine or system."

"Motivation to learn could be attained by a motivating socio-technical design of assistance systems, but suitable and context-specific design strategies are lacking."

"A systematic literature review of motivation theories in education, at the workplace, and in system design was carried out."

"The resulting 16 theories were integrated into a conceptual model of motivating assistance system design in industrial production."

"Learning motivation results from the satisfaction of the needs for autonomy, competence, and relatedness, which in turn is mediated through the design of the system (including interface, task, and behavior)."

"Strategies for motivational design are derived from the model, and an example from the discrete processing industry is used to illustrate how the model could be applied to design assistance systems in this domain."

Introduction

"In many contemporary work environments, motivation cannot be taken for granted, and operators often lack the motivation to learn more about their machine or system (Schult and others [168])."

"Until now no theory exists which adequately supports the development of strategies and methods to design assistance systems in ways that instill or increase motivation to learn."

"In industries that rely on the operation of advanced technologies, the motivation to learn is of utmost importance as performance in these settings depends on operator skills and knowledge (Adler [169]; Walton and Sussman [170])."

"Cyber-physical production systems provide ample opportunities to support learning: Given their processing capabilities and the availability of large amounts of data, appropriately designed assistance systems could serve as digital learning environments."

"We use a theory-driven approach to derive practical strategies that system designers can use to create assistance systems which help increase operators' motivation to learn more about their machine or system."

Method

"To develop strategies of motivational assistance system design, different motivation theories from three major application areas were gathered: motivation in education, at the workplace, and in system design."

"The search strings TI(work* OR education* OR learn* OR industr*) AND TI motivat* AND (theor* OR framework OR model OR design OR approach) and (system* OR "assistance system" OR design*) AND motivat* and TI((motivation*) AND (design OR human factors)) AND theory resulted in 8886 research papers."

"Due to the fact that motivation research in the field of system design is a rather new topic and specific theories are not represented as well as in the other two fields, we decided to broaden the search string to include theories on Human Factors in general that mention motivation in the title."

"Step four resulted in a total of 16 theories: five theories on motivation in education, seven theories on motivation at the workplace, and four theories on motivation in system design."

Motivation Theories

"The main themes of the general motivation theories are the presence and satisfaction of innate needs, the importance of intrinsically motivated action, and the quality of the goals an individual pursues."

"Main elements of the presented theories on motivation in education are the value of a task or its anticipated reward versus punishment, beliefs in one's own abilities, the presence of attributions and their role in shaping future behavior, and the presence of a larger social context in which an individual's actions are embedded."

"A similar concept has been suggested by Expectancy-Value Theory, which defines a task as needing some personal importance to be relevant and motivating, as well as by Locke and Latham who stated that a meaningful reason can enhance goal acceptance (Gagné and Deci [171])."

"While there has been little empirical evaluation of the theory per se, its main concepts of motivational affordances and need satisfaction are deeply grounded in the research on gamification and SDT."

Model Development

"If the task and behavior components of the system are designed with the JCM core characteristics in mind, this will impact motivation, engagement, and well-being via mediation through the critical psychological states—experienced meaningfulness of work, felt responsibility for the work outcomes, and knowledge of actual work results."

"The core characteristics proposed in JCM which determine the engagement in a task and therefore the overall outcome of motivation are moderated by person variables such as employees' growth need strength, skills, knowledge, and goals."

"The system layer acts on the person variables through mediation via need satisfaction, motivation, and learning outcomes: A motivating interface may increase employees' growth need strength or enhance their ability to execute some task."

"This also becomes apparent in the way the spheres are laid out: The context, person, and job factors start from the left and end in the system sphere, which mediates operators' need satisfaction with the interface at the rightmost point."

Deriving Strategies

"To introduce design elements that help motivate employees to learn more about a machine or system, different strategies can be derived from the theoretical assumptions of the model presented in the previous section."

"As domain characteristics have profound impacts on operators' work (Müller and Oehm [172]), designers should first assess any specific parameters and constraints present in the industry the assistance system will be designed for."

"Skills and knowledge are assessed regarding the industry-specific requirements from the work domain layer, while growth need strength as a moderator variable of JCM is closely related to the factors on the job characteristics layer."

"The theoretical model for motivational assistance system design in industrial production (MADIP) lets designers derive many strategies to create an environment that increases the motivation to learn."

"After assessing the work domain constraints as well as the operator- and job-specific variables, strategies pertaining to the actual system can be developed."

Discussion

"To provide a basis for future research and to derive strategies, a theoretical model of motivational assistance system design was introduced."

"While some of the theories included in the model are rather new, they mainly integrate the concepts of need satisfaction and qualitative differences in motivation from SDT, which have been empirically evaluated in various domains and settings."

"To perform a valid test, it would be necessary to design at least two assistance systems (one according to the model and one without it), implement them in the field, and evaluate their impacts on changes in operator motivation in a longitudinal study."

"To further increase the practical usefulness of the model, specific situated motivational affordances in the context of assistance system design should be investigated."

"Although the present study is mainly concerned with enhancing motivation to learn through the design of assistance systems, there are several other strategies organizations can use to instill such motivation."

Different Underlying Motivations and Abilities Predict Student Versus Teacher Persistence in an Online Course [285]

This is a machine-generated summary of:

Higashi, Ross M.; Schunn, Christian D.; Flot, Jesse B.: Different underlying motivations and abilities predict student versus teacher persistence in an online course [285].

Published in: Educational technology research and development (2017).

Link to original: https://doi.org/10.1007/s11423-017-9528-z.

If you want to cite the papers, please refer to the original.

For technical reasons we could not place the page where the original quote is coming from.

Abstract-Summary
"A study of 172 K-12 students and 114 K-12 teachers taking the same free, online, summertime programming course finds that student and teacher populations have different underlying motivational models that predict rates of persistence in the course despite having generally similar motivational levels."

"Student persistence is predicted by prior programming knowledge, intrinsic interest in the subject matter, and mastery approach goals."

"Future design of free choice learning environments can take these factors into account to increase rates of user persistence for different target user populations."

Background
"Achievement Goal Theory (AGT) originated in the 1980s with the primary distinction between learners with mastery goals and performance goals."

"Learners who report high levels of Mastery Approach goals tend to display many characteristics and behaviors that are beneficial to learning, including increased persistence when facing difficulty (Elliot and others [173])."

"Empirically, the relationship between Mastery Approach goals and high persistence is well-documented, Performance Approach generally contributes to persistence (though not always learning outcomes), and the relationship between Performance Avoidance or Mastery Avoidance and persistence is less straightforward (Elliot and others [173]; Hulleman and Senko [174])."

"As our study's objectives are to investigate the effects of participant motivation on persistence in MOOCs, we employ the slightly older but better understood three-goal model (mastery approach, performance approach, and performance avoidance), acknowledging that it comes with known limitations."

Predictions About Persistence in Online Courses

"We draw attention to another heterogeneity of interest: K-12 students and teachers may both be enrolled in a given course as learners."

"Such a difference might not manifest as a difference in course performance or mean levels of specific motivational constructs, but rather as a differential factor in predicting persistence: if most teachers are subject to this pressure, then teachers with more such motivation may persist at a higher rate; meanwhile, students—not as subject to the same pressure—may show a null or negative relationship between Performance Avoidance motivation and persistence."

"The main research questions to be answered in our study are: (1) whether Achievement Goal Theory and Expectancy Value Theory constructs predict learner persistence in a free online course, and (2) whether the set of predictive constructs is the same or different between K-12 students and teachers enrolled as learners in the same course."

Methods

"The beginner-level programming topics (e.g. movement commands, loops) covered in the course were considered common content that both students and teachers would need to know, so only a single, general-audience course section was made available, open to everyone."

"Given our research questions, we selected only K-12 teachers and K-12 students who answered all the pre-survey items (i.e., excluding the small number of adult learners who were not teachers), resulting in a sample of 172 K-12 students and 114 K-12 teachers."

"Mastery Avoidance items were omitted from our instruments due to concerns about linguistic complexity for younger students (e.g., "One of my goals is to avoid learning less than I could about programming," could be problematic for elementary school age students), to address concerns about test fatigue, and the weaker prior research connection of this construct to learner persistence."

Results and Discussion

"The resulting student model included Mastery Approach motivation and pre-test score."

"To the best student model, the best-fitting teacher model included Performance Approach motivation as a positive factor, and Performance Avoidance motivation as a negative factor (i.e., a completely different set of two predictors)."

"To examine this possibility, we conducted two complementary follow-up analyses on the ($n = 33$) teachers and ($n = 60$) students who had completed the surveys at pre and post: one analysis examined whether there were differential pre-post mean shifts in construct values for teachers versus students, and a second analysis tested for differential relative order stability (or test–retest reliability) within each group."

"There was no significant shift in most predictors in the model, except for intrinsic interest, which increased for both groups ($p = 0.055$ for teachers, $p < 0.001$ for students)."

General Discussion

"There were also different factors, however, predicting student and teacher persistence: (1) Pre-test score predicts persistence for students but not teachers; (2) Performance Approach and Performance Avoidance motivation explain teacher persistence, while student persistence is better modeled using Mastery Approach; and (3) Identity is a significant predictor for persistence in teachers, but not students."

"The demand for teachers to appear knowledgeable about subject material they will later teach might drive them to persist through the learning activities, whereas for students the learning itself may the primary goal of such an experience (i.e., a difference in the importance of extrinsic versus intrinsic goals)."

"Current Achievement Goal Theory does not explain how two sub-populations of learners with the same overall levels of Mastery Approach, Performance Approach, and Performance Avoidance motivation can support two different models of persistence."

Conclusions

"Our study examined the ability of motivational constructs from the educational motivation literature to predict user persistence in a MOOC-like online learning environment."

"Persistence among K-12 students was predicted by prior knowledge, mastery approach orientation, and intrinsic interest."

"Persistence among teachers was predicted by performance approach orientation, performance avoidance orientation, intrinsic interest, and identity."

[Section 7]
"Free online courses, including Massively Open Online Courses (MOOCs), are an increasingly popular form of online instruction offering for large audiences."

"Since the cost to offer a course to diverse audience scales only with the costs of internet bandwidth, institutions offering free online courses very seldom institute a limit on the number of online consumers of the course material."

"Individuals' role relationships to the content, e.g. as teachers versus as students, could play a role in determining persistence as well; teaching status has been examined as a predictor of MOOC performance (DeBoer and others [175]), though not course persistence."

"We examine persistence in an online STEM course using two motivational frameworks drawn from the educational research literature—Achievement Goal Theory and Expectancy-Value Theory—which connect motivation and persistence in learning spaces."

Promoting Students' Self-Determined Motivation in Maths: Results of a 1-year Classroom Intervention [286]

This is a machine-generated summary of:

Brandenberger, Claudia C.; Hagenauer, Gerda; Hascher, Tina: Promoting students' self-determined motivation in maths: results of a 1-year classroom intervention [286].

Published in: European Journal of Psychology of Education (2017).

Link to original: https://doi.org/10.1007/s10212-017-0336-y.

If you want to cite the papers, please refer to the original.

For technical reasons we could not place the page where the original quote is coming from.

Abstract-Summary
"To counteract this negative development, a multicomponent intervention study based on the main ideas of self-determination theory, students' emotions in relation to maths, and the concept of self-regulation was developed and implemented in Swiss seventh-grade maths classes with basic demands."

"The study applied a quasi-experimental design comparing two experimental groups (combined student/teacher group: teachers and students participated in the intervention; student-only group: only students participated in the intervention) and a control group."

"Intervention effects were evaluated longitudinally (pre-post test) via self-reports of 348 students."

"Within-group results (t tests) showed no significant changes for students in the control group, a significant decrease in intrinsic and identified motivation for students in the student-only group, and a significant increase in intrinsic motivation and self-concept for students in the combined student/teacher group."

Introduction

"Given the crucial role of self-determined motivation in maths with regard to achievement, and that of achievement with regard to school careers, maintaining and fostering students' motivation and their beliefs in their own abilities must be viewed as a particularly important educational objective (Pekrun and others [176])."

"The aim of the present study was to evaluate whether seventh-grade students' self-determined motivation and self-concept (as a motivational self-belief) in maths could be maintained or even fostered through a classroom-based, multi-component intervention study based on the main ideas of the self-determination theory of motivation (SDT; Deci and Ryan [177]), students' emotions in relation to maths (Pekrun [178]) and the concept of self-regulation (Perels and others [179])."

"Learning environments that facilitate the satisfaction of these basic needs have a positive impact on students' self-determined motivation."

"If interventions aim at increasing the fulfilment of basic needs, (1) positive effects on self-determined motivation, emotions and academic self-concept are expected and (2) learning and emotion theories have to be addressed to foster basic needs."

The Present Study

"In response to the limitations of previous intervention research, the present study tested whether students' self-determined motivation and self-concept could be maintained or fostered by (a) a combined student/teacher intervention (with both students and their teachers participating in the intervention) and/or (b) a student-only intervention in the classroom setting, compared to a control group."

"To evaluate the effects of the intervention, we assessed students' self-determined motivation and self-concept before (at the beginning of seventh grade, pre-measurement) and after (at the end of the school year, post-measurement) the intervention."

"Based on the idea of the co-construction of learning (Fend [180]) and the fact that teachers play a crucial role in their students' motivation and beliefs (Martin

[181]), our main hypotheses addressing the intervention effects are as follows: we expected (4a) both intervention groups to show positive effects compared to the control group, with (4b) the combined student/teacher intervention being the most effective with regard to self-determined motivation and self-concept."

Method

"22 maths teachers and their 348 seventh-grade students in basic-demand maths classes from 17 secondary schools in the German-speaking part of the canton of Bern in Switzerland (mean age at the beginning of the study = 12.75, SD = 0.64) voluntarily participated and completed both surveys in seventh grade; of these, 179 were female (51.4%) and 169 were male (48.6%), and 117 (33.6%) had a migration background and 231 (66.4%) did not."

"To test hypotheses 1–3, effect sizes were calculated for within-group changes between pre- and post-test outcome measures using paired sample t tests and Cohen's d. Multivariate analyses of covariance (MANCOVA) controlling for the pre-test variables (self-concept and motivation, maths achievement and students' perception of basic need support from their maths teachers) as well as gender and migration background were conducted in order to examine differences in the development of students' self-determined motivation and self-concept between the three groups (hypotheses 4a and 4b)."

Results

"There were no significant pre-test differences between the groups prior to the intervention in any of the scales except for self-concept ($F(2, 345) = 3.678$, $p = 0.026$), with post hoc comparisons indicating that all three groups were significantly different from one another."

"For identified motivation, although the mean scores for students in the student/teacher group indicate an increase from pre- to post-test, the results fall short of significance ($t(133) = -1.660$, $p = 0.099$)."

"For self-concept, the results reveal an increase in the mean scores from pre- to post-measurement for students in all groups, but this development is significant only for students in the combined student/teacher intervention."

"Prior to the intervention, students with a migration background reported significantly higher intrinsic ($M_{with} = 3.18$ (0.87), $M_{without} = 2.96$ (0.93), $p = 0.037$), identified ($M_{with} = 4.21$ (0.80), $M_{without} = 3.95$ (0.84), $p = 0.005$) and external motivation ($M_{with} = 3.28$ (0.66), $M_{without} = 3.06$ (0.69), $p = 0.004$."

Discussion

"We investigated the effectiveness of a joint student/teacher intervention and a student-only intervention compared to a control group for the development of students' self-determined motivation and their self-concept during the first year of secondary education in Switzerland (grade 7)."

"To evaluate the effects of the interventions, we assessed students' maths-related motivation and their self-concept before (at the beginning of seventh grade) and after (at the end of the school year) the intervention."

"The most important finding of this study was that the combined student/teacher intervention showed the most positive effect on students' self-determined motivation (hypothesis 4b)."

"Despite the small effect sizes, our results suggest that it is possible to support seventh-grade students' self-determined motivation in maths through a teacher and student intervention based on the ideas of SDT, emotion- and self-regulation-theory."

Conclusion

"Key elements of the intervention that seem to have contributed to increase students' motivation and self-concept include aspects of self-determination theory (basic needs, SDT) and stage-environment-fit theory (SEFT), as well as inputs on emotions, learning strategies and self-regulation strategies."

"The effects of our study also show that the deterioration of self-determined motivation across adolescence is not an inevitable developmental trend; rather, it can be influenced by teachers who shape the learning environment professionally and by students as they develop learning competencies and reflect on their attitudes towards learning."

"The need for competence can be fulfilled when students develop skills that facilitate self-regulated learning and emotion regulation and teachers foster students' self-concept by providing informational, positive feedback, introducing optimally challenging learning activities, explaining how mistakes can be informational and stressing to students how much they have improved."

Unraveling the Motivational Effects and Challenges of Web-Based Collaborative Inquiry Learning Across Different Groups of Learners [287]

This is a machine-generated summary of:

Raes, Annelies; Schellens, Tammy: Unraveling the motivational effects and challenges of web-based collaborative inquiry learning across different groups of learners [287].

Published in: Educational technology research and development (2015).

Link to original: https://doi.org/10.1007/s11423-015-9381-x.

If you want to cite the papers, please refer to the original.

For technical reasons we could not place the page where the original quote is coming from.

Abstract-Summary

"This study deals with the implementation of a web-based collaborative inquiry (WISE) project in secondary science education and unravels the contribution and challenges of this learning approach to foster students' motivation to learn science, and its relation with student and class-level characteristics."

"Students' motivation was quantitatively studied based on the self-determination theory, and it was hypothesized that web-based collaborative inquiry can be considered as a need-supportive environment, which in turn can foster autonomous motivation and decrease controlled motivation."

"These students were significantly more positive about web-based inquiry in science education compared to students from a science track."

"We describe how the learning environment can be improved to satisfy students' basic needs and improve good quality motivation for science learning."

Introduction

"An important question in this regard is if web-based collaborative inquiry can improve students' motivation to learn science, and more importantly, how web-based collaborative inquiry should be implemented to motivate different groups of students to learn science."

"By specifying the contextual environments and need satisfaction that foster optimal learning, the self-determination theory is a relevant framework for unraveling the motivational effects and challenges of web-based collaborative inquiry in authentic classroom contexts."

"This study tries to fill the gaps in existing research by focusing on the motivational effects and challenges of web-based collaborative learning from an SDT perspective including the three basic needs: autonomy, competence, and relatedness."

"This study meets the research gap of limited research that gains insight into the interplay between individual differences and the way students experience a technology-enhanced intervention (Kim and Hannafin [182])."

The Self-determination Perspective

"This study put forth web-based collaborative inquiry by means of the Web-based Inquiry Science Environment (further referred to as WISE) (Slotta and Linn [183]) as a particular learning approach that can be perceived as need-supportive because of the features that may foster good quality motivation, that is, increased autonomous motivation and decreased controlled motivation for science learning."

"Although, as described above, the WISE learning environment has several characteristics to meet students' basic needs, it is rather unclear how students experience a WISE intervention regarding autonomy, competence, and relatedness support."

"What are the effects of the implementation of WISE on student motivation for science learning (autonomous and controlled motivation)?"

"Regarding research questions 1 and 3, based on previous research, it is hypothesized that web-based collaborative inquiry can increase students' autonomous motivation to learn science and decrease students' controlled motivation due to its perceived-need supportive characteristics."

Method

"Because the intervention had to be carried out according to a set of instructional principles and a strict protocol, but the science teachers did not have the time to go through a training period beforehand, it was decided to involve 34 Master's students to conduct and support the implementation of the WISE project."

"The Master's students served as the actual teachers during the project, whereas the regular classroom teachers predominantly observed the learning processes."

"Master's students experienced the learning environment from the teacher's perspective by developing a curriculum project using the underlying knowledge integration approach (see below)."

"Students' motivation for science learning was measured quantitatively based on the SDT perspective by means of an adapted version of the Academic Self-Regulation Questionnaire originally developed by Ryan and Connell 18, yet redesigned by Vansteenkiste and others [184]."

"The following procedure was used to analyze the effects of student and class-level characteristics on students' motivation for science."

Results

"A main effect of academic track regarding students' autonomous motivation for science prior to the intervention was found ($\chi^2 = 23.47$, df = 1, $p < 0.001$), indicating that general track students' autonomous motivation for science is significantly lower than the motivation for science reported by science track students."

"Based on the fixed part of the four-level unconditional null model, we can state that before the intervention students' controlled motivation for science learning across all students, groups, and classes was 1.56 (0.05) (=intercept β_0) and that no significant change was found after the intervention [=slope β_1, i.e. 0.08 (0.05)]."

"A main effect of academic track regarding students' controlled motivation for science prior to the intervention was found ($\chi^2 = 10.62$ df $= 1$, $p < 0.01$), indicating that general track students' controlled motivation for science is significantly higher than the motivation for science reported by science track students."

Discussion

"We can state that the hypothesis of an increased autonomous motivation for science learning is not entirely confirmed, but only holds for general track students."

"General track students' autonomous motivation prior to the intervention was significantly lower compared to science track students, and their controlled motivation was significantly higher which indicates a lower quality motivation profile."

"Within the group of students who were reluctant towards using WISE in the future (i.e., more students from a science track), most students indicated that they experienced a lower learning gain and that they understand it much better if the teacher explains the content and do not have to investigate everything by themselves."

"Future research should try to include the real classroom teacher to investigate the motivational and learning effects of teachers classroom-based interventions and additional research should also include the measurement of students' perceived need satisfaction and study the effects of teacher-centered interventions."

Identifying the Factors Leading to Success: How an Innovative Science Curriculum Cultivates Student Motivation [288]

This is a machine-generated summary of:

Scogin, Stephen C.: Identifying the Factors Leading to Success: How an Innovative Science Curriculum Cultivates Student Motivation [288].

Published in: Journal of Science Education and Technology (2016).

Link to original: https://doi.org/10.1007/s10956-015-9600-6.

If you want to cite the papers, please refer to the original.

For technical reasons we could not place the page where the original quote is coming from.

Abstract-Summary
"The purpose of this study was to identify specific factors contributing to the program's effectiveness in engaging students."

"Sage, Newbury Park [185] was used to develop a conceptual model identifying the central phenomenon, causal conditions, intervening conditions, strategies, contexts, and student outcomes of the project."

"Student motivation was determined to be the central phenomenon explaining the success of the program, with student empowerment, online mentor interaction, and authenticity of the scientific experiences serving as causal conditions."

"Teachers contributed to student motivation by giving students more freedom, challenging students to take projects deeper, encouraging, and scaffolding."

"Scientists contributed to student motivation by providing explanations, asking questions, encouraging, and offering themselves as partners in the inquiry process."

"The findings of this study provide relevant information on how to develop curriculum, use technology, and train practitioners and mentors to utilize strategies and actions that improve learners' motivation to engage in authentic science in the classroom."

Introduction
"PlantingScience (PS), developed in 2005 by the Botanical Society of America (BSA), is a prestigious Science Prize for Online Resources in Education (SPORE) award-winning program recognized by the American Association for the Advancement of Science (AAAS) for its complex design engaging classroom teachers, students, and scientist-mentors in an innovative, computer-supported science-learning environment (Hemingway and others [186])."

"Used internationally by over 12,000 students at the time of this study, PS provided advanced technology tools to mix scientific inquiry, classroom instruction, and online mentoring by practicing scientists and advanced science graduate students."

"Inspired by the challenge, the BSA developed PS with three overarching objectives: (1) provide opportunities for scientific inquiry experiences, (2) integrate plants into the science curriculum, and (3) facilitate online mentoring between scientists and students (Hemingway [187])."

Background

"Programs of this nature are not new, as students have attended summer camps, weekend research events, and after-school science mentoring programs at local colleges and universities for many years (e.g., Yale Science Outreach, Mentoring for Science at Harvard)."

"One example program developed in the 1990s was an online mentoring project linking a geology graduate student with students studying earthquakes in an Earth Science class (O'Neill and others [188])."

"Similar to CoVis, PS partners practicing scientists or graduate students with classroom science learners."

"According to the Science article touting PS as a successful SPORE winning program (Hemingway and others [186]), "Talking online with a scientist is exciting and motivating to students."

"PS is a complex program blending traditional classroom interaction with progressive methods such as inquiry-based learning, online mentoring, and scientist-student partnerships."

Methodology

"These strategies were utilized in the following manners: (1) Data were collected from extended fieldwork including PS classroom observations, teacher interviews, focus group discussions, and scientist-student asynchronous dialogues. (2) Data were collected in various ways, including direct observations, personal interviews, focus group conversations, and textual dialogues. (3) Researchers from both inside and outside the project served as peer reviewers at various stages of the process."

"Teachers and scientist-mentors used diverse strategies and actions while doing PS, and students responded in various ways."

"One of the most obvious ways Dan and Kelly orchestrated the PS project was by providing time on the computers for students to communicate with scientist-mentors."

"When students were apathetic in the beginning, as evidenced by comments such as, "Botany isn't my favorite subject in school," and "Science is not my strongest class," teachers and scientist-mentors used different strategies than when the students came into the project motivated and excited."

Implications

"PS provided a unique opportunity for students to develop their own research questions and strategies, but only because teachers allowed that to happen."

"In spite of all the challenges, this study bears witness that teachers can successfully orchestrate the complexities of PS in ways fostering student motivation."

"While using programs like PS can be extremely challenging, the evidence presented in this study implies PS works in part because of the contributions of dedicated teachers who are willing to let students work in complex learning environments."

"While teachers directed and orchestrated the classroom environment, scientists focused on mentoring and giving their student-teams individualized attention."

"By developing more robust curriculum that includes a mentoring component, the field of science education stands to make great strides in providing greater learning experiences for students in motivationally supportive contexts."

Limitations

"The sample in this study represented a small proportion of the overall population of participants who used PS through the years."

"A limitation of the study was the ability to extrapolate findings to the entire population."

"While researchers involved in the PS evaluation project spent time in other classrooms across the country and with other focus groups composed of additional scientists and teachers involved in PS, the sheer volume of qualitative data precipitated bounding the sample to its current size."

Conclusion

"The findings provided new understandings about a program (i.e., PS) that was positively influencing student motivation and engagement in science education."

"This study uncovered specific factors contributing to student motivation, specifically revealing how teachers, scientists, and curriculum developers played a role in the process."

"While this study does not make the claim that online mentoring programs involving scientists are the key to turning around science education, the concrete factors contributing to the popularity of PS uncovered in this study are germane to ongoing discussions about increasing student motivation in science education."

"Intense, field-based investigations of successful programs like PS provide relevant information on how to develop curriculum, use technology, and train practitioners and mentors to utilize strategies and actions that improve learners' motivation to engage in authentic science in the classroom."

Charting the Motivation, Self-efficacy Beliefs, Language Learning Strategies, and Achievement of Multilingual University Students Learning Arabic as a Foreign Language [289]

This is a machine-generated summary of:

Calafato, Raees: Charting the motivation, self-efficacy beliefs, language learning strategies, and achievement of multilingual university students learning Arabic as a foreign language [289].

Published in: Asian-Pacific Journal of Second and Foreign Language Education (2023).

Link to original: https://doi.org/10.1186/s40862-023-00194-5.

If you want to cite the papers, please refer to the original.

For technical reasons we could not place the page where the original quote is coming from.

Abstract-Summary

"Research on the relationship between students' language learning motivation (LLM), language learning strategies (LLS), self-efficacy beliefs (SEB), and achievement in non-European languages has been both limited and overwhelmingly cross-sectional, often with little attention paid to their multilingualism."

"Combining complex dynamic systems and sociocultural perspectives, this article reports on a study that explored changes in the LLM, LLS, and SEB trajectories of two multilingual students taking Arabic at a university in Norway over two semesters, including how these trajectories, alongside their multilingual competence, related to their achievement in the course."

"Data were gathered through weekly semi-structured interviews and a rating scale-based log that the participants kept of their LLM and SEB, as well as their exam scores over two semesters."

"The results indicated that, although their LLS remained fairly consistent or grew more diversified, the participants were not completely successful in maintaining or boosting their LLM, SEB, or achievement."

Introduction

"Because motivation can be object-bound (e.g., students' motivation for learning Arabic can differ from that of other languages such as English or Spanish; see Calafato and Tang [190]; Calafato [191], Dewaele and Proietti Ergün [192], Zheng and others [193]) and influenced by context (e.g., students in some countries may be motivated to learn Arabic in a way that qualitatively differs from those residing in other countries), there is a need for studies on learning Arabic as a foreign language (but also other non-European languages) from contexts outside those mentioned above, for example, in Europe, which was the focus of this study."

"The study, which adopted an integrated dynamic systems and sociocultural approach (Yashima & Arano [194]) as its primary framework, contributes to our understanding of the interactions between students' LLM, LLS, SEB, multilingual competence (i.e., their knowledge of multiple languages and associated experiences), and achievement as these concern the learning of Arabic as a foreign language in tertiary education in Europe."

A dynamic Systems and Sociocultural Framework for LLM

"Ramirez-Arellano and others [195], meanwhile, see interactions between motivation, affect, cognition, and achievement as both causal and reciprocal, underlining the need to study these variables (as represented by LLM, SEB, LLS, and multilingual competence in this study) together, rather than separately, when researching language learners."

"These variations in research findings could be a result of, as already stated, the many different approaches to categorizing LLS adopted by researchers, the specificities of the English learning context (as opposed to Arabic or other languages), as well as the snapshot-in-time, questionnaire-based nature of most studies on LLM, LLS, and SEB, where interactions between participants and researchers are likely to remain restricted and may, therefore, not fully reveal the experiences of participants."

"This study sought to contribute to research on the interactions between LLM, LLS, SEB, multilingual competence, and achievement vis-à-vis the learning of Arabic as a foreign language by exploring these variables over two semesters (i.e., an academic year)."

Methods

"The first interview with each participant lasted 60 min and elicited sociobiographical information (place of birth, early life, education, etc.), reasons for studying Arabic, desired proficiency in the language, current Arabic language ability, prior experiences with learning Arabic and other languages, travel to the Middle East, multilingual competence, level of motivation to learn Arabic, LLS, course materials, activities, and structure, support from parents, and class environment (e.g., relationship with teachers and classmates)."

"In terms of theoretical knowledge, this meant taking participants' responses and organizing these in relation to need states when it came to examining their LLM (e.g., learning Arabic to improve career-specific competences, wanting to help integrate refugees from Iraq and Syria into Norwegian society, interest in Arabic calligraphy, need to learn vocabulary rapidly, etc.), including for need strength, desire, value, and/or importance (Reeve & Lee [196])."

Results

"She said that for them, it (i.e., studying Arabic) would have been easier to understand if she were Arab or from a purely Norwegian-speaking household because then she would have had a more monolingual or heritage language background and it would make sense that she wanted to learn other languages."

"When asked if her motivation to learn Arabic had changed, she said that she remained very interested in the Arabic language and culture and that a strong motivator for her would be to go and live in the Middle East for a semester or so and use the language daily in authentic situations."

"She said that she was trying the Pimsleur Arabic course and was working with a teacher to learn a song in Arabic, which motivated her more to learn the language."

Discussion

"Before proceeding to a discussion of the findings (these have been organized under three subsections based on the study's research questions), it is important to state here that this study is one of the first to have investigated LLM regarding Arabic as a foreign language in Europe by combining dynamic systems and sociocultural approaches to cover multiple variables such as participants' SEB, LLS, and multilingual competence across different timescales."

"In the ontogenetic domain, her multilingual competence and desire to integrate Arabic into her identity had a positive impact on her LLM and SEB in the fall semester, though this impact turned negative in the spring semester and served to demotivate Talia and led her to doubt her abilities as a multilingual learner."

Conclusion

"For teachers, whether in schools or tertiary education, the findings of this study indicate the importance of keeping track of not only learners' achievement over time, which is formally done through tests, exams, and tasks, but also their LLM and SEB, which are not necessarily systematically or regularly assessed and can be difficult to observe in the classroom."

"For researchers, the study's findings underline the risks of conducting cross-sectional LLM research to the extent that such research might not be relevant the day or week after and thus could not be used by teachers or educational institutions to implement appropriate measures in support of students' language learning."

"This study also shed light on multilingual competence, which is strongly being promoted in the current education landscape, as it applies to learning distant languages (i.e., from those in one's repertoire), and how it relates to LLM, LLS, SEB, and achievement."

Teachers' Mastery Goals: Using a Self-Report Survey to Study the Relations Between Teaching Practices and Students' Motivation for Science Learning [290]

This is a machine-generated summary of:

Vedder-Weiss, Dana; Fortus, David: Teachers' Mastery Goals: Using a Self-Report Survey to Study the Relations between Teaching Practices and Students' Motivation for Science Learning [290].

Published in: Research in Science Education (2017).

Link to original: https://doi.org/10.1007/s11165-016-9565-3.

If you want to cite the papers, please refer to the original.

For technical reasons we could not place the page where the original quote is coming from.

Abstract-Summary

"Employing achievement goal theory (Ames Journal of Educational psychology, 84(3), 261–271 [197]), we explored science teachers' instruction and its relation to students' motivation for science learning and school culture."

"We then used this survey and hierarchical linear modeling (HLM) analyses to study the relations between 35 science teachers' mastery goals in each of the TARGETS dimensions, the decline in their grade-level 5–8 students' (N = 1.356) classroom and continuing motivation for science learning, and their schools' mastery goal structure."

"The findings suggest that adolescents' declining motivation for science learning results in part from a decreasing emphasis on mastery goals by schools and science teachers."

Achievement Goal Theory

"The relations between students' motivation for science learning (and for learning in other domains) and teachers' self-reported mastery goals have only seldom been studied (Miller and Murdock [198])."

"To provide a better understanding of science teachers' mastery goals and their impact on students' motivation for science learning, in this study we aimed to (a) employ the TARGETS framework to explore the relations between science teachers' mastery goals, their students' perceptions of these goals and their classroom, and continuing motivation for science learning and (b) examine the relations between science teachers' mastery goals and their schools' mastery structures."

"Are science teachers' mastery goals (self-reported as well as perceived by students) related to their students' classroom motivation for learning science (mastery goal orientation and classroom engagement)?"

"Which of science teachers' TARGETS mastery-emphasizing practices are related to their students' motivation for learning science (mastery goal orientation, classroom and extracurricular engagement)?"

Methods

"HLM allowed us to test the relation between teachers' self-reported mastery goals and their students' perceptions of the teachers' goals."

"The student-level measurement was created by a single student score and allowed us to test which of the TARGETS dimension was related to students' perceptions of their teacher's mastery goals and which of them was associated with age-related changes in students' perceptions (second research question)."

"We tested the relation between science teachers' self-reported mastery goals and their school's mastery structure in two ways: (a) by correlating teachers' reports of their own goals with their reports of the school structure and (b) by correlating teachers' reports of their own goals with their students' aggregated perceptions of the school's structure."

Results

"The HLM analysis also revealed that the mean personal mastery orientation of each class was significantly related their science teacher's self-reported mastery goals ($\gamma_{01} = 0.35$, $p < 0.05$) as well as to the students' perceptions of their teacher's mastery goals ($\gamma_{01} = 0.68$, $p < 0.001$)."

"Students' mean classroom engagement was not significantly related to their science teacher's self-reported mastery goals, but it was related to the perceived mastery goals of the teacher by the students ($\gamma_{01} = 0.50$, $p < 0.001$)."

"Students' mean perception of their science teacher's mastery goals was related to their teachers' self-reported mastery goals ($\gamma_{01} = 0.40$, $p < 0.05$)."

"Among the different TARGETS dimensions, only Authority/Autonomy was significantly related to students' mastery goals; the decline in students' mastery goals ($\gamma_{10} = -1.23$) was reduced as teachers' support of student autonomy increased ($\gamma_{11} = 0.28$, $p < 0.01$)."

Discussion

"Addressing the first research question, the HLM analyses suggest that students' motivation for science learning in and out of school is related to their science teacher's mastery goals (as reported by either teacher or students)."

"Science teachers' usage of mastery-emphasizing practices from the Task dimension was related to students' engagement in science learning in and out of school and also to the students' perceptions of the teachers' mastery goals, particularly its change with age."

"The Authority/Autonomy dimension also was not related to students' perceptions of their science teachers' mastery goals."

"Unpacking mastery structure into the TARGETS dimensions, as we did in this study, strengthened our assumption that studying the influence of science teachers' goals on students' motivation requires one to distinguish between the different TARGETS dimensions, especially if one wishes to give practical recommendations."

Conclusion

"This study, which used teachers' self-report data, offers solid quantitative support of previous findings (Vedder-Weiss and Fortus [199]), which were based only on students' reports and suggested that adolescents' declining motivation for science learning (in and out of school) results, at least partly, from declines in schools and science teachers' mastery goals."

"Illustrating the utilization of this survey offered new insights into the relative relations between the different TARGETS dimensions and students' motivation for science learning."

"Educators and policy makers wishing to support adolescents' classroom and continuing motivation for science learning and prevent its decline with age should consider the reinforcement of practices that emphasize mastery goals; in particular, they should consider the reinforcement of practices that relate to the design of tasks and the autonomy offered to students."

[Section 6]

"Framed by achievement goal theory, in this study, we developed a self-report survey assessing science teachers' usage of practices that emphasize mastery goals."

"We used a surveyed teachers in different school cultures to advance our understanding of the relations between school goal structure, science teachers' goals, and the decline in students' classroom and continuing motivation for science learning."

"This article is organized as follows: First, we review the literature on achievement goal theory and the constructs involved in this study, including classroom and school goal structure, the TARGETS framework, science classroom engagement, and continuing motivation."

"We then present descriptive statistics and results regarding the relations between science teacher's mastery goals, their schools' goal structures, and their students' motivation."

"We conclude the article with a discussion of the insights afforded by unpacking goal structure by considering the teachers' perspectives and by situating the study in various school cultures."

Investigating the Effects of the Flipped Classroom Model on Omani EFL Learners' Motivation Level in English Speaking Performance [291]

This is a machine-generated summary of:

Abdullah, Mohamad Yahya; Hussin, Supyan; Ismail, Kemboja: Investigating the effects of the flipped classroom model on Omani EFL learners' motivation level in English speaking performance [291].

Published in: Education and Information Technologies (2019).

Link to original: https://doi.org/10.1007/s10639-019-09911-5.

If you want to cite the papers, please refer to the original.

For technical reasons we could not place the page where the original quote is coming from.

Abstract-Summary
"The current study aims to examine the effect of implementing the Flipped Classroom Model (FCM) on the motivation level of Omani EFL (English as a Foreign Language) learners to speak English."

"In the quantitative dimension, a Motivation in English Speaking Performance Questionnaire (MESPQ) was distributed in three phases: before, during, and after the application of FCM over a period of 15 weeks."

"The findings showed that using FCM as a teaching approach helped to develop a creative, engaging, and motivating climate in the EFL speaking class."

"Significant differences ($p < 0.001$) in students' motivation levels before and during the implementation of FCM indicated that FCM contributed significantly to increasing students' motivation in English speaking performance over time."

"Besides, the Omani EFL students showed significantly greater inclination to be involved in the speaking activities over time, including their interest in applying FCM in all taught courses."

"Further studies on FCM could focus specifically on its effect on learners' achievement."

Introduction

"The paucity of opportunities to use English outside the classroom also causes EFL learners to lose interest and motivation in learning English (Leman [200])."

"Although a lot of work has been reported on different aspects of learning motivation, very little work has been carried out to examine the impact of the flipped classroom model on EFL learners' motivation level."

"No empirical study has been conducted in Oman on the influence of implementing the flipped classroom model on EFL learners' motivation to interact in English."

"The current study aims at investigating the effect of utilizing the flipped classroom model (FCM) on Omani EFL learners' motivation to interact in English; and more specifically, the effects of FCM on the dimensions of motivation including attitude towards speaking English, perception of speaking performance evaluation, and perception of speaking activities."

Literature Review: Motivation in the Language Learning Process

"They suggested that the instrumental motivation should get more attention in the EFL context because EFL learners, studying the target language within their own culture, obviously lacked the social context for experiencing integrative motivation."

"It is accepted that teachers play a key role in building learner motivation and attitudes towards language learning."

"According to Lightbown and Spada (2013), cited in Badrkoohi and Maftoon [201], EFL teachers must design their teaching based on their learners' needs and characteristics as this could contribute to motivating EFL learners to overcome the language learning obstacles and weaknesses that hinder them from learning the target language."

"It is thus clear from the vast amount of research on motivation that the level of EFL learners' motivation is based on their attitudes about English language learning which in turn influences their learning success (Gardner [202]; Dörnyei [203]; Ghorbandordinejad [204])."

Flipped Classroom Model

"In the flipped language classroom, students have the opportunity to practise expressing themselves after learning relevant and useful aspects of the target language on their own from online resources and their experiences in their real world (Reeve [205])."

"The second part of the flipped classroom which includes interaction with the teacher and with other students involves students in critical thinking processes to allow them to construct their knowledge collaboratively and evolve their language skills (Kurt [206]; Üğüten and Balci [207]; Wu and others [208])."

"Basal [209] showed that the flipped classroom model increases the opportunity to overcome the limitations of the class, enables students to prepare in advance, and enhances classroom participation."

"Kurt's experimental data highlighted the higher level of self-efficacy beliefs and better learning outcomes as the main findings for the implementation of the flipped classroom model in the education programme of a pre-service English teacher."

Methodology

"This study used a mixed methods approach to obtain qualitative and quantitative data aimed at answering the research questions developed from the stated objectives."

"In the qualitative dimension, reflective journal forms were collected to obtain more in-depth information."

"The quantitative and qualitative data obtained from these research instruments were triangulated to arrive at valid and reliable answers to the research questions of this study."

Design of the Study: Implementation of the Flipped Classroom Model

"This enabled students to learn in their own time and at their own pace before the class session, freeing up class time for practice, discussion, activities, and interaction (Bergmann and Sams [210]; Kvashnina and Martynko [211]; Kurt [206])."

"Students become the determining dynamic of the class sessions for discussing different topics with their teacher and peers and performing a variety of student-led activities."

"This type of in-class activity was expected to help students to be more self-regulating, motivated, and be aware of the assessment criteria."

"Prior to the class, all students were expected to have watched the video lectures on their own."

"Of the lecture, the teacher would give a brief introduction to the content of the lecture and then call on the groups of the peer teaching and the peer evaluation panel to assume their roles according to previous instructions."

Participants

"According to the nature of the current study, a purposeful sampling technique, specifically quasi-experimental procedure was employed in this study."

"Quasi-experimental design can be used "when individuals are not randomly assigned" (Creswell [212], p. 271)."

"The population from which the sample of ($n = 27$) respondents was drawn for the present study comprised the undergraduate EFL students enrolled in the compulsory course of Advanced Communication Skills (ENGL 154) at BUC."

Data Collection Instruments

"To examine the influence of FCM on the variable of motivation in students' English speaking performance, two main instruments were used: a questionnaire, and a reflective journal."

"This 13-item questionnaire (Motivation in English Speaking Performance Questionnaire, or MESPQ), was designed based on previous ESL/EFL motivation research questionnaires by Gardner [213], Dörnyei and others [214], Ryan [215] and Lamb [216]."

"The component of Perception of Speaking Performance Evaluation contained two items such as "I am motivated to speak English better because of classmates' feedback"."

"The last component comprised three items such as "I am motivated to speak English in group work competition"."

"The main part of the reflective journal form consisted of a set of questions which aimed to address various points relevant to the objectives and research questions of the study such as their thoughts and reactions toward the application of FCM, their perceptions about motivation along with their EFL speaking performance."

Data analysis Methods

"Of all, to ensure that all variables satisfied the assumption of normality distribution, quantitative data for univariate and multivariate normality was assessed with statistical tests."

"The statistical tests of skewness and kurtosis showed that all values between − 2.0 and + 2.0 for all variables met the criteria of the normal distribution."

"The test of one-way repeated measured ANOVA was carried out to assess the effectiveness of implementing FCM on the variable of motivation in English speaking performance at three separate times (pre-, midway, and post the application of FCM)."

"One-way repeated measure multivariate analysis of variance (MANOVA) was conducted to determine whether there were differences among the mean scores of the dimensions within the variable of motivation over time (pre-, midway, and post FCM application)."

Findings: Quantitative Data

"In order to compare the mean scores of students' motivation in their English speaking performance throughout the application of FCM in their EFL speaking classroom, a one-way repeated measure ANOVA was conducted to assess whether there were differences among these mean scores over time with a statistical test at time 1 (prior to FCM implementation), time 2 (after 7 weeks), and time 3 (after 14 weeks, post implementation)."

"In order to evaluate the differences in the mean scores of (N = 27) students' motivation in English speaking performance including the subscales of attitude to speaking English, perception of speaking performance evaluation, and perception of speaking activities, a one-way repeated measure MANOVA was conducted to assess whether there were differences among these three subscales over the period (Pre-test, before the implementation of FCM; Post-test, after 7 weeks form utilizing the FCM; Post-test2, after 14 weeks i.e. at the end of the implementation) as the three subscales were significantly correlated."

Qualitative Data

"The current study sought to find out the effects of implementing FCM on the variable of motivation in English speaking performance for Omani EFL learners."

"The participants (N = 27) were asked to write their reflections about the application of FCM in their EFL speaking classroom by responding to a set of trigger-questions on each of the three phases."

"Students' willingness to speak English was seen to have increased over time."

"Students expressed the desire to continue practising different collaborative speaking tasks inside and outside the classroom to allow them to practise speaking in English more efficiently."

"This growing willingness to be involved in any speaking opportunities can stem from their progressive motivation throughout the implementation of FCM."

"The progressive improvement in students' motivation level in English speaking performance is obvious in both quantitative and qualitative findings."

Discussion and Conclusion
"The current study aimed to examine the effect of implementing FCM on the motivation level in the English speaking performance of a small sample of Omani EFL learners."

"Most of the previous studies have determined that the flipped classroom environment promotes student involvement in the learning process and increases their satisfaction in learning as a result."

"Based on both reported quantitative and qualitative findings, the engaging, encouraging, and interactive flipped classroom environment of the current study contributed effectively to the gradual improvement that occurred in the students' motivation levels in their EFL speaking classroom."

"The quantitative results showed that the overall students' motivation in English speaking performance started increasing over the period of FCM application."

"Students' perceptions have positively and significantly affected their motivation and English speaking performance as a result."

Limitations
"Although the sample size is small and was specific to one group of EFL students at BUC, the approach taken in the current study, and the in-depth analyses of both the quantitative and qualitative data enhance the trustworthiness, reliability, and validity of the findings."

"This study is the first empirical study conducted in Oman to examine the effect of implementing the flipped classroom model on EFL learners' motivation level in their English speaking performance."

"It is hoped that more research will follow to further improve our understanding and application of the flipped classroom method."

The Effect of Using a Game-Based Translation Learning App on Enhancing College EFL Learners' Motivation and Learning Experience [292]

This is a machine-generated summary of:

Chen, Yulin: The effect of using a game-based translation learning app on enhancing college EFL learners' motivation and learning experience [292].

Published in: Education and Information Technologies (2022).

Link to original: https://doi.org/10.1007/s10639-022-11174-6.

Copyright of the summarized publication:

If you want to cite the papers, please refer to the original.

For technical reasons we could not place the page where the original quote is coming from.

Abstract-Summary
"This study investigated the motivation of Taiwanese undergraduate EFL students in learning English-to-Chinese translation and the intention to use a digital game-based learning app called CHEN-slate."

"The responses to a questionnaire, designed based on theories including Dörnyei's L2 motivational self-system and the technology acceptance model, were analyzed quantitatively and qualitatively to identify the most prevalent type of motivation among students to learn English-to-Chinese translation (one-way repeated-measures ANOVA), students' perception toward the use of CHEN-slate, and the relationship between students' intention to use CHEN-slate and their learning motivation (linear regression analysis)."

"The findings of this study also indicate that students have positive attitudes toward the adoption of CHEN-slate and have high intention to use the app facilitate their learning process."

"A significant positive relationship was found between students' intention to use the app and their translation learning experience."

"This study offers pedagogical implications for instructors in translation courses to enhance students' learning motivation and effectively use game-based learning apps in translation courses."

Introduction
"Translation has long played a vital role in daily life and the process of learning foreign languages (Malmkjaer [217]; Riccardi [218])."

"Translation has long been a part of language teaching and learning."

"Chang and others [219] pointed out that translation is necessary for learning a foreign language and investigating the relationship between two languages."

"Translation can also be presented as a teachable craft, and although it is closely related to language learning itself, it is a distinct and separate skill (Azizinezhad [220])."

"Although translation is believed to play a major role in foreign language research, the theory of language teaching in the twentieth century ignored the role of translation (Bahremand [221])."

"Given the key role of translation in effective communication, idea exchange, and cultural communication, teachers should not ignore the role of translation in foreign language learning (Bahremand [221])."

Literature Review

"Dörnyei proposed an L2 motivational self-system, which includes intrinsic and extrinsic motivation, to encourage language learners to learn a TL (target language)."

"The ideal L2 self is considered to be promotion focus and major concerns about learners' growth, achievement, and goal achievement, which are considered factors motivating them to increase their efforts in their learning process (Higgins [222])."

"The L2 motivational self-system of Dörnyei [203, 223] was adapted and modified as a motivational self-system comprising three components, namely ideal self, ought-to self, and translation learning experience."

"Although many studies have investigated the motivation of EFL learners, few have evaluated learning motivation based on the L2 motivational self-system in terms of EFL translation courses."

"Is a relationship present between the intention of Taiwanese undergraduate students to use CHEN-slate in English–Chinese translation courses and their motivation to learn English–Chinese translation?"

Method

"The second section of the questionnaire investigated students' motivation to learn English–Chinese translation."

"The third section of the questionnaire was based on the TAM proposed by Davis [224] and surveyed students' intention to use the interactive app (CHEN-slate) in translation courses."

"One purpose of conducting semi-structured interviews was to verify the quantitative findings regarding Taiwanese students' motivation to learn English–Chinese translation as well as the possible influence of using CHEN-slate in translation courses on their learning experience."

"One-way repeated-measures ANOVA was used to identify the students' most prevalent type of motivation to learn English–Chinese translation, the result of which addressed the first research question in this study."

"Linear regression and correlation analysis were used to identify the relationships between students' motivation to learn English–Chinese translation and their intention to use CHEN-slate in translation courses."

Results

"To answer the first research question, one-way repeated-measures ANOVA and semi-structured interview were conducted to find out what the most prevalent type of motivation of students for learning translation is."

"The results of one-way repeated-measures ANOVA with the sphericity assumption indicated a significant difference between the effects of the three types of learning motivation on students' overall learning motivation ($F[2, 134]$ 20.87, $p < 0.05$ 0.00, $np^2 = 0.24$)."

"The aforementioned findings answer the first research question and indicate that Taiwanese undergraduate EFL students are more motivated to learn English–Chinese translation in terms of the ideal self and translation learning experience than in terms of the ought-to self."

"All the interviewed students indicated that they wanted to develop their translation skills and showed an intrinsic interest in learning translation."

Discussion

"The first major study finding answers the first research question on the theoretical framework of motivation, indicating that ideal self- and translation-learning experiences play larger roles in motivating Taiwanese EFL undergraduate students to learn English-to-Chinese translation than the ought-to self-motivation."

"The present study found that positive classroom experiences, such as positive experiences with the teacher, learning materials, peer collaboration, and using CHEN-slate to learn play a significant role in encouraging target students to learn English–Chinese translation."

"The results reveal the perceptions of students regarding the use of CHEN-slate in translation courses as well as the relationship among students' intention to use the app and their learning motivation and posttest results."

"A significant relationship was observed between the students' intention to use the app and their motivation to learn English–Chinese translation."

"Liu and others [225] also indicated that information feedback, game quality, and autonomous motivation (ideal self) significantly influence students' learning effectiveness and their intention to continue to use DGBL systems."

Conclusions

"The results of this study indicate that among the three components of the motivational self-system, ideal self and translation learning experience played an integral role in motivating Taiwanese undergraduate students to learn English–Chinese translation."

"In terms of enhancing the translation learning experience of students, the study findings indicate that efficient teaching techniques and the use of CHEN-slate as a supportive tool in translation courses presented a positive learning experience to train students' translation skills."

"The study findings also support the role of CHEN-slate in translation classes in motivating students to learn English–Chinese translation, benefiting their translation learning experience, and enhancing their translation skills, as revealed by the posttest scores."

"The app could enhance students' translation learning process in various manners."

"The main purpose of developing the app was to provide an alternative learning tool for students to enhance their translation learning experience."

Validation of the Mathematics Motivation Questionnaire (MMQ) for Secondary School Students [293]

This is a machine-generated summary of:

Fiorella, Logan; Yoon, So Yoon; Atit, Kinnari; Power, Jason R.; Panther, Grace; Sorby, Sheryl; Uttal, David H.; Veurink, Norma: Validation of the Mathematics Motivation Questionnaire (MMQ) for secondary school students [293].

Published in: International Journal of STEM Education (2021).

Link to original: https://doi.org/10.1186/s40594-021-00307-x.

If you want to cite the papers, please refer to the original.

For technical reasons we could not place the page where the original quote is coming from.

Abstract-Summary

"We focus on the assessment of mathematics motivation among secondary school students."

"We provide validity and reliability evidence for the Mathematics Motivation Questionnaire (MMQ)—adapted from the Science Motivation Questionnaire designed for college students—using data from 2551 secondary students from seven states across the United States."

"To assess criterion validity, analyses using a subset of the data that included students' mathematics standardized scores (n = 536) indicated that intrinsic value, self-regulation, and self-efficacy were significantly positively correlated with mathematics achievement, whereas test anxiety was significantly negatively correlated with mathematics achievement."

"The MMQ provides a reliable, valid, and feasible measure of the specific factors underlying mathematics motivation among secondary students."

Theoretical Framework

"The Attitude Towards Mathematics Inventory (ATMI; Tapia & Marsh [227]) has four constructs of enjoyment, motivation, self-confidence, and value, indicated by 40 items, which has content validity evidence from high school students."

"The original MSLQ (Pintrich & de Groot [228]) has 44 items for 5 constructs (intrinsic value, self-efficacy, test anxiety, cognitive strategy use, and self-regulation)."

"The revised MSLQ is (a) not specifically contextualized for motivation in mathematics learning, (b) too long for secondary students, and (c) contains constructs that are not quite relevant to our purpose of study to assess multifaceted mathematics motivation."

"Past research indicates mathematics achievement is influenced by (a) beliefs about one's competence (e.g., self-efficacy); (b) the perceived value of mathematics (e.g., intrinsic value and utility value); (c) self-regulated strategies to learn and perform, and (d) feelings of anxiety toward learning or being assessed in mathematics."

Method

"An exploratory factor analysis (EFA) was conducted by randomly splitting the data in half (n = 1275) to identify the underlying factor structure and irrelevant items that did not fit into any factors in the MMQ."

"After identifying the factor structure and relevant items, we conducted confirmatory factor analyses (CFAs) using the other half of the data (n = 1276) to confirm and refine the factor structure identified through the EFA."

"As we finalized a factor structure and items for the MMQ, we utilized nonlinear SEM methods to calculate the internal consistency reliability coefficient of ω for ordinal data, using Mplus and SAS, following the two-step approaches in Yang and Green [229]."

"For criterion validity evidence, we calculated Pearson correlation coefficients between the scores averaged from the items loaded for the identified factors and the state standardized mathematics achievement test scores (AERA, APA, & NCME [230])."

Results

"The negative correlation coefficients were due to items intended to assess mathematics test anxiety, which were worded such that high scores indicate more anxiety."

"Item 18 ("I hate taking the math tests"), which was designed to be an indicator of test anxiety, was dropped because even a student without test anxiety may hate taking a math test."

"The five factors were renamed based on the motivation theories and self-regulation: intrinsic value, self-regulation, self-efficacy, utility value, and test anxiety."

"The modification indices indicated that test anxiety Item 15 ("I think about how my math grade will affect my overall grade point average") might be a source of misfits (i.e., potential indicators of error covariances)."

"Some items had large MIs (over 4.0) including anxiety Item 13 ("I worry about failing math tests.") cross-loading on other factors (intrinsic value, self-regulation, self-efficacy, and utility value), indicating areas for improvement in the model fits."

Discussion

"This study contributes to our understanding of what motivates students to persist and learn in STEM disciplines (Atit and others [231]; Jiang and others [232]; Wang [233]), particularly the factors underlying mathematics motivation among secondary school students."

"According to expectancy-value theory (Wigfield and others [234]), motivation depends on students' expectancies for success (closely related to self-efficacy) and their perceived task values (intrinsic, utility, and attainment value)."

"This factor structure has several distinctions from the SMQ factor structure identified by Glynn and others [235], which also identified five factors from 30 items using principal component analysis: intrinsic motivation and personal relevance

(10 items), self-efficacy and assessment anxiety (9 items), self-determination (4 items), career motivation (2 items), and grade motivation (5 items)."

"The MMQ can be used to investigate the relationship between students' mathematics motivation and their STEM achievement, while also considering other plausible factors that may affect students' performance."

Conclusion

"The finalized MMQ provides a feasible measure for capturing secondary students' motivation to learn in mathematics, presenting sound validity and reliability evidence and revealing varied motivation by demographic background."

"In revealing the multidimensional nature of mathematics motivation, the MMQ provides implications for the development of future motivation-based interventions in STEM (Rosenzweig & Wigfield [226]), such as those designed to promote self-regulation and self-efficacy or to reduce test anxiety."

[Section 6]

"Designing instruction that motivates students to learn is a fundamental challenge among educators in science, technology, engineering, and mathematics (STEM) disciplines (Hernandez and others [236]; Rosenzweig & Wigfield [226])."

"Many secondary school students show declining mathematics motivation and achievement, in part due to differences in the school context and instructional practices, as well as the increased complexity of the learning material (Eccles and others [237])."

"Understanding the factors that contribute to secondary school students' mathematics motivation can contribute toward increasing student participation in STEM disciplines."

"In response, we created the Mathematics Motivation Questionnaire (MMQ) for secondary students, which was adapted from the popular and well-validated Science Motivation Questionnaire (SMQ; Glynn and others [235]) for college students."

"This study examined the construct validity, reliability, and criterion validity evidence of the MMQ applicable for secondary school students to understand the motivational factors underlying learning in mathematics."

Role of Social Representations in Student Motivation for Acquiring Further Education [294]

This is a machine-generated summary of:

Koshkin, Andrey Petrovich; Abramov, Ruslan Agarunovich; Rozhina, Ekaterina Yurevna; Novikov, Andrey Vadimovich: Role of Social Representations in Student Motivation for Acquiring Further Education [294].

Published in: Interchange (2018).

Link to original: https://doi.org/10.1007/s10780-018-9328-3.

If you want to cite the papers, please refer to the original.

For technical reasons we could not place the page where the original quote is coming from.

Abstract-Summary
"Perspective, it is extremely important to determine relevant social factors majorly impacting the students' representation about further education and its role in building and shaping up the motivation to continue education."

"This study focuses on the analysis of social representation in the processes of motivation of students to acquire further vocational education in course of university studies."

"Attitude of Russian students to further education was investigated by means of six scales: (1) information awareness of further education programs; (2) social representation about further education (in general, regarding the contents, conditions, cost); (3) representation about required program characteristics and expectations; (4) wish to receive further education in present and in future, as well as bars thereto."

"Study results prove that social representations have little effect on students' intention to receive further vocational education in course of university studies, but the effect is significant in a long-term perspective."

Introduction
"Positive social representations about the need of mastering skills and knowledge, expectations of new opportunities and a chance to meet their educational needs also have to become important factors in students' motivation to start FVE courses (Ball and others [238])."

"This group also includes students of higher educational institutions who are attending programs of professional retraining in course of their main studies, but obtain a diploma after graduation from university (i.e., after they receive their first education)."

"Certain social representations of students about education should influence their behavior and create certain incentives and motivations for continuation of education."

"Based on scientific approaches and theories, a hypothesis has been formulated: social representations are one of the most significant factors in motivating students for further vocational education."

"In order to prove this hypothesis in our study, we hereby establish a goal to define a role of social representations and other factors having a major effect on students' aspiration for further education."

Methodology

"In course of presented research a method of sociological questionnaire survey was used."

"Existing representations of students about further vocational education and their desire to pursue one of these program were studied."

"Survey results were analyzed with the use of factor analysis, K-means cluster analysis, and regression analysis."

"Structure-wise, the questionnaire included five sections: (1) information awareness of further vocational education; (2) social representations about further education (in general, contents, price, timetable); (3) representation about required characteristics of further vocational education programs and expectations therefrom; (4) desire to pursue further vocational education in present and future, as well as hindering factors; (5) social and demographic unit."

Results

"Students have four types (clusters) of representations about further vocational education: (1) those considering FVE to be generally useful for job search, but not wanting to spend personal time for its acquisition (30%); (2) those who don't want to put any efforts into education acquisition, saying that it is an employer's liability (44%); (3) those considering FVE to be just a waste of their time (4%); (4) those considering that the objective of further education is a diploma confirming their skills (22%)."

"It turned out that students of the second (75%) and the third (96%) clusters considered the most effective source of information in further education programs to be "social networks", unlike 57% of the first cluster representatives ($p < 0.05$)."

"High importance of such factors as expectation of easy studies, need for good facilities and understanding the utility of education for employment is primarily defining students' aspirations to obtain FVE in future."

Discussion

"The majority of social representation studies in the education field, especially in the area of additional and further education programs, pay attention to low information awareness of further education opportunities among students."

"According to the study of Merenkov and Sushchenko [239] conducted on the basis of sociological survey among 623 students from 17 universities of Ural Federal District, most of Russian further education programs are distinguished by inefficiency of existing system of informing students of new courses and programs."

"FVE program leader should pay much more attention to informing the students via university or educational institution web-sites send personalized emails."

"Among other things, active interaction with lecturers plays a very important role in students' motivation for continuing education and studies in FVE programs."

"Perception of further education by society and the student group itself are the main factors affecting the motivation for further studies."

Conclusion

"This study determined the role of social representations compared to other factors mainly affecting the aspiration of students to receive further vocational education."

"It was identified that social representations about further vocational education were the most significant factor in students' motivation for FVE acquisition only in a long-term perspective."

"It appears that positive social representations have a rather slight impact on students' wish to advance and obtain further education in course of main university studies."

"It's worth mentioning that FVE leaders and university administration should not only build positive social representations about further education in students, but also pay more attention to the questions of motivations for acquisition of further education."

Improving College Student Engagement and Motivation in a Gamified Learning Environment: The Pilot Study in China [295]

This is a machine-generated summary of:

Lv, Jing: Improving college student engagement and motivation in a gamified learning environment: the pilot study in China [295].

Published in: Current Psychology (2023).

Link to original: https://doi.org/10.1007/s12144-023-04884-8.

If you want to cite the papers, please refer to the original.

For technical reasons we could not place the page where the original quote is coming from.

Abstract-Summary

"The paper addresses the impact of interactive learning technologies, using gamification as an example, on students' motivation, engagement, and academic achievements."

"game improved motivation (mean = 4.14, SD = 0.77) and engagement (mean = 3.54, SD = 1.05)."

"The resulting data suggested the proposed model's effectiveness in teaching Chinese grammar with the game-based interactive learning technology."

Introduction

"Game-based learning (GBL) is an essential and innovative teaching method that can enhance students' language learning outcomes while simultaneously fostering collaboration and engagement (Amr [240]; Belkhouche and others [241])."

"There is a direct correlation between student motivational indicators, engagement, and academic achievements, which significantly impact the outcomes of language learning endeavors (Xie [242])."

"This might be attributed to motivational and engagement indicators directly affecting students' willingness and persistence to learn and practice the language (Mayer & Johnson [243])."

"The research findings designed to test the effects of GBL on students' motivational indicators and student achievements in learning the Arabic language validated the hypothesis that game-based learning improved these parameters (Kotob & Ibrahim [244]; Wang & Tahir [245])."

"The paper addresses the impact of interactive learning technologies, using gamification as an example, on students' motivation, engagement, and academic achievements in a college-level Chinese grammar course."

Methods and Resources

"The study participants were divided into an experimental group and a control group, where the Chinese grammar course topics were taught using GBL and the traditional method, respectively."

"The participants studied a Chinese grammar course."

"The instructor prepared a paper-and-pencil test for the students in the control group before the lesson, which is a well-established and validated approach to in-class testing."

"Participants receive personalized recommendations on questions related to their scores, points, performance compared to leaders, and correct answers (if available)."

"To evaluate the attitudes of participants towards the implemented experiential method, a questionnaire employing a Likert scale was administered."

"Student's t-test and one-way ANOVA were used to determine statistical differences between the groups."

"Limitations associated with the age of the participants in the study impede the attainment of statistical generalizations in this research."

Results

"Item A7 exhibits the second-highest mean agreement score (4.41), indicating a highly significant level of agreement."

"Item A4 ("I felt satisfied while taking the test") demonstrates the third highest level of agreement, with a very strong average rating of 4.31."

"Items A3 and A1 exhibited a "very strong" level of agreement."

"Students' responses to item A2 ("I felt inspired and enthusiastic during the exam") indicate a "very strong" rating with an average value of 4.20."

"The experimental group exhibited a very strong level of student engagement in learning Chinese grammar."

“Students’ responses to question A16 ("I used Kahoot during the lecture because I thought I did well on the test") ranked third in magnitude with an average value of 3.66.”

Discussion

“A study testing the GBL’s effect on student motivational indicators and achievements in learning the language confirmed that GBL has improved these indicators (Kotob & Ibrahim [244]).”

“enhances student learning, with the greatest impact on the dynamic pattern, engagement, motivation, and improvement.”

“The findings concerning GBL’s effect on university students’ engagement in learning Chinese grammar suggest that students’ engagement in learning grammar was strong as a result of using Kahoot.”

“The measurement of the impact of GBL on students’ motivational indicators and proficiency in learning the language confirmed that GBL improved students’ academic performance (Kotob & Ibrahim [244]).”

“This implied that GBL positively affected learning, improving academic achievements among students (Licorish and others [246]).”

“Students’ achievement on the pre-test in Chinese language learning management was 53.00%, with 71.43% scored on the post-test.”

“Students’ satisfaction with the management of Chinese language learning with Kahoot!".

Conclusions

“This study demonstrated that an interactive gaming learning environment positively influences motivational indicators, engagement, and students’ achievement.”

“The research findings provide valuable insights into the improvement and development of effective educational gaming tools utilized in the classroom to enhance student motivation and participation.”

“It is crucial to expand the research to study the experience of GBL implementation.”

“Since the current study was pilot in nature, there is a need for future research that is enhanced and comprehensive.”

Virtual Tutor and Pupil Interaction: A Study of Empathic Feedback as Extrinsic Motivation for Learning [296]

This is a machine-generated summary of:

Oker, Ali; Pecune, Florian; Declercq, Christelle: Virtual tutor and pupil interaction: A study of empathic feedback as extrinsic motivation for learning [296].

Published in: Education and Information Technologies (2020).

Link to original: https://doi.org/10.1007/s10639-020-10123-5.

If you want to cite the papers, please refer to the original.

For technical reasons we could not place the page where the original quote is coming from.

Abstract-Summary

"Using an embodied conversational agent platform, we tested human–computer interactions with 22 children aged 9–11 years."

"Children performed several numeracy exercises set by two different virtual agents."

"One agent provided solely verbal feedback (unimodal), while the other one combined facial expressions based on real muscle contractions with its verbal feedback (bimodal)."

"Results showed that the overall duration of agent-pupil interactions varied, children found the bimodal agent more empathic, and produced significantly more correct answers."

"There was a positive correlation between accuracy and mean reaction times for correct answers with the bimodal agent."

"The lack of a correlation for the unimodal agent is discussed in the light of empathy and motivation in social cognition."

Feedback and Motivation

"There is considerable interest in how different feedback modalities, particularly formative feedback proposed by a virtual agent can motivate pupils to engage in a learning process."

"According to these authors, the primary function of a teacher's verbal behavior is to provide new learning content to students, whereas the primary function of nonverbal behavior is to develop students' desires and affects regarding this new content and instill in them the motivation to learn more about the subject."

"Students have implicit knowledge of their teachers' emotional states in class, and this has an impact on their motivation to learn (Sutton and Wheatley [247])."

"The use of facial expressions to provide this feedback leads teachers to be perceived of as more empathic, thereby increasing students' motivation."

"Formative feedback is an important part of the learning process and according to literature and facial expressions involved tutor's feedback is to key to communicate empathic intent."

Facial Expressions of Empathic Virtual Agents

"The facial expression of the virtual agent alone does not convey empathic engagement."

"We therefore calculated participants' reaction times when they answered questions either from an agent who provided feedback accompanied by emotional facial expressions rendered by FACS or from another agent who provided solely verbal feedback, in order to see whether these differences might indicate empathic engagement."

"Regarding qualitative answers, in order to see whether the contextualization of our experiment (exercises to be performed by the pupils, human–computer interaction, feedback regarding correct and incorrect responses) led to the virtual agents being perceived of as empathic (Oker and others [248]), we also administered a questionnaire on perceived social support designed for pre-adolescent children (Gordon [249])."

"Our main prediction was that even when pupils are revising an already learned lesson, feedback from an empathic agent combined with facial expressions generated using FACS improves performances in terms of correct answers and reaction times, reflecting empathic engagement and motivation."

Method

"A unimodal agent provided coherent, unimodal (verbal) feedback about the pupils' answers."

"A bimodal agent provided coherent, bimodal (verbal and facial expressions based on the FACS taxonomy) feedback about the pupils' answers."

"We wanted to make sure whether the facial expressions rendered by virtual agents were understood by younger pupils than our target subjects in order to validate the stimulus used in the following experiment."

"The virtual agent started the experiment by thanking the participants and explaining what they would have to do: "First of all, we will study a subject together that I know you have already been taught in class."

"Alice, our unimodal or bimodal agent, verbally asked pupils to perform mental arithmetic (e.g., 2×2)."

"Pupils completed a questionnaire assessing the bimodal or unimodal agent's empathy depending their group."

Results

"We calculated the mean correct response latencies, accuracy rates and the mean number of trials completed for each group."

"We calculated the accuracy rate, number of trial blocks, and mean correct response latency (in ms) for each participant in each group (bimodal and unimodal)."

"The ANOVA did, however, reveal a significant effect for the accuracy rate, $F(1, 21) = 4.878$, Bonferroni-corrected $p < 0.04$, $\eta^2_p = 0.20$, as the mean proportion of correct answers was higher for the empathic (i.e., bimodal) agent (84% vs. 74%)."

"We observed a positive correlation of moderate strength (Pearson's r) between the correct response rate and the mean correct response latency, though only for the bimodal group ($r = 0.616$, $p < 0.05$)."

"The higher accuracy rate in the bimodal condition was positively correlated with the mean correct response latency."

Discussion

"We believe that if the virtual tutor is perceived of as empathic by the children, this will have positive repercussions such as more accurate responses to an exercise (reflecting motivation), and longer interaction durations (reflecting engagement with agent)."

"While there are many ways of manipulating the empathic nature of virtual agents, in the present study we only focused on the use of coherent facial expressions to complement verbal feedback."

"Again, some researchers claim that children only display accurate facial expression recognition if these expressions are accompanied by coherent verbal messages (for a review, see Herba and Phillips [250])."

"Our results indicate that practical interventions based on this approach (for a review, see Sugai and others [251]) should emphasize the use of verbal feedback combined with coherent facial expressions to motivate children."

[Section 6]

"We conducted an experiment among children aged 9–11 years old, specifically to study how the Facial Action Coding System (FACS; Ekman and Friesen [252]; Ekman and others [253]) can be used to create empathic embodied conversational agents (ECAs) to be used as virtual intelligent tutors in the classroom."

"Our aim is to highlight the added value of the feedback provided by empathic embodied conversational agents in learning process even though pupils were already exposed to the educational content compared to non-empathic feedback."

"We investigated whether: 1) virtual tutors need to present coherent verbal and facial expressions in order to be perceived of as empathic and thereby boost pupils' extrinsic motivation to study and make the interaction last longer."

"2) the empathic nature of virtual agents can be efficiently perceived by children if the tutor's facial expressions are rendered according to human facial muscle contractions."

Learning Strategies as a Mediator for Motivation and a Sense of Achievement Among Students Who Study in MOOCs [297]

This is a machine-generated summary of:

Magen-Nagar, Noga; Cohen, Lizi: Learning strategies as a mediator for motivation and a sense of achievement among students who study in MOOCs [297].

Published in: Education and Information Technologies (2016).

Link to original: https://doi.org/10.1007/s10639-016-9492-y.

If you want to cite the papers, please refer to the original.

For technical reasons we could not place the page where the original quote is coming from.

Abstract-Summary
"The goal of the research was to examine the contribution of learning strategies as a mediator for motivation and a sense of achievement in a Massive Open Online Course (MOOC), among students who participate in a unique program in Israel, called "Academy Online – MOOCs in the Israeli Education System"."

"At the end of the course, the students answered a Motivated Strategies for Learning Questionnaire (MSLQ), and a Sense of Achievement in a MOOC questionnaire, which was based on the principles of the pedagogical model of the program."

"The findings of the SEM path analysis show that the variable 'Learning Strategies' is a significant mediator for motivation and a sense of academic achievement in a MOOC for both groups, and a stronger mediator for the astronomy-related sciences students."

"It may be concluded that using a variety of learning strategies in a curriculum which integrates a MOOC affects the sense of academic achievement."

Background

"In recent years, the trend indicates increased integration of MOOCs as part of high-school students' curriculum (Jackson [254]; Protalinski [255])."

"It appears that integrating MOOCs in 'Blended Learning', especially in mathematics and science, helps preventing students from dropping out of high-school."

"In order to integrate MOOCs in high-school in an informed way, there should be re-thinking and renewed curriculum and teaching–learning strategies design in a way suitable for the twenty-first century (Brahimi and Sarirete [256])."

"In order for MOOCs to be an integral part of the high-school curriculum in an informed way, teachers must be given opportunities to experience MOOCs as part of their professional training (Vivian and others [257])."

"The education system in Israel has joined the global trend and has developed a program for high-school students, called "Academy Online – MOOCs in the Israeli Education System"."

Literature Review

"The goals of the program are: To integrate into the education system an innovative teaching–learning strategy of MOOC type, based on online academic learning, for students of science and technology; To provide for Israeli students a global level learning in collaboration with learners throughout the world; to challenge students in Robotics or Astronomy internship program, to deepen their subject knowledge in those fields."

"The goal of the current study is to examine the contribution of learning strategies as a mediator for motivation and sense of academic achievement in a MOOC among high-school students who study science and technology (astronomy-related sciences program and mechatronics-robotics course)."

"The hypothesis of the study is: The higher the student's motivation, the more he applies learning strategies, and as a result, his sense of academic achievement in MOOC is higher."

Method

"The sense of academic achievement among mechatronics-robotics students is explained by learning strategies and motivation in about half of the explained variance (50%), as opposed to about a third of the explained variance (30%) among the astronomy-related sciences students."

"That among students of astronomy-related sciences and mechatronics-robotics, learning strategies have a positive, significant, medium strength effect on the sense of academic achievement ($\beta = 0.36$, $\beta = 0.34$, $p < 0.001$, respectively)."

"Testing the significance of the mediation of the variable 'learning strategies' found the total effect between motivation and the sense of academic achievement without learning strategies among the astronomy-related sciences students is $\beta = 0.57$***, and among mechatronics-robotics students $\beta = 0.60$***."

"It can be concluded that the variable 'learning strategies' is a significant mediator between motivation and the sense of academic achievement in a MOOC for both groups, and stronger among the astronomy-related sciences students."

Discussion

"It appears that the intervention program was focused on providing learning strategies, and according to the research findings, it is likely that the relevant experience in a variety of learning strategies has assisted the learners to understand and move forward in the course assignments, and thus turned them into skilled learners with a high sense of achievement in MOOC learning."

"These findings may indicate that an informed use of learning strategies, based on motivational orientations, such as intrinsic orientation, beliefs about the course's value, and self-efficacy, allows creating a deep and wide picture about the characteristics of the learning and the sense of achievement of students who study in a MOOC."

"The research findings lead to the conclusion that learning strategies are a significant mediator between motivation and a sense of academic achievement in MOOC among students."

Work Context Influences on Older Workers' Motivation for Continuing Education [298]

This is a machine-generated summary of:

Thieme, Paula; Brusch, Michael; Büsch, Victoria; Roßnagel, Christian Stamov: Work context influences on older workers' motivation for continuing education [298].

Published in: Zeitschrift für Erziehungswissenschaft (2015).

Link to original: https://doi.org/10.1007/s11618-014-0600-8.

If you want to cite the papers, please refer to the original.

For technical reasons we could not place the page where the original quote is coming from.

Abstract-Summary

"Work ability can be maintained through continuing education."

"Understanding and possibly even increasing older employees' motivation to participate in continuing education is vital to achieving the European Union's aim of individuals' "lifelong learning" to cope with challenges of technological and organisational change."

"This empirical study analyses the influence of work context factors on motivation to participate in continuing education."

"Data source is the "Survey on continuing in employment after reaching pensionable age", commissioned in 2008 by the Bundesinstitut für Bevölkerungsforschung (Federal Institute for Population Research)."

"Results show weak but significant influences of some work context factors."

Introduction

"The working age population has been calculated to decrease by 6.5 million until the year 2025 (Bundesagentur für Arbeit [258]), whilst worker mean age will rise to 41.7 years (Börsch-Supan and Wilke [259], p. 41)."

"One strategy to avoid inordinate contribution burdens for younger workers is to decrease the number of pensioners by prolonging working life (Bundesministerium des Innern [260])."

"Against this backdrop, the question of individual motivation for prolonging one's working life arises."

"A national representative poll on the attitudes towards prolonged working life by the German Federal Institute of Population Research indicated that 47.3% of people between 55–64 years were prepared to work past retirement age, a number regarded as rather high (Büsch and others [261])."

"We were interested in indicators of motivation to participate in continuing education (MPCE) as one of the support factors for work after age 65."

Theoretical Background, Current State of Research, and Hypotheses

"There are both empirical and practical reasons for our interest in work context influences on older workers' continuing education motivation."

"Socio-emotional selectivity theory (SST) is also helpful in modelling work context influences on older workers' motivation to partake in continuing education."

"We assume that The relationship between company size and motivation to participate in continuing education is moderated by age such that motivation is higher in larger companies for younger, but not older workers."

"We propose The relationship between working hours and motivation to participate in continuing education is moderated by age such that motivation is higher in younger, but not older workers who work more hours."

"We hypothesise that, Older workers who participated in continuing education in the past 3 years feel more participation motivation than those who have not participated."

Method

"We considered work context factors an important part of the various contextual factors that may foster a worker's motivation to participate in continuing education."

"The survey provides our model's dependent variable (DV), motivation to participate in continuing education (MPCE), measured by three items: "Continually learning new things is very important in my life", "I shall always strive to continually train" and "I like to attend continuing training classes" rated on a 5-point Likert-type ordinal scale (with 1 = "fully applicable" to 5 = "not at all applicable")."

"We also recorded sex, age, working hours, company size, security of employment, tenure, temporary work contract, task level, desire to work past retirement age and continuing education participation in the last 3 years."

"Respondents gave their gender (m/w), age (in completed life years) and categorised themselves as part-time (15–35 h/week), full-time (35 h/week or more), marginally employed (less than 15 h) or unemployed to provide working hours as per contract."

Results

"We investigate work context-related influences on older workers' MPCE using multiple regression analyses on a sample of 1500 workers between 55–64 years old."

"Older workers on higher task levels indicated lower MPCE."

"Contrary to our assumptions, workers who wish to work past their retirement age feel less MPCE."

"Our hypothesis of older workers who envisage working past their retirement age also feel higher MPCE received partial support."

"A similar effect may pertain to our measures of MPCE, where items do not directly relate to job-related continuing education, so the effect of work context variables may be somewhat diluted by other influences."

"Controlling for the factors analysed in our regression analysis, we can conclude that past participation represents the strongest predictive influence on older workers' MPCE."

"Psychological working conditions, task level and organisational tenure play a role for predicting MPCE."

Discussion
"The contribution of the present study is in identifying work context-related motivational differences between older and younger workers."

"We hypothesised that work context factors influence MPCE differently in older workers than in younger workers."

"Drawing on our theoretical background, older workers on higher task levels may have already achieved a satisfactory career level and therefore attach less importance to their career goals, leaving them with less motivation to engage in continuing education."

"As work motivation does not generally decline with age but becomes increasingly task-specific, depending on changing personal life goals (see Stamov Roßnagel and Hertel [262]), MPCE might also develop along shifting areas of personal interest and relevance."

"Motivated older workers should signal their interest in continuing education rather than resign or wait for training to be offered."

Techniques for Increasing Educational Motivation and the Need to Assess Students' Knowledge: The Effectiveness of Educational Digital Games in Learning English Grammatical Material [299]

This is a machine-generated summary of:

Shakhmalova, Irina; Zotova, Natalia: Techniques for Increasing Educational Motivation and the Need to Assess Students' Knowledge: The Effectiveness of Educational Digital Games in Learning English Grammatical Material [299].

Published in: Journal of Psycholinguistic Research (2023).

Link to original: https://doi.org/10.1007/s10936-023-09983-y.

If you want to cite the papers, please refer to the original.

For technical reasons we could not place the page where the original quote is coming from.

Abstract-Summary
"Educational digital games can be an effective way to teach English grammatical material because they provide an interactive and engaging learning experience."

"The purpose of this study is to clarify how playing digital games affects students' motivation and performance in university-level English grammar classes. [The North-Eastern Federal Institute of MK Ammosova in Neryungri] used a quasi-experimental study, testing, respondent survey, and statistical data analysis method for this purpose."

"114 Fourth-year students were the participants, and they were split into the experimental and control groups at random."

"Students in the experimental group were given a learning format that included the use of digital games designed to teach English grammar (Quizlet and Kahoot!)."

"The traditional teaching strategies offered by the university curriculum were used with the students in the control group (written assignments, textbooks, presentations, and tests)."

"The students in the experimental group performed better."

"These results suggest that digital games are a more productive and effective tool for teaching English grammar than traditional games."

"Students were also highly motivated, as they found digital games to be both entertaining and effective for language acquisition."

Introduction
"Educational digital games can help learners to practice their grammar skills in a fun and engaging way, making the learning experience less intimidating and more enjoyable."

"It is important to note that not all educational digital games are equally effective, and the effectiveness of a game may depend on a variety of factors, such as the game's design, the learners' prior knowledge, and their motivation levels."

"There is a need to explore the potential of educational digital games as a tool for improving learners' understanding and application of English grammatical material."

"The problem is to determine the extent to which educational digital games can effectively teach English grammar and to identify the factors that contribute to their effectiveness."

"By investigating the effectiveness of educational digital games, the study can help to identify the strengths and limitations of this teaching tool, and provide insights into how it can be best utilized in language learning."

Literature Review

"The use of digital games increases students' eagerness to participate in learning (Kim [263])."

"The effectiveness of learning is greatly improved by students' active participation in games."

"An effective strategy to raise student learning outcomes and motivation is to design learning environments based on the educational qualities of games."

"Playing games to learn a foreign language can help students become fluent in it in natural conversational settings."

"An experiential Digital Game-Based Learning (DGBL) approach can effectively boost student motivation."

"This technology tool demonstrates the value of language games for enhancing the teaching and learning of grammar and assists students in their grammar exercises (Baharudin & Yunus [264])."

"Games in foreign language classes get students involved in learning, which in turn gives them a good foundation for learning grammar (Leaning [265])."

"Describe the impact of DGBL on students' motivation to learn English grammar in higher education."

Methods and Materials

"The way that games were incorporated into the curriculum of the students was as follows: twice a week, the teachers administered a test using games after the students had finished the tasks in the practical classes and gone over the theoretical material on the grammar sections in the lectures."

"The traditional methods of the university curriculum (written assignments, textbooks, presentations, tests) were applied to the students in the control group."

"A survey was administered to the students in the experimental group."

"The survey questions were designed to ascertain the students' motivation level and attitude toward online games throughout the course of the study."

"It looked into how students reacted to the use of digital games during lectures on English grammar and how that affected their desire to retain information through yes-or-no questions."

Results

"The majority (43%) of students had a "moderate" score, according to the results of the technical skills task."

"23% of students ("good") were aware of additional software and game functions."

"An experimental programme for teaching English grammar to students using the digital games Quizlet and Kahoot!".

"The game design allowed students to focus more on their gaps and errors while also serving to reinforce the material they had already learned."

"Question 1: How have digital games affected your motivation to learn English?"

"The responses show positive feedback from students in relation to the use of games."

"The responses show positive feedback from students in relation to the use of games."

"Students' motivation to learn English grammar was "very high" as a result of the use of digital games."

"This means that students' level of motivation to learn was also high."

Discussion

"This study looked at how playfulness in learning affects students' motivation."

"Students with high gameplay experience were six times more likely to have high learning motivation than students with low gameplay experience, according to the findings."

"Gender characteristics were not considered in the design of the current study, but the findings support the high level of motivation seen in students who used digital games in their learning as opposed to those who received teaching."

"Another study used bingo games to investigate the effects of game-based learning on motivation, knowledge sharing, and critical thinking ability during learning (Chang & Yeh [266])."

"The findings revealed that while bingo games might have a bigger influence on boosting learning motivation, Socrative might promote more knowledge sharing and critical thinking."

"Of the use of digital games, students' motivation to learn English grammar was high."

Conclusions

"Students in the experimental group received instruction using digital games (Quizlet and Kahoot!) designed to teach English grammar."

"Traditional methods were used to teach grammar to students in the control group."

"The test results revealed that the experimental group of students had significantly more improved grammar knowledge."

"These findings demonstrate that digital games are a more productive and effective tool for teaching English grammar than traditional games."

"The number of students with "poor" scores decreased by 6%."

"The survey results revealed that digital games increased students' motivation to learn."

"Of using digital games, students' motivation to learn English grammar was high."

"They have practical value for teaching as there has been a development and evaluation of an educational program for teaching English grammar through the use of the online games Quizlet and Kahoot!".

Bibliography

1. Ryan, R. M., & Deci, E. L. (2000). Intrinsic and extrinsic motivations: Classic definitions and new directions. Contemporary Educational Psychology, 25(1), 54–67. https://doi.org/10.1006/ceps.1999.1020
2. Ryan, R. M., & Deci, E. L. (2017a). Self-determination theory: Basic psychological needs in motivation, development, and wellness. Guilford Publications.
3. Ryan, R. M., & Deci, E. L. (2017b). Self-determination theory: Basic psychological needs in motivation, development, and wellness. Guilford Publications.
4. Boersma, F, J., & Chapman, J, W. (1992). Manual for the perception of ability scale for students. Los Angeles: Western psychological services. Bond, T. G., & fox, C. M. (2015). Applying the Rasch model: Fundamental measurement in the human sciences (3rd ed.). Mahwah, NJ: L. Erlbaum.
5. Marsh, H. W., Craven, R. G., Hinkley, J. W., & Debus, R. L. (2003). Evaluation of the big-two-factor theory of academic motivation orientations: An evaluation of jingle-jangle fallacies. Multivariate Behavioral Research, 38(2), 189–224.
6. Peklaj, C., & Bucik, N. (2003). Vprasalnik bralne motivacije za mlajse ucence [Reading motivation questionnaire for young students]. Bralna motivacija kot dejavnik učinkovitega učenja in izobraževanja. Raziskovalno poročilo [Reading motivation as a factor of effective learning and education. Research report]. Ljubljana: Znanstveni inštitut Filozofske fakultete.
7. Nicholls, J. G., Cobb, P., Wood, T., Yackel, E., & Patashnick, M. (1990). Assessing students' theories of success in mathematics: Individual and classroom differences. Journal for Research in Mathematical Education, 21, 109–122.
8. Duda, J. L., & Nicholls, J. G. (1992). Dimensions of achievement motivation in schoolwork and sport. Journal of Educational Psychology, 84(3), 290–299.
9. Deci, E. L., Eghrari, H., Patrick, B. C., & Leone, D. R. (1994). Facilitating internalization: The self-determination theory perspective. Journal of Personality, 62(1), 119–142.
10. Battin-Pearson, S., Newcomb, M. D., Abbott, R. D., Hill, K. G., Catalano, R. F., & Hawkins, D. J. (2000). Predictors of early high school dropouts: a test of five theories. Journal of Educational Psychology, 92, 568–582.
11. Harde, P. L., & Reeve, J. (2003). A motivational model of rural students: intentions to persist in, versus drop out of high school. Journal of Educational Psychology, 95(2), 347–356.
12. Thornburgh, N. (2006). Dropout nation. Time, 167(16), 30–40.

13. Vallerand, R. J., Fortier, M. S., & Guay, F. (1997). Self-determination and persistence in a real-life setting: Toward a motivational model of high school dropout. Journal of Personality and Social Psychology, 72, 1161–1176. https://doi.org/10.1037/0022-3514.72.5.1161 .
14. Wasonga, T., Chirstman, D. E., & Kilmer, L. (2003). Ethnicity, gender, and age: predicting resilience among urban high school students. American Secondary Education, 32(1), 62–74.
15. Bukatko, D., & Kaehler, M. W. (1998). Child development: A thematic approach (3rd ed.). New York: Houghton Mifflin Company.
16. Vansteenkiste, M., Soenens, B., Verstuyf, J., & Lens, W. (2009). 'What is the usefulness of your schoolwork?' The differential effects of intrinsic and extrinsic goal framing on optimal learning. Theory and Research in Education, 7, 155–164.
17. Msila, V. (2005). The education exodus: the flight from township schools. African Education Review, 2(2), 173–188.
18. Ryan, R. M., & Connell, J. P. (1989). Perceived locus of causality and internalization: Examining reasons for acting in two domains. Journal of Personality and Social Psychology, 57, 749–761. https://doi.org/10.1037/0022-3514.57.5.749 .
19. Bridgeland, J. M., Dilulio, J. J., & Morison, K. B. (2006). The silent epidemic: perspectives of high school dropouts. A report by Civic Enterprises in association with Peter D. Hart Research Associates for the Bill and Melinda Gates Foundation.
20. Jimerson, S. R., Egeland, B., Sroufe, L. A., & Carlson, B. (2000). A prospective, longitudinal study of high school dropouts: examining multiple predictors across development. Journal of School Psychology, 38, 525–549.
21. Suh, S., Suh, J., & Houston, I. (2007). Predictors of categorical at-risk high school dropouts. Journal of Counselling and Development, 85(2), 196–203.
22. Brown, T. C., & Latham, G. P. (2006). The effect of training in verbal self-guidance on performance effectiveness in a MBA program. Canadian Journal of Behavioural Science/Revue, 38, 1–11. https://doi.org/10.1037/h0087266 .
23. Seijts, G. H., Latham, G. P., Tasa, K., & Latham, B. W. (2004). Goal setting and goal orientation: An integration of two different yet related literatures. Academy of Management Journal, 47, 227–239. https://doi.org/10.2307/20159574 .
24. Anseel, F., Beatty, A. S., Shen, W., Lievens, F., & Sackett, P. R. (2015). How are we doing after 30 years? A meta-analytic review of the antecedents and outcomes of feedback-seeking behavior. Journal of Management,41(1), 318–348.
25. Ashford, S. J. (1986). Feedback seeking in individual adaptation: A resource perspective. Academy of Management Journal,29(3), 465–487.
26. Ashford, S. J., De Stobbeleir, K., & Nujella, M. (2016). To seek or not to seek: Is that the only question? Recent developments in feedback-seeking literature. The Annual Review of Organizational Psychology and Organizational Behavior,3, 213–239.
27. Park, G., Schmidt, A. M., Scheu, C., & DeShon, R. P. (2007). A process model of goal orientation and feedback seeking. Human Performance,20(2), 119–145.
28. Williams, J. R., Miller, C. E., Steelman, L. A., & Levy, P. E. (1999). Increasing feedback seeking in public contexts: It takes two (or more) to tango. Journal of Applied Psychology,84(6), 969–976.
29. Levy, P. E., Cober, R. T., & Miller, T. (2002). The effect of transformational and transactional leadership perceptions on feedback-seeking intentions. Journal of Applied Social Psychology,32(8), 1703–1720.
30. VandeWalle, D., Ganesan, S., Challagalla, G. N., & Brown, S. P. (2000). An integrated model of feedback-seeking behavior: Disposition, context, and cognition. Journal of Applied Psychology,85(6), 996–1003.
31. Muthén, L. K., & Muthén, B. O. (1998–2015). Mplus user's guide (7th ed.). Los Angeles, CA: Muthén & Muthén.
32. Hofer, B. K. (2000). Dimensionality and disciplinary differences in personal epistemology. Contemporary Educational Psychology, 25(4), 378–405. https://doi.org/10.1006/ceps.1999.1026

33. Rarujanai, K., Teo, E. W., Ngien Siong, C., Ling, A., & Kuan, G. (2022). Understanding players' sportspersonship attitude, expectancy-related beliefs, and subjective task values in field Hockey: An integrated approach. International Journal of Environmental Research and Public Health, 19(8). https://doi.org/10.3390/ijerph19084819
34. Kim, C., & Keller, J. M. (2010). Motivation, volition, and belief change strategies to improve mathematics learning. Journal of Computer Assisted Learning, 26(5), 407–420. https://doi.org/10.1111/j.1365-2729.2010.00356.x
35. Habók, A., Magyar, A., Németh, M. B., & Csapó, B. (2020). Motivation and self-related beliefs as predictors of academic achievement in reading and mathematics: Structural equation models of longitudinal data. International Journal of Educational Research, 103, 101634. https://doi.org/10.1016/j.ijer.2020.101634
36. Voica, C., Singer, F. M., & Stan, E. (2020). How are motivation and self-efficacy interacting in problem-solving and problem-posing? Educational Studies in Mathematics, 105(3), 487–517. https://doi.org/10.1007/s10649-020-10005-0
37. Eynde, P. O., & De Corte, E. (2003). Students' Mathematics-Related Belief Systems: Design and Analysis of Questionnaire. Paper Presented at the Annual Meeting of the American Educational Research Association, 1–14.
38. Duncan, G. J., & Magnuson, K. A. (2005). Can family socioeconomic resources account for racial and ethnic test score gaps? The Future of Children,15(1), 35–54.
39. Grolnick, W. S., Friendly, R. W., & Bellas, V. M. (2009). Parenting and children's motivation at school. In K. R. Wentzel & A. Wigfield (Eds.), Handbook of motivation at school (pp. 279–300). New York: Routledge.
40. Usher, A., & Kober, N. (2012). What roles do parent involvement, family background, and culture play in student motivation? Retrieved from the Center on the Education Policy website. https://www.cep-dc.org/cfcontent_file.cfm?Attachment=UsherKober%5FBackground4%5FMotivation%5F5%2E22%2E12%2Epdf . Accessed 2 Jan 2019.
41. Willms, J. (2003). Student engagement at school a sense of belonging and participation results from PISA 2000. Paris: OECD
42. Abraham, J., & Barker, K. (2015b). Exploring gender difference in motivation, engagement and enrolment behaviour of senior secondary physics students in New South Wales. Research in Science Education, 45(1), 59–73. https://doi.org/10.1007/s11165-014-9413-2 .
43. Martin, A. J. (2003). The student motivation scale: Further testing of an instrument that measures school students' motivation. Australian Journal of Education, 47(1), 88–106. https://doi.org/10.1177/000494410304700107 .
44. Abraham, J., & Barker, K. (2014). Sustaining young people's enrolment intentions in relation to physics: Development and validation of a tool. The Australian Journal of Educational & Developmental Psychology, 14, 93–116.
45. Lyons, T. (2006). The puzzle of falling enrolments in physics and chemistry courses: Putting some pieces together. Research in Science Education, 36(3), 285–311. https://doi.org/10.1007/s11165-005-9008-z .
46. Jones, F. L., & McMillan, J. (2001). Scoring occupational categories for social research: A review of current practice, with Australian examples. Work, Employment and Society, 15(3), 539–563. https://doi.org/10.1177/09500170122119147 .
47. Bodkin-Andrews, G., O'Rourke, V., Grant, R., Denson, N., & Craven, R. (2010). Validating racism and cultural respect: Testing the psychometric properties and educational impact of perceived discrimination and multiculturation for Indigenous and non-Indigenous students. Educational Research and Evaluation, 16(6), 471–493. https://doi.org/10.1080/13803611.2010.550497 .
48. Kelloway, K. E. (1998). Using LISREL for structural equation modelling. Thousand Oaks: SAGE Publications.
49. Ainley, J., Kos, J., & Nicholas, M. (2008). Participation in science, mathematics and technology in Australian education (Research Monograph No 63). Retrieved from Camberwell, Victoria, Australia:

50. Hazari, Z., Sadler, P. M., & Tai, R. H. (2008). Gender differences in the high school and affective experiences of introductory college physics students. The Physics Teacher, 46(7), 423–427. https://doi.org/10.1119/1.2981292 .
51. Kennedy, J., Lyons, T., & Quinn, F. (2014). The continuing decline of science and mathematics enrolments in Australian high schools. Teaching Science, 60(2), 34.
52. Stokking, K. M. (2000). Predicting the choice of physics in secondary education. International Journal of Science Education, 22(12), 1261–1283.
53. Taasoobshirazi, G., & Carr, M. (2008). Gender differences in science: An expertise perspective. Educational Psychology Review, 20(2), 149–169. https://doi.org/10.1007/s10648-007-9067-y .
54. Whitelegg, E., Murphy, P., & Hart, C. (2007). Girls and physics: Dilemmas and tensions. In R. Pintó & D. Couso (Eds.), Contributions from science education research (pp. 27–36). Dordrecht: Springer Netherlands.
55. Häussler, P., & Hoffmann, L. (2002). An intervention study to enhance girls' interest, self-concept, and achievement in physics classes. Journal of Research in Science Teaching, 39(9), 870–888. https://doi.org/10.1002/tea.10048 .
56. Hamre, B. K., & Pianta, R. C. (2001). Early teacher-child relationships and the trajectory of children's school outcomes through eighth grade. Child Development, 72, 625–638.
57. Jang, H., Kim, E. J., & Reeve, J. (2012). Longitudinal test of self-determination theory's motivation mediation model in a naturally occurring classroom context. Journal of Educational Psychology, 104, 1175–1188.
58. Jang, H., Reeve, J., & Deci, E. L. (2010). Engaging students in learning activities: It is not autonomy support or structure but autonomy support and structure. Journal of Educational Psychology, 102, 588–600.
59. Raftery-Helmer, J. N., & Grolnick, W. G. (2016). Children's coping with academic failure: Relations with contextual and motivational resources supporting competence. The Journal of Early Adolescence 36(8), 1017–1041.
60. Swanson, J., Valiente, C., Lemery-Chalfant, K., & O'Brien, T. C. (2011). Predicting early adolescents' academic achievement, social competence, and physical health from parenting, ego resilience, and engagement coping. The Journal of Early Adolescence, 31, 548–576.
61. Zimmer-Gembeck, M. J., & Locke, E. M. (2007). The socialization of adolescent coping behaviors. Relationships with families and teachers. Journal of Adolescence, 30, 1–16.
62. Barrington, B. L., & Hendricks, B. (1989). Differentiating characteristics of high school graduates, dropouts, and nongraduates. The Journal of Educational Research, 82, 309–319.
63. Cairns, R. B., Cairns, B. D., & Neckerman, H. J. (1989). Early school dropout: Configurations and determinants. Child Development, 60, 1437–1452.
64. Ekstrom, R. B., Goertz, M. E., Pollack, J. M., & Rock, D. A. (1986). Who drops out of high school and why? Findings from a national study. Teacher College Record, 87, 356–373.
65. Grissom, J. B., & Shepard, L. A. (1989). Repeating and dropping out of school. In L. A. Shepard & M. L. Smith (Eds.), Flunking grades: Research and policies on retention. (pp. 16–33). London Falmer Press.
66. Wehlage, G. G., & Rutter, R. A. (1986). Dropping out: How much do schools contribute to the problem? Teachers College Record, 87, 374–392.
67. Deci, E. L., & Ryan, R. M. (1985). Intrinsic motivation and self-determination in human behavior. New York: Plenum Press.
68. Skinner, E. A., & Wellborn, J. G. (1994). Coping during childhood and adolescence: A motivational perspective. In D. Featherman, R. Lerner & M. Perlmutter (Eds.), Life-span development and behavior (Vol. 12, pp. 91–133). Hillsdale, NJ: Erlbaum.
69. Skinner, E. A., & Wellborn, J. G. (1997). Children's coping in the academic domain. In S. A. Wolchik & I. N. Sandler (Eds.), Handbook of children's coping: Linking theory and intervention (pp. 387–422). New York, NY: Plenum Press.
70. Skinner, E. A., Edge, K., Altman, J., & Sherwood, H. (2003). Searching for the structure of coping: A review and critique of category systems for classifying ways of coping. Psychological Bulletin, 129, 216–269.

71. Reschly, A. L., Huebner, E. S., Appleton, J. J., & Antaramian, S. (2008). Engagement as flourishing: The contribution of positive emotions and coping to adolescents' engagement at school and with learning. Psychology in the Schools, 45, 419–431.
72. Sugai, G., Horner, R. H., & Sprague, J. R. (1999). Functional-assessment-based behavior support planning: Research to practice to research. Behavioral Disorders, 24(3), 253–257.
73. Mathur, S.R., Kavale, K.A., Quinn, M.M., Forness, S.R. & Rutherford, R.B. (1998). Social skills interventions with students with emotional and behavioral problems: A quantitative synthesis of single-subject research. Behavioral Disorders, 193–201.
74. Cannella, G. S. (1997). Deconstructing Early Childhood Education: Social Justice and Revolution. Michigan: Peter Lang Inc.
75. Weiner, H. (1965). Conditioning history and maladaptive human operant behavior. Psychological Reports, 17(3), 935–942.
76. Wolpe, J., & Rowan, V. C. (1988). Panic disorder: A product of classical conditioning. Behaviour Research and Therapy, 26(6), 441–450.
77. Carver, C. S., & Scheier, M. F. (1981). Attention and self-regulation. New York: Springer.
78. Duval, T. S., Duval, V. H., & Mulilis, J. P. (1992). Effects of self-focus, discrepancy between self and standard, and outcome expectancy favorability on the tendency to match self to standard or to withdraw. Journal of Personality and Social Psychology, 62, 340–348.
79. Brehm, J. W., & Self, E. A. (1989). The intensity of motivation. Annual Review of Psychology, 40, 109–131.
80. Brehm, J. W., Wright, R. A., Solomon, S., Silka, L., & Greenberg, J. (1983). Perceived difficulty, energization, and the magnitude of goal valence. Journal of Experimental Social Psychology, 19, 21–48.
81. Duval, T. S., & Wicklund, R. A. (1972). A theory of objective self awareness. New York: Academic Press.
82. Liebling, B. A., & Shaver, P. (1973). Evaluation, self-awareness, and task performance. Journal of Experimental Social Psychology, 9, 297–306.
83. McDonald, P. J. (1980). Reactions to objective self-awareness. Journal of Research in Personality, 14, 250–260.
84. Wicklund, R. A., & Duval, T. S. (1971). Opinion change and performance facilitation as a result of objective self-awareness. Journal of Experimental Social Psychology, 7, 319–342.
85. Burgio, K. L., Merluzzi, T. V., & Pryor, J. B. (1986). Effects of performance expectancy and self-focused attention on social interaction. Journal of Personality and Social Psychology, 50, 1216–1221.
86. Carver, C. S., Blaney, P. H., & Scheier, M. F. (1979a). Focus of attention, chronic expectancy, and responses to a feared stimulus. Journal of Personality and Social Psychology, 37, 1186–1195.
87. Carver, C. S., Blaney, P. H., & Scheier, M. F. (1979b). Reassertion and giving up: The interactive role of self-directed attention and outcome expectancy. Journal of Personality and Social Psychology, 37, 1859–1870.
88. Duval, T. S., & Silvia, P. J. (2002). Self-awareness, probability of improvement, and the self-serving bias. Journal of Personality and Social Psychology, 82, 49–61.
89. Silvia, P. J., & Duval, T. S. (2001b). Predicting the interpersonal targets of self-serving attributions. Journal of Experimental Social Psychology, 37, 333–340.
90. Wright, R. A., Killebrew, K., & Pimpalapure, D. (2002). Cardiovascular incentive effects where a challenge is unfixed: Demonstrations involving social evaluation, evaluator status, and monetary reward. Psychophysiology, 39, 188–197.
91. Gendolla, G. H. E., Richter, M., & Silvia, P. J. (2008). Self-focus and task difficulty effects on cardiovascular reactivity. Psychophysiology, 45, 653–662.
92. Gendolla, G. H. E., & Richter, M. (2006). Cardiovascular reactivity during performance under social observation: The moderating role of task difficulty. International Journal of Psychophysiology, 62, 185–192.

93. Scheier, M. F., Carver, C. S., & Gibbons, F. X. (1979). Self-directed attention, awareness of bodily states, and suggestibility. Journal of Personality and Social Psychology, 37, 1576–1588.
94. Blackwell, L. S., Trzesniewski, K. H., & Dweck, C. S. (2007). Implicit theories of intelligence predict achievement across an adolescent transition: A longitudinal study and an intervention. Child Development, 78, 246–263.
95. Midgley, C., Maehr, M. L., Hruda, L. Z., Anderman, E., Anderman, L., Freeman, K. E., et al. (2000). Manual for the Patterns of Adaptive Learning Scales (PALS). Ann Arbor: University of Michigan. Retrieved July 10, 2007, from http://www.umich.edu/~pals/pals/ .
96. Blackwell, L. S. (2002). Psychological mediators of student achievement during the transition to junior high school: The role of implicit theories. Unpublished doctoral dissertation, Columbia University, New York.
97. Bryan RR, Glynn SM, Kittleson JM: Motivation, achievement, and advanced placement intent of high school students learning science. Sci Educ. 2011, 95: 1049-1063. https://doi.org/10.1002/sce.20462.
98. Sanfeliz M, Stalzer M: Science motivation in the multicultural classroom. Sci Teach. 2003, 70: 64-66.
99. Glynn SM, Brickman P, Armstrong N, Taasoobshirazi G: Science motivation questionnaire II: validation with science majors and nonscience majors. J Res Sci Teach. 2011, 48: 1159-1176. https://doi.org/10.1002/tea.20442.
100. Boiche JCS, Sarrazin PGS, Grouzet FME, Pelletier LG, Chanal J: Students' motivational profiles and achievement outcomes in physical education: a self-determination perspective. J Educ Psychol. 2008, 100: 688-701.
101. Landry, E., Amara, N., & Rherrand, I. (2006). Why are some university researchers more likely to create spin-offs than others? Evidence from Canadian universities. Research Policy, 35(10), 1599–1615.
102. Meyer, M. (2003). Academic entrepreneurs or entrepreneurial academics? Research-based ventures and public support mechanisms. R&D Management, 33(2), 107–115.
103. Chiesa, V., & Piccaluga, A. (2000). Exploitation and diffusion of public research: the chase of academic spin-offs companies in Italy. R & D Management, 30(4), 329–339.
104. O'Shea, R., Allen, T., Morse, K., O'Gorman, C., & Roche, F. (2004). Universities and technology transfer: a review of academic entrepreneurship literature. Irish Journal of Management, 25(2), 11–29.
105. Searle, C. (2006). Academic capitalism and university incentives for faculty entrepreneurship. Journal of Technology Transfer, 31, 227–239.
106. Doutriaux, J., & Peterman, D. (1982). Technology transfer and academic entrepreneurship. In: Proceedings of Frontiers of Entrepreneurship Research, Babson College Conference, 430–448.
107. Louis, K., Blumenthal, D., Gluck, M., & Stoto, M. (1989). Entrepreneurs in academe: behaviours among life scientist. Administrative Science Quarterly, 34(1), 110–131.
108. Samsom, K., Gurdon, M. (1990). Entrepreneurial scientist: organizational performance in scientist-started high technology firms. In: Proceedings of Frontiers of Entrepreneurship Research, Babson College Conference, 437–451
109. Doutriaux, J. (1991). University culture, spin-off strategy, and success of academic entrepreneurs at Canadian universities. In: Proceedings of Frontiers of Entrepreneurship Research, Babson College Conference, 406–421.
110. Doutriaux, J., & Dew, G. (1992). Motivation of academic entrepreneurs and spin-off development: analysis of regional and university effects through case studies. In: Proceedings of Frontiers of Entrepreneurship Research, Babson College Conference, 231–232.
111. Chrisman, J., Hynes, T., & Fraser, S. (1995). Faculty entrepreneurship and economic development: the case of the University of Calgary. Journal of Business Venturing, 10(4), 267–281.
112. Henrekson, M., & Rosenberg, N. (2000). Incentives for academic entrepreneurship and economic performance: Sweden and the United States. The Journal of Technology Transfer, 26(3), 207–231.

113. Shane, S., & Khurana, R. (2003). Bringing individuals back in: the effects of career experience on new firm founding. Industrial and Corporate Change, 12(3), 519–543.
114. Shane, S. (2004). Academic entrepreneurship: University Spinoffs and wealth creation. Massachusetts: Edwar Elgar.
115. Lehrer, M., & Asakawa, K. (2004). Pushing scientists into the marketplace: promoting science entrepreneurship. California Management Review, 46(3), 55–76.
116. Ding, W., & Stuart, T. (2006). When do scientists become entrepreneurs? The social structural antecedents of commercial activity in the academic life sciences. American Journal of Sociology, 112(1), 97–144.
117. Zhang, J. (2006). A study of academic entrepreneurs using venture capital data. Institute for the Study of Labor (IZA), IZA Discussion Papers No. 2992.
118. Jones-Evans, D. (1998). Universities, technology transfer and spin-off activities—academia entrepreneurship in different European Regions. Targeted Socio-economic Research Project No. 1042. Final Report. University of Glamorgan, Business School, and European Commission. http://www.lambertreview.org.uk/pdffiles/uni/utheschoolforbusinandregiondevelreport.pdf .
119. Klofsten, M., & Jones-Evans, D. (2000). Comparing academic entrepreneurship in Europe. Small Business Economics, 14(4), 199–309.
120. Laukannen, M. (2003). Exploring academic entrepreneurship: drivers and tensions of university based business. Journal of Small Business and Enterprise Development, 10(4), 372–382.
121. Vohora, A., Wright, M., & Lockett, A. (2004). Critical junctures in the development of university high technology spinout companies. Research Policy, 33(1), 147–175.
122. Mosey, S., & Wright, M. (2007). From human capital to social capital: a longitudinal study of technology-based academic entrepreneurs. Entrepreneurship, Theory and Practice, 31(6), 909–935.
123. Shane, S., Kolvereid, L., & Westhead, P. (1991). An exploratory examination of the reasons leading to new firm formation across country and gender. Journal of Business Venturing, 6, 431–446.
124. Birley, S., & Westhead, P. (1994). A taxonomy of business start-up reasons and their impact on firm growth and size. Journal of Business Venturing, 9, 7–31.
125. Dubini, P. (1988). The influence of motivations and environment on business start-ups: some hints for public policies. Journal of Business Venturing, 4, 11–26.
126. Carter, N. M., Gartner, W. B., Shaver, K. G., & Gatewood, E. J. (2003). The career reasons of nascent entrepreneurs. Journal of Business Venturing, 18, 13–39.
127. Shane, S., Locke, E., & Collins, C. (2003). Entrepreneurial motivation. Human Resource Management Review, 13(2), 257–279.
128. Segal, G., Borgia, D., & Schoenfeld, J. (2005). The motivation to become an entrepreneur. International Journal of Entrepreneurial Behaviour and Research, 11(1), 42–57.
129. Weatherston, J. (1995). Academic Entrepreneurs: Is a spin-off Company too risky?. International Council for Small Business (ICSB). http://www.sbaer.uca.edu/research/icsb/1995/pdf/20.pdf .
130. Barrett MS, Bond N (2014) Connecting through music: the contribution of a music programme to fostering positive youth development. Res Stud Music Educ 37(1):37–54
131. Cogdill SH (2014) Applying research in motivation and learning to music education: what the experts say update. Appl Res Music Educ 33(2):49–57
132. Deci E, Ryan R (1985) The general causality orientations scale: self-determination in personality. J Res Pers 19(2):109–134
133. Deci, E. L., & Ryan, R. M. (2000). The "what" and "why" of goal pursuits: Human needs and the self-determination of behavior. Psychological Inquiry, 11, 227–268. https://doi.org/10.1207/S15327965PLI1104_01 .
134. Deci E, Ryan R (2014) The importance of universal psychological needs for under-standing motivation in the workplace. In: Gagné M (ed) The Oxford handbook of work engagement, motivation, and self-determination theory. Oxford University Press, Oxford, pp 13–32

135. Deci E, Ryan R (2017) Self-Determination Theory. Basic psychological needs in motivation, development, and wellness. Guildford Press, New York
136. Berliner P (1994) Thinking in jazz: the infinite art of improvisation. University of Chicago Press, Chicago
137. Gatien G (2009) Categories and music transmission. Action Crit Theory Music Educ 8(2):94–119
138. Gridley M (2003) Jazz styles: history and analysis, 8th edn. Prentice Hall, New Jersey
139. Schreier M (2012) Qualitative content analysis in practice. SAGE Publications, Thousand Oaks
140. Virkkula E, Nissilä S-P (2014) In-service teachers' learning through integrating theory and practice. SAGE Open J 4(4):1–8
141. Küpers E, van Dijk M, McPherson G, van Geert P (2014) A dynamic mode that links skill acquisition with self-determination in instrumental music lessons. Musicae Scientiae 18(1):17–34
142. Radel R, Pelletier L, Baxter D, Fournier M, Sarrazin P (2013) The paradoxical effect of controlling context on intrinsic motivation in another activity. Learn Instr 29:95–102
143. Green L (2001) How popular musicians learn: a way ahead for music education. Ashgate, London
144. Green L (2008) Music, informal learning and the school: a new classroom pedagogy. Ashgate, Aldershot
145. Burwell K (2012) Apprenticeship in music: a contextual study for instrumental teaching and learning. Int J Music Educ 31(3):276–291
146. Evans P (2015) Self-determination theory: an approach to motivation in music education. Musicae Scientiae 19(1):65–83
147. Lebler D (2007) Student-as-master? Reflections on a learning innovation in popular mu-sic pedagogy. Int J Music Educ 25(3):205–221
148. Brière, N. M., Vallerand, R. J., Blais, M. R., & Pelletier, L. G. (1995). Développement et validation d'une mesure de motivation intrinsèque, extrinsèque et d'amotivation en contexte sportif : L'Échelle de Motivation dans les Sports (EMS). International Journal of Sport Psychology, 26, 465–489. On the development and validation of the French form of the Sport Motivation Scale.
149. Ratelle, C. F., Guay, F., Vallerand, R. J., Larose, S., & Senécal, C. (2007). Autonomous, controlled, and amotivated types of academic motivation: A person-oriented analysis. Journal of Educational Psychology, 99, 734–746. https://doi.org/10.1037/0022-0663.99.4.734 .
150. Kowal, J., & Fortier, M. S. (1999). Motivational determinants of flow: Contributions from self-determination theory. The Journal of Social Psychology, 139(3), 355–368. https://doi.org/10.1080/00224549909598391
151. Pelletier, L. G., Fortier, M. S., Vallerand, R. J., Tuson, K. M., Brière, N. M., & Blais, M. R. (1995). Toward a new measure of intrinsic motivation, extrinsic motivation, and amotivation in sports: The Sport Motivation Scale (SMS). Journal of Sport & Exercise Psychology, 17, 35–53.
152. Pelletier, L. G., Fortier, M. S., Vallerand, R. J., & Briere, N. M. (2001). Associations among perceived autonomy support, forms of self-regulation, and persistence: A prospective study. Motivation and Emotion, 25, 279–306. https://doi.org/10.1023/A:1014805132406 .
153. Sarrazin, P., Vallerand, R. J., Guillet, E., Pelletier, L. G., & Cury, F. (2002). Motivation and dropout in female handballers: A 21-month prospective study. European Journal of Social Psychology, 57, 749–761.
154. Vallerand, R. J. (1997). Toward a hierarchical model of intrinsic and extrinsic motivation. In M. P. Zanna (Ed.), Advances in experimental social psychology (pp. 271–360). New York: Academic Press.
155. Harwood, C., Cumming, J., & Fletcher, D. (2004). Motivational profiles and psychological skills usage within elite youth sport. Journal of Applied Sport Psychology, 16, 318–332. https://doi.org/10.1080/10413200490517986 .

156. Hodge, K., Allen, J. B., & Smellie, L. (2008). Motivation in masters sport: Achievement and social goals. Psychology of Sport and Exercise, 9, 157–176. https://doi.org/10.1016/j.psychsport.2007.03.002 .
157. McNeill, M. C., & Wang, C. K. J. (2005). Psychological profiles of elite school sports players in Singapore. Psychology of Sport and Exercise, 6, 117–128. https://doi.org/10.1016/j.psychsport.2003.10.004 .
158. Cumming, J., & Hall, C. (2004). The influence of goal orientation on self-efficacy in exercise. Journal of Applied Social Psychology, 34, 747–763. https://doi.org/10.1111/j.1559-1816.2004.tb02568.x .
159. Matsumoto, H., & Takenaka, K. (2004). Motivational profiles and stages of exercise behavior change. International Journal of Sport and Health Science, 2, 89–96.
160. Wang, C. K. J., & Biddle, S. J. H. (2001). Young people's motivational profiles in physical activity: A cluster analysis. Journal of Sport & Exercise Psychology, 23, 1–22.
161. Ntoumanis, N. (2002). Motivational clusters in a sample of British physical education classes. Psychology of Sport and Exercise, 3, 177–194. https://doi.org/10.1016/S1469-0292(01)00020-6 .
162. Wang, C. K. J., Chatzisarantis, N. L. D., Spray, C. M., & Biddle, S. J. H. (2002). Achievement goal profiles in school physical education: Differences in self-determination, sport ability beliefs, and physical activity. The British Journal of Educational Psychology, 72, 433–445. https://doi.org/10.1348/000709902320634401 .
163. Vlachopoulos, S. P., Karageorghis, C. I., & Terry, P. C. (2000). Motivation profiles in sport: A self-determination theory perspective. Research Quarterly for Exercise and Sport, 71, 387–397.
164. Vallerand, R. J. (2007a). A hierarchical model of intrinsic and extrinsic motivation for sport and physical activity. In M. S. Hagger & N. L. D. Chatzisarantis (Eds.), Intrinsic motivation and self-determination in exercise and sport (pp. 255–279). Champaign: Human Kinetics.
165. Cresswell, S. L., & Eklund, R. C. (2005a). Motivation and burnout among top amateur rugby players. Medicine and Science in Sports and Exercise, 37, 469–477. https://doi.org/10.1249/01.MSS.0000155398.71387.C2 .
166. Cresswell, S. L., & Eklund, R. C. (2005b). Motivation and burnout in professional rugby players. Research Quarterly for Exercise and Sport, 76, 370–376.
167. Lemyre, N. P., Treasure, D. C., & Roberts, G. C. (2006). Influence of variability in motivation and affect on elite athlete burnout susceptibility. Journal of Sport & Exercise Psychology, 28, 32–48.
168. Schult A, Beck E, Majschak J-P (2015) Steigerung der Effizienz von Verarbeitungs- und Verpackungsanlagen durch Wirkungsgradanalysen. Pharma + Food 18:66–68
169. Adler PS (1988) Managing flexible automation. Calif Manag Rev 30:34–56. https://doi.org/10.2307/41166513
170. Walton RE, Sussman GI (1987) People policies for new machines. Harv Bus Rev 65:98–106
171. Gagné M, Deci EL (2005) Self-determination theory and work motivation. J Organ Behav 26:331–362. https://doi.org/10.1002/job.322
172. Müller R, Oehm L (2019) Process industries versus discrete processing: how system characteristics affect operator tasks. Cogn Technol Work 21:337–356. https://doi.org/10.1007/s10111-018-0511-1
173. Elliot, A. J., McGregor, H. A., & Gable, S. (1999). Achievement goals, study strategies, and exam performance: A mediational analysis. Journal of Educational Psychology, 91, 549–563.
174. Hulleman, C. S., & Senko, C. (2010). Up around the bend: Forecasts for achievement goal theory and research in 2020. In The decade ahead: Theoritical perspectives on motivation and achievement (Vol. 16, pp. 71–104). Emerald Group Publishing Limited.
175. DeBoer, J., Stump, G. S., Seaton, D., & Breslow, L. (2013). Diversity in MOOC students' backgrounds and behaviors in relationship to performance in 6.002x. In the Proceedings of the Sixth Learning International Networks Consortium Conference.
176. Pekrun, R., vom Hofe, R., Blum, W., Frenzel, A. C., Goetz, T., & Wartha, S. (2007). Development of mathematical competencies in adolescence. The PALMA longitudinal study. In

M. Prenzel (Ed.), Studies on the educational quality of schools. The final report on the DFG priority Programme (pp. 17–37). Waxmann: Münster.

177. Deci, E. L., & Ryan, R. M. (2002). Overview of self-determination theory: An organismic dialectical perspective. Handbook of self-determination research, 3–33.
178. Pekrun, R. (2014). Emotions and learning. Educational practices series-24. Switzerland: UNESCO International Bureau of Education.
179. Perels, F., Dignath, C., & Schmitz, B. (2009). Is it possible to improve mathematical achievement by means of self-regulation strategies? Evaluation of an intervention in regular maths classes. European Journal of Psychology of Education, 24, 17–31.
180. Fend, H. (1998). Qualität im Bildungswesen. Schulforschung zu Systembedingungen, Schulprofilen und Lehrerleistung [quality in education. School research on system conditions, school profiles and teacher performance]. Weinheim und München, Juventa.
181. Martin, A. J. (2006). The relationship between teachers' perceptions of student motivation and engagement and teachers' enjoyment of and confidence in teaching. Asia-Pacific Journal of Teacher Education, 34, 73–93.
182. Kim, M. C., & Hannafin, M. J. (2011). Scaffolding problem solving in technology-enhanced learning environments (TELEs): Bridging research and theory with practice. Computers & Education, 56(2), 403–417. https://doi.org/10.1016/j.compedu.2010.08.024 .
183. Slotta, J. D., & Linn, M. C. (2009). WISE science, web-based inquiry in the classroom. New York: Teachers College Press.
184. Vansteenkiste, M., Sierens, E., Soenens, B., Luyckx, K., & Lens, W. (2009). Motivational profiles from a self-determination perspective: The quality of motivation matters. Journal of Educational Psychology, 101(3), 671–688. https://doi.org/10.1037/A0015083 .
185. Toplis R (2011) Students' attitudes in science. Educ Sci 243:22–23
186. Hemingway C, Dahl W, Haufler C, Stuessy C (2011) Building botanical literacy. Science 331(6024):1535–1536
187. Hemingway C (2008) Annual report to NSF. Retrieved from the Botanical Society of America website: http://botany.org/PlantingScience/Reports/DRK12_submitted08report.pdf
188. O'Neill DK, Wagner R, Gomez LM (1996) Online mentors: experimenting in science class. Educ Leadersh 54:39–42
189. Teng, M. F., Wang, C., & Wu, J. G. (2021). Metacognitive Strategies, Language Learning motivation, self-efficacy belief, and English achievement during remote learning: A structural equation modelling approach. RELC Journal. https://doi.org/10.1177/00336882211040268
190. Calafato, R., & Tang, F. (2019). The status of Arabic, superdiversity, and language learning motivation among non-Arab expats in the Gulf. Lingua, 219, 24–38.
191. Calafato, R. (2021a). "I'm a salesman and my client is China": Language learning motivation, multicultural attitudes, and multilingualism among university students in Kazakhstan and Uzbekistan. System, 103, 102645.
192. Dewaele, J. M., & Proietti Ergün, A. L. (2020). How different are the relations between enjoyment, anxiety, attitudes/motivation and course marks in pupils' Italian and English as foreign languages? Journal of the European Second Language Association, 4(1), 45–57.
193. Zheng, Y., Lu, X., & Ren, W. (2019). Profiling Chinese university students' motivation to learn multiple languages. Journal of Multilingual and Multicultural Development, 40(7), 590–604.
194. Yashima, T., & Arano, K. (2015). Understanding EFL learners' motivational dynamics: A three-level model from a dynamic systems and sociocultural perspective. In Z. Dörnyei, P. D. MacIntyre, & A. Henry (Eds.), Motivational dynamics in language learning (pp. 285–314). Multilingual Matters.
195. Ramirez-Arellano, A., Acosta-Gonzaga, E., Bory-Reyes, J., & Hernández-Simón, L. M. (2018). Factors affecting student learning performance: A causal model in higher blended education. Journal of Computer Assisted Learning, 34(6), 807–815.
196. Reeve, J., & Lee, W. (2019). A neuroscientific perspective on basic psychological needs. Journal of Personality, 87(1), 102–114.

197. Ames, C. (1992a). Achievement goals and the classroom motivational climate. In D. H. Schunk & J. L. Meece (Eds.), Student perceptions in the classroom (pp. 327–348). Hillsdale, NJ Hove and London: Lawrence Arlbaum.
198. Miller, A. D., & Murdock, T. B. (2007). Modeling latent true scores to determine the utility of aggregate student perceptions as classroom indicators in HLM: The case of classroom goal structures. Contemporary Educational Psychology, 32(1), 83–104..
199. Vedder-Weiss, D., & Fortus, D. (2013). School, teacher, peer's and parents' goals emphases and adolescents' motivation to learn science in and out of school. Journal of Research in Science Teaching, 50(8), 952–988.
200. Leman, F. B. 2003. Problems in learning English in rural secondary schools: A case study (Unpublished MA thesis, Faculty of Education, Universiti Kebangsaan Malaysia).
201. Badrkoohi, A., & Maftoon, P. (2017). Individual differences and L2 motivation: The case of EFL learners. International Journal of Applied Linguistics and English Literature, 6(3), 199–209.
202. Gardner, R. C. (2001). Integrative motivation and second language acquisition. In Z. Dörnyei & R. Schmidt (Eds.), Motivation and second language acquisition (p. 1_20). Honolulu: University of Hawaii Press.
203. Dörnyei, Z. (2009). The L2 motivational self system. In Z. Dörnyei & E. Ushioda (Eds.), Motivation, language identity and the L2 self (pp. 9–42). Bristol: Multilingual Matters.
204. Ghorbandordinejad, F. (2014). Motivation and autonomy as predictors of foreign language proficiency among the Iranian students of distance vs. traditional universities. International Journal of Applied Linguistics and English Literature, 3(6), 113–123.
205. Reeve, J. (2009). Why teachers adopt a controlling motivating style toward students and how they can become more autonomy supportive. Educational Psychologist, 44(3), 159–175.
206. Kurt, G. (2017). Implementing the flipped classroom in teacher education: Evidence from Turkey. Journal of Educational Technology & Society, 20(1), 211–221.
207. Üğüten, S. D., & Balci, Ö. (2017). Flipped learning. Journal of Suleyman Demirel University Institute of Social Sciences, 26(1), 253–264.
208. Wu, W. C. V., Hsieh, J. S. C., & Yang, J. C. (2017). Creating an online learning community in a flipped classroom to enhance EFL learners' oral proficiency. Journal of Educational Technology & Society, 20(2), 142–157.
209. Basal, A. (2015). The implementation of a flipped classroom in foreign language teaching. Turkish Online Journal of Distance Education, 16(4), 28–37.
210. Bergmann, J., & Sams, A. (2012). Flip your classroom: Reach every student in every class every day. Washington, DC: Internal Society for Technology in Education.
211. Kvashnina, O. S., & Martynko, E. A. (2016). Analyzing the potential of flipped classroom in ESL teaching. International Journal of Emerging Technologies in Learning (iJET), 11(03), 71–73.
212. Creswell, J. W. (2014). Research design: Qualitative, quantitative, and mixed methods approaches (4th ed.). Los Angeles: Sage publications, Inc..
213. Gardner, R. C. (1985). Social psychology and second language learning: The role of attitude and motivation. London: Edward Arnold.
214. Dörnyei, Z., Csizér, K., & Nemeth, N. (2006). Motivation, language attitudes and globalization: A Hungarian perspective. Clevedon: Multilingual Matters.
215. Ryan, S. (2009). Self and identity in L2 motivation in Japan: The ideal L2 self and Japanese learners of English. In Z. Dörnyei & E. Ushioda (Eds.), Motivation, language identity and the L2 self (pp. 120–143). Bristol: Multilingual Matters.
216. Lamb, M. (2012). A self system perspective on young adolescents' motivation to learn English in urban and rural settings. Language Learning, 62(4), 997–1023.
217. Malmkjaer, K. (Ed.). (1998). Translation and Language Teaching: Language Teaching and Translation. St. Jerome Publishing.
218. Riccardi, A. (Ed.). (2002). Translation studies: Perspectives on an emerging discipline. Cambridge University Press.

219. Chang, P. C., Yu, Y. K., Li, T. C., & Peng, M. Y. (1998). Practical translation: An introductory coursebook. Bookman Books.
220. Azizinezhad, M. (2006). Is translation teachable?" dalam. Translation Journal, 10(2).
221. Bahremand, A. (2015). The concept of translation in different teaching approaches and methods. Journal of Social Sciences and Humanities Research, 3(01), 6–10.
222. Higgins, E. T. (1998). Promotion and prevention: Regulatory focus as a motivational principle. Advances in Experimental Social Psychology, 30, 1–46.
223. Dörnyei, Z. (2005). The Psychology of the Language Learner: Individual Differences in Second Language Acquisition. Lawrence Erlbaum Associates. https://doi.org/10.1017/S0272263105370288
224. Davis FD (1989) Perceived usefulness, perceived ease of use, and user acceptance of information technology. MIS Q 13:319–340. https://doi.org/10.2307/249008
225. Liu, Y. C., Wang, W. T., & Lee, T. L. (2021). An integrated view of information feedback, game quality, and autonomous motivation for evaluating game-based learning effectiveness. Journal of Educational Computing Research, 59(1), 3–40.
226. Rosenzweig, E. Q., & Wigfield, A. (2016). STEM motivation interventions for adolescents: A promising start, but further to go. Educational Psychologist, 51(2), 146–163.
227. Tapia, M., & Marsh, G. (2004). An instrument to measure mathematics attitudes. Academic Exchange Quarterly, 8, 16–21.
228. Pintrich PR, De Groot EV. Motivational and self-regulated learning component of classroom academic performance. J Educ Psychol. 1990;82:33–40.
229. Yang, Y., & Green, S. B. (2015). Evaluation of structural equation modeling estimates of reliability for scales with ordered categorical items. European Journal of Research Methods for the Behavioral and Social Sciences, 11(1), 23–34.
230. American Educational Research Association [AERA], American Psychological Association [APA], National Council on Measurement in Education [NCME]. (2014). Standards for educational and psychological testing. American Educational Research Association.
231. Atit, K., Power, J. R., Veurink, N., Uttal, D. H., Sorby, S., Panther, G., Msall, C., Fiorella, L., & Carr, M. (2020). Examining the role of spatial skills and mathematics of motivation on middle school achievement. International Journal of STEM Education. https://doi.org/10.1186/s40594-020-00234-3
232. Jiang, S., Simpkins, S. D., & Eccles, J. S. (2020). Individuals' math and science motivation and their subsequent STEM choices and achievement in high school and college: A longitudinal study of gender and college generation status differences. Developmental Psychology, 56(11), 2137–2151.
233. Wang, X. (2013). Why students choose STEM majors: Motivation, high school learning, and postsecondary context of support. American Educational Research Journal, 50(5), 1081–1121.
234. Wigfield, A., Tonks, S. M., & Klauda, S. L. (2016). Expectancy-value theory. In K. R. Wentzel & D. B. Miele (Eds.), Handbook of motivation of school (2nd ed., pp. 55–74). Routledge.
235. Glynn SM, Taasoobshirazi G, Brickman P: Science motivation questionnaire: construct validation with nonscience majors. J Res Sci Teach. 2009, 46: 127-146. https://doi.org/10.1002/tea.20267.
236. Hernandez, P. R., Schultz, PW., Estrada, M., Woodcock, A., & Chance, R. C. (2013). Sustaining optimal motivation: A longitudinal analysis of interventions to broaden participation of underrepresented students in STEM. Journal of Educational Psychology, 105(1). https://doi.org/10.1037/a0029691 .
237. Eccles, J. S., Wigfield, A., Midgley, C., Reuman, D., Iver, D. M., & Feldlaufer, H. (1993). Negative effects of traditional middle schools on students' motivation. The Elementary School Journal, 93(5), 553–574.
238. Ball, S., Maguire, M., & Macrae, S. (2000). Choice, pathways and transitions post-16: New youth, new economies in the global city. London: Routledge.

239. Merenkov, A. V., & Sushchenko, A. D. (2016). Potrebnosti studentov vuzov v dopolnitel'nom obrazovanii: osobennosti formirovaniya i realizacii. Voprosy obrazovaniya. https://doi.org/10.17323/1814-9545-2016-3-204-223 .
240. Amr, K. (2012). Learning through games: Essential features of an educational game. Doctoral dissertation. Syracuse University.
241. Belkhouche, B., Al Darei, N., Ali, S., Al Mandhari, S., & Al Mehairi, M. (2014). Learning Arabic through games. In Eighth Annual Conference on Computer Games Multimedia and Allied Technologies (March 2014) (pp. 1–6). CGAT.
242. Xie, X. (2003). Objects management and English achievements Teacher's role in inspiring students' motivation and objects management. Exploration of Medical Education, 1, 1–7.
243. Mayer, R., & Johnson, C. (2010). Adding instructional features that promote learning in a game-like environment. Journal of Educational Computing Research, 42(3), 241–265. https://doi.org/10.2190/ec.42.3.a
244. Kotob, M., & Ibrahim, A. (2019). Gamification: The effect on students' motivation and achievement in language learning. Journal of Applied Linguistics and Language Research, 6(1), 177–198.
245. Wang, A. I., & Tahir, R. (2020). The effect of using Kahoot! for learning–A literature review. Computers & Education, 149, 103818. https://doi.org/10.1016/j.compedu.2020.103818
246. Licorish, S. A., George, J. L., Owen, H. E., & Daniel, B. (2017). Go Kahoot!" enriching classroom engagement, motivation and learning experience with games. In Proceedings of the 25th international conference on computers in education (pp. 755–764). New Zealand: Asia-Pacific Society for Computers in Education.
247. Sutton, R. E., & Wheatley, K. F. (2003). Teachers' emotions and teaching: A review of the literature and directions for future research. Educational Psychology Review, 15(4), 327–358.
248. Oker, A., Courgeon, M., Prigent, E., Eyharabide, V., Bazin, N., Urbach, M., et al. (2015a). A virtual reality study of help recognition and metacognition with an affective agent. International Journal of Synthetic Emotions, 6(1), 60–73.
249. Gordon, A. T. (2011). Assessing social support in children: Development and initial validation of the social support questionnaire for children. LSU Doctoral Dissertations, 3529.
250. Herba, C., & Phillips, M. (2004). Annotation: Development of facial expression recognition from childhood to adolescence: Behavioural and neurological perspectives. Journal of Child Psychology and Psychiatry, 45(7), 1185–1198.
251. Sugai, G., Horner, R. H., Dunlap, G., Hieneman, M., Lewis, T. J., Nelson, C. M., et al. (2000). Applying positive behavior support and functional behavioral assessment in schools. Journal of Positive Behavior Interventions, 2(3), 131–143.
252. Ekman, P., & Friesen, W. V. (1978). Manual for the facial action coding system. Palo Alto: Consulting Psychologists Press.
253. Ekman, P., Friesen, W. V., & Hager, J. C. (2002). Facial action coding system: The manual on CD ROM. A Human Face, Salt Lake City, 77–254.
254. Jackson, N. M. (2013). MOOCs go to K12: Higher ed trend expands to high schools. District Administration. Retrieved May 17, 2015 from http://www.districtadministration.com/article/moocs-go-k12-higher-ed-trend-expands-high-schools .
255. Protalinski, E. (2013). Coursera moves beyond universities and into K-12 with 12 new institutions and 28 free courses for teachers. Retrieved May 13, 2015 from http://www.thenextweb.com/insider/2013/05/01/coursera-moves-beyond-universities-by-partnering-with-12-institutions-to-offer-28-free-courses-for-teachers/#!t4vwT .
256. Brahimi, T., & Sarirete, A. (2015). Learning outside the classroom through MOOCs. Computers in Human Behavior, 51, 601–609.
257. Vivian, R., Falkner, K., & Falkner, N. (2014). Addressing the challenges of a new digital technologies curriculum: MOOCs as a scalable solution for teacher professional development. Research in Learning Technology, 22, 1–19.
258. Bundesagentur für Arbeit. (2011). Perspektive 2025 – Fachkräfte für Deutschland. Nuremberg. http://www.arbeitsagentur.de/Perspektive-2025 . Accessed 26 May 2014.

259. Börsch-Supan, A., & Wilke, C. (2009). Zur mittel- und langfristigen Entwicklung der Erwerbstätigkeit in Deutschland. Zeitschrift für Arbeitsmarktforschung, 42(1), 29–48.
260. Bundesministerium des Inneren. (2011). Demografiebericht der Bundesregierung. Berlin. http://www.bmi.bund.de/SharedDocs/Downloads/DE/Broschueren/2012/demografiebericht.pdf?__blob=publicationFile . Accessed 26 May 2014.
261. Büsch, V., Dorbritz, J., Heien, T., & Micheel, F. (2010). Weiterbeschäftigung im Rentenalter. Wünsche – Bedingungen – Möglichkeiten. Materialien zur Bevölkerungswissenschaft, 129. Wiesbaden: Bundesinstitut für Bevölkerungsforschung.
262. Stamov Roßnagel, C., & Hertel, G. (2010). Older workers' motivation: Against the myth of general decline. Management Decision, 48(6), 894–906.
263. Kim, B. (2015). Designing gamification in the right way. Library Technology Reports, 51(2), 29–35.
264. Baharudin, N. A. A., & Yunus, M. M. (2018). GRAMMODO: A platform in learning grammar among ESL learners in Malaysia. Innovation for Sustainable Growth, 106–108.
265. Leaning, M. (2015). A study of the use of games and gamification to enhance student engagement, experience and achievement on a theory-based course of an undergraduate media degree. Journal of Media Practice, 16(2), 155–170. https://doi.org/10.1080/14682753.2015.1041807
266. Chang, W. L., & Yeh, Y. C. (2021). A blended design of game-based learning for motivation, knowledge sharing, and critical thinking enhancement. Technology, Pedagogy and Education, 30(2), 271–285. https://doi.org/10.1080/1475939x.2021.1885482
267. Wang, Yurou; Wind, Stefanie A. Psychometric evaluation of a new internalization of learning motivation (ILM) scale. Current Psychology (2020). https://doi.org/10.1007/s12144-020-00909-8
268. Grobler, Adelene A.; Knight, Mirriam R.; Lens, Willy; Lacante, Marlies Motivational predictors of successful transition from grades 11 to 12 in South Africa. European Journal of Psychology of Education (2014). https://doi.org/10.1007/s10212-014-0223-8
269. Dishon-Berkovits, Miriam A study of motivational influences on academic achievement. Social Psychology of Education (2014). https://doi.org/10.1007/s11218-014-9257-7
270. Papi, Mostafa; Bondarenko, Anna Vitalyevna; Wawire, Brenda; Jiang, Chen; Zhou, Shiyao Feedback-seeking behavior in second language writing: motivational mechanisms. Reading and Writing (2019). https://doi.org/10.1007/s11145-019-09971-6
271. Cho, Eunsoo; Toste, Jessica R.; Lee, Minhye; Ju, Unhee Motivational predictors of struggling readers' reading comprehension: the effects of mindset, achievement goals, and engagement. Reading and Writing (2018). https://doi.org/10.1007/s11145-018-9908-8
272. Hidayatullah, Achmad; Csíkos, Csaba The Role of Students' Beliefs, Parents' Educational Level, and The Mediating Role of Attitude and Motivation in Students' Mathematics Achievement. The Asia-Pacific Education Researcher (2023). https://doi.org/10.1007/s40299-023-00724-2
273. Islam, Md. Nurul; Chakrabarty, Sumi A Study of the Relationship between Familial Factors and Academic Motivation and Achievement in High School Students. Psychological Studies (2019). https://doi.org/10.1007/s12646-019-00500-7
274. Abraham, Jessy; Barker, Katrina Motivation and Engagement with Physics: a Comparative Study of Females in Single-Sex and Co-educational Classrooms. Research in Science Education (2018). https://doi.org/10.1007/s11165-018-9770-3
275. Raftery-Helmer, Jacquelyn N.; Grolnick, Wendy S. Parent and teacher effects on academic coping and classroom engagement: Testing a motivational model. Motivation and Emotion (2018). https://doi.org/10.1007/s11031-018-9676-4
276. PourAli, PariNaz; SeifNaraghi, Maryam; Naderi, Ezatollah Humanistic Education and students' Educational Motivation in Tehran Primary Schools. International Journal of Mental Health and Addiction (2016). https://doi.org/10.1007/s11469-016-9703-1
277. Silvia, Paul J.; McCord, Daniel M.; Gendolla, Guido H. E. Self-focused attention, performance expectancies, and the intensity of effort: Do people try harder for harder goals?. Motivation and Emotion (2010). https://doi.org/10.1007/s11031-010-9192-7

278. Davis, William E.; Kelley, Nicholas J.; Kim, Jinhyung; Tang, David; Hicks, Joshua A. Motivating the academic mind: High-level construal of academic goals enhances goal meaningfulness, motivation, and self-concordance. Motivation and Emotion (2015). https://doi.org/10.1007/s11031-015-9522-x
279. Jones, Brett D.; Wilkins, Jesse L. M.; Long, Margaret H.; Wang, Feihong Testing a motivational model of achievement: How students' mathematical beliefs and interests are related to their achievement. European Journal of Psychology of Education (2011). https://doi.org/10.1007/s10212-011-0062-9
280. Campos-Sánchez, Antonio; López-Núñez, Juan Antonio; Carriel, Víctor; Martín-Piedra, Miguel-Ángel; Sola, Tomás; Alaminos, Miguel Motivational component profiles in university students learning histology: a comparative study between genders and different health science curricula. BMC Medical Education (2014). https://doi.org/10.1186/1472-6920-14-46
281. Morales-Gualdrón, Silvia T.; Gutiérrez-Gracia, Antonio; Roig Dobón, Salvador The entrepreneurial motivation in academia: a multidimensional construct. International Entrepreneurship and Management Journal (2009). https://doi.org/10.1007/s11365-009-0113-8
282. Virkkula, Esa Evaluating motivational characteristics in vocational music education within the perspective of self-determination theory. Empirical Research in Vocational Education and Training (2020). https://doi.org/10.1186/s40461-020-00098-5
283. Gillet, Nicolas; Vallerand, Robert J.; Rosnet, Elisabeth Motivational clusters and performance in a real-life setting. Motivation and Emotion (2009). https://doi.org/10.1007/s11031-008-9115-z
284. Dostert, Janik; Müller, Romy Motivational assistance system design for industrial production: from motivation theories to design strategies. Cognition, Technology & Work (2020). https://doi.org/10.1007/s10111-020-00643-y
285. Higashi, Ross M.; Schunn, Christian D.; Flot, Jesse B. Different underlying motivations and abilities predict student versus teacher persistence in an online course. Educational technology research and development (2017). https://doi.org/10.1007/s11423-017-9528-z
286. Brandenberger, Claudia C.; Hagenauer, Gerda; Hascher, Tina Promoting students' self-determined motivation in maths: results of a 1-year classroom intervention. European Journal of Psychology of Education (2017). https://doi.org/10.1007/s10212-017-0336-y
287. Raes, Annelies; Schellens, Tammy Unraveling the motivational effects and challenges of web-based collaborative inquiry learning across different groups of learners. Educational technology research and development (2015). https://doi.org/10.1007/s11423-015-9381-x
288. Scogin, Stephen C. Identifying the Factors Leading to Success: How an Innovative Science Curriculum Cultivates Student Motivation. Journal of Science Education and Technology (2016). https://doi.org/10.1007/s10956-015-9600-6
289. Calafato, Raees Charting the motivation, self-efficacy beliefs, language learning strategies, and achievement of multilingual university students learning Arabic as a foreign language. Asian-Pacific Journal of Second and Foreign Language Education (2023). https://doi.org/10.1186/s40862-023-00194-5
290. Vedder-Weiss, Dana; Fortus, David Teachers' Mastery Goals: Using a Self-Report Survey to Study the Relations between Teaching Practices and Students' Motivation for Science Learning. Research in Science Education (2017). https://doi.org/10.1007/s11165-016-9565-3
291. Abdullah, Mohamad Yahya; Hussin, Supyan; Ismail, Kemboja Investigating the effects of the flipped classroom model on Omani EFL learners' motivation level in English speaking performance. Education and Information Technologies (2019). https://doi.org/10.1007/s10639-019-09911-5
292. Chen, Yulin The effect of using a game-based translation learning app on enhancing college EFL learners' motivation and learning experience. Education and Information Technologies (2022). https://doi.org/10.1007/s10639-022-11174-6

293. Fiorella, Logan; Yoon, So Yoon; Atit, Kinnari; Power, Jason R.; Panther, Grace; Sorby, Sheryl; Uttal, David H.; Veurink, Norma Validation of the Mathematics Motivation Questionnaire (MMQ) for secondary school students. International Journal of STEM Education (2021). https://doi.org/10.1186/s40594-021-00307-x
294. Koshkin, Andrey Petrovich; Abramov, Ruslan Agarunovich; Rozhina, Ekaterina Yurevna; Novikov, Andrey Vadimovich Role of Social Representations in Student Motivation for Acquiring Further Education. Interchange (2018). https://doi.org/10.1007/s10780-018-9328-3
295. Lv, Jing Improving college student engagement and motivation in a gamified learning environment: the pilot study in China. Current Psychology (2023). https://doi.org/10.1007/s12144-023-04884-8
296. Oker, Ali; Pecune, Florian; Declercq, Christelle Virtual tutor and pupil interaction: A study of empathic feedback as extrinsic motivation for learning. Education and Information Technologies (2020). https://doi.org/10.1007/s10639-020-10123-5
297. Magen-Nagar, Noga; Cohen, Lizi Learning strategies as a mediator for motivation and a sense of achievement among students who study in MOOCs. Education and Information Technologies (2016). https://doi.org/10.1007/s10639-016-9492-y
298. Thieme, Paula; Brusch, Michael; Büsch, Victoria; Roßnagel, Christian Stamov Work context influences on older workers' motivation for continuing education. Zeitschrift für Erziehungswissenschaft (2015). https://doi.org/10.1007/s11618-014-0600-8
299. Shakhmalova, Irina; Zotova, Natalia Techniques for Increasing Educational Motivation and the Need to Assess Students' Knowledge: The Effectiveness of Educational Digital Games in Learning English Grammatical Material. Journal of Psycholinguistic Research (2023). https://doi.org/10.1007/s10936-023-09983-y

Self-regulated Learning

4

Introduction by the Editor

Self-regulated learning (SRL) is a process in which learners take control of their own learning, managing their own cognitive, motivational, and behavioral processes to achieve academic success. It is an approach that has gained considerable attention in educational psychology because of its profound implications for improving academic performance and lifelong learning skills. Self-regulated learning also refers to the process by which learners set goals for their learning, monitor their progress, and regulate their cognitive, motivational, and behavioral strategies to achieve these goals. Motivation is the driving force that initiates, sustains and guides goal-directed behavior. Understanding the interplay between self-regulated learning and motivation is essential for fostering effective learning environments and promoting lifelong learning. The research articles in this section explore the relationship between self-regulated learning and motivation, examining how motivational factors influence self-regulated learning and how self-regulated learning strategies can enhance motivation.

Self-regulated Learning and Motivation

Self-regulated learning and motivation are inextricably linked concepts that play a pivotal role in academic success and personal growth (Mega et al., 2014). In an era where lifelong learning is becoming increasingly important, the ability to take ownership of one's learning process and cultivate a strong sense of motivation is essential for navigating the ever-changing landscape of knowledge and skills. Exploring the significance of self-regulated learning and motivation, their interconnections, and their impact on academic achievement and personal development are prevalent in the literature. Self-regulated learning has been observed to be a complex and multifaceted process, encompassing various dimensions and aspects that involve actively monitoring and managing one's cognitive, metacognitive, and motivational processes to achieve desired learning outcomes. It encompasses

M. S. Khine (ed.), *Motivation Science*, https://doi.org/10.1007/978-981-97-9247-4_4

a range of strategies and behaviors, such as goal-setting, self-monitoring, self-evaluation, and strategic planning, that enable individuals to take control of their learning experiences (Zimmerman & Schunk, 2012). By employing these strategies, self-regulated learners are better equipped to navigate the complexities of the learning process, adapt to new challenges, and persist in the face of obstacles.

One of the key advantages of self-regulated learning is its ability to foster metacognitive awareness, which is the understanding and regulation of one's own thought processes. Metacognitive awareness enables learners to reflect on their learning strategies, identify areas of strength and weakness, and make informed decisions about how to approach and modify their learning approaches (Abdelrahman, 2020). This increased self-awareness not only improves academic performance but also promotes a growth mindset, encouraging learners to view challenges as opportunities for growth and development. Motivation serves as the driving force that encourages individuals to engage in and persist with learning activities. It encompasses intrinsic and extrinsic factors that influence an individual's desire to learn, level of commitment, and willingness to invest time and effort in the learning process. Intrinsic motivation, which stems from an inherent interest or enjoyment in the learning activity itself, is particularly powerful because it fosters determination to learn.

The relationship between self-regulated learning and motivation is reciprocal and mutually beneficial. When learners self-regulate, they engage in cognitive and metacognitive activities to actively monitor and control their learning. Self-regulated learning strategies can enhance motivation by promoting a sense of autonomy, competence, and relatedness which are fundamental psychological needs that foster intrinsic motivation. When learners feel in control of their learning process, experience a sense of mastery, and perceive their efforts as meaningful, they are more likely to remain motivated and engaged. Conversely, motivation plays a crucial role in sustaining the effortful and persistent behaviors required for effective self-regulated learning. Highly motivated individuals are more likely to set challenges, monitor their progress, and adapt their strategies when faced with difficulties.

In an academic setting, emphasizing the significance of self-regulated learning and motivation is crucial. Research has consistently shown that students who demonstrate higher levels of self-regulation and motivation tend to achieve better academic outcomes, such as higher grades, better retention of knowledge, and greater overall academic success (Zhang et al., 2021). These skills not only contribute to academic achievement but also prepare students for the challenges of lifelong learning, enabling them to navigate to their future and adapt to new learning contexts. Beyond the academic field, self-regulated learning and motivation have far-reaching implications for personal and professional growth. Self-regulated learners are better equipped to identify their learning needs, set meaningful goals, and engage in continuous learning opportunities, positioning themselves as valuable assets in a dynamic workforce.

In addition, self-regulated learning and motivation are integral to personal growth and well-being. By cultivating these skills, individuals can pursue their passions, explore new interests, and engage in activities that enrich their lives. Self-regulation enables individuals to manage their time and resources effectively, while motivation provides the impetus to persevere in the face of challenges and setbacks, ultimately fostering a sense of achievement and fulfillment. In developing self-regulated learning and motivation, individual differences, such as prior knowledge, self-efficacy beliefs, and cultural background, can influence an individual's propensity for self-regulation and motivation should also be considered. Besides, environmental factors such as teaching practices, learning resources, and support systems can either facilitate or hinder the development of these skills. To meet these challenges, educators and institutions have a crucial role to play in creating learning environments that foster self-regulated learning and motivation. This can be achieved through a variety of strategies, such as explicit instruction in self-regulation strategies, providing opportunities for self-directed learning, fostering a mastery-oriented classroom climate, and providing timely and constructive feedback.

In summary, self-regulated learning and motivation are essential components of academic success, personal growth, and lifelong learning. By cultivating these skills, individuals can take ownership of their learning journey, persevere through challenges, and continually adapt to the ever-changing demands of an increasingly complex world. While the development of self-regulated learning and motivation is a multifaceted process influenced by several factors, educational institutions, and educators have a key role to play in creating supportive learning environments that foster these critical skills. Ultimately, investing in self-regulated learning and motivation not only improves academic performance but also empowers individuals to become lifelong learners capable of meeting the challenges and opportunities that lie ahead.

Some of the research papers included in this section are the effects of motivation, academic stress, and age in predicting self-directed learning readiness. psychological ownership, and knowledge behaviors during a pandemic: role of approach motivation, motivational effects on self-regulated learning with different tasks, exploring the fluctuation of motivation and use of self-regulatory processes during learning with hypermedia, learning English as a second language: motivation, self-regulation and learning environment perceptions, and understanding undergraduate science students' multiple motivations for research. How do motivation and self-efficacy interact in problem-solving and problem-posing, motivation, pedagogy, and gender: examining the multifaceted and dynamic situational responses of women and men in college STEM courses, and the role of affective and motivational factors in designing personalized learning environments are also included in this section.

References

Abdelrahman, R. M. (2020). Metacognitive awareness and academic motivation and their impact on academic achievement of Ajman University students. *Heliyon*, *6*(9).

Mega, C., Ronconi, L., & De Beni, R. (2014). What makes a good student? How emotions, self-regulated learning, and motivation contribute to academic achievement. *Journal of Educational Psychology*, *106*(1), 121.

Zheng, B., Chang, C., Lin, C. H., & Zhang, Y. (2021). Self-efficacy, academic motivation, and self-regulation: how do they predict academic achievement for medical students? *Medical Science Educator*, *31*, 125–130.

Zimmerman, B. J., & Schunk, D. H. (2012). Motivation: An essential dimension of self-regulated learning. In *Motivation and self-regulated learning* (pp. 1–30). Routledge.

Machine Generated Summaries

Disclaimer: The summaries in this chapter were generated from Springer Nature publications using extractive AI auto-summarization: An extraction-based summarizer aims to identify the most important sentences of a text using an algorithm and uses those original sentences to create the auto-summary (unlike generative AI). As the constituted sentences are machine selected, they may not fully reflect the body of the work, so we strongly advise that the original content is read and cited. The auto generated summaries were curated by the editor to meet Springer Nature publication standards. To cite this content, please refer to the original papers.

Machine generated keywords: student, learner, online, course, regulation, affective, anxiety, design, assessment, learn, feeling, education, learn environment, continuum, psychological.

Composition and Consistency of the Desired Affective State: The Role of Personality and Motivation [141]

This is a machine-generated summary of:

Augustine, Adam A; Hemenover, Scott H.; Larsen, Randy J.; Shulman, Tirza E.: Composition and consistency of the desired affective state: The role of personality and motivation [141].

Published in: Motivation and Emotion (2010).

Link to original: https://doi.org/10.1007/s11031-010-9162-0.

If you want to cite the papers, please refer to the original.

For technical reasons we could not place the page where the original quote is coming from.

Abstract-Summary

"Personality and motivations/goals were related to the content of the desired affective state."

"Extraversion, neuroticism, agreeableness, and conscientiousness were related to the content of the desired affective state."

"Our results suggest that the content of the desired affective state may be largely dependent on personality, motivation, and, potentially, an interaction between personality and motivation."

Models of Affect Regulation and Affect

"The desired affective state (DAS) is a primary component of several of the prominent models of affect regulation that have surfaced in the extant literature."

"In both Bonanno's [1] and Larsen's [2] model of affect regulation, the goal of the affect-regulatory process is to reach the DAS."

"Those studies that do exist have largely focused on individual differences in the DAS as organized by the circumplex model of affect (see Larsen and Diener [3, 4]; Russell [5]; Rusting and Larsen [6])."

"In the circumplex model of affect, self-reported affect is organized around two dimensions, pleasantness and activation."

"Research regarding the DAS has yielded individual differences in both the desired activation and valence of affect."

The Desired Affective State

"Rusting and Larsen [6] found that the level of affect activation one typically experiences is related to the amount of affect activation that one desires, although the direction of this relationship is dependent on personality."

"To these findings, Kampfe and Mitte [7] found that neuroticism positively predicted a desire for unactivated unpleasant affects and negatively predicted desires for activated pleasant affect; this pattern is more consistent with the actual affective experience of the neurotic."

"In line with these findings for neuroticism, Tamir [8] found that those high in neuroticism desire trait-consistent negative affect (i.e., higher negative affect) when driven by performance goals."

"Individuals and cultures (i.e., European Americans) that value influencing others (making others suit one's personal needs) desire more activated pleasant affect."

"When under a utilitarian (using affect) motivation, the content of the DAS may be different than when operating under a hedonic (experience) motivation (Tamir and others [9])."

Overview of the Current Research

"Motives/goals will predict the content of the DAS."

"Given prior research (Rusting and Larsen [6]), we hypothesize that extraversion and neuroticism will predict desired activation levels, with extraverts desiring trait-consistent and neurotics trait-inconsistent affect activation."

"Given the links between these two traits and an increased frequency of apparently successful efforts to achieve pleasant affect, we hypothesize that conscientiousness and agreeableness, will predict a desire for the valence of affective states; those high in these constructs will desire affective states with a more pleasant tone."

"Experienced affect is related to higher-level motivations/goals (i.e., power, achievement, etc.; Nelissen and others [10]), desired affect may be linked with higher-level goals."

"Using this free response data gathered from Study 1, a set of seven higher-level motives/goals for the DAS will be derived."

"Using an experience sampling design, in Study 2 we explore differences in the DAS based on these seven higher-level goals."

Study 1

"Participants completed the DAS questionnaires on four separate occasions; each report was made approximately 25 days after the prior report."

"At Times 2–4, participants completed the DAS questionnaire."

"The responses to these questions were not used to calculate participants' DAS (this data was gathered using the DAS questionnaire described previously), but were used for the generation of a questionnaire for Study 2."

"At level 2, we predicted the mean value of the DAS for each person using the equation: $\beta_{0j} = \gamma_{00} + \gamma_{01}(\text{Extraversion}) + \gamma_{02}(\text{Agreeableness}) + \gamma_{03}(\text{Conscientiousness}) + \gamma_{04}(\text{Neuroticism}) + \gamma_{05}(\text{Openness}) + u_{0j}$."

"Consistent with our hypotheses, conscientiousness and agreeableness predict trait-consistent affective desires; those high in these constructs wish to experience more pleasant (i.e., AP, P) and less unpleasant affect (i.e., U, AU), regardless of activation level."

"Study 1 demonstrated that the DAS is related to the personality traits of extraversion, neuroticism, agreeableness, and conscientiousness."

Study 2

"Differences in the content of the DAS, while under varying motivations, should be indicative of the motivation in question (i.e. those with a hedonic motive desire more pleasant affects, those with a relaxation motive desire more low activation affects, etc.)."

"Those with a hedonic motive desire more high activation affect than those with a relaxation motive."

"Those with perfect balance, productivity, and relaxation motives desire less activated pleasant affect than those with a hedonic motive."

"Those with a self-verification motive desire more activated pleasant affect than those with a relaxation or perfect balance motive."

"Individuals with hedonic and productivity motives desire more unactivated pleasant affect than those with relaxation motives."

"Those with relaxation motives desire more low activation affect than those with perfect balance, hedonic, productivity, or avoid anhedonic motives."

"Consistent with our predictions, the affects desired were in line with the DAS motive."

General Discussion

"In Study 2, the broad motivations for desired affect that were gathered as a part of Study 1 showed relationships with the DAS."

"Observed differences in the content of the DAS between motives reflected the nature of those motives (i.e., individuals operating under a hedonic motive desired more pleasant affects)."

"Given this, the findings that broad motivations influence the DAS, findings that indicate that higher order values/motivations influence experienced affect (Nelissen and others [10]), and findings that motives predict temporal variance in the DAS (Study 2), it seems clear that motives have an influence on desired affect."

"The extent to which certain personality and motivational constructs predict the DAS may depend on their relationships with actual affective experience and controlling actual affect degrades the ability to accurately examine desired affect."

[Section 7]

"An ever increasing body of literature focuses on affect regulation, or "processes by which individuals influence which emotions they have, when they have them, and how they experience and express these emotions" (Gross [11], p. 275)."

"It is this goal, the desired affective state, which is the topic of the current research."

"In one longitudinal and one experience sampling we explore the consistency and composition of the desired affective state, as well as personality and motivational factors that may predict the content of one's desired affective state."

Does the Visual Appeal of Instructional Media Affect Learners' Motivation Toward Learning? [142]

This is a machine-generated summary of:

Tomita, Kei: Does the Visual Appeal of Instructional Media Affect Learners' Motivation Toward Learning? [142].

Published in: TechTrends (2017).

Link to original: https://doi.org/10.1007/s11528-017-0213-1.

If you want to cite the papers, please refer to the original.

For technical reasons we could not place the page where the original quote is coming from.

Abstract-Summary

"The author investigated whether the perception of the visual appeal of instructional handouts affects learners' self-reported motivation to learn from the handouts."

"The appealing handout was perceived as motivating when it was seen after the minimalist handout."

"The group that saw the minimalist handout first did not perceive the appealing handout as motivating."

"The result implies that not only the design but also the learning context influences learners' self-reported motivation to engage with the handout."

Theoretical Framework

"There are two primary theories framing this study: cognitive load theory (CLT) and Dewey's [12] concept of aesthetics."

"Rather, aesthetics describes a quality of experience that "possesses internal integration and fulfillment" (Dewey [12], p. 46)."

"Parrish [13] mentions that an aesthetic experience happens as a result of the completion of tasks which require "deep engagement, and willingness to follow through to completion" (p. 5)."

"The aesthetic experience could be described as an experience of satisfaction after people face, engage in, and resolve a challenge."

"Berleant [14] emphasizes the importance of people's engagement in order for them to have an aesthetic experience."

"Csikszentmihalyi and Robinson [15] report that engagement caused by the visual design of an object with which people are interacting is a salient aspect of the aesthetic experience."

The Gap in Field Knowledge

"While Mayer's [16] Coherence Principle, based on CLT, attempts to get immediate learning outcomes by improving the efficiency of the design of instructional materials, considering the appealing aspect of instructional materials could foster learners' motivation (Keller and Burkman [17]; Rieber [18])."

"Such enhanced motivation could engage learners and bring them an aesthetic learning experience, which could help them form a positive attitude toward future learning (Parrish [13, 19]; Uhrmacher [20])."

"Balancing the considerations for cognition and emotion might be challenging, it is nevertheless important when people are designing instructional media."

Method

"After the three primary questions, in order to get more insight regarding how the students perceived each handout, students were asked to fill out the AttrakDiff questionnaire consisting of 28 items measuring four qualities of the visual design."

"The data were analyzed to ascertain 1) the students' perceptions of minimalist and appealing visual design and, 2) whether the perceptions of the instructional handouts actually affect their self-reported motivation to learn from the handouts."

"In order to investigate if there were significant differences in students' self-reported motivation due to the difference of the visual design of training handouts as well as to the ordering effect, the responses to the statement "How likely are you to agree with the statement, 'The visual design of the PDF handout helps to motivate me toward learning'?"."

Results

"There was no significant effect of the visual design difference: $F(1, 28) = 1.724$, $p = 0.800$."

"There was no significant effect of the visual design difference: $F(1, 28) = 0.000$, $p = 1.00$."

"There was no significant difference in the scores of pragmatic quality."

"There was no significant difference in the scores of hedonic quality-simulation."

"The hedonic quality-identity score was significantly higher in the case of appealing handouts."

"The result is consistent with the ANOVA which showed that attractiveness score for the appealing handout was significantly higher in the group seeing the minimalist handout first."

"Paying attention to the words "boring" and "easy," the result could be considered consistent with the ANOVA which showed that the self-reported motivation score for the "appealing" handout was significantly higher in the group seeing the "appealing" handout first."

Discussion
"The interaction was also significant for the students' self-reported attractiveness, and the appealing handout was considered significantly more motivating in the case of the group that saw the appealing handout first."

"A possible interpretation of such results is that the perceived attractiveness does not always increase students' motivation to learn from the handout."

"One student who saw the minimalist handout first commented about the appealing handout, "the color did attract me at the first glance."

"While the open-ended questions were intended to elicit responses about the visual characteristics of the handouts, multiple students commented that they were not motivated to learn the handouts since they were not interested in the topic of the handout, the basics of Microsoft Excel."

"Some students mentioned that they would be motivated to learn regardless of the visual design of the handout if they were interested in the subject matter."

Conclusion
"The major findings resulting from this study were that 1) students' perceptions of the visual design of handouts were affected by the order in which they viewed the handouts, and 2) the perceived attractiveness of the visual design of the handouts did not increase motivation to learn from the handouts."

"Since this study did not observe a significant influence of the perceived attractiveness of the handout on students' motivation to learn from the handout, researchers and designers attempting to engage students using attractive visuals should be aware of the complexity of the issue."

"While the visual design difference itself was not a significant factor in determining student motivation to learn from the handout, with the help of the ordering effect, students perceived one design as more motivating than the other."

[Section 8]
"In the actual practice of instructional material design, balancing the considerations for cognition and emotion is a complex task."

"While authors like Mayer [16] and Winn [21] emphasize that designers should refrain from using visuals for the purpose of attracting learners' interest, there are scholars who suggest the importance of using visuals to influence people's emotions."

"This study investigated whether the perception of the visual appeal of an instructional handout actually affects learners' self-reported motivation to learn from the handout."

"Because this study tried to investigate how students perceive the visual design, it chose to analyze students' self-reported motivation."

Effects of Motivation, Academic Stress and Age in Predicting Self-Directed Learning Readiness (SDLR): Focused on Online College Students [143]

This is a machine-generated summary of:

Heo, JeongChul; Han, Sumi: Effects of motivation, academic stress and age in predicting self-directed learning readiness (SDLR): Focused on online college students [143].

Published in: Education and Information Technologies (2017).

Link to original: https://doi.org/10.1007/s10639-017-9585-2.

If you want to cite the papers, please refer to the original.

For technical reasons we could not place the page where the original quote is coming from.

Abstract-Summary
"The purpose of this study is to determine whether the self-directed learning readiness (SDLR) among online students might be significantly predicted by motivation, academic stress, and age."

"The Pearson correlation is used to examine whether there are significant correlations among motivation, academic stress, age, and SDLR."

"Results show that motivation and academic stress have an influence on the level of SDLR."

"The multiple-regression is used to analyze any influence on three predictors (motivation, academic stress, and age) in predicting SDLR."

"Motivation and academic stress turn out as significant predictors which are influential in predicting the level of SDLR."

"The variable of the age among online college students is not influential in predicting the level of SDLR."

Introduction

"There is no doubt that learning online will continue to have a critical impact on the majority of higher education students via Open Course Ware (OCW) and Massive Open Online Courses (MOOCs) (e.g., Udemy, Coursera, Open edX, Future Learn, etc.)."

"Self-directed students are independent and autonomous in their distance learning and are positively motivated to learn at their own paces completing their online courses successfully."

"SDL should be very important educational concept for successful online learning and lifelong education to instructors as well as learners."

"Online instructors will be able to consider and use core educational predictors and demographical variables which can be influential in predicting SDLR in order to encourage online learners to be self-directed and autonomous in their e-learning circumstances."

Research Question

"To accomplish the purpose of this study, research questions are as follows: Research question one: Is there a significant correlation between motivation and SDLR among college students?"

"Research question two: Is there a significant correlation between academic stress and SDLR among college students?"

"Research question tree: Is there a significant correlation between age and SDLR among online college students?"

"Research question four: Which of the possible three predictor variables (motivation, academic stress, and age) is the most influential in predicting SDLR?"

Method

"The original SDLRs of Guglielmino [22] includes 58 items in a five-point likert type scale and consists of eight subscales (e.g., openness to learning opportunities, self-concept as an effective learner, independence in learning, responsibility for one's own learning, love of learning, positive orientation to the future, creativity, and ability to use basic study skills and problem-solving skills)."

"The short form of Guglielmino's SDLRs produced a Cronbach's alpha = 0.89 reliability coefficient."

"The short form of motivation uses 4 items that are selected from modified short form of IMMS."

"The modified SAS has 15 items with a six-point likert type scale with three realms of stress and one sources of stress for the students' self."

"The short form of SAS in three realms of stress by one source of stress produced a reliability coefficient (Cronbach's alpha = 0.92) in this study."

Results

"Motivation had significant correlation with five subscales of SDLR (openness to learning opportunities: $r = 459$ $p < 0.01$, self-concept as an effective learner: $r = 0.412$ $p < 0.01$, independence in learning: $r = 0.392$ $p < 0.01$, responsibility for one's own learning: $r = 0.417$ $p < 0.01$, love of learning: $r = 0.557$ $p < 0.01$)."

"The academic stress had significantly negative correlation with five subscales of SDLR (openness to learning opportunities: $r = -0.470$ $p < 0.01$, self-concept as an effective learner: $r = -0.423$ $p < 0.01$, independence in learning: $r = -0.523$ $p < 0.01$, responsibility for one's own learning: $r = -0.296$ $p < 0.01$, love of learning: $r = -0.451$ $p < 0.01$)."

"The result of this study revealed that motivation and academic stress can significantly have an effect on predicting SDLR."

Conclusion and Discussion

"With the results of the first research question, there was significantly positive relationship between online students' motivation and level of SDLR."

"After analyzing the second research question, the results revealed that there was a negative correlation between academic stress and level of SDLR."

"The result for the significant correlation between academic stress and SDLR is congruent with previous researches that academic stress and burnout can reduce the level of SDLR (Choi [23]; Kim & Lee [24])."

"The results identified that online students with low academic stress had a high level of SDLR."

"After analyzing the third research question, there was no significant correlation between age and level of SDLR."

Effects of Predictors (Motivation, Academic Stress, and Age) on Predicting SDLR

"Then, the predictors of motivation and academic stress had a significant influence on SDLR for online college students in this study."

"Online students need to be positively motivated and reduce the level of academic stress for high level of SDLR on their online courses."

"The results showed that the level of SDLR can be predicted by the predictor variables (e.g., motivation and academic stress)."

"The variable of age among online students was not influential in predicting the level of SDLR."

Suggestions and Implications

"According to findings in this study, there was significant positive relationship between a set of variables (i.e., online students' motivation and academic stress) and level of SDLR."

"A set of core concepts (e.g., motivation and academic stress) have significantly influence on the level of SDLR."

"Online educators will be able to strategically consider and use a variety of core educational variables which can enhance positive motivation and reduce academic stress among online learners."

"Then, online learners will be likely to strongly be self-directed and autonomous increasing their motivation and reducing the level of academic stress in their online education."

"Further studies will be needed to analyze the direct and indirect effects among a variety of core educational concepts (e.g., satisfaction, learning experience, academic achievement, academic outcomes, as well as motivation, academic stress) through path analysis or structural equation modeling for developing successful online education."

Psychological Ownership and Knowledge Behaviors During a Pandemic: Role of Approach Motivation [144]

This is a machine-generated summary of:

Batool, Uzma; Raziq, Muhammad Mustafa; Obaid, Asfia; Sumbal, Muhammad Saleem Ullah Khan: Psychological ownership and knowledge behaviors during a pandemic: role of approach motivation [144].

Published in: Current Psychology (2022).

Link to original: https://doi.org/10.1007/s12144-022-03450-y.

If you want to cite the papers, please refer to the original.

For technical reasons we could not place the page where the original quote is coming from.

Abstract-Summary
"The purpose of this article is to understand the relationship between psychological ownership, knowledge sharing, knowledge hiding and employee motivation in knowledge intensive organizations."

"We take employee motivation in terms of approach motivation and avoidance motivation and examine moderating role of the former in case of the psychological ownership – knowledge sharing relationship, and the latter in case of the psychological ownership – knowledge hiding relationship."

"Results show that stronger feelings of psychological ownership lead to both positive work behavior (i.e., knowledge sharing) as well as negative work behavior (i.e., knowledge hiding)."

"Approach motivation positively moderates the positive relationship between psychological ownership and knowledge sharing."

"Research has ignored the role of different types of employee motivation, particularly approach motivation vis-à-vis knowledge behaviors."

"Managers may incorporate reward practices to motivate employees towards knowledge sharing, and nurture an organizational climate, which discourages knowledge hiding."

Introduction
"At one end, job-based psychological ownership encourages employees to enhance ownership bond towards job and be more proactive, whereas on the other end, employees with ownership feelings tend to be more possessive resulting in negative work behaviors such as knowledge hiding (Wang and others [25]; Peng & Pierce [26]) opine that, employees hide knowledge when they feel the ownership of the knowledge they hold."

"Employees with stronger feelings of psychological ownership hide knowledge from their co-workers specifically when avoidance motivation is high (Wang and others [25])."

"Researchers have paid noticeable attention (both empirical and theoretical) to employee's responses toward knowledge sharing and knowledge hiding (Gagné and others [27]; Pan and others [28]), what motivates knowledge sharing resulting from psychological ownership is unexplored."

"Results show that stronger feelings of psychological ownership lead to both positive work behaviors (i.e., knowledge sharing) as well as negative work behaviors (i.e., knowledge hiding)."

Theory and Hypotheses Development

"Hameed and others [29] suggest that individuals with high level of psychological ownership show humane spirit and promote knowledge sharing among employees."

"Based on these, it can be hypothesized that: H2: Psychological ownership positively associates with knowledge sharing behavior of employees."

"Even employees with high psychological ownership may hesitate to exchange knowledge when requested."

"Employees experiencing high collective psychological ownership will share knowledge with coworkers if their approach motivation is high."

"Approach motivation amplifies positive influence of psychological ownership on knowledge sharing."

"It is hypothesized that: H3: Avoidance motivation moderates the positive effect of collective psychological ownership on knowledge hiding such that the influence is stronger when avoidance motivation is high."

"H4: Approach motivation moderates the positive Influence of collective psychological ownership on knowledge sharing such that the effect is stronger when approach motivation is high."

Method

"We targeted population of knowledge workers working in high-tech organizations and educational institutes in Pakistan."

"Knowledge management in these kind of organizations is usually considered as the most substantial element for the firm's sustainability and growth (Xia and others [30])."

"To assess knowledge hiding behavior, measures were adopted from Connelly and others [31]."

"Knowledge hiding included its dimensions such as rationalized hiding, playing dumb and evasive hiding and sample items included: "agreed to help him/her but never really intended to"; "pretended that I did not know the information"; and "explained that I would like to tell him/her, but was not supposed to"."

"Sample items included: "I usually help my colleagues solve work related problems"."

"Prior studies have found that respondent's demographic variables including gender and age may impact worker's knowledge hiding behaviors (Zhao and others [32])."

Results

"To check for common method variance, Harman's single factor test (Harman [33]) was used, and results indicated that data were not affected."

"To check constructs suitability, we estimated internal consistency, discriminant validity, and convergent validity."

"Consistent with the predefined standards our data reinforced the convergent and discriminant validity of scales."

"The results show that collective psychological ownership considerably boosts knowledge hiding (H1: $\beta = 0.724$; $t = 5.958$, $p < 0.000$), and knowledge sharing (H2: $\beta = -0.580$; $t = 9.535$, $p < 0.000$)."

"Moderating effect of avoidance motivation is insignificant, whereas approach motivation significantly moderates the relationship of psychological ownership and knowledge sharing (H4: $\beta = 0.11$; $t = 2.023$, $p < 0.044$)."

Discussion

"The aim of this study was to examine relationships among psychological ownership, knowledge sharing, and knowledge hiding."

"Out study demonstrates that psychological ownership positively associates with both knowledge hiding and knowledge sharing behaviors, and these two contrasting behaviors are linked with an individual's motivation."

"Our study makes clarity that whether employee with psychological ownership share knowledge or hide knowledge depends on one's motivation."

"This study adds to the existing literature on knowledge management at team level by demonstrating psychological ownership direct positive impact on knowledge sharing and knowledge hiding."

"This study suggests that psychological ownership can result in negative work outcomes like knowledge hiding."

"While there is also sufficient literature on psychological ownership and knowledge behaviors, however, future studies should be undertaken to consider different forms of psychological ownership and their impact on work behaviors."

Conclusions

"This study demonstrates the double-edged effects of psychological ownership on knowledge behaviors."

"Based on theory of psychological ownership, our hypothesized model revealed that stronger feelings of possession lead to positive work behavior (i.e., knowledge sharing) as well as the negative work behavior (i.e., knowledge hiding) simultaneously."

"Consistent with previous literature, this study reveals that employee's intentions to share knowledge or hide knowledge while experiencing psychological ownership depend on their inner self and motivation."

Motivational Effects on Self-regulated Learning with Different Tasks [145]

This is a machine-generated summary of:

Vollmeyer, Regina; Rheinberg, Falko: Motivational Effects on Self-Regulated Learning with Different Tasks [145].

Published in: Educational Psychology Review (2006).

Link to original: https://doi.org/10.1007/s10648-006-9017-0.

If you want to cite the papers, please refer to the original.

For technical reasons we could not place the page where the original quote is coming from.

Abstract-Summary
"In our cognitive motivational process model (Vollmeyer & Rheinberg, Zeitschrift für Pädagogische Psychologie, 12:11–23, 1998) we assume that initial motivation affects performance via motivation during learning and learning strategies."

"We discuss methods with which the model can be tested."

"Initial motivation with its four factors challenge, probability of success, interest, and anxiety was measured with the Questionnaire on Current Motivation (QCM; Rheinberg, Vollmeyer, & Burns, Diagnostica, 47:57–66, 2001)."

"As an indicator for the functional state we assessed flow with the FKS (Rheinberg, Vollmeyer, & Engeser, Diagnostik von Motivation und Selbstkonzept [Diagnosis of Motivation and Self-Concept], Hogrefe, Göttingen, Germany, 261–279, 2003)."

Introduction

"The aim of the model was (1) to specify factors of initial motivation (theoretical assumptions and results are presented in the section "Initial motivation"), (2) to collect possible mediators for the influence of initial motivation on performance (theoretical assumptions are presented in the section "Mediators for the influence of initial motivation on performance"), and finally (3) to emphasize different learning outcomes."

"(Vollmeyer & Rheinberg [34]), we tested the assumption that high initial motivation led learners to spend more time on a learning task (persistence) and hence to acquire more knowledge."

"From our cognitive-motivational model we derived the following hypotheses: for our learning task, a linear system, we tested the hypothesis that initial motivation (interest, challenge, probability of success, anxiety) would influence performance via the mediating variables of strategy systematicity and motivational state during learning."

"During the learning period (Rounds 1 to 3) the mediating variables (i.e., strategy systematicity, motivational state) should affect knowledge acquisition."

Discussion

"Motivated learners spent more time with the learning material."

"Already with our initial motivation measure we could demonstrate how important is the motivation learners start with when working with new material."

"When we tested combinations of motivational factors, we could identify learners who were anxious and challenged."

"These learners worked differently with the learning material."

"We had the impression that learners reacted with annoyance when they had to answer all the questions, especially when their levels on the measured factors did not change while they were working with the learning material."

"Beside learners' annoyance we noticed that learners were less differentiated on the motivational factors compared to their answers on initial motivation."

"Beside all problems that we encountered when studying self-regulated learning it seems to be a promising way to model how initial motivation affects learning through mediating variables."

Exploring the Fluctuation of Motivation and Use of Self-regulatory Processes During Learning with Hypermedia [146]

This is a machine-generated summary of:

Moos, Daniel C.; Azevedo, Roger: Exploring the fluctuation of motivation and use of self-regulatory processes during learning with hypermedia [146].

Published in: Instructional Science (2007).

Link to original: https://doi.org/10.1007/s11251-007-9028-3.

If you want to cite the papers, please refer to the original.

For technical reasons we could not place the page where the original quote is coming from.

Abstract-Summary

"Results indicated that while participants in both conditions gained declarative knowledge, participants who received conceptual scaffolds during learning demonstrated deeper understanding of the circulatory system on the post-test."

"In terms of self-regulatory processes, the results indicated that participants in the CS condition used significantly more planning processes during learning than participants in the NS condition."

"Participants in both conditions significantly decreased their use of strategies as they progressed through the learning task."

"Regarding motivation while learning with hypermedia, results indicated that participants in both conditions reported significantly increased levels of interest as they progressed through the learning task."

Introduction

"This line of research has found that students' use of certain self-regulatory processes during learning with hypermedia may affect their knowledge development, especially in the absence of scaffolds (e.g., Azevedo and Cromely [35])."

"In order to address the difficulties some students have in self-regulating their learning with hypermedia, research has examined the potential benefits of providing students with scaffolds during learning (e.g., Azevedo et al. [36])."

"In order to address difficulties students have in developing conceptual knowledge when learning with hypermedia, research has examined the potential benefit of providing them with scaffolds."

"While some research has found that this type of scaffold fosters knowledge development (e.g., Jonassen et al. [37]; Hartley and Sydes [38]), other research has found limited benefit of providing students with guiding questions during learning with hypermedia."

Method

"While all participants had access to a global learning goal (make sure you lean about the different parts and their purpose, how they work both individually and together, and how they support the human body) during learning with hypermedia, participants randomly assigned to the CS condition received the following five guiding questions during learning with hypermedia: (1) What are the most important things the circulatory system does to keep us alive? (2) How do the parts of the circulatory system do those important things you just mentioned? (3) When blood leaves the right side of the heart it goes to one place, and when the blood leaves the left side of the heart it goes to a different place."

"As with the scoring of the participants' use of SRL processes during learning, the participants' responses to the four motivation questions were gathered at different time episodes: Prior to learning with hypermedia (pre-task portion of OMQ), Time 1 of the learning task (8 min), Time 2 of the learning task (16 min), Time 3 of the learning task (24 min), and immediately following the learning task (post-task portion of OMQ)."

Results

"Fifteen (83%) of the participants in the NS condition and 15 (79%) of the participants in the CS condition had a low mental model on the pre-test."

"This result indicated that the distribution of students in the NS condition and CS condition statistically significantly differ on their mental model scores for the post-test."

"It can be inferred that while a majority of participants from the CS and NS condition had low mental models of the circulatory system on the pre-test, there were more participants from the CS condition that had a high mental model of the circulatory system on the post-test."

"The participants' SRL was analyzed in separate repeated measures ANOVAs with the frequencies of coded SRL processes at four different times during the 30-min learning task (Time 1: 0–8 min, Time 2: 8–16 min, Time 3: 16–24 min, and Time 4: 24–30 min) as a within-subjects factor, and scaffolding conditions (No Scaffolding [NS] and Conceptual Scaffolding [CS]) as a between-subjects factor."

Discussion

"Examining the individual raw frequencies of strategies indicates that participants' selection of new information sources in the hypermedia environment may explain why their strategy use fluctuated over the 30 min learning task."

"To potentially expanding our theoretical understanding of cognitive and motivational processes involved in learning with hypermedia, this study also has methodological implications for future research."

"To the theoretical and methodological contributions, this study also extends previous research that has examined the impact of conceptual scaffolds on how students learn with hypermedia (Hannafin et al. [39])."

"While this study offers theoretical and methodological contributions to research examining the impact of conceptual scaffolds on cognitive and motivational processes during learning, there are some limitations that need to be addressed."

"The relationship between conceptual scaffolds, motivation, and cognitive processes may be mediated by the particular developmental group (undergraduates), context (learning with hypermedia), and/or task (learning about the circulatory system)."

Learning English as a Second Language at the University Level in Jordan: Motivation, Self-Regulation and Learning Environment Perceptions [147]

This is a machine-generated summary of:

Alzubaidi, Eman; Aldridge, Jill M.; Khine, Myint Swe: Learning English as a second language at the university level in Jordan: motivation, self-regulation and learning environment perceptions [147].

Published in: Learning Environments Research (2014).

Link to original: https://doi.org/10.1007/s10984-014-9169-7.

If you want to cite the papers, please refer to the original.

For technical reasons we could not place the page where the original quote is coming from.

Abstract-Summary

"The overarching aim of this study was to investigate students' perceptions of the learning environment and whether these influenced their motivation and self-regulation in learning English as a second language at the university level in Jordan."

"The collection of data involved the administration of two surveys: one to assess students' perceptions of the learning environment and another to assess students' motivation and self-regulation in learning English as a second language."

"Simple correlation and multiple regression analysis were used to examine the influence of students' perceptions of the learning environment of English language classes on self-reports of motivation and self-regulation."

"Statistically significant ($p < 0.05$) simple correlations were found between students' perceptions of their learning environment and their motivation and self-regulation."

Background

"As there are no studies to date that have explored the nature of the classroom environment among university students in Jordan, this study made the first attempt to examine students' perceptions of their language learning environments and how these relate to student motivation and self-regulation in Jordan."

"Zimmerman [40] emphasised that self-regulated learning involves the degree to which students meta-cognitively, motivationally and behaviourally participate in the learning process."

"According to Boekaerts and Cascallar [41], students' self-regulation in academic settings is an important construct that can influence students' engagement in learning and subsequently achievement in school."

"They have also indicated that the students' successful engagement in learning is dependent upon the level of motivation and self-regulation in the learning process."

"This research built on and extended past research related to students' motivation and self-regulation in learning English as a second language."

Methods

"The study involved the administration of the What Is Happening In this Class? (WIHIC) questionnaire (to assess students' perceptions of their English language classrooms) and the Engagement in English Language Learning and

Self-Regulation (EELLS) survey (to assess students' motivation and self-regulation in English language classes)."

"To assess students' engagement and self-regulation in English language, the Engagement in English Language Learning and Self-Regulation (EELLS) survey was used."

"The EELLS was adapted from the Student Adaptive Learning Engagement Survey (SALES), which was originally developed by Velayutham and others [42] to assess students' engagement and self-regulation in science classes."

"The EELLS includes four scales considered to be important to the engagement of students in English language classes, namely, Learning Goal Orientation, Self-Efficacy, Task Value and Self-Regulation."

"To investigate whether relationships existed between students' perceptions of the learning environment and student motivation (Task Value, Learning Goal Orientation and Self-Efficacy) and self-regulation, we analysed the data from 994 students using simple correlation and multiple regression analyses."

Results

"The data collected from 994 university-level student in Jordan were analysed in various ways to examine: the factor structure; internal consistency reliability; discriminant validity; and, ability to differentiate between schools."

"As with the WIHIC, the data collected from 994 university-level students were analysed to examine reliability and validity of the EELLS, in terms of the: factor structure; internal consistency reliability; discriminant validity; and ability to differentiate between faculties."

"To examine the ability of each scale of the Arabic version of the EELLS to differentiate between perceptions of students in different schools, an analysis of variance (ANOVA), with school membership as the main effect, was used."

"Examination of the standardised regression weight (β) suggest that six of the seven learning environment scales independently accounted for statistically significant ($p < 0.01$) amounts of variance in students' Self-efficacy."

Discussion and Conclusion

"These results provide strong support for the reliability and validity of the Arabic versions of both WIHIC and EELLS, making available two convenient tools that can be used by researchers and teachers to gather information about students' perceptions of their learning environment and motivation and self-regulation in university-level English classes in Jordan."

"Simple correlation and multiple regression analyses, using the data collected from 994 university-level English students in 13 schools, were used to determine whether associations exist between the seven scales of the WIHIC and the four scales of the EELLS (Learning Goal Orientation; Task Value; Self-Efficacy; and Self-Regulation)."

"The strong, positive relationships between Equity and the four motivation and self-regulation outcomes indicate that students who perceive themselves to be treated equally by the teacher are more likely to be motivated and use self-regulatory practices in learning English as a second language."

Motivation Under the Microscope: Understanding Undergraduate Science Students' Multiple Motivations for Research [148]

This is a machine-generated summary of:

Smith, Jessi L.; Deemer, Eric D.; Thoman, Dustin B.; Zazworsky, Lisa: Motivation under the microscope: Understanding undergraduate science students' multiple motivations for research [148].

Published in: Motivation and Emotion (2013).

Link to original: https://doi.org/10.1007/s11031-013-9388-8.

If you want to cite the papers, please refer to the original.

For technical reasons we could not place the page where the original quote is coming from.

Abstract-Summary

"Based on a tripartite model that conceptualizes research motivations as intrinsic, extrinsic, and failure avoidant, analyses revealed five distinct research motivational profiles which were described as Unmotivated; Neutral Engagement, Ternary-Driven; Emerging Engagement; and High Engagement."

"Results showed students were optimally motivated toward science when highly endorsing both intrinsic and extrinsic motivations, with low failure avoidance."

"These findings contribute to the literature on self-determination theory, intrinsic motivation and multiple goals, and these data create a framework for understanding undergraduate science laboratory experiences that may aid in efforts to broaden the participation of students in science."

Introduction

"The aim of the current project was to document the motivational profiles of students enrolled in scientific laboratory classes at three U.S. universities and examine the extent to which such motivational profiles are associated with science class experiences, science identity, and future research intentions."

"We aim to examine students' multiple motivations (IR, ER, and FA) for scientific research and use a person-centered analytic approach to capture a more qualitative and nuanced understanding of the distinct ways in which these three motivations combine."

"Then, our aim was to understand the relationship among multiple motivations that drive undergraduate science students in research experiences and discover how each motivational profile is associated with students' interest and identity with science, as well as their intentions to pursue scientific research."

Method

"A total of 1,151 undergraduate students participated in the study, however, several cases were removed from the data set due to excessive missing data."

"At the midpoint of each semester, students were asked to complete an online survey that inquired into their perceptions of the laboratory class they were enrolled in, as well as their science identity, class experience, and motivations toward scientific research."

"Science class experiences were measured using a portion of the 23-item Subjective Science Attitude Change Measures (SSACM; Stake and Mares [43]), which originally consisted of the following scales: (a) science motivation (6 items); (b) science confidence (6 items); (c) science knowledge (6 items); and (e) new social niche (5 items)."

"Three items ($\alpha = 0.93$) were used in the present study to measure student intent to engage in undergraduate research."

Results

"We present a latent profile analysis (LPA) that describes the number and shape of research motivation profile-groups, based on RMS scores, which emerge in our sample."

"The model is then extended, such that categorical profile-group membership is associated with science identity, science class experience, and future research intentions (conceptually similar to MANOVA, but results in an omnibus Chi square rather than an F statistic)."

"Because of the variability in FA, students in this profile group reported the second-highest scores on science identity, science class experiences, and future research intentions, leading us to define this group as "emerging engagement"."

"This last analysis tests whether gender differentially predicts science outcomes (identity, attitudes, and research intentions) across the 5 profile-groups."

"In this full mixture model, LPA group membership was influenced by all study variables and within each group, gender was a predictor of each auxiliary variable: science identity, class experiences, and research intentions."

Discussion

"This was the only profile-group that was characterized by consistently low levels of science identity, science class experiences, and future motivation for research."

"We hope our focus on understanding the research motivation profiles among students in science laboratory classes (physics, chemistry, and biology) is a step forward in advancing research on broadening the scientific workforce."

"Testing the role of autonomy, competence, and relatedness need support in understanding students' scientific research motivations would be an important next step for future investigations, which we expand upon below."

"Ternary-driven motivation might then yield to "emerging engagement" as students gain more research experience over time and worry relatively less about demonstrating incompetence (e.g., Shim and others [44])."

"Even in countries such as India, with comparatively high levels of undergraduate students pursuing science research majors and careers, student motivation has declined over time and fewer and fewer students are opting to persist in science (e.g., Shaikh [45])."

Summary

"These findings contribute to the literature on multiple-goals using advanced person-centered statistical techniques."

"We revealed five types of tripartite research motivational profiles Unmotivated; Neutral Engagement, Ternary-Driven; Emerging Engagement; and High Engagement and validated each profile with important science educational variables."

"We situated our data in the science laboratory to contribute to understanding science laboratory experiences that may aid in future efforts to recruit and retain a robust future of scientific scholars."

Scaffolding via Group-Dynamic Assessment to Positively Affect Motivation, Learning Anxiety, and Willingness to Communicate: A Case Study of High School Students [149]

This is a machine-generated summary of:

Azizi, Zeinab; Farid Khafaga, Ayman: Scaffolding via Group-dynamic Assessment to Positively Affect Motivation, Learning Anxiety, and Willingness to Communicate: A Case Study of High School Students [149].

Published in: Journal of Psycholinguistic Research (2023).

Link to original: https://doi.org/10.1007/s10936-023-09935-6.

If you want to cite the papers, please refer to the original.

For technical reasons we could not place the page where the original quote is coming from.

Abstract-Summary
"Although a mass of studies has demonstrated the effectiveness of scaffolding through group-dynamic assessment (G-DA) in fostering English as a foreign language (EFL) learners' language skills and components, its contributions to developing psychological constructs, such as motivation, learning anxiety (LA), and willingness to communicate (WTC) have remained largely unexplored in the context of Iranian high schools."

"This study purported to disclose the contributions of G-DA to Iranian high school students' motivation, LA, and WTC."

"The results of the independent samples t-tests evidenced that EG's motivation significantly improved compared to CG after the interventions."

"The findings revealed that the G-DA-based instruction significantly contributed to relieving EG's LA."

Introduction

"Every new teaching approach practiced in L2 the classroom affects and is affected concurrently by students' psychological constructs, such as motivation, learning anxiety (LA), and willingness to communicate (WTC)."

"Given the importance of new teaching approaches like G-DA in affecting L2 learners' motivation, LA, and WTC, it is essential to explore its contributions to L2 learners' motivation, LA, and WTC."

"The review of the available literature reveals that the contributions of G-DA to L2 learners' motivation, LA, and WTC has received scant attention in the high school contexts of Iran."

"This study attempts to respond to this lacuna by exploring the contributions of G-DA to Iranian high school students' motivation, LA, and WTC."

"The results of this study can enrich the literature to offer further understanding of the applicability and effectiveness of G-DA in developing Iranian high school students' motivation, LA, and WTC."

Literature Review

"As Poehner [46] stresses, in G-DA all L2 learners' ZPD is considered."

"In simple terms, motivation is an integration of internal and external forces that L2 learners put in to start, direct, and control the learning tasks in hand (Deci and others [47])."

"Instrumental motivation deals with L2 learners' eagerness and efforts to satisfy external needs by learning the L2 (Gardner [48])."

"Much previous research has explored the contributions of DA and G-DA to psychological constructs in L2 learners."

"Ritonga and others [49] inspected the extent to which interventionist DA, interactionist DA, and non-DA effectively influence EFL learners' speaking accuracy and fluency, foreign language classroom anxiety, and foreign language learning motivation."

"Chen and others [50] studied the effects of a DA-based speech recognition learning system on fostering L2 learners' speaking skills and speaking learning anxiety."

Methodology

"The current study adopted the true-experimental design to explore the contributions of G-DA to cultivating Iranian high school students' motivation, LA, and WTC."

"The researchers selected the high school students as they were available to them."

"The OQPT was the first instrument employed to homogenize the high school students with regard to general English language proficiency."

"Regarding reliability, the researchers administered it to 35 high school students who were similar to the participants of the main study in respect of gender, grade, and language proficiency."

"They invited 35 high school students who were similar to the participants of the main study in respect of gender, grade, and language proficiency to fill out the instruments."

"The last step was administering AMTBQ, FLCAS, and WTCQ at the end of the interventions to measure the high school students' motivation, LA, and WTC."

Results

"To disclose if this difference was statistically significant, the researchers used an independent samples t-test."

"There was a big difference between the two groups' Ms. The researchers ran an independent samples t-test to uncover if this big difference was statistically significant."

"Having ensured that the required assumptions were met, the researchers used an independent samples t-test."

"The researchers ran an independent samples t-test to see if this difference was statistically significant."

"As it is clear, there existed a big difference between the two groups' Ms. The researchers employed an independent samples t-test to clarify if this big difference was statistically significant."

"Having ensured that the required assumptions were met, the researchers employed an independent samples t-test."

"There existed a big difference between the two groups' Ms. Accordingly, the researchers utilized an independent samples t-test to clarify if this big difference was statistically significant."

Discussion

"The findings evidenced that G-DA-based instruction significantly contributed to the high school students' motivation, LA, and WTC at the end of the intervention."

"The G-DA-based instruction could create a learning in which the high school students' motivation significantly prompted, their LA anxiety significantly relieved, and their WTC significantly upheld."

"As posited by this theory, it may be argued that the G-DA-based instruction might positively affect the high school students' self-regulation and self-efficacy."

"In light of this theory, it can be argued that G-DA-based instruction might create a learning setting in which the high school students might have reached the possible selves."

"As a result of the positive effects of the G-DA-based instruction, the high school students may have become aware of the discrepancies between preferred future self-guides and the conceived plausibility of those guides, thereby enhancing their motivation, relieving their LA, and maintaining their WTC (Dörnyei and others [51])."

Conclusions and Implications
"This study examined the contributions of G-DA to high school students' motivation, LA, and WTC in Iran."

"As G-DA could positively affect the psychological constructs, it can be considered as an alternative approach to boost high school students' achievements."

"Fifthly, as high school student's English learning is affected by psychological constructs, EFL teachers can know that the implementation of alternative approaches like G-DA can pay the way for their cultivation."

"Because the interventions lasted 16 sessions, the contributions of G-DA to the high school students' motivation, LA, and WTC were not disclosed in the long term."

"A longitudinal study should be conducted to reveal its effects on high school students' motivation, LA, and WTC over a long period of time."

"Owing to the present study's quantitative design, it could not disclose how G-DA affected the high school students' motivation, LA, and WTC."

Putting the Pieces Together: Reviewing the Structural Conceptualization of Motivation Within SDT [150]

This is a machine-generated summary of:

Howard, Joshua L.; Gagné, Marylene; Morin, Alexandre J. S.: Putting the pieces together: reviewing the structural conceptualization of motivation within SDT [150].

Published in: Motivation and Emotion (2020).

Link to original: https://doi.org/10.1007/s11031-020-09838-2.

If you want to cite the papers, please refer to the original.

For technical reasons we could not place the page where the original quote is coming from.

Abstract-Summary

"Self-determination theory presents a multi-dimensional approach to human motivation in which motivation is driven by a range of classifiably distinct regulation styles."

"The lack of explicit integration between these two aspects in the conceptualization of motivation has recently led to debates and a flux of research on the structure of motivation as defined in self-determination theory."

"A novel solution is proposed, namely a semi-radex structure of motivation, in which types of motivation are predictably ordered by degree of self-determination while also maintaining their unique contribution as distinct factors."

Behavioral Regulations According to Self-determination Theory

"Internalization, defined as an "active assimilation of behavioral regulations that are originally alien or external to the self" (Ryan [52], p. 405), is what can transform extrinsic motivation into something closer, yet different from, intrinsic motivation."

"Identified regulation is the next most highly internalized form of extrinsic motivation and involves the recognition of the inherent instrumental value of the behavior."

"Whereas intrinsic motivation is related to behaviors that are valued in and of themselves, identified regulation is related to behaviors that are expected to lead to valued outcomes."

"Introjected regulation is conceptualized as a partially internalized type of extrinsic motivation (Deci and Ryan [53]), or a type of motivation with a moderate level of perceived locus of causality."

"External regulation is the least internalized form of extrinsic regulation, with the most externally focused locus of causality, and refers to behaviors that are coerced by the promise (or threat) of external rewards (or punishments)."

The Current Structure of Motivation in SDT

"A study examining the full range of regulations in predicting well-being may find a positive association between identified regulation and the outcome, however it is not possible to distinguish if this result is due to a unique characteristic of identified regulation (i.e., meaningfulness) or to the global degree of self-determination."

"Studies relying on a single score reflecting participants global levels of self-determination (such as the RAI) will be able to examine how the degree of self-determination associates with covariates but will be unable to consider the unique contributions of each regulation subscale (i.e., thereby treating motivation as a single-dimensional construct)."

"This exemplifies a major pitfall of the current conceptualization of motivation as the lack of clarity and integration of these two aspects (degree of self-determination and specific regulation types) allows for divergent assumptions which come with theoretical and practical implications that are much too strong to be left unchecked."

Research Evidence for the Continuum Hypothesis of Motivation and the Unique Quality of Each Behavioral Regulation

"These approaches have demonstrated support for a multidimensional representation of motivation as encompassing multiple types of behavioral regulations, themselves displaying a pattern of correlation matching the expected continuum structure of self-determination."

"Rather than focusing on factor analysis, correlations, or regressions among conceptually distinct behavioral regulations, Chemolli and Gagné [54] conducted an even more stringent test of the one-dimensional continuum structure of motivation by testing whether a single dimension could be assumed to provide an adequate representation of responses to multidimensional motivation measures."

"Loadings on the G-factor should follow the continuum hypothesis, with items representing more self-determined forms of motivation (e.g., intrinsic motivation) showing strong positive loadings and items representing less self-determined motivation forms (e.g., external regulation) showing weaker loadings."

"The observed pattern of factor loadings on the G-factor did not follow the SDT's continuum hypothesis, but rather seemed to simply reflect participants' global level of motivation across all types of behavioral regulations."

A proposition: The Semi-Radex Structure

"A semi-radex conceptualization is supported by the entirety of reviewed evidence including the predictable ordering noticed in correlation tables, factor analyses finding distinct regulations types, Rasch analysis suggesting that a single dimension in not sufficient, bifactor modeling which identifies both self-determination and regulation-specific factors, multidimensional scaling revealing a strong but not sufficient central dimension, and the commonly observed shapes of motivation profiles."

"The application of the bifactor-ESEM framework provides a way to simultaneously estimates participants' location on the underlying self-determination continuum estimated from their responses obtained across all motivation items (i.e., the G-factor), together with their specific levels on each of the specific types of behavioral regulation left unexplained by this G-factor (i.e., the S-factors)."

"This approach provides a way to simultaneously account for both of the core aspects of SDT representation of human motivation: The continuum (i.e., the global self-determination component) and specificity (i.e., the unique quality of each type of regulation)."

Conclusion

"Motivation as defined by SDT, which has traditionally concurrently been described as unique elements and described using a continuum of self-determination, can now be integrated and described as a semi-radex."

"In this semi-radex, motivational regulations both retain unique properties while being ordered in a predictable manner."

"As resolving an ongoing debate concerning the simplex structure often cited in the literature, this semi-radex structure refines SDT theoretically by providing an integrative and nuanced conceptualization of motivation."

"This will help foster more precise theorizing and testing of the two key elements of SDT alongside one another, and furthermore allow for easier integration of research findings."

[Section 6]

"Even though SDT explicitly notes the importance of recognizing the conceptually distinct nature of these various types of behavioral regulations (i.e., a multidimensional approach), the continuum hypothesis might be erroneously taken to suggest that the regulations themselves do not really represent distinct factors, but merely points along a single continuum of self-determination (i.e., a one-dimensional approach)."

"These operationalizations do not clearly align with SDTs multidimensional conceptualization of motivation as encompassing a series of qualitatively distinct types of behavioral regulations, thereby raising a variety of questions concerning how best to conceptualize and assess human motivation."

"Although both components of SDTs conceptualization of human motivation (e.g., the one-dimensional continuum hypothesis and the multidimensional nature of behavioral regulations) caries valuable information, they both provide an incomplete picture of human motivation when considered on their own."

Self-control Motivationally Reconsidered: "Acting" Self-controlled Is Different to "Being Good" at Self-control [151]

This is a machine-generated summary of:

Grund, Axel; Carstens, Christoph-Alexander: Self-control motivationally reconsidered: "Acting" self-controlled is different to "being good" at self-control [151].

Published in: Motivation and Emotion (2018).

Link to original: https://doi.org/10.1007/s11031-018-9721-3.

If you want to cite the papers, please refer to the original.

For technical reasons we could not place the page where the original quote is coming from.

Abstract-Summary
"We scrutinize this understanding by differentiating between an ability self-concept of self-control strength and experiential acts of self-control."

"In cross-sectional Study 1 (N = 228), we compared a typicality measure of experiential acts of imperative self-control with the widely used Self-Control Scale (Tangney and others in J Pers 72:271–322 [55])."

"Multilevel analyses showed that between-person differences (Level 2) in self-control strength were generally unrelated to experiential acts of self-control in everyday life."

"We found a positive Level 2 effect for acting self-controlled."

"With regard to momentary affect, both between- and within differences (Level 1) in acting self-controlled served as substantial predictors, in addition to momentary self-determination."

"Other context-dependent effects (i.e., studying vs. leisure time) further emphasize the need to consider motivational interpretations of self-control (strength)."

Introduction
"Acting self-controlled is viewed as effortful and psychologically costly, as has been demonstrated in numerous studies on ego-depletion (e.g. Baumeister and others [59]; Milyavskaya and Inzlicht [60]; Muraven and others [61]; see Carter and others [62], for an extensive critique on the ego-depletion effect)."

"We also want to scrutinize the self-evident interpretation that the benefits of self-control strength as an interindividual difference variable are actually achieved by acts of self-control in everyday life."

"Whereas the latter captures individuals' perceptions about how good they are at self-control, the former asks to what extent they typically (have to) act self-controlled in everyday life (ASC)."

"Our main assumption is that individuals, who say that they are "good" at self-control, actually have to act self-controlled less often because of their motivational basis, particularly in achievement contexts."

The Resource Model of Self-control (Strength)

"There needs to be at least a little something to "overcome", otherwise, self-control would be of no use."

"The notion of self-control further implies that the resistance lies within the person, not outside the person (cf. Sokolowski [63])."

"Because inhibitory self-control is psychologically exhausting, people need to recover from acting self-controlled, and therefore fail in a subsequent task that requires controlled action."

"People who ostensibly exercise inhibitory self-control at task one are expected to perform worse on the second task, which is interpreted in terms of a general self-control resource or strength that depleted after initial use."

"Self-control is thought to be extendable by training (Baumeister and others [64]; Muraven [65])."

"Inter individual differences in the amount of self-control strength have been proposed (e.g., Tangney and others [55])."

Self-control Motivationally Reconsidered

"“procrastinators", in direct opposition to individuals who report being high in self-control, seem to favor postmodern well-being values over modern achievement values (Grund and Fries [66]), which again renders the interpretation of a self-control failure quite problematic."

"Introjected regulation refers to non-self-determined behavior where the instrumental consequences of an action are administered not by others (i.e., external regulation) but "by the individuals to themselves" (Deci and Ryan [67], p. 236)."

"In that sense, introjected regulation shares common ground with self-controlled action, as both involve aversive, intrapsychic conflict."

"If only goal-maintenance is achieved at the cost of self-integrity, such a mode of regulation is termed self-control."

"Self-regulation resembles a "democratic" form of regulation, where several internal and external stimuli are acknowledged in order to integrate and satisfy both immediate and long-term needs of the individual (cf. Rheinberg and Engeser [68])."

Summary and Research Questions

"We expected opposing relationships for both measures with regard to the amount of intrapsychic conflict individuals typically experience, and therefore, a negative relationship between both measures of self-control."

"We expected that acting self-controlled (ASC) would go along with high perceptions of intrapsychic conflict and low well-being because acting self-controlled is effortful and emotionally costly, and should eventually deteriorate well-being when serving as chronic modus of action (e.g., Kuhl and Fuhrmann [56]; Kuhl and Beckmann [57]; Ryan and others [58])."

"We wanted to further understand both perspectives of self-control by shedding light on potential motivational processes that may account for the positive effects of self-ascribed self-control strength, especially with regard to academic achievements and the assumed negative effects of acting self-controlled."

"We expected that acting non-self-determined is strongly associated with acting self-controlled, similar to our measure of intrapsychic conflict in Study 1."

Study 1: A Complementary Measure of Self-control

"Mixed with the items of acting self-controlled, we asked participants to answer four items that we developed to assess the extent of participants' general intrapsychic conflict experience."

"Whereas participants' self-ascribed self-control strength (SCS) and acts of self-control (ASC) fell around the theoretical midpoint of the respective 5-point scales, participants reported relatively high levels in life satisfaction and positive affect, whereas they reported relatively low levels in intrapsychic conflict and negative affect."

"Given the substantial overlap between SCS and ASC as well as possible confounding effects for intrapsychic conflict, gender and age, we applied stepwise multiple regression analyses in order to further investigate the possible effects of both self-control conceptions on psychological well-being."

"Our aim was to compare two measures of self-control, either to assess participants' ability concept of self-control strength (SCS) or to determine whether the execution of imperative, effortful self-control is something typical for them (ASC)."

Study 2: Self-control in Everyday Life
"Participants' self-ascribed self-control strength (SCS), acting self-controlled (ASC), and intrapsychic conflict experiences were assessed using the same measures as in Study 1."

"We expected that the effect of SCS (Level 2) on acts of self-control (Level 1) depends on the context in which the engagement is embedded in (i.e., academic vs. leisure related activities), reflecting a general preference for achievement over well-being values and related behaviors (cf. Grund and Senker [69])."

"These motivational preferences should be reflected in how these engagements are experienced (acts of self-control) and the actual behavioral patterns of the participants (frequency of engagement in academic and leisure-related activities)."

"The higher participants' self-ascribed self-control strength, the more they participated in academic relative to leisure-related activities during the experience-sampling period."

"When we entered intrapsychic conflict experiences (Level 2) and self-determination (Level 1) as additional predictors in the model (Model 7), the effect for both ASC measures diminished substantially, but remained significant for momentarily self-controlled action."

General Discussion
"Our major point was that it is important for future research to differentiate between acting self-controlled, as a specific modus of action regulation, triggered by intrapsychic conflict and accompanied by perceptions of effort and aversion, and self-ascribed self-control strength, resembling a self-concept of one's self-control capacities that relates to a host of desirable outcomes."

"We are aware that such a behavioral pattern could also be interpreted in line with the notion of self-control as an effortful, suppressing process or skill (i.e., being only able to choose studying over leisure time because one possesses self-control strength)."

"Given the prevalent finding that self-ascribed self-control strength is not only positively related to affective and cognitive well-being in general but also to feeling happier in the moment (e.g., Hofmann and others [70]; Grund and others [71]), it is also tempting to reason that acting self-controlled depicts an desirable modus of self-regulation in terms of psychological well-being and goal attainment."

Conclusion
"Given the present data we cannot conclude that they achieve this by the effortful execution of inhibitory self-control."

"This calls into question the nature of the self-control concept and its underlying processes."

"Imperative self-control is only indicated when inner resistances need to be overcome, which seems to be the critical aspect when it comes to psychological well-being."

"Research on interindividual differences in self-control strength may profit from the inclusion of relevant motivational variables that provide alternative explanations for the prevalent and impressive effects."

Student's Social Interaction in Inquiry-Based Science Education: How Experiences of Flow Can Increase Motivation and Achievement [152]

This is a machine-generated summary of:

Ellwood, Robin; Abrams, Eleanor: Student's social interaction in inquiry-based science education: how experiences of flow can increase motivation and achievement [152].

Published in: Cultural Studies of Science Education (2017).

Link to original: https://doi.org/10.1007/s11422-016-9769-x.

If you want to cite the papers, please refer to the original.

For technical reasons we could not place the page where the original quote is coming from.

Abstract-Summary

"This research investigated how student social interactions within two approaches to an inquiry-based science curriculum could be related to student motivation and achievement outcomes."

"Data sources included formal and informal participant interviews, participant journal reflections, curriculum artifacts including quizzes, worksheets, and student-generated research posters, digital video and audio recordings, photographs, and researcher field notes."

"The results of this research illustrate how setting conditions that promote focused concentration and communicative interactions can be positively related to student motivation and achievement outcomes in inquiry-based science."

"Off-Campus participants also designed and implemented a more cognitively complex research project, provided more in-depth analyses of their research results, and expanded their perceptions of what it means to act like a scientist to a greater extent than participants in the On-Campus case."

"Student interactions that promoted the criteria necessary for initiating flow, which included having clearly defined goals, receiving immediate feedback, and maintaining a balance between challenges and skills, fostered enhanced student motivation and achievement outcomes."

"Implications for science teaching and future research include shifting the current focus in inquiry-based science from a continuum that progresses from teacher-directed to open inquiry experiences to a continuum that also deliberately includes and promotes the necessary criteria for establishing flow."

"Attending to Flow Theory and incorporating student experiences with flow into inquiry-based science lessons will enhance student motivation and achievement outcomes in science and bolster the success of inquiry-based science."

Flow Theory

"Csikszentmihalyi described individuals being in states of flow when they have "a sense that one's skills are adequate to cope with the challenges at hand, a goal-directed, rule-bound action system that provides clear clues as to how well one is performing…concentration is intense…and the sense of time becomes distorted" (Csikszentmihalyi [72], p. 71)."

"Flow occurs when one can balance the challenges presented with the skills required to complete the challenges; individuals willingly engage in a task (Csikszentmihalyi [73])."

"Csikszentmihalyi identified three necessary conditions to create dimensions of flow: clear goals throughout the activity or process, immediate feedback, and a balance between challenges and skills (Csikszentmihalyi [74])."

"Additional research, such as that conducted by Stefan Engeser [75], has also shown that individuals who enter into states of flow have enhanced performance in their artistic, athletic, and cognitive abilities as well as increased enjoyment, perseverance, and dedication toward the activities in which they are engaged."

Inquiry-Based Science Education

"Very few research studies have embedded researchers in the classrooms to study the entire student experience, or focused on student social interactions and their influence on student ability to act like scientists and IBSE outcomes."

"The term "acting like a scientist," as typically depicted within science education textbooks, resources, and teacher interpretations, describes acting like a scientist with going through procedural steps of "the scientific method;" students develop research questions, develop and implement experiments, collect and analyze data, and report their findings to others (Wong and Hodson [76])."

"Our study investigated the influences of student social interactions, particularly those reflective of acting like a scientist as described by scientists, on the outcomes of IBSE experiences and strives to broaden the educational perspective of "acting like a scientist" to include the contextual and human process of meaning making within the scientific process."

Problem Statement and Purpose of Study

"Our research investigated whether two groups of middle school aged students engaged in and learning from two contextually different IBSE projects engaged in social interactions and collaborations similarly to the scientists in the Wong and Hodson study [76] and the Schwartz and Lederman study [77]."

"We also examined if those student social interactions led to more authentically acting like scientists, enhanced engagement, motivation and/or achievement."

"The specific research question asked was: Within two approaches to inquiry-based science, how do student social interactions relate to student motivation and achievement?"

"We required an understanding of whether, and if so how, social interactions may be related to student motivation and achievement, thus we focused our attention on any student interactions that seemed to influence student progress in, and motivation concerning, the class and/or student projects."

Materials and Methods

"In order for us to gain access to data that would allow us to answer our research question, and assess whether student social interactions influenced IBSE outcomes and flow experiences, we developed a researcher observation sheet following recommendations from James Paul Gee's [78]."

"Although each of these categories is relevant to understanding student experiences with IBSE, in order to manage the considerable amount of data we anticipated would be collected through this research and to focus our observations on social interactions that were the most relevant to the situation and outcomes within an IBSE context, our research isolated four of Gee's categories in developing our researcher observation sheet."

"In order to understand the influence of student social interactions and flow within each case, and to develop accurate case profiles for each case, we developed a researcher observation sheet that was modeled after Gee's seven categories for developing constructed meaning and understanding of situations."

Data Analysis

"Conducting coding matrix queries within NVIVO9 facilitated our ability to collapse the initial 425 codes into five emergent themes based on parameters of frequency of occurrence and apparent impact."

"Codes that either frequently appeared across data sources or had noticeable impact on students were used to begin collapsing original codes into themes."

"Using the parameters of frequency of occurrence and apparent impact established codes that reflected the student experience."

"Outside coding consensus, with the same researcher, was again established after we identified five emergent themes relevant to flow and social interaction, thus inter-rater reliability was established."

"Once emergent themes were established, a detailed, thick description narrative was created that depicted student experiences and perspectives within each of the two cases."

"Flow' was supported by the five themes that emerged from our analysis of how the student social interactions that took place within each case were of influence."

Results

"Opportunities experienced by students to act like scientists, which included engaging in discussions about project ideas, plans and implementation processes as well as data collection and possible explanations for the data they obtained, and discussions about how to deal with challenges that arose through the project, influenced student motivation and their efforts to analyze and express their understandings of their research."

"As students in both cases began to develop their intended research questions and explore potential research sites in Phase Two of the project, participants in the Off-Campus case were allotted more time to discuss their observations and project ideas."

"The time that Off-Campus participants had been allotted to develop their research question had allowed them to discuss and refine their ideas based on group, peer, and teacher feedback and had situated them within the optimal zone for maintaining flow; they would be challenged within manageable levels, thus not becoming overwhelmed or bored by the activity."

Discussion and Conclusions

"Our research illustrates and explains how student social interactions and emotions affect the likelihood of entering into states of flow and how they have influence on student achievement and motivation outcomes within IBSE experiences."

"Fostering student interactions in IBSE that promote conditions necessary for initiating flow, which include having clearly defined goals, receiving immediate feedback, and maintaining a balance between challenges faced and skills required, can enhance overall student motivation and achievement outcomes in IBSE."

"The results of this research suggest that shifting our attention from an IBSE continuum that focuses on a progression from teacher-directed to open-inquiry experiences based on the level of student responsibility within the lesson to one that also focuses on providing the conditions necessary for flow to emerge will foster student abilities to authentically act like scientists and will enhance motivation and achievement outcomes in IBSE."

[Section 8]
"Inquiry-based science education (IBSE) is an educational practice that can foster student ability to learn scientific concepts, experience "acting like a scientist," and understand the nature of science (Chang and Song-Ling [79])."

"In our attempt to discover how student social interactions influence IBSE outcomes, we discovered that student experiences with Flow ultimately led to elevated outcomes in motivation and achievement in IBSE."

"The results of our research show that developing and implementing IBSE experiences that set conditions necessary to promote flow experiences, where students have opportunity to engage in discussions about short and long-term project goals, give and receive regular and specific feedback about project plans and progress, as well as discussing challenges that arise and how to overcome the challenges, led to elevated achievement and motivation outcomes."

"Our research reveals how educational practitioners can bolster student motivation and achievement in inquiry-based science education and provide a means for achieving more consistently positive results from IBSE."

How Are Motivation and Self-efficacy Interacting in Problem-Solving and Problem-Posing? [153]

This is a machine-generated summary of:

Voica, Cristian; Singer, Florence Mihaela; Stan, Emil: How are motivation and self-efficacy interacting in problem-solving and problem-posing? [153].

Published in: Educational Studies in Mathematics (2020).

Link to original: https://doi.org/10.1007/s10649-020-10005-0.

If you want to cite the papers, please refer to the original.

For technical reasons we could not place the page where the original quote is coming from.

Abstract-Summary

"Affects are intuitively accepted as having a role in the key stages that determine success in problem-solving (PS) and problem-posing (PP)."

"While in the PS context, several students expressed reluctance to report emotional feelings or frustrations over failing to find a solution for the given problems, in the PP context, a phenomenon of adaptation to one's own cognitive possibilities appeared and the posed problems created a state of comfort/ enjoyment in students."

"It seems that intrinsic motivation is influenced by factors that are different in PP vs PS tasks, with an effect on perceived self-efficacy."

"The PP activity instilled a robust sense of coping efficacy, which made the students perceive their work as successful, to a large extent."

"In PS, the initially perceived self-efficacy triggered motivation to persevere with the solving, while in PP, the initial motivation generated by a novel task evolved into students' perceptions of self-efficacy and confidence in one's own capabilities."

Introduction

"A more comprehensive view for understanding students' problem-posing and solving might include, as Goldin [80] stated at some point, cognitive, affective, conative (motivational), social, and behavioral aspects."

"The present research is motivated by the need to explore students' epistemic affects in the context of complex mathematics problem-solving and posing."

"In this interdisciplinary study, the theoretical framework we have worked with is situated at the confluence of mathematical problem-solving and problem-posing, cognitive psychology, psychology of affects, and social cognition in a large sense."

"When one solves a knowledge-based specific task, epistemic affects interfere, among which, motivation and self-efficacy."

"In order to study students' motivation and self-efficacy in solving a specific task, it is necessary to analyze the data from three perspectives: cognition, affects, and the social context in which the respective students are situated."

Method

"In the instructions for the two cognitive tasks (concerning PS and PP, respectively), we encouraged students to present in detail their attempts to solve or pose a problem, even if they were not successful."

"Students were urged to express themselves freely; for a more accurate description of their emotions, attitudes, and beliefs, we suggested that they periodically note their affective states in relation to their attempts to solve or pose problems and, eventually, to turn to other people, asking them to make objective observations, "from outside," on their behavior."

"For the PS task, students used the following strategies: framing the problem in a typology; anticipating/grasping a result and directing the solution to verify it; formulating a related, simpler problem; exploring particular cases and trying to generalize the methods of solving these cases; and resorting to methods or tools that are not specific to mathematics."

Results and Discussion

"The simplicity of the problems (or at least the impression that things are this way) seems to motivate these students to persevere in solving: in all cases where students mention "simplicity," there are repeated attempts to solve all problems, spread over several pages; in some cases, even complete correct solutions were given."

"I have to be able to solve them!") Therefore, her motivation is strengthened by a social context (Denisa prepares students from Olympic teams and describes in her report how she encourages them to be patient and persevere in solving given problems)."

"We have seen that, in the PS context, the students of our sample have expressed self-efficacy in relation to their own attempts to solve the given problems."

Conclusions

"An analysis of the (intrinsic) motivation manifested by the students in solving the tasks also shows differences between the PP context and the PS context."

"The data we have show that, in both contexts (i.e., PS, and PP contexts), negative affective states involve de-motivating of students and a low level of self-efficacy, while positive emotional states are a strong motivational factor: this is consistent with the findings of other studies (e.g., Caprara & Gerbino [81]; MacIntyre & Vincze [82])."

"The obtained data indicate that, in a PP context, the students in our sample expressed mainly positive emotions, states that can be put in correspondence with the higher success rate in solving the task (i.e., the proposal of coherent and consistent problems)."

Motivation, Pedagogy, and Gender: Examining the Multifaceted and Dynamic Situational Responses of Women and Men in College STEM Courses [154]

This is a machine-generated summary of:

Stolk, Jonathan D.; Gross, Michael D.; Zastavker, Yevgeniya V.: Motivation, pedagogy, and gender: examining the multifaceted and dynamic situational responses of women and men in college STEM courses [154].

Published in: International Journal of STEM Education (2021).

Link to original: https://doi.org/10.1186/s40594-021-00283-2.

If you want to cite the papers, please refer to the original.

For technical reasons we could not place the page where the original quote is coming from.

Abstract-Summary

"Most studies of learner motivation, however, are conducted outside of STEM and are focused at the contextual level, which may describe why students attend college or choose a degree program, but not why they engage in classroom activities."

"There is little research that meaningfully connects learner motivations with gender identity and course pedagogy."

"This study addresses these gaps by examining the interconnections among course pedagogy, gender, and situational-level motivations, which reveal why learners engage in different course activities and how engagement may vary over time."

"This detailed perspective on learner motivations is essential for instructors to gain insights into how their pedagogical and course design choices influence students' motivational responses and to more effectively develop interventions that support positive forms of motivation among all students."

"Participants in the study are undergraduate students enrolled in 72 introductory-level STEM courses across 11 institutions, and the dataset includes over 5000 unique responses to the Situational Motivation Scale, a Self-Determination Theory-based instrument that was administered weekly in each course."

"Cluster distributions by gender and pedagogy indicate significant differences in lecture-based learning courses, with women reporting less self-determined motivations compared to men."

"Motivational response profiles of women and men are both more similar, and more positive overall, in courses that employ active learning."

"The gender analyses raise questions about how instructors may more effectively promote the positive motivations of all students through their course design decisions."

Introduction
"While our understandings of women and men's self-efficacy, goals, and values has progressed over the past two decades, studies on the orientations or types of motivations women and men express in learning situations are quite limited."

"Most of the work linking motivation and gender lies outside of STEM, and most studies that examine gendered expression of intrinsic versus extrinsic motivations are conducted at the contextual level (e.g., why students attend school or participate in a degree program), not at the situational or course activity level."

"Since few studies have directly linked women and men's situational motivations to STEM pedagogies, we have little information on how instructors may more effectively select instructional modes and create experiences that engage all students in STEM learning."

"This study begins to address these research gaps, by characterizing college students' controlled versus autonomous motivations in STEM learning activities, and exploring the connections among gender, pedagogy, and situational motivations in STEM learning environments."

Methods
"For courses classified as lecture-based learning, greater than 75% of student responses described course activities as "lecture.""

"The motivation dataset is weighted toward lecture-based pedagogies, with 3894 responses (76.3%) from lecture-based learning courses and 1211 responses (23.7%) from active learning courses."

"To analyze students' situational motivations, this study applied quantitative clustering, a technique that has proven useful in characterizing motivations (Braten and Olaussen [83]; Csizer and Dornyei [84]; Ratelle and others [85]), as it enables examination of the motivational response 'shape' and reveals how individuals simultaneously express a combination of different motivation types."

"Each cluster is described in terms of mean values from the four motivation subscales of the SIMS instrument."

"Cluster distributions for students' motivational responses were compared across lecture-based and active pedagogies and across self-reported gender identities."

"For individuals with at least five SIMS responses (49.3% of participants), we examined the percentages of students who reported n different motivation response profiles (clusters)."

Results and Discussion

"The AU cluster represents the most positive situational motivation response profile, characterized by high identified regulation (IR) and intrinsic motivation (IM) levels, and low external regulation (ER) and amotivation (AM) levels."

"Students with the ID-EX response likely perceive usefulness or importance in their course activities, but they couple this perceived value more strongly with extrinsic pressure or obligation (external regulation) than internalized interest and enjoyment (intrinsic motivation)."

"As the most controlled and least self-determined situational motivation response, we expect AM to appear when students experience high levels of stress and extrinsic pressure, feelings of uncertainty about what they are doing, and no sense of enjoyment, interest, or value in the learning."

"Analyses of the motivational response profiles by pedagogy using the Mann–Whitney U test showed a significant difference in the cluster distribution for active ($N = 1213$) and lecture-based ($N = 3789$) learning ($U = 3{,}077{,}924$, $p < 0.001$), with more positive situational motivations in active learning environments compared to lecture-based learning environments."

Conclusions

"Situational motivations are more positive in courses that employ active pedagogies, as compared to lecture-based courses."

"Motivational response profiles of women and men are both more similar and more positive overall in STEM courses that employ active pedagogies such as project- and discussion-based learning."

"While individuals' situational motivational responses in STEM courses may be stable or dynamic, this study shows that the majority of students report multiple unique motivation clusters that span three or more cluster profiles."

"The gender-based analyses raise questions about how instructors may more effectively promote the positive motivational engagement of all students through their course design and implementation decisions."

The Role of Affective and Motivational Factors in Designing Personalized Learning Environments [155]

This is a machine-generated summary of:

Kim, ChanMin: The role of affective and motivational factors in designing personalized learning environments [155].

Published in: Educational technology research and development (2012).

Link to original: https://doi.org/10.1007/s11423-012-9253-6.

If you want to cite the papers, please refer to the original.

For technical reasons we could not place the page where the original quote is coming from.

Abstract-Summary

"Guidelines for designing virtual change agents (VCAs) are proposed to support students' affective and motivational needs in order to promote personalized learning in online remedial mathematics courses."

"Automated, dynamic, and personalized support is emphasized in the guidelines through maximizing interactions between VCAs and individual students."

"The strategies that VCAs convey throughout the interactions are constructed to support emotion regulation and motivation based on theories and prior research on emotions and motivation."

Introduction

"The purpose of this paper is to report the development of design guidelines that address students' affective and motivational factors in order to promote personalized learning and provide meaningful feedback in online remedial mathematics courses."

"Emotional and motivational problems (e.g., math anxiety and task avoidance) potentially confound and complicate efforts to examine student performance in remedial courses (Ironsmith and others [86])."

"Students experience negative emotions (e.g., shame, anger, frustration) due to their past failure on a college placement test, which required them to take the remedial mathematics courses."

"This is why both emotions and motivation need to be understood and optimized to promote student learning and performance."

"Both student motivation and emotions are critical to success in college mathematics courses (Schiefele and Csikszentmihalyi [87])."

"There is a need to support students' emotion regulation while improving motivation in remedial mathematics education."

Design Guidelines for Virtual Change Agents
"The strategies were constructed to facilitate students' positive reappraisal of a course activity or its outcome (e.g., assignments in an online math course) by maximizing their controllability of the four emotion process components described by Gross [88]: (a) situation, (b) attention, (c) appraisal, and (d) responses."

"The limited effects of the interventions in those studies led to the inclusion of strategies promoting students' perception of task value (i.e., Design Guideline Corollary 1.1), also based on theoretical foundations that explain reciprocal impacts of emotions and motivation and both should be considered for students' reappraisal of the situation."

"As learned from the previous studies as well as the synthesis of theoretical foundations, the integrated use of all the strategies and interactions proposed in the design guidelines should be tested for the purpose of improving students' emotional and motivational experiences."

Discussion
"Design guidelines for VCAs were proposed to address students' affective and motivational factors to promote personalized learning in online remedial mathematics courses."

"Understanding the common factors that influence both emotions and motivation (e.g., task value, controllability, social interactions) are important in designing interventions that promote student success in such challenging courses."

"The design guidelines are meaningful in that they guide the creation of interventions to promote both motivation and emotion regulation unlike other existing models and theories."

"These do not provide guidelines for how to design interventions for emotion regulation in the contexts of online learning."

"The design guidelines presented in this paper maximize the possibility of studying and supporting students' affective and motivational factors within the learning and performance contexts."

Predict or Describe? How Learning Analytics Dashboard Design Influences Motivation and Statistics Anxiety in an Online Statistics Course [156]

This is a machine-generated summary of:

Valle, Natercia; Antonenko, Pavlo; Valle, Denis; Sommer, Max; Huggins-Manley, Anne Corinne; Dawson, Kara; Kim, Dongho; Baiser, Benjamin: Predict or describe? How learning analytics dashboard design influences motivation and statistics anxiety in an online statistics course [156].

Published in: Educational technology research and development (2021).

Link to original: https://doi.org/10.1007/s11423-021-09998-z.

If you want to cite the papers, please refer to the original.

For technical reasons we could not place the page where the original quote is coming from.

Abstract-Summary
"Based on the achievement goal theory, this experimental study explored the influence of predictive and descriptive learning analytics dashboards on graduate students' motivation and statistics anxiety in an online graduate-level statistics course."

"Results indicate that, compared to the control group, the predictive dashboard significantly reduced learners' interpretation anxiety and had an effect on intrinsic goal orientation that depended on learners' lower or higher initial levels of intrinsic goal orientation."

"To the control group, both predictive and descriptive dashboards reduced worth of anxiety (negative attitudes towards statistics) for learners who started the course with higher levels of worth anxiety."

"Thematic analysis revealed that learners who adopted a more performance-avoidance goal orientation approach demonstrated higher levels of anxiety regardless of the dashboard used."

Introduction

"Statistics anxiety may be compounded in online learning environments due to lack of immediate feedback from instructors and the requirements for great levels of metacognitive skills as well as proficiency in technology, especially when statistical software is used (Hsu and others [91]; Zimmerman [92])."

"This study investigates how to better support learners' affective and cognitive outcomes in online learning environments by using an important affordance of online contexts: learning analytics."

"We designed and tested two learning analytics dashboards – predictive self-referenced vs. descriptive norm-referenced (Biesinger & Crippen [93]; Koenka & Anderman [94]) – to explore how different design and pedagogical features influence learners' motivation, statistics anxiety, and learning performance in an introductory online statistics course."

"Despite the increasing interest in the potential of learning analytics to support learners' reflection, behavioral changes, motivation, and learning performance (Gašević and others [95]; Park & Jo [96]), empirical studies investigating the influence of learning analytics dashboards on learners' cognitive and affective outcomes are still limited (Bodily & Verbert [97]; Jivet and others [98]), particularly in authentic settings such as online statistics courses and that are based on an experimental design."

Theoretical Background

"Statistics anxiety can be described as the anxiety learners experience when dealing with statistics "in any form and at any level" (Onwuegbuzie [99])."

"Statistics anxiety is reflected by the mental disorganization, worry, and tension that learners feel when they have to work with statistics content or problems (Zeidner [100])."

"There is overwhelming evidence that motivation influences learners' statistics anxiety, learning experience, and performance in statistics courses (Lavasani and others [90]; Onwuegbuzie & Seaman [101]; Onwuegbuzie & Wilson [102])."

"Despite the interactions between motivation and anxiety, there is a lack of research on how theories of motivation can be used to develop learning analytics tools, such as learning analytics dashboards, to support learners' motivation (Bodily & Verbert [97]; Lonn and others [103]) and learning performance in anxiety-inducive courses such as statistics."

Motivation

"From a traditional perspective of achievement goal theory, learners can adopt two types of goal orientations: performance or mastery goals."

"A mastery goal orientation focuses on individual progress and learning, while a performance goal orientation focuses on relative ability and comparison with other learners (Ames [104]; Pintrich [105])."

"This approach is referred to as the multiple goal perspective in contrast to the traditional normative goal theory, which proposes that mastery and performance goal orientations are mutually exclusive (Harackiewicz and others [106])."

"The learning analytics dashboards used in this study subscribe to the multiple goal perspective, which combines performance and mastery goal orientation elements."

"The need for more evidence-based research is critical as educators design, implement, and test learning analytics dashboards and their influence on students' motivation, interest, self-regulation, awareness, self-efficacy, and learning outcomes (Bodily & Verbert [97]; Jivet and others [98]; Valle and others, in press)."

Conceptual Framework

"Mastery goal orientation is reflected in the emphasis on the content to be learned, specifically the names of topics shown in the graph's x-axes, which are used in both the predictive and descriptive dashboards."

"Performance goal orientation is reflected in the design of the performance feedback presented to learners, which is either norm-referenced in the descriptive dashboard or self-referenced in the predictive dashboard."

"RQ4 – How do learners perceive their learning experience, motivation, and statistics anxiety while using predictive or descriptive learning analytics dashboards?"

Materials and Methods

"This study used quantitative and qualitative data (Creswell and others [107]; Johnson & Onwuegbuzie [108]) to provide a more comprehensive account of the use of predictive and descriptive dashboards and their influence on learners' statistics anxiety, motivation, and learning performance outcomes in an online statistics course."

"Following the IRB approved protocol, we collected the following quantitative information: number of dashboard views, prior content knowledge scores, prior experience with programming languages, quiz results, final grades, and pre and post results (beginning and end of the semester, respectively) from the Motivation Strategies for Learning Questionnaire scale – MSLQ (Pintrich and others [109, 110]) and from the Statistical Anxiety Rating Scale – STARS (Cruise and others [111])."

"The predictive and descriptive learning analytics dashboards used as the intervention in this study were developed using R software (version R 3.5.2 – ©2004–2016) and Shiny™, a web development framework that enables the visual display of results from R. The design of the dashboards was based on the multiple goal orientation perspective (Harackiewicz and others [106]), where elements from the performance approach (e.g., focus on predicted letter grades) and the mastery goal approach (e.g., focus on the content or skills to be learned) orientations are combined to support learners' motivation."

Results

"To what extent does the use of a predictive or descriptive learning analytics dashboard in an online statistics course influence learners' motivation?"

"To what extent does the use of a predictive or descriptive learning analytics dashboard in an online statistics course influence learners' statistics anxiety?"

"The model indicated that learners in the predictive dashboard had on average significantly lower interpretation anxiety scores at the end of the course than those in the control group."

"To what extent does the use of a predictive or descriptive learning analytics dashboard in an online statistics course influence learners' performance?"

"How do learners perceive their learning experience, motivation, and statistics anxiety while using the predictive or descriptive learning analytics dashboards?"

"Focusing on what I'm learning Although we did not find robust results regarding differences in learning outcomes between groups (control, descriptive, and predictive dashboards), most participants reflected on their performance as they described the course and its resources."

Discussion

"We describe an experimental study conducted over two semesters based on the multiple goal orientation theory to illuminate how and whether mastery and performance goal orientation approaches embedded in learner-centered learning analytics dashboards can influence learners' motivation, anxiety, and learning performance."

"Both the descriptive and predictive dashboard reduced worth of statistics anxiety in comparison to the control group, indicating that dashboard designs that combine mastery goal and performance goal orientation elements can be employed to reduce learners' negative attitudes towards statistics, particularly among learners who start the course with higher levels of worth of statistics anxiety."

"Despite the combination of mastery and performance goal orientation elements in the design of the predictive and descriptive dashboards and the influence they had on learners' intrinsic goal orientation, interpretation anxiety, and worth of statistics anxiety, we did not find statistically significant differences between groups in relation to final exam or total quiz scores."

Conclusions
"Our results suggest that the combination of performance and mastery learning components used in the predictive dashboard had a positive influence on learners' interpretation anxiety by reducing it in comparison to the control group."

"The predictive dashboard also supported intrinsic motivation for those who started the course with higher levels of intrinsic motivation; however, it had a detrimental influence on motivation among less motivated learners, which challenges the assumption that self-referenced approaches are necessarily more motivational as suggested by previous studies (Jonathan and others [112]; Martin & Elliot [113])."

"Both the descriptive and predictive dashboards reduced negative attitudes towards statistics (worth of statistics anxiety), especially among learners who started the course with higher levels of aversion towards statistics."

Implications and Future Directions
"Predictive dashboards are more likely to reduce interpretation anxiety possibly because of the self-referenced elements used, which contextualized learners' individual progression in the course without comparisons with their class."

"Future research could explore whether the inclusion of other theory-based elements such as motivation scaffolding prompts would also help reduce learners' interpretation anxiety."

"Predictive and descriptive dashboards combining performance and mastery goal orientation elements can be used to reduce learners' negative attitudes towards statistics courses."

"We hope that the theory-based approach we used in the conceptualization, development, and implementation of learner-centered learning analytics dashboards will motivate other researchers to explore the potential of the multiple goal orientation perspective to support learner's motivation in anxiety-driven courses such as statistics (Dela Rosa & Bernardo [114]; Harackiewicz and others [106]; Pintrich [115])."

Motivation for Creativity in Architectural Design and Engineering Design Students: Implications for Design Education [157]

This is a machine-generated summary of:

Casakin, Hernan; Kreitler, Shulamith: Motivation for creativity in architectural design and engineering design students: implications for design education [157].

Published in: International Journal of Technology and Design Education (2009).

Link to original: https://doi.org/10.1007/s10798-009-9103-y.

If you want to cite the papers, please refer to the original.

For technical reasons we could not place the page where the original quote is coming from.

Abstract-Summary
"The goals were to assess motivation for creativity in architectural design and engineering design students based on the Cognitive Orientation theory which defines motivation as a function of a set of belief types, themes, and groupings identified as relevant for the development of creativity."

"Differences between the two groups of students were expected in some of the scores of the belief types, themes, and groupings."

"Participants were 112 students (52 from architecture, and 60 from engineering) who were administered the questionnaire of the Cognitive Orientation of Creativity."

"Major characteristics of the architectural design students were an emphasis on the inner world, inner-directedness, and a development of the self, and of the engineering design students being receptive to the environment, and demanding from oneself despite potential difficulties."

Introduction
"The development of creativity is a primary concern of design problem solving (e.g., Christiaans [116]; Cross [117]; Gero [118]), and of design education (Casakin and others [119]; Sim and Duffy [120])."

"It is the major requisite for solving of design problems in disciplines like architecture and engineering, in which students have to make use of their creative abilities."

"Based on design studio experience, it was observed that more often than not teachers and students assess the creativity of the design outcomes with little concern for the motivational aspects that have led to them."

"Gaining a better insight in regard to the main aspects characterizing motivation for creativity in the groups of students from these two related but at the same time differently oriented domains is the main challenge of this investigation, whose implications are expected to be of particular relevance for design education."

Motivation as a Driving Force for Design Creativity and Design Education

"In architectural design, Kreitler and Casakin [121] studied the role of motivational disposition for creativity anchored in the Cognitive Orientation theory."

"Despite the strong impact that motivation has in creative design domains such as architecture and engineering, designers, i.e., teachers and students alike, are mostly unaware of this impact and rarely consider motivation when they assess creativity."

"In regard to design creativity, the relation between the outputs and the motivational disposition is not one-to-one so that the same motivational disposition may be responsible for various outputs in domains, such as curiosity (Kreitler and Kreitler [122]) or planning (Kreitler and Kreitler [123])."

"Of the downgrading or overlooking of motivation, observation shows that teachers and students dealing with design creativity tend to concentrate more on the mechanics of creativity than on the motor that triggers the stimulus necessary for getting involved in creative acts."

Cognitive Orientation Theory and Method: Motivation for Creativity

"The conceptual and methodological approach founded on the cognitive orientation (CO) theory (Kreitler and Kreitler [124]; Kreitler [125]) has been applied for studying motivation for creativity in design, in the domains of architecture and engineering."

"The motivational disposition is assessed by means of a CO questionnaire which includes statements referring to the themes corresponding to the four belief types, that the respondent is requested to accept or reject."

"The validity of CO measures of motivational tendencies was proved in previous studies assessing CO motivations concerning a broad range of different behaviors and domains of application (Kreitler and others [126]; Kreitler and Kreitler [123, 127, 128, 122])."

"We expected that there would be differences in motivation for design creativity between architecture and engineering students, and that the differences would be manifested in the structure of the motivation—namely, the belief types and themes of the motivation—rather than in the overall level of creativity."

Method

"The participants were 112 students of two groups."

"The first group included 52 students of architectural design (35 men and 17 women), from a school of architecture in a university center in central Israel."

"They were recruited randomly from the population of students in the first to fifth years of studies."

"The second group included 60 students of engineering design (45 men, 15 women) from a major university in central Israel."

"All participants were administered the Cognitive Orientation Questionnaire of Creativity (COQ-CR; Kreitler and Kreitler [129]), which was designed to provide a measure of the strength and structure of motivation for creativity."

"The COQ-CR yielded four main scores for each of the belief types, and eleven scores for each of the groupings of themes."

"The COQ-CR was administered to both architectural and engineering students, either in small group sessions or individually."

Results

"It shows that the first and strongest factor (accounting for 48.4% of the variance) found in the group of architectural design students is saturated on the following groupings: Self development; emphasizing one's uniqueness; self-expression; demanding from oneself; contributing to society, in addition to emphasis on the inner world."

"Architectural students scored higher than engineering students in the following groupings: (1) self development [investing in oneself, promoting oneself, guarding oneself]; (2) emphasis on the inner world [identifying, knowing, developing and expressing thinking, feeling and imagination]; (3) inner-directedness [emphasis on one's desires, will and decision, self confidence in one's ability to succeed] (5) underscoring one's uniqueness [emphasis on oneself as an individual unique in one's talents and one's way of perceiving, behaving and being, not necessarily accompanied by nonconformity]; (8) functioning under conditions of uncertainty [readiness to act under conditions of uncertainty concerning the results, with no control over the circumstances, which may resemble risk-taking]; (10) self expression [concern with using one's talents and expressing oneself and one's feelings with authenticity]; (11) non-functionality [readiness to act even if functionality is not clearly evident from the start]."

Discussion

"This difference suggests that functionality plays a different role in the case of the students of the two groups: in the case of engineers, motivation for creativity is characterized by being pro-functionality (i.e., in favor of functional concerns to support the design activity), but in the architectural students motivation for creativity is promoted rather by adopting a non-functional approach (i.e., not caring about functional concerns as a main issue for the design activity)."

"Further findings based on the comparison between the two groups of students indicated differences in two of the four belief types supporting creativity: the scores of beliefs about goals and beliefs about self were higher in the group of architectural students than in the group of engineering students."

"It appears that in contrast to the engineering group, the motivation for creativity in the group of architectural students is guided primarily by beliefs that concern oneself: the actual self image and in addition the wishful or ideal self image as reflected in one's personal goals."

Conclusions
"The investigation dealt with the study of motivation for creativity in architectural and engineering students."

"A measure of such complexity allows for multiple combinations of themes to structure motivation for creativity in students with different profiles."

"Considering domain differences in engineering and architecture, and individual differences in the personality and cognitive background of students, applying the CO theory, which is basically and originally a theoretical framework for changing behavioral manifestations, can contribute to extending the potential for promoting individual creativity in each particular domain."

"Findings derived from the current investigation provide teachers with a concrete instrument for promoting motivation, considered as a prime generator or motor of creativity for engineering and architectural design education."

"This insight may enable educators to promote creativity of students in the different domains so that they may best serve their profession by manifesting the creativity which is best adapted to the requirements of architecture or engineering."

Predicting Research Productivity in STEM Faculty: The Role of Self-determined Motivation [158]

This is a machine-generated summary of:

Stupnisky, Robert H.; Larivière, Vincent; Hall, Nathan C.; Omojiba, Oluwamakinde: Predicting Research Productivity in STEM Faculty: The Role of Self-determined Motivation [158].

Published in: Research in Higher Education (2022).

Link to original: https://doi.org/10.1007/s11162-022-09718-3.

If you want to cite the papers, please refer to the original.

For technical reasons we could not place the page where the original quote is coming from.

Abstract-Summary

"How are university faculty members in STEM disciplines motivated to conduct research, and how does motivation predict their success?"

"The current study assessed how multiple types of self-determined motivation predict research productivity in a sample of 651 faculty from 10 US institutions."

"Using structural equation modeling, the basic psychological needs of autonomy and competence predicted autonomous motivation (enjoyment, value) that, in turn, was the strongest predictor of self-reported research productivity."

Does Self-determined Motivation Predict STEM Faculty Research Success and Productivity?

"Following from this research, Stupnisky and others [132] surveyed 105 pre-tenure faculty from two Midwestern doctoral US universities and found faculty who perceived their basic psychological needs of autonomy and competence were being satisfied also tended to report higher levels of intrinsic motivation that, in turn, predicted greater perceived and expected success in research."

"Stupnisky and others [131] studied 1846 US faculty from 19 institutions and found satisfaction of autonomy and competence needs to predict autonomous motivation (a combination of enjoyment and value) that was positively related to self-reported research productivity, beyond demographic and position details."

"To address gaps in prior research, this study examined all SDT basic psychological needs and motivation types, collected a large sample of STEM faculty across multiple disciplines and institutions, utilized multi-item measures of constructs including self-reported research success and bibliometric indictors of research productivity, and employed robust statistical analyses."

"How does motivation predict faculty research success (self-reported, bibliometric)?"

Methods

"Four social-environmental factors reported by faculty as important to their success were measured using four-item scales from Stupnisky and others (133, 131; 1 = Strongly disagree to 5 = Strongly agree): personal balance (e.g., "I have been able to balance my work and home/personal life"), professional balance ("I have been able to balance my teaching, research, and service work"), clear expectations ("I have come to understand what the expectations are for me at work"), and collegiality ("There is someone in my department who I can ask for advice and guidance")."

"Whereas typical linear regression models assume a normal distribution of the outcome variable (e.g., self-reported research success on a Likert scale), count variables are often positively skewed yielding a Poisson distribution (i.e., most values are relatively low, few are high)."

Results

"Faculty reported high mean levels of autonomy, competence, and relatedness for research, and in turn, reported relatively high levels of autonomous motivation for research compared to negative introjected, external, or amotivation."

"Full professors had the highest levels of autonomy, competence, autonomous motivation, and perceived success, as well as the lowest negative introjected and external motivation."

"Step 1 of the hierarchical negative binomial regression included all success covariates and basic psychological needs, with results showing research percent (contract) to be a positive predictor of publications and citations, and competence and relatedness to predict publications."

"In Step 2c, all motivation types were entered simultaneously, and although the predictive significance of autonomous and amotivation decreased, they remained among the most important predictors of publications and citations."

"We also observed evidence of mediation as autonomy was positively correlated with publications and citations but became nonsignificant or significantly negative with the inclusion of autonomous motivation in the models."

Discussion

"This study further improved upon the past studies on faculty motivation by (1) examining motivation beyond intrinsic (i.e., autonomous) motivation to consider the roles of introjected and external motivation, as well as amotivation; (2) using a large, multi-institution sample of faculty members across a variety of STEM disciplines; (3) utilizing multi-item scales and latent variables in structural equation models to improve the reliability and validity of findings; (4) measuring research success with self-report and bibliometrics; and (5) controlling for other common, non-motivation variables related to faculty success."

"The only factor that was comparatively as impactful as autonomous motivation for faculty research success in our latent variable analyses was having good professional balance (i.e., between teaching, research, and service)."

"The current study results should be informative to higher education institutions, particularly those striving to increase scholarly productivity, as to specific strengths and deficits in faculty motivation for research that contribute to measurable gains in research activity."

How Basic Psychological Needs and Motivation Affect Vitality and Lifelong Learning Adaptability of Pharmacists: A Structural Equation Model [159]

This is a machine-generated summary of:

Tjin A Tsoi, Sharon L. N. M.; de Boer, Anthonius; Croiset, Gerda; Koster, Andries S.; van der Burgt, Stéphanie; Kusurkar, Rashmi A.: How basic psychological needs and motivation affect vitality and lifelong learning adaptability of pharmacists: a structural equation model [159].

Published in: Advances in Health Sciences Education (2018).

Link to original: https://doi.org/10.1007/s10459-018-9812-7.

If you want to cite the papers, please refer to the original.

For technical reasons we could not place the page where the original quote is coming from.

Abstract-Summary
"Continuing education (CE) and continuing professional development (CPD) are needed to secure safe and good quality healthcare."

"The aim of the study was to investigate the hypothesized associations and their directions between pharmacists' basic psychological needs in CE, their academic motivation, well-being, learning outcomes."

"Data were collected through four questionnaires measuring: academic motivation, basic psychological needs (BPN), vitality and lifelong learning adaptability of pharmacists in the CE/CPD learning context."

"Frustration of BPN is positively directly related to the less desirable type of academic motivation, controlled motivation (0.88) and negatively directly related to vitality (− 1.61) and negatively indirectly related to learning outcomes in CE."

"Further research should be conducted to discover how we can prevent these needs from being frustrated in order to design a motivating, vitalizing and sustainable CE/CPD system for pharmacists and other healthcare professionals."

"Basic psychological needs are very important predictors for well-being and learning outcomes."

"Further research should be conducted to discover how we can prevent these needs from being frustrated in order to design a motivating, vitalizing and sustainable CE/CPD system for pharmacists and other healthcare professionals."

Introduction

"Most of these studies in the education domain were conducted in structured settings such as high schools and universities and emphasize the relation between motivation and learning outcomes like study strategies, study performance and students' well-being (Vansteenkiste and others [134]; Guay and others [135]; Kusurkar and others [136])."

"We were interested in studying the motivation of healthcare professionals in CE and CPD and its effect on their well-being and learning outcomes."

"According to SDT, tools for re-energizing pharmacists can be discovered by understanding how BPN and motivation in CE/CPD relate to well-being and learning outcomes."

"The objective of the present study is to investigate a complete route (consisting of an upper "bright" and a lower "dark" side) from predictors to outcome with the possible associations between pharmacists' BPN in CE, their academic motivation, well-being and learning outcomes."

"Frustrating BPN is associated with a decrease in AM (directly), increase in CM (directly), decrease in well-being (directly and indirectly through Academic Motivation) and decrease in learning outcomes (directly and indirectly)."

Methods

"This study was conducted among 425 Dutch pharmacists working in a community or hospital pharmacy that participated in CE courses organized by the Netherlands Centre for Post-Academic Education in Pharmacy."

"The regulations for CPs were revised in 2015, and Dutch pharmacists are now required to participate in CE according to the CanMEDS competency framework."

"The translated questionnaires were assessed by a group of five experts in the field consisting of pharmacists and educators in pharmacy."

"The different dimensions of academic motivation like identified and external regulation were measured by items such as "because I believe CE will improve my competencies as a professional" and "to prove to myself that I am capable of succeeding in CE", respectively."

"To assess basic psychological need satisfaction (BPNsat) and frustration (BPN-frus) of the pharmacists in the context of CE, we used a 6-factor 24-item scale that was validated across four cultures (Chen and others [137])."

Results
"Significant negative correlations were found between BPNsat and BPNfrus (− 0.373, $p < 0.01$), between CM and Vitality (− 0.225, $p < 0.05$) and between BPNfrus and Vitality (− 0.610, $p < 0.01$)."

"Statistically significant positive correlations were found between BPNsat and AM (0.234, $p < 0.05$), BPNsat and Vitality (0.398, $p < 0.05$), BPNsat and LLA (0.297, $p < 0.01$)."

"The goodness of fit (step 4) was determined based on the model fit indices for this model: $\chi^2 = 3152.05$ (df = 1866.00, p = 0.00), CFI = 0.65, TLI = 0.63, RMSEA = 0.08 and SRMR = 0.11."

"It did, however, directly affect the CM significantly positively (0.88), directly affect vitality significantly negatively (− 1.61) and indirectly affect LLA through vitality negatively."

Discussion
"The measurements (AM, CM, BPNfrus, BPNsat, vitality and LLA) used in this study were collected through validated questionnaires originating from SDT research and research in professionalism assessment among pharmacists."

"Our findings suggest that frustration of BPN could be an important negative predictor for vitality/well-being (directly) and LLA (indirectly) in pharmacists."

"Satisfying or frustrating BPN as possible predictors for psychological functioning and learning outcomes are supported by earlier studies conducted in different contexts with workers, athletes and across different cultures within SDT (Chen and others [137]; Ryan and Deci [138]; Ryan and others [139]; Bartholomew and others [140])."

"Our findings suggest that preventing BPN frustration is important for the well-being and learning outcomes of pharmacists."

"Because our findings suggest that BPN play an important role as predictors for psychological functioning and good learning outcomes, questions for future research are: "What are the pharmacists' needs for CE?"."

Conclusion
"Our findings, supported by SDT literature, suggest that BPN are important predictors for well-being and learning outcomes, because frustration of BPN is negatively related to vitality (directly) and to learning outcomes in CE (indirectly)."

"Further research should be conducted to discover how we can prevent these needs from being frustrated in order to design a motivating, vitalizing and sustainable CE/CPD system for pharmacists and other healthcare professionals."

Bibliography

1. Bonanno, G. A. (2001). Emotion self-regulation. In G. A. Bonanno & T. J. Mayne (Eds.), Emotions: Current issues and future directions (pp. 251–285). New York, NY: Guilford.
2. Larsen, R. J. (2000). Toward a science of mood regulation. Psychological Inquiry, 11, 129–141.
3. Larsen, R. J., & Diener, E. (1985). A multitrait-multimethod examination of affect structure: Hedonic level and emotional intensity. Personality and Individual Differences, 6, 631–636.
4. Larsen, R. J., & Diener, E. (1992). Promises and problems with the circumplex model of emotion. Review of Personality and Social Psychology, 13, 25–59.
5. Russell, J. A. (1980). A circumplex model of affect. Journal of Personality and Social Psychology, 39(6), 1161–1178.
6. Rusting, C. L., & Larsen, R. J. (1995). Moods as sources of stimulation: Relationships between personality and desired mood states. Personality and Individual Differences, 18, 321–329.
7. Kampfe, N., & Mitte, K. (2009). What you wish is what you get? The meaning of individual variability in desired affect and affective discrepancy. Journal of Research in Personality, 43, 409–418.
8. Tamir, M. (2005). Don't worry be happy? Neuroticism, trait-consistent affect regulation, and performance. Journal of Personality and Social Psychology, 89, 449–461.
9. Tamir, M., Chiu, C. Y., & Gross, J. J. (2007). Business or pleasure? Utilitarian versus hedonic considerations in emotion regulation. Emotion, 7, 546–554.
10. Nelissen, R. M. A., Dijker, A. J. M., & de Vries, N. K. (2007). Emotions and goals: Assessing relations between values and emotions. Cognition and Emotion, 21, 902–911.
11. Gross, J. J. (1998). The emerging field of emotion regulation: An integrative review. Review of General Psychology, 2, 271–299.
12. Dewey, J. (2005). Art as experience. New York: Berkley Publishing Group (Original work published 1934).
13. Parrish, P. (2009). Aesthetic principles for instructional design. Educational Technology Research and Development, 57(4), 511–582. https://doi.org/10.1007/s11423-007-9060-7 .
14. Berleant, A. (2010). Art and Engagement (Reprint edition). Temple University Press.
15. Csikszentmihalyi, M., & Robinson, R. E. (1991). The art of seeing: An interpretation of the aesthetic encounter. Los Angeles: J. Paul Getty Museum.
16. Mayer, R. E. (2009). Multimedia learning (2nd ed.). Cambridge University Press. https://doi.org/10.1017/CBO9780511811678
17. Keller, J. M., & Burkman, E. (1993). Motivation principles. In M. L. Fleming & W. H. Levie (Eds.), Instructional message design: Principles from the behavioral and cognitive sciences (2nd ed., pp. 3–53). Englewood Cliffs: Educational Technology Publications.
18. Rieber, L. P. (1994). Computers, graphics & learning. Madison, WI: Brown & Benchmark.
19. Parrish, P. (2005). Embracing the aesthetics of instructional design. Educational Technology, 45(2), 16–25.
20. Uhrmacher, P. B. (2009). Toward a theory of aesthetic learning experiences. Curriculum Inquiry, 39(5), 613–636. https://doi.org/10.1111/j.1467-873X.2009.00462.x .
21. Winn, W. (1993). Perception principles. In M. L. Fleming & W. H. Levie (Eds.), Instructional message design: Principles from the behavioral and cognitive sciences (2nd ed., pp. 55–126). Englewood Cliffs, NJ: Educational Technology Publications.

22. Guglielmino, L. M. (1997). Reliability and validity of the self-directed learning readiness scale and the learning preference assessment. Expanding Horizons in Self-directed Learning, 209–222.
23. Choi, I. (2014). The study of self-regulated learning related variables in web-based blended learning: with a focus on school adjustment behavior, academic burnout, self-determination, and participation in e-learning. Open Education Research, 22(2), 237–260.
24. Kim, H., & Lee, J. (2014). Effects of cyber universities career maturity on academic stress: mediator effects of self-directed learning readiness. The Korea Contents Association, 14(11), 486–496.
25. Wang, L., Law, K. S., Zhang, M. J., Li, Y. N., & Liang, Y. (2019). It's mine! Psychological ownership of one's job explains positive and negative workplace outcomes of job engagement. Journal of Applied Psychology, 104 No(2), 229.
26. Peng, H., & Pierce, J. (2015). Job-and organization-based psychological ownership: Relationship and outcomes. Journal of Managerial Psychology, 30(2), 151–168. https://doi.org/10.1108/JMP-07-2012-0201
27. Gagné, M., Tian, A. W., Soo, C., Zhang, B., Ho, K. S. B., & Hosszu, K. (2019). Different motivations for knowledge sharing and hiding: The role of motivating work design. Journal of Organizational Behavior, 40 No(7), 783–799.
28. Pan, W., Zhang, Q., Teo, T. S., & Lim, V. K. (2018). The dark triad and knowledge hiding. International Journal of Information Management, 42, 36–48.
29. Hameed, Z., Khan, I. U., Sheikh, Z., Islam, T., Rasheed, M. I., & Naeem, R. M. (2019). Organizational justice and knowledge sharing behavior: The role of psychological ownership and perceived organizational support. Personnel Review, 48(3), 748–773. https://doi.org/10.1108/PR-07-2017-0217
30. Xia, Q., Yan, S., Zhang, Y., & Chen, B. (2019). The curvilinear relationship between knowledge leadership and knowledge hiding: the moderating role of psychological ownership. Leadership & Organization Development Journal, 40(6), 669–683. https://doi.org/10.1108/LODJ-10-2018-0362
31. Connelly, C. E., Zweig, D., Webster, J., & Trougakos, J. P. (2012). Knowledge hiding in organizations. Journal of Organizational Behavior, 33 No(1), 64–88.
32. Zhao, H., Xia, Q., He, P., Sheard, G., & Wan, P. (2016). Workplace ostracism and knowledge hiding in service organizations. International Journal of Hospitality Management, 59, 84–94.
33. Harman, D. (1967). A single factor test of common method variance. Journal of Psychology, 35(1967), 359–378.
34. Vollmeyer, R., & Rheinberg, F. (2000). Does motivation affect performance via persistence? Learning and Instruction, 10, 293–309.
35. Azevedo, R., & Cromley, J. G. (2004). Does training on self-regulated learning facilitate students' learning with hypermedia? Journal of Educational Psychology, 96(3), 523–535.
36. Azevedo, R., Cromley, J. G., & Seibert, D. (2004a). Does adaptive scaffolding facilitate students' ability to regulate their learning with hypermedia. Contemporary Educational Psychology, 29, 344–370.
37. Jonassen, D., Hartley, J., & Trueman, M. (1986). The effects of learner-generated versus text-provided headings on immediate and delayed recall and comprehension: An exploratory study. Human Learning: Journal of Practical Research & Application, 5(3), 139–150.
38. Hartley, J., & Sydes, M. (1997). Are structured abstracts easier to read than traditional ones? Journal of Research in Reading, 20(2), 122–136.
39. Hannafin, M., Land, S., & Oliver, K. (1999). Open learning environments: Foundation, methods, and models. In C. M. Reigeluth (Ed.), Instructional-design theories and models: A new paradigm of instructional theory (Vol. II, pp. 115–140). Mahwah, NJ: Erlbaum.
40. Zimmerman, B. J. (2008). Investigating self-regulation and motivation: Historical background, methodological developments, and future prospects. American Educational Research Journal, 45(1), 166–183.

41. Boekaerts, M., & Cascallar, E. (2006). How far have we moved toward the integration of theory and practice in self-regulation? Educational Psychology Review, 18, 199–210. https://doi.org/10.1007/s10648-006-9013-4 .
42. Velayutham, S., Aldridge, J. M., & Fraser, B. J. (2011). Development and validation of an instrument to measure students' motivation and self-regulation in science learning. International Journal of Science Education, 15, 2159–2179.
43. Stake, J. E., & Mares, K. R. (2001). Science enrichment programs for gifted high school girls and boys: Predictors of program impact on science confidence and motivation. Journal of Research in Science Teaching, 38, 1065–1088. https://doi.org/10.1002/tea.10001 .
44. Shim, S. S., Ryan, A. M., & Anderson, C. J. (2008). Achievement goals and achievement during early adolescence: Examining time-varying predictor and outcome variables in growth-curve analysis. Journal of Educational Psychology, 100, 655–671. https://doi.org/10.1037/0022-0663.100.3.655 .
45. Shaikh, M. R. (2013). Socio-demographic profile and career choice of science students: A study of science students of University of Pune, Pune India. International Journal on Integrating Technology in Education, 2(1), 1–8. https://doi.org/10.5121/ijite . 2013.2101.
46. Poehner, M. E. (2009). Group dynamic assessment: mediation for the L2 classroom. TESOL Quarterly, 43(3), 471–491. https://doi.org/10.1002/j.1545-7249.2009.tb00245.x
47. Deci, E. L., & Ryan, R. M. (2008). Self-determination theory: A macrotheory of human motivation, development, and health. Canadian Psychology/Psychologie Canadienne, 49, 182–185. https://doi.org/10.1037/a0012801 .
48. Gardner, R. C. (1985). Social psychology and second language learning: the role of attitudes and motivation. Edward Arnold.
49. Ritonga, M., Farhangi, F., Ajanil, B., & Farid Khafaga, A. (2022). Interventionist vs. interactionist models of dynamic assessment (DA) in the EFL classroom: impacts on speaking accuracy and fluency (SAF), foreign language classroom anxiety (FLCA), and foreign language learning motivation (FLLM). Language Testing in Asia, 12(1), 1–21. https://doi.org/10.1186/s40468-022-00195-0
50. Chen, C. H., Koong, C. S., & Liao, C. (2022). Influences of integrating dynamic assessment into a speech recognition learning design to support students' English speaking skills, learning anxiety and cognitive load. Educational Technology & Society, 25(1), 1–14. https://www.jstor.org/stable/48647026 .
51. Dörnyei, Z., Csizér, K., & Nemeth, N. (2006). Motivation, language attitudes and globalization: A Hungarian perspective. Clevedon: Multilingual Matters.
52. Ryan, R. M. (1995). Psychological needs and the facilitation of integrative processes. Journal of Personality, 63, 397–427.
53. Deci, E. L., & Ryan, R. M. (2000). The" what" and" why" of goal pursuits: Human needs and the self-determination of behavior. Psychological Inquiry, 11, 227–268.
54. Chemolli, E., & Gagné, M. (2014). Evidence against the continuum structure underlying motivation measures derived from self-determination theory. Psychological Assessment, 26, 575–585.
55. Tangney, J. P., Baumeister, R. F., & Boone, A. L. (2004). High self-control predicts good adjustment, less pathology, better grades, and interpersonal success. Journal of Personality, 72(2), 271–322. https://doi.org/10.1111/j.0022-3506.2004.00263.x
56. Kuhl, J., & Fuhrmann, A. (1998). Decomposing self-regulation and self-control: The volitional components checklist. In J. Heckhausen & C. Dweck (Eds.), Life span perspectives on motivation and control (pp. 15–49). Mahwah: Erlbaum.
57. Kuhl, J., & Beckmann, J. (1994). Alienation: Ignoring one's preferences. In J. Kuhl & J. Beckmann (Eds.), Volition and personality: Action versus state orientation (pp. 375–390). Göttingen: Hogrefe.
58. Ryan, R. M., Koestner, R., & Deci, E. L. (1991). Ego-involved persistence: When free-choice behavior is not intrinsically motivated. Motivation and Emotion, 15, 185–205.

59. Baumeister, R. F., Bratslavsky, E., Muraven, M., & Tice, D. M. (1998). Ego depletion: Is the active self a limited resource? Journal of Personality and Social Psychology, 74, 1252–1265. https://doi.org/10.1037/0022-3514.74.5.1252 .
60. Milyavskaya, M., & Inzlicht, M. (2017). What's so great about self-control? Examining the importance of effortful self-control and temptation in predicting real-life depletion and goal attainment. Social Psychological and Personality Science, 8, 603–611. https://doi.org/10.1177/1948550616679237 .
61. Muraven, M., Tice, D. M., & Baumeister, R. F. (1998). Self-control as a limited resource: Regulatory depletion patterns. Journal of Personality and Social Psychology, 74, 774–789. https://doi.org/10.1037/0022-3514.74.3.774 .
62. Carter, E. C., Kofler, L. M., Forster, D. E., & McCullough, M. E. (2015). A series of meta-analytic tests of the depletion effect: Self-control does not seem to rely on a limited resource. Journal of Experimental Psychology: General, 144, 796–815. https://doi.org/10.1037/xge0000083 .
63. Sokolowski, K. (1997). Sequentielle und Imperative Konzepte des Willens [Sequential and imperative theories of volition]. Psychologische Beiträge, 39, 346–369.
64. Baumeister, R. F., Gailliot, M., DeWall, C. N., & Oaten, M. (2006). Self-regulation and personality: How interventions increase regulatory success, and how depletion moderates the effects of traits on behavior. Journal of Personality, 74, 1773–1801. https://doi.org/10.1111/j.1467-6494.2006.00428.x .
65. Muraven, M. (2010). Building self-control strength: Practicing self-control leads to improved self-control performance. Journal of Experimental Social Psychology, 46, 465–468. https://doi.org/10.1016/j.jesp.2009.12.011 .
66. Grund, A., & Fries, S. (2018). Understanding procrastination: A motivational approach. Personality and Individual Differences, 121, 120–130. https://doi.org/10.1016/j.paid.2017.09.035 .
67. Deci, E. L., & Ryan, R. M. (2000). The "what" and "why" of goal pursuits: Human needs and the self-determination of behavior. Psychological Inquiry, 11, 227–268. https://doi.org/10.1207/S15327965PLI1104_01 .
68. Rheinberg, F., & Engeser, S. (2010). Motive training and motivational competence. In O. C. Schultheiss & J. C. Brunstein (Eds.), Implicit Motives (pp. 510–548). Oxford: University Press.
69. Grund, A., & Senker, K. (2018). Motivational foundations of self-control and mindfulness and their role in study–leisure conflicts. Manuscript submitted for publication.
70. Hofmann, W., Luhmann, M., Fisher, R. R., Vohs, K. D., & Baumeister, R. F. (2014). Yes, but are they happy? Effects of trait self-control on affective well-being and life satisfaction. Journal of Personality, 82(4), 265–277. https://doi.org/10.1111/jopy.12050
71. Grund, A., Grunschel, C., Bruhn, D., & Fries, S. (2015). Torn between want and should: An experience-sampling study on motivational conflict, well-being, self-control, and mindfulness. Motivation and Emotion, 39, 506–520.
72. Csikszentmihalyi, M. (1990). Flow: The psychology of optimal experience. New York, NY: Harper and Row. https://doi.org/10.1007/978-94-017-9088-8_14 .
73. Csikszentmihalyi, M. (1999). Flow: The psychology of optimal Experience. New York, NY: Harper and Row. https://doi.org/10.5860/CHOICE.28-0597 .
74. Csikszentmihalyi, M. (1996). Creativity: Flow and the psychology of discovery and invention. New York, NY: Harper and Collins. https://doi.org/10.1177/001698629704100309 .
75. Engeser, S. (2012). Advances in flow research. New York, NY: Springer. https://doi.org/10.1007/978-1-4614-2359-1 .
76. Wong, S. L., & Hodson, D. (2008). From the horse's mouth: What scientists say about scientific investigation and scientific knowledge. Science Education, 93(1), 109–130. https://doi.org/10.1002/sce.20290 .
77. Schwartz, R., & Lederman, N. (2008). What scientists say: Scientists' views of nature of science and relation to science context. International Journal of Science Education, 30(6), 727–771. https://doi.org/10.1080/09500690701225801 .

78. Gee, J. P. (2005). An introduction to discourse analysis theory and method. New York, NY: Routledge. https://doi.org/10.4324/9781315819679 .
79. Chang, C.-Y., & Song-ling, M. (1999). Comparison of Taiwan science students' outcomes with inquiry-group versus traditional instruction. Journal of Educational Research, 92(6), 1–12. https://doi.org/10.1080/00220679909597617 .
80. Goldin, G. A. (2017). Mathematical creativity and giftedness: Perspectives in response. ZDM Mathematics Education, 49(1), 147–157.
81. Caprara, G. V., & Gerbino, M. (2001). Affective perceived self-efficacy: The capacity to regulate negative affect and to express positive affect. In G. V. Caprara (Ed.), Self-efficacy assessment, (pp. 35–50). Trento, Italy: Edizioni Erickson.
82. MacIntyre, P. D., & Vincze, L. (2017). Positive and negative emotions underlie motivation for L2 learning. Studies in Second Language Learning and Teaching, 7, 61–88.
83. Bråten, I., & Olaussen, B. S. (2005). Profiling individual differences in student motivation: A longitudinal cluster-analytic study in different academic contexts. Contemporary Educational Psychology, 30(3), 359–396. https://doi.org/10.1016/j.cedpsych.2005.01.003 .
84. Csizer, K., & Dornyei, Z. (2005). Language learners' motivational profiles and their motivated learning behavior. Language Learning, 55(4), 613–659. https://doi.org/10.1111/j.0023-8333.2005.00319.x .
85. Ratelle, C. F., Guay, F., Vallerand, R. J., Larose, S., & Senécal, C. (2007). Autonomous, controlled, and amotivated types of academic motivation: A person-oriented analysis. Journal of Educational Psychology, 99, 734–746. https://doi.org/10.1037/0022-0663.99.4.734 .
86. Ironsmith, M., Marva, J., Harju, B., & Eppler, M. (2003). Motivation and performance in college students enrolled in self-paced versus lecture-format remedial mathematics courses. Journal of Instructional Psychology, 30(4), 276–284.
87. Schiefele, U., & Csikszentmihalyi, M. (1995). Motivation and ability as factors in mathematics experience and achievement. Journal for Research in Mathematics Education, 26(2), 163–181.
88. Gross, J. J. (2008). Emotion regulation. In M. Lewis, J. M. Haviland-Jones, & L. F. Barrett (Eds.), Handbook of emotions (3rd ed., pp. 497–512). New York, NY: Guilford.
89. Elliot, A. J., & McGregor, H. A. (1999). Test anxiety and the hierarchical model of approach and avoidance achievement motivation. Journal of Personality and Social Psychology, 76(4), 628.
90. Lavasani, M. G., Weisani, M., & Shariati, F. (2014). The role of achievement goals, academic motivation in statistics anxiety: Testing a causal model. Procedia-Social and Behavioral Sciences, 114, 933–938.
91. Hsu, M. K., Wang, S. W., & Chiu, K. K. (2009). Computer attitude, statistics anxiety and self-efficacy on statistical software adoption behavior: An empirical study of online MBA learners. Computers in Human Behavior, 25(2), 412–420.
92. Zimmerman, B. J. (2008). Investigating self-regulation and motivation: Historical background, methodological developments, and future prospects. American Educational Research Journal, 45, 166–183. https://doi.org/10.3102/0002831207312909
93. Biesinger, K., & Crippen, K. (2010). The effects of feedback protocol on self-regulated learning in a web-based worked example learning environment. Computers & Education, 55(4), 1470–1482.
94. Koenka, A. C., & Anderman, E. M. (2019). Personalized feedback as a strategy for improving motivation and performance among middle school students. Middle School Journal, 50(5), 15–22.
95. Gašević, D., Dawson, S., & Siemens, G. (2015). Let's not forget: Learning analytics are about learning. TechTrends, 59(1), 64–71.
96. Park, Y., & Jo, I.-H. (2019). Factors that affect the success of learning analytics dashboards. Educational Technology Research and Development, 67(6), 1547–1571.
97. Bodily, R., & Verbert, K. (2017). Review of research on student-facing learning analytics dashboards and educational recommender systems. IEEE Transactions on Learning Technologies, 10(4), 405–418.

98. Jivet, I., Scheffel, M., Specht, M., & Drachsler, H. (2018). License to evaluate: Preparing learning analytics dashboards for educational practice. Proceedings of the 8th International Conference on Learning Analytics and Knowledge, 31–40.
99. Onwuegbuzie, A. J. (2004). Academic procrastination and statistics anxiety. Assessment and Evaluation in Higher Education, 29, 3–19. https://doi.org/10.1080/0260293042000160384 .
100. Zeidner, M. (1990). Does test anxiety bias scholastic aptitude test performance by gender and sociocultural group? Journal of Personality Assessment, 55(1–2), 145–160.
101. Onwuegbuzie, A. J., & Seaman, M. A. (1995). The effect of time constraints and statistics test anxiety on test performance in a statistics course. The Journal of Experimental Education, 63(2), 115–124.
102. Onwuegbuzie, A. J., & Wilson, V. A. (2003). Statistics Anxiety: Nature, etiology, antecedents, effects, and treatments—A comprehensive review of the literature. Teaching in Higher Education, 8(2), 195–209.
103. Lonn, S., Aguilar, S. J., & Teasley, S. D. (2015). Investigating student motivation in the context of a learning analytics intervention during a summer bridge program. Computers in Human Behavior, 47, 90–97.
104. Ames, C. (1992). Classroom goals, structures and student motivation. Journal of Educational Psychology, 84(3), 261–271.
105. Pintrich, P. R. (2000b). The role of goal orientation in self-regulated learning. In M. Boekaerts, P. R. Pintrich, & M. Zeidner (Eds.), Handbook of self-regulation (pp. 451–502). Academic Press.
106. Harackiewicz, J. M., Barron, K. E., Pintrich, P. R., Elliot, A. J., & Thrash, T. M. (2002). Revision of achievement goal theory: Necessary and illuminating. Journal of Educational Psychology, 94, 638–645.
107. Creswell, J. W., Plano Clark, V. L., Gutmann, M. L., & Hanson, W. E. (2003). Advanced mixed methods research designs. In A. Tashakkori & C. Teddlie (Eds.), Handbook of mixed methods in social and behavioral research (pp. 209–240). Sage.
108. Johnson, R. B., & Onwuegbuzie, A. J. (2004). Mixed methods research: A research paradigm whose time has come. Educational Researcher, 33(7), 14–26.
109. Pintrich, P. R., Smith, D. A. F., Garcia, T., & McKeachie, W. J. (1991). A manual for the use of the Motivated Strategies for Learning Questionnaire (MSLQ). Regents of the University of Michigan.
110. Pintrich, P. R., Smith, D. A. F., Garcia, T., & Mckeachie, W. J. (1993). Reliability and predictive validity of the motivated strategies for learning questionnaire (MSLQ). Educational and Psychological Measurement, 53(3), 801–813. https://doi.org/10.1177/0013164493053003024
111. Cruise, R. J., Cash, R. W., & Bolton, D. L. (1985). Development and validation of an instrument to measure statistical anxiety. American Statistical Association, 92–97.
112. Jonathan, C., Tan, J. P.-L., Koh, E., Caleon, I. S., & Tay, S. H. (2017). Enhancing students' critical reading fluency, engagement and self-efficacy using self-referenced learning analytics dashboard visualizations. In W. Chen, J.-C. Yang, A. F. Mohd Ayub, S. L. Wong, & A. Mitrovic (Eds.), Proceedings of the 25th International Conference on Computers in Education (pp. 457–462). Asia-Pacific Society for Computers in Education.
113. Martin, A. J., & Elliot, A. J. (2016). The role of personal best (PB) goal setting in students' academic achievement gains. Learning and Individual Differences, 45, 222–227.
114. Dela Rosa, E. D., & Bernardo, A. B. I. (2013). Testing multiple goals theory in an Asian context: Filipino university students' motivation and academic achievement. International Journal of School & Educational Psychology, 1(1), 47–57.
115. Pintrich, P. R. (2000). Multiple goals, multiple pathways: The role of goal orientation in learning and achievement. Journal of Educational Psychology, 92(3), 545–555.
116. Christiaans, H. (2002). Creativity as a design criterion. Creativity Research Journal, 14(1), 41–54.
117. Cross, N. (1997). Descriptive models of creative design: Application to an example. Design Studies, 18, 427–455.

118. Gero, J. S. (2000). Creativity, emergence and evolution in design. Knowledge Based Systems, 9(7), 435–448.
119. Casakin, H., Davidovitch, N., & Milgram, R. M. (2009). Creative thinking as a predictor of creative problem-solving in architectural design. Psychology of Aesthetics, Creativity, and the Arts (Accepted).
120. Sim, S. K., & Duffy, A. H. B. (2004). Knowledge transformers—a link between learning and creativity. AI EDAM: Artificial Intelligence for Engineering Design, Analysis, and Manufacturing, 18(3), 271–279.
121. Kreitler, S., & Casakin, H. (2009). Self-perceived creativity: The perspective of design. The European Journal of Psychological Assessment, 25(3), 194–203.
122. Kreitler, S., & Kreitler, H. (1994). Motivational and cognitive determinants of exploration. In H. Keller, K. Schneider, & B. Henderson (Eds.), Curiosity and exploration (pp. 259–284). New York: Springer.
123. Kreitler, S., & Kreitler, H. (1987a). Plans and planning: Their motivational and cognitive antecedents. In S. L. Friedman, E. K. Scholnick, & R. R. Cocking (Eds.), Blueprints for thinking: The role of planning in cognitive development (pp. 110–178). New York: Cambridge University Press.
124. Kreitler, H., & Kreitler, S. (1982). The theory of cognitive orientation: Widening the scope of behavior prediction. Progress in Experimental Personality Research, 11, 101–169.
125. Kreitler, S. (2004). The cognitive guidance of behavior. In J. T. Jost, M. R. Banaji, & P. A. Prentice (Eds.), Perspectivism in social psychology: The Yin and Yang of scientific progress (pp. 113–126). Washington, DC: American Psychological Association.
126. Kreitler, S., Chaitchik, S., Kreitler, H., & Weissler, K. (1994). Who will attend tests for the early detection of breast cancer? Psychology and Health, 9, 463–483.
127. Kreitler, S., & Kreitler, H. (1987b). The motivational and cognitive determinants of individual planning. Genetic, Social and General Psychology Monographs, 113, 81–107.
128. Kreitler, S., & Kreitler, H. (1988a). The cognitive approach to motivation in retarded individuals. In N. W. Bray (Ed.), International review of research in mental retardation (pp. 61–123). San Diego, CA: Academic Press. 15.
129. Kreitler, S., & Kreitler, H. (1990a). The cognitive foundations of personality traits. New York: Plenum.
130. Ryan, R. M., & Deci, E. L. (2017). Self-determination theory: Basic psychological needs in motivation, development, and wellness. Guilford Publications.
131. Stupnisky, R. H., BrckaLorenz, A., & Nelson Laird, T. F (2019a). How does faculty research motivation type relate to success? A test of self-determination theory. International Journal of Educational Research, Special Edition on Faculty Motivation, 98, 25–35. https://doi.org/10.1016/j.ijer.2019.08.007
132. Stupnisky, R. H., Hall, N. C., Daniels, L. M., & Mensah, E. (2017). Testing a model of pretenure faculty members' teaching and research success: Motivation as a mediator of balance, expectations, and collegiality. The Journal of Higher Education, 88(3), 376–400. https://doi.org/10.1080/00221546.2016.1272317
133. Stupnisky, R. H., Weaver-Hightower, M., & Kartoshkina, Y. (2015). Exploring and testing predictors of new faculty success: A mixed method study. Studies in Higher Education, 40(2), 368–390. https://doi.org/10.1080/03075079.2013.842220 .
134. Vansteenkiste, M., Simons, J., Lens, W., Sheldon, K. M., & Deci, E. L. (2004). Motivating learning, performance, and persistence: The synergistic effects of intrinsic goal contents and autonomy-supportive contexts. Journal of Personality and Social Psychology, 87, 246–260.
135. Guay F, Ratelle CF, Chanal J (2008) Optimal learning in optimal contexts: the role of self-determination in education. Can Psychol 49:233–240
136. Kusurkar RA, Croiset G, Mann KV, Custers E, Ten Cate O: Have motivation theories guided the development and reform of medical education curricula? A review of the literature. Acad Med. 2012, 87: 735-743. https://doi.org/10.1097/ACM.0b013e318253cc0e.
137. Chen B et al (2015) Basic psychological need satisfaction, need frustration, and need strength across four cultures. Motiv Emot 39:216–236. https://doi.org/10.1007/s11031-014-9450-1

138. Ryan, R. M., & Deci, E. L. (2000b). The darker and brighter sides of human existence: Basic psychological needs as a unifying concept. Psychological Inquiry, 11(4), 319–338.
139. Ryan, R. M., Bernstein, J. H., & Brown, K. W. (2010). Weekends, work, and well-being: Psychological need satisfactions and day of the week effects on mood, vitality, and physical symptoms. Journal of Social and Clinical Psychology, 29(1), 95–122.
140. Bartholomew, K. J., Ntoumanis, N., Ryan, R. M., Bosch, J. A., & Thøgersen-Ntoumani, C. (2011). Self-determination theory and diminished functioning: The role of interpersonal control and psychological need thwarting. Personality and Social Psychology Bulletin, 37, 1459–1473. https://doi.org/10.1177/0146167211413125 .
141. Augustine, Adam A; Hemenover, Scott H.; Larsen, Randy J.; Shulman, Tirza E. Composition and consistency of the desired affective state: The role of personality and motivation. Motivation and Emotion (2010). https://doi.org/10.1007/s11031-010-9162-0
142. Tomita, Kei Does the Visual Appeal of Instructional Media Affect Learners' Motivation Toward Learning?. TechTrends (2017). https://doi.org/10.1007/s11528-017-0213-1
143. Heo, JeongChul; Han, Sumi Effects of motivation, academic stress and age in predicting self-directed learning readiness (SDLR): Focused on online college students. Education and Information Technologies (2017). https://doi.org/10.1007/s10639-017-9585-2
144. Batool, Uzma; Raziq, Muhammad Mustafa; Obaid, Asfia; Sumbal, Muhammad Saleem Ullah Khan Psychological ownership and knowledge behaviors during a pandemic: role of approach motivation. Current Psychology (2022). https://doi.org/10.1007/s12144-022-03450-y
145. Vollmeyer, Regina; Rheinberg, Falko Motivational Effects on Self-Regulated Learning with Different Tasks. Educational Psychology Review (2006). https://doi.org/10.1007/s10648-006-9017-0
146. Moos, Daniel C.; Azevedo, Roger Exploring the fluctuation of motivation and use of self-regulatory processes during learning with hypermedia. Instructional Science (2007). https://doi.org/10.1007/s11251-007-9028-3
147. Alzubaidi, Eman; Aldridge, Jill M.; Khine, Myint Swe Learning English as a second language at the university level in Jordan: motivation, self-regulation and learning environment perceptions. Learning Environments Research (2014). https://doi.org/10.1007/s10984-014-9169-7
148. Smith, Jessi L.; Deemer, Eric D.; Thoman, Dustin B.; Zazworsky, Lisa Motivation under the microscope: Understanding undergraduate science students' multiple motivations for research. Motivation and Emotion (2013). https://doi.org/10.1007/s11031-013-9388-8
149. Azizi, Zeinab; Farid Khafaga, Ayman Scaffolding via Group-dynamic Assessment to Positively Affect Motivation, Learning Anxiety, and Willingness to Communicate: A Case Study of High School Students. Journal of Psycholinguistic Research (2023). https://doi.org/10.1007/s10936-023-09935-6
150. Howard, Joshua L.; Gagné, Marylene; Morin, Alexandre J. S. Putting the pieces together: reviewing the structural conceptualization of motivation within SDT. Motivation and Emotion (2020). https://doi.org/10.1007/s11031-020-09838-2
151. Grund, Axel; Carstens, Christoph-Alexander Self-control motivationally reconsidered: "Acting" self-controlled is different to "being good" at self-control. Motivation and Emotion (2018). https://doi.org/10.1007/s11031-018-9721-3
152. Ellwood, Robin; Abrams, Eleanor Student's social interaction in inquiry-based science education: how experiences of flow can increase motivation and achievement. Cultural Studies of Science Education (2017). https://doi.org/10.1007/s11422-016-9769-x
153. Voica, Cristian; Singer, Florence Mihaela; Stan, Emil How are motivation and self-efficacy interacting in problem-solving and problem-posing?. Educational Studies in Mathematics (2020). https://doi.org/10.1007/s10649-020-10005-0
154. Stolk, Jonathan D.; Gross, Michael D.; Zastavker, Yevgeniya V. Motivation, pedagogy, and gender: examining the multifaceted and dynamic situational responses of women and men in college STEM courses. International Journal of STEM Education (2021). https://doi.org/10.1186/s40594-021-00283-2

155. Kim, ChanMin The role of affective and motivational factors in designing personalized learning environments. Educational technology research and development (2012). https://doi.org/10.1007/s11423-012-9253-6
156. Valle, Natercia; Antonenko, Pavlo; Valle, Denis; Sommer, Max; Huggins-Manley, Anne Corinne; Dawson, Kara; Kim, Dongho; Baiser, Benjamin Predict or describe? How learning analytics dashboard design influences motivation and statistics anxiety in an online statistics course. Educational technology research and development (2021). https://doi.org/10.1007/s11423-021-09998-z
157. Casakin, Hernan; Kreitler, Shulamith Motivation for creativity in architectural design and engineering design students: implications for design education. International Journal of Technology and Design Education (2009). https://doi.org/10.1007/s10798-009-9103-y
158. Stupnisky, Robert H.; Larivière, Vincent; Hall, Nathan C.; Omojiba, Oluwamakinde Predicting Research Productivity in STEM Faculty: The Role of Self-determined Motivation. Research in Higher Education (2022). https://doi.org/10.1007/s11162-022-09718-3
159. Tjin A Tsoi, Sharon L. N. M.; de Boer, Anthonius; Croiset, Gerda; Koster, Andries S.; van der Burgt, Stéphanie; Kusurkar, Rashmi A. How basic psychological needs and motivation affect vitality and lifelong learning adaptability of pharmacists: a structural equation model. Advances in Health Sciences Education (2018). https://doi.org/10.1007/s10459-018-9812-7

Motivation and Technology 5

Introduction by the Editor

Motivation is a multifaceted construct that has intrigued scholars and practitioners for centuries. As technology continues to evolve and permeate every aspect of our lives, its impact on motivation warrants a thorough examination. The advent of technology has revolutionized the way we approach and understand motivation. In education, technology has transformed traditional pedagogical methods, providing new tools and platforms that improve student engagement and learning outcomes. Digital learning environments, online courses, and interactive software have introduced new ways of motivating students. For example, gamification, the integration of game mechanics into learning activities, has shown great promise in increasing student motivation (Alsawaier, 2018). By incorporating elements such as points, badges, and leaderboards, gamified learning platforms can make educational activities more fun and rewarding, thereby increasing intrinsic motivation. Research has shown that gamification not only increases student engagement but also fosters a sense of achievement and progress, which are critical components of intrinsic motivation (Garcia-Lopez & Acosta-Gonzaga, 2023). Studies in this section consider the complex relationship between motivation and technology, exploring how technological advances influence intrinsic and extrinsic motivation in a variety of settings, including education, workplace productivity, and personal development.

Motivation and Technology
In a rapidly advancing technological environment, the impact of technology on human motivation has become an area of intense scrutiny and scholarly discourse. Motivation, a complex psychological construct that drives behavior and decision-making, is central to our understanding of human achievement, well-being, and personal growth. As technology permeates every aspect of our lives, it is imperative to examine its role in shaping motivation across various domains, from

M. S. Khine (ed.), *Motivation Science*, https://doi.org/10.1007/978-981-97-9247-4_5

education and work to personal pursuits. This introduction delves into the multifaceted relationship between technology and motivation, exploring its potential to enhance engagement, facilitate self-regulation, and foster intrinsic motivation while acknowledging potential risks and challenges.

One of the most compelling ways in which technology influences motivation is through its ability to create engaging and immersive experiences. Today's technological advances have given rise to interactive platforms, gamified learning environments, and augmented and virtual reality applications. These tools have the power to engage users and transform traditionally passive activities into interactive and dynamic experiences (Oliveira et al. 2022). In education, for example, game-based learning platforms that integrate elements of competition, rewards, and storytelling have shown promising results in motivating and engaging students. By tapping into the intrinsic human drives for challenge, curiosity, and achievement, these interactive technologies can transform otherwise mundane tasks into engaging experiences, leading to increased motivation and sustained focus (Faust, 2021).

Technology has also emerged as a powerful tool to facilitate self-regulation and goal-setting, two critical components of motivation. Wearable devices, fitness trackers, and mobile applications provide individuals with real-time feedback on their progress, enabling them to set measurable goals and monitor their achievements. This ability to track and quantify progress can be a powerful motivator, allowing individuals to celebrate milestones, identify areas for improvement, and adjust their strategies accordingly. Technology has also opened up new avenues for collaboration and social interaction, which can profoundly influence motivation. Online communities, social media platforms, and collaborative tools enable individuals to connect, share ideas, and work together toward common goals (Vidergor, 2021). This sense of belonging and social support can serve as a powerful motivator, particularly in contexts where individuals might feel isolated or lacking in motivation.

While gamification elements such as rewards, badges, and leaderboards can initially motivate individuals, an over-reliance on these external motivators can undermine intrinsic motivation in the long run. Individuals may become more focused on earning points or outperforming others, rather than engaging in tasks for their intrinsic value and personal growth. This shift towards extrinsic motivation can lead to a decrease in overall well-being, creativity, and sustained engagement. The relationship between technology and motivation is multifaceted and complex, with both opportunities and challenges to consider. Technology offers the potential to enhance engagement through interactive and immersive experiences, facilitate self-regulation and goal-setting, personalize learning environments, and foster collaboration and social support. These aspects can profoundly influence motivation, leading to increased persistence, achievement, and personal growth. However, technology also presents risks, such as distractions, extrinsic motivation overload, and social comparison pressures, which can undermine intrinsic motivation and overall well-being.

The role of technology in motivating students is critical to fostering engagement, persistence, and achievement in educational settings. By using technology to personalize learning experiences, provide engaging resources, and support collaborative interactions, educators can enhance students' intrinsic motivation and promote positive learning outcomes. While there are challenges related to digital distraction, equity, privacy, and ethical use of technology, addressing these considerations can help educators harness the full potential of technology to effectively motivate students. Going forward, the use of innovative technologies and evidence-based practices in education can empower educators to create inspiring and supportive learning environments that motivate students to reach their full potential and succeed in academic endeavors.

The research papers included in this section are, taking part in technology education: elements in students' motivation, self-regulation, motivation, and outcomes in HyFlex classrooms, a study of learning and motivation in a new media enriched environment for middle school science, motivational engagement, and video gaming: a mixed methods study, and leveraging learning experience design: digital media approaches to influence motivational traits that support student learning behaviors. Other studies that focus on motivation and technology are, the design of motivational agents and avatars, new approach to evaluating the motivation and satisfaction of university students using advanced visual technologies, predictive effect of digital literacy, self-control, and motivation on the academic achievement in the science, technology and society learning area, and eye-tracking and artificial intelligence to enhance motivation and learning.

References

Alsawaier, R. S. (2018). The effect of gamification on motivation and engagement. *The International Journal of Information and Learning Technology*, *35*(1), 56–79.

Faust, A. (2021). *The Effects of Gamification on Motivation and Performance.* Berlin, Germany: Springer Gabler.

García-López, I. M., Acosta-Gonzaga, E., & Ruiz-Ledesma, E. F. (2023). Investigating the impact of gamification on student motivation, engagement, and performance. *Education Sciences*, *13*(8), 813.

Oliveira, W., Hamari, J., Joaquim, S., Toda, A. M., Palomino, P. T., Vassileva, J., & Isotani, S. (2022). The effects of personalized gamification on students' flow experience, motivation, and enjoyment. *Smart Learning Environments*, *9*(1), 16.

Vidergor, H. E. (2021). Effects of digital escape room on gameful experience, collaboration, and motivation of elementary school students. *Computers & Education*, *166*, 104156.

Machine Generated Summaries

Disclaimer: The summaries in this chapter were generated from Springer Nature publications using extractive AI auto-summarization: An extraction-based summarizer aims to identify the most important sentences of a text using an algorithm and uses those original sentences to create the auto-summary (unlike generative AI). As the constituted sentences are machine selected, they may not fully reflect the body of the work, so we strongly advise that the original content is read and cited. The auto generated summaries were curated by the editor to meet Springer Nature publication standards. To cite this content, please refer to the original papers.

Machine generated keywords: student, technology, stem, classroom, science, course, middle school, middle, engagement, teacher, digital, learn, science technology, technological, online

Taking Part in Technology Education: Elements in Students' Motivation [165]

This is a machine-generated summary of:

Autio, Ossi; Hietanoro, Jenni; Ruismäki, Heikki: Taking part in technology education: elements in students' motivation [165]

Published in: International Journal of Technology and Design Education (2010)

Link to original: https://doi.org/10.1007/s10798-010-9124-6

If you want to cite the papers, please refer to the original.

For technical reasons we could not place the page where the original quote is coming from.

Abstract-Summary

"The purpose of this study was to determine the elements motivating comprehensive school students to study technology education."

"We tried to discover how students' motivation towards technology education developed over the period leading up to their school experience and the effect this might have on their future involvement with technology."

"The study group consisted of four 15 to 16-year-old students, each representing a totally different case of motivation towards technology education."

"In choosing individuals for the study the main criteria were gender and negative or positive motivation towards technology education."

Introduction

"The main problem in Finland is that technological content is mostly taught in technical craft lessons."

"Although the national curriculum states that both technical and textile crafts are compulsory for both boys and girls in grades 3–6, for various practical reasons, such as scheduling and the number of teachers employed, pupils are forced to select only one of the craft classes."

"Even when pupils can choose shared craft education, studies show that about 70% of the girls would still choose only textiles, while about 80% of the boys would choose technical crafts (Autio [1])."

"Järvinen and Autio [2] found that 63% of primary school teacher education students (n = 118) were not willing to teach technical craft after they have finished their studies."

"This attitude towards technology has a negative effect when selecting subjects such as physics in upper secondary school and on considerations of studying in technical universities and science departments in universities."

Theoretical Background

"Motivation is one lens with which to investigate factors that contribute to students' interest, engagement and persistence in learning activities (Gilman and Anderman [3])."

"According to them, self-determination theory (SDT) posits the existence of three major types of motivational constructs namely intrinsic motivation, extrinsic motivation, and amotivation."

"Intrinsic motivation refers to engaging in activities for themselves, out of pleasure, fun, and enjoyment."

"Extrinsic motivation refers to engaging in activities for outcomes that are separate from the activity."

"To intrinsic and extrinsic motivation, a third motivational construct is amotivation, which occurs when amotivated individuals do not perceive contingencies between their actions and subsequent outcomes."

"We classified the elements motivating comprehensive school students to study technology education according to the Peltonen and Ruohotie [4] model of school motivation."

Empirical Research
"The aim of this research was to examine comprehensive school students' motivation in technology education and to determine the elements motivating comprehensive school students to study technology education."

"Besides that we tried to find out how students' motivation towards technology education was formed during their life."

"How is the motivation in technology education transformed throughout comprehensive school students' life?"

"What are the main elements motivating comprehensive school students to study technology education?"

"The analysis was carried out through determining out the interesting and essential elements motivating students in technology education."

"The motivation curves indicate each person's motivation in technology education during their whole life."

"The study group consisted of four 15 to 16-year-old students, each representing a totally different case of motivation in technology education."

"In the choice of individuals to be tested the main consideration was given to gender and to negative and positive motivation in technology education."

Results
"The motivation curves were first drawn by the test participants, but they were later discussed in more detail during the theme interviews."

"Each test participant drew themselves a motivation curve and the value between 0 and 100% was based on their personal feeling and on their own opinion of their motivation."

"The factors may have had negative or positive effects on a participant's motivation."

"The significance of the factors is based on the participants' direct comments, which were documented during the interviews."

A Boy in Technology Education: Subject 1
"His motivation in technology education developed steadily throughout his school years."

"The first noticeable increase in motivation was found when technology education classes started in primary school, and, for the first time, he received sound instruction."

"After finishing comprehensive school, Subject 1 thinks that his activity in technology education will decrease a little, but that his motivation towards technology in general will remain positive."

"It seems that Subject 1 has moved from his early childhood's amotivation to a level where behavior in technology education is internalized in the individual's self and value system."

"Subject 1 became familiar with technology education in comprehensive school."

"Subject 1 also responded positively to technology education; even at this lower level of the comprehensive school, craft and technology became his favorite subject."

"In his opinion, technology education is a comfortable counterbalance to academic subjects."

A Girl in Technology Education: Subject 2

"Subject 2 has been interested in technology education since early childhood."

"The first increase in motivation was found when craft and technology classes started in primary school, and she learned something valuable about technology education."

"Especially in secondary school the motivation again increased because Subject 2 liked to work with machines, and there were more materials and interesting projects to choose from."

"After finishing secondary school, Subject 2 thinks that her activity in technology education will decrease, but that her attitude towards technology in general will remain positive, but diminished."

"From the interview we can conclude that Subject 2's motivation in her early childhood was based on external or introjected regulation and her grandfather was a highly valued role model."

"Subject 2 remembers the work of making the guitar as the most agreeable project of all."

A Boy not in Technology Education: Subject 3

"He represents a boy who has not chosen any technology education lessons in secondary school."

"The first remarkable increase in interest and motivation came when technology education started in primary school, and for the first time he learned some valuable technical skills."

"After finishing school, Subject 3 thinks that he will not have any activities in technology education because he will be concentrating on his academic career."

"His motivation to engage in technology education may well reduce close to zero after his school years."

"According to Subject 3, technology education is not a significant matter in his life."

"Subject 3's thoughts regarding technology education reflect those values and attitudes that come from home."

"He places value neither on the craft nor on vocational education in the field of technology."

"During his first school years, however, Subject 3 liked technology education."

A Girl not in Technology Education: Subject 4

"She represented a girl who has not chosen any technology classes in secondary school."

"She returned to the same school, but in the 8th grade there was no compulsory technology education and as mentioned she did not take any electives in technology education."

"The highest point in her motivation curve was when she was in nursery school and interested in all craft and technology activities."

"Already in primary school the motivation started to decrease because there was much more compulsory work without any freedom of choice."

"The learning is motivated by curiosity but the school will suppress the motivation."

"According to Ryan and Deci intrinsic motivation will decrease in the first eight school years."

"Technology education has been very unpleasant to her throughout her school years."

"The tools and classrooms were in good order at school so she does not believe that these affected her negative motivation."

Summary of All Test Participants

"According to Deci and Ryan [5], one way to achieve intrinsic motivation is to expand the feeling of autonomy among students."

"The entire classroom environment appeared to be important for motivation; according to the test participants, the classroom in technology education should have enough space for everybody, enough materials, and tools in good order."

"According to Stipek [6], it is even more important to pay attention to provide an optimal and suitable learning environment than to concentrate on students' personal problems in terms of motivation."

"It seems that if we ask students to do too difficult tasks in technology education with limited competence, the motivation is based only on extrinsic forms."

"Social relations: e.g., teacher, teacher–student interaction, classroom atmosphere, and parents were also found to be important elements in motivation but not as essential as those elements in subject content, learning environment and personality."

Discussion

"The motivation of most of the subjects, in technology education was relatively positive at least during some parts of their educational career."

"Students who typically enroll in technology education are attracted to the types of projects they will be engaged in (Weber and Custer [7])."

"It seems that several other school subjects have more motivational problems than technology education."

"It appears to be the case that some students do not place a value either on the craft or on vocational education in the field of technology education in Finnish schools."

"Motivation in technology education can be significantly improved by developing special programmes (Mammes [8]), where teachers are aware of the differing interests of both genders and consider ways of making the environment and the subject attractive to all (Silverman and Pritchard [9])."

"Most of the students valued only the product at first, but later on internal feedback turned out to be one of the key elements in their motivation."

"Motivation has been viewed as the primary determinant of students' learning and school success."

Self-regulation, Motivation, and Outcomes in HyFlex Classrooms [166]

This is a machine-generated summary of:

Athens, Wendy: Self-regulation, motivation, and outcomes in HyFlex classrooms [166]

Published in: Educational technology research and development (2023)

Link to original: https://doi.org/10.1007/s11423-023-10243-y

If you want to cite the papers, please refer to the original.

For technical reasons we could not place the page where the original quote is coming from.

Abstract-Summary

"At a public, 4-year, open access university, students were given the choice of how they would attend class: face-to-face, online, or livestream (synchronous session in Microsoft Teams®)."

"The 876 students in this study registered in course sections designated as face-to-face delivery yet were provided attendance flexibility due to the pandemic."

"This unique situation provided a research opportunity to explore the self-regulatory, motivational, and contextual factors that affected students' attendance choices as well as their academic outcomes (pass/withdrawal) and perceptions of satisfaction."

"Self-regulatory and motivational factors were explored in relation to attendance decisions."

"Besides COVID-related issues and work/life balance, a sizable percentage (13%) of students explained their attendance decisions in terms of quality of their learning and in this way demonstrated self-regulation."

"Motivation was an issue for 17% of the students, typically expressed as not attending in the way they learned best or skipping class."

Introduction

"At a public, 4-year, open access university, students were given the choice of how they would attend class during Fall 2020 and Spring 2021 terms: either face-to-face, online, or livestream (synchronous session in Microsoft Teams®)."

"To enable HyFlex delivery, the institution adapted nearly 300 classrooms with ceiling-mounted single cameras and microphones, automated recording of whiteboards into Kaltura®, live-streamed using Microsoft Teams® via a podium computer, and stored online content in Canvas®."

"RQ2 How did students demonstrate self-regulation in their attendance choices in HyFlex classrooms?"

"RQ3 What was the impact of motivation on attendance choice in HyFlex classrooms?"

Literature Review

"A self-regulated learner would move from a noisy to a quiet study space or in the case of HyFlex, plan ahead and attend in the mode most conducive to their learning."

"Nicol and MacFarlane-Dick [11] provide seven principles of good feedback to facilitate self-regulation that include instructor-student dialogue around learning, reflection on feedback, and positive beliefs."

"Self-regulation is important for online students who must manage much of their learning due to reduced instructor presence (Hung and others [12]; Kebritchi and others [13]; Kizilcec and others [14]; Wang and others [15])."

"Wang and others [15] found that online students who employ SRL strategies, e.g. time management, goal setting, and metacognition were more motivated in their studies."

"A study by Rotgans and Schmidt [16] with 1166 first year polytechnic students used the MSLQ and observed achievement-oriented classroom behaviors (engagement in group work, presentations, and persistence in self-directed learning) to test the connection between motivation and academic achievement."

Methods

"Commitment entailed providing access to the students in particular course sections, completing pre- and post-surveys about their experiences teaching in HyFlex classrooms, and participating in focus groups."

"For those students who opted into the study, data collection included pre- and post- surveys and the important Friday pulse survey which asked, "How did you attend class this week and why?" (Malczyk [10])."

"Only 12% of students who opted into the study in Fall 2020 and 7% in Spring 2021 consistently responded to the Friday pulse survey on a week by week basis (at least four responses), which weakened reliability of the results."

"The student post-survey revealed student perceptions of best and worst aspects of the HyFlex delivery, feelings of connectedness to instructors and peers, and overall satisfaction."

Results

"The sample pool consisted of students who had registered for face-to-face classes which were now delivered via HyFlex, allowing student choice of attendance."

"Of the students who expressed low motivation, a freshman Psychology major who lived 27 miles from campus, worked 20 h per week, and utilized all four modes of attendance stated, "The most challenging aspect was that I had the option to livestream or watch a recording later but while those options are convenient and useful it makes it easier to procrastinate and to not wake up on time to attend in person.""

"A sophomore who lived five miles from campus and attended either face-to-face or online said "I didn't care for my teacher's teaching methods, it was hard to find motivation to go if I woke up late (my class was in the morning)."

Discussion

"This research study explored how and why students attended class (face-to-face, online, or livestream) in HyFlex classrooms, as well as their academic success (pass/withdrawal) and perception of satisfaction."

"Highly satisfied students commented: "The instructor did a fantastic job of including everyone and even having us work with each other in groups occasionally," "I prefer to work and learn from home."

"Based on the work of Raes and others [17], who found that those students who never attended F2F had lower "relatedness" to instructor and peers and lower intrinsic motivation, an interesting future research study could measure the satisfaction and connectedness of remote only students compared to those who attend in person."

"HyFlex provides full flexibility, but in this study some instructors required students to gain permission for remote attendance, which became more common practice in the second term when there were no restrictions on in-person attendance."

Conclusion

"This unique situation provided a public, open access higher education institution an opportunity to explore student attendance behaviors in HyFlex classrooms relative to academic outcomes (pass/withdrawal) and perceptions of satisfaction."

"HyFlex classrooms enabled students daily choice of how they would attend class, either face-to-face, online, or livestream."

"The need for instructor preparation is paramount in terms of highly organizing course materials, providing explicit avenues for communication to occur, developing active learning exercises that span location boundaries and build community, and mastering the technology."

"The fact that a higher percentage of failures were among first year students emphasizes the importance of explicit self-regulatory skill development in conjunction with HyFlex instruction, within an encouraging and supportive learning environment (Bower and others [18])."

A Study of Learning and Motivation in a New Media Enriched Environment for Middle School Science [167]

This is a machine-generated summary of:

Liu, Min; Horton, Lucas; Olmanson, Justin; Toprac, Paul: A study of learning and motivation in a new media enriched environment for middle school science [167]

Published in: Educational technology research and development (2011)

Link to original: https://doi.org/10.1007/s11423-011-9192-7

If you want to cite the papers, please refer to the original.

For technical reasons we could not place the page where the original quote is coming from.

Abstract-Summary
"This study examines middle school students' learning and motivation as they engaged in a new media enriched problem-based learning (PBL) environment for middle school science."

"Using a mixed-method design with both quantitative and qualitative data, we investigated the effect of a new media environment on sixth graders' science learning, their motivation, and the relationship between students' motivation and their science learning."

Theoretical Framework
"Current literature on achievement motivation suggests that there are two complementary ways to motivate students: through improving the student's belief in her probability of success—known as expectancy, and increasing the subjective value of the task as it relates to areas such as interestingness or intrinsic value."

"Research shows that expectancy is highly correlated with motivation to engage in learning tasks and self-efficacy influences actual achievement that then reciprocally shapes self-perceptions of competency (Pintrich and Schunk [19])."

"Numerous studies have demonstrated that cooperative learning and group activities, such as those presented in problem-based learning environments, have a positive effect on students' interest, engagement, and motivation (Shernoff and others [20])."

"In our previous research (Liu and others [21]), we investigated student engagement with a new media enriched PBL environment by exploring the factors that students considered motivating."

"Our current study continues that line of inquiry and investigates changes in sixth graders' science learning and connections between their motivation and learning after using a new media enhanced PBL environment, Alien Rescue."

Method

"The goal of Alien Rescue is to engage sixth-grade students in solving a complex problem that requires them to use the tools, procedures, and knowledge of space and planetary science to learn about our solar system and processes of scientific inquiry."

"These students were from "Regular Education" classes and they used Alien Rescue as a self-paced curriculum unit for 3 weeks in their daily 45-min science classes."

"Learning performance in this study was measured by students' understanding of the various scientific concepts introduced in Alien Rescue."

"The science knowledge test was administered before and after students used Alien Rescue to measure any change."

"Upon completion of the curriculum, sixth graders were also asked to respond in writing to six open-ended questions: "How would you describe Alien Rescue to a friend?" ""

Results

"The correct responses in the science knowledge test increased significantly from pretest to posttest for both male and female students and a small but significant difference was observed between males and females ($M_{male} = 83.53$; $M_{female} = 79.36$)."

"The multiple regression analysis, examining the relationship between students' motivation scores and their science knowledge posttest scores, controlling for the effect of the pretest scores, showed a significant moderate R^2 of 0.26, $F(2, 129) = 23.17$, $p < 0.01$."

"Students' motivation score was found to significantly predict their science knowledge test scores: $b = 5.43$, $t(129) = 2.6$, $p < 0.01$."

"There is a tendency that "effort/importance" is also positively related to the science knowledge posttest scores, suggesting that the students who worked harder learned more than those who did not, but this calls for further research."

Discussion

"In addressing the first research question about the effect of a new media PBL environment on sixth graders' science learning, the findings based upon both quantitative and qualitative data showed that both male and female students significantly increased their science knowledge from pretest to posttest after using Alien Rescue, with female students' "unsure" responses decreasing more than male students."

"The gain in students' knowledge supports previous research examining the effectiveness of Alien Rescue as a new media enhanced PBL environment by showing a positive impact on sixth graders' science learning (Liu and Bera [22]; Liu and others [23, 24]), and is consistent with the encouraging results from other new media infused learning environments (Barab and others [25]; Ketelhut [26])."

"Alien Rescue uses a problem-based learning approach that motivates students to learn in a self-directed manner."

"Students' responses to "What did you like about Alien Rescue? [and why]" corroborated this finding as 63% of the reasons related to probes, aliens, and other new media features."

[Section 5]
"Research has also shown that instructional context strongly affects students' motivation."

"Educators and schools that do not utilize these new media based technologies run the risk of failing to leverage the digital literacies of their students, creating contexts that may lead to a "switching off" (Dudfield [27]), disengagement from classroom activities, or missed opportunities to maximize student potential (Judson [28])."

"We are interested in better understanding how curricula delivered via new media environments engage students in learning activities."

"We examine middle school students' learning and motivation as they experience space science curriculum via a new media enriched problem-based learning (PBL) environment."

Support, Belonging, Motivation, and Engagement in the College Classroom: A Mixed Method Study [168]

This is a machine-generated summary of:

Zumbrunn, Sharon; McKim, Courtney; Buhs, Eric; Hawley, Leslie R.: Support, belonging, motivation, and engagement in the college classroom: a mixed method study [168]

Published in: Instructional Science (2014)

Link to original: https://doi.org/10.1007/s11251-014-9310-0

Copyright of the summarized publication:

If you want to cite the papers, please refer to the original.

For technical reasons we could not place the page where the original quote is coming from.

Abstract-Summary
"This explanatory sequential mixed methods study examined how belonging perceptions, academic motivation, and engagement might mediate the relationship between academic contextual characteristics and achievement using structural equation modeling and qualitative follow-up interviews with college students from a large, Midwestern university."

"In line with the Self-System Model of Classroom Support for Motivation (Connell and Wellborn, in: Gunnar and Sroufe (eds.) Minnesota Symposium on Child Psychology: Self-processes and Development [29]), Model 1 hypothesized student belonging and motivation to be directly predicted by supportive classroom environment perceptions, and to directly predict engagement, which was hypothesized to predict achievement."

"Model 2 elaborated on the traditional self-system model and hypothesized student belonging to mediate the relationship between supportive classroom environment perceptions and student motivation."

"Supportive classroom environment perceptions predicted students' belonging beliefs, which in turn predicted students' motivation, engagement, and achievement in the course."

Links Between Student Belonging, Engagement, and Achievement
"The self-system model of motivational development (Connell and Wellborn [29]) emphasizes the complex dynamics of classroom environments and describes processes by which aspects of the social context impact student self-beliefs and subsequent academic engagement and achievement."

"Previous research findings have consistently supported the utility of the self-system model in academic settings (Buhs [30]; Connell and others [31]; Hamre and Pianta [32]; Van Ryzin and others [33]), with many studies emphasizing the predictive role of student belonging in academic and socio-emotional adjustment."

"In a study of fifth and sixth grade students, for example, Furrer and Skinner [34] showed that students' perceptions of relatedness to their teacher were associated with engagement in classroom activities."

"In related studies with middle school students, Wentzel [35, 36] found that students' relationships with teachers predicted academic engagement and performance."

Supportive Classroom Environments and Student Belonging as Antecedents of Motivation, Engagement, and Achievement

"Two motivational variables that have been examined and linked to student belonging are self-efficacy and task value."

"Their findings suggested a positive relationship between freshmen students' feelings of class belonging and their subsequent academic self-efficacy and task value."

"These findings and models suggesting that belonging/relatedness may precede student perceptions of self-efficacy and value indicate a possible elaboration of the self-system model."

"Walker and others [37], for example, examined relationships between motivational variables and engagement with college students and found that value of academics and self-efficacy uniquely predicted cognitive engagement."

"Evaluation of these models is important because most investigations to date have not included evaluations of specific process models that evaluate the potential predictive roles of supportive classroom characteristics, belonging, motivation, engagement, and achievement, and these relationships might reveal important aspects of college student development and retention."

Quantitative Phase Method

"Student participants completed the following quantitative instruments: (1) a demographic questionnaire developed for this study; (2) an adapted version of the Psychological Sense of School Membership Scale (PSSM; Goodenow [38]); (3) the task value and academic self-efficacy subscales from the Motivated Strategies for Learning Questionnaire (MSLQ; Garcia and Pintrich [39]); and (4) selected scales from the Student Evaluation of Teaching questionnaire (SET; Barnes and others [40])."

"Participants rated items on a 5-point Likert-type scale ranging from "Not at all true" to "Extremely true," and scores were created by taking the mean of all items."

"The combined scale was composed of 14 items (e.g., "The instructor seems to care whether students learn the material," "The instructor conveys material in a way that is easy to understand," and "The instructor expects academic excellence from students;" $\alpha = .93$)."

"This scale included 12 items ($\alpha = .91$) designed to measure students' engagement [e.g., "(The student) actively participates," and "(The student) comes to class"]."

"Instructors rated students on a 5-point Likert scale ranging from "Does not apply" to "Certainly applies.""

Quantitative Results

"Additional direct paths from belonging, self-efficacy, and task value to achievement were included to examine whether or not engagement mediated links to our achievement outcome as suggested by Connell and colleagues (e.g., Connell and Wellborn [29])."

"Model 2 was created to test the alternative hypotheses presented that revised the prior assumptions from Model 1 and indicated that instructor academic and social support is a likely predictor of belonging and that belonging is antecedent to self-efficacy and task value."

"This allowed us to examine whether or not belonging might mediate linkages from instructor academic and social support to self-efficacy and task value."

"Non-significant paths from instructor academic and social support to self-efficacy, from belonging to engagement, from task value to engagement and from task value to achievement were removed in this subsequent analysis."

Qualitative Analytic Procedures

"Researchers then met to discuss their initial reactions and discuss potential categories that represented key statements from the participants, a process known as basic thematic categorization (Saldaña [41])."

"Three major theme categories emerged from the data: perceptions of belonging; perceptions of peer acceptance and support; and perceptions of instructor acceptance and support."

"An initial coding list was co-constructed by the two readers using a sub-set of participant responses."

"After independent initial coding was complete, the researchers met again to discuss a final coding list, which considered connections, contrasts, and comparisons among the theme categories (Saldaña [41])."

"Transcripts were then re-read in entirety and coded using final sub-category codes."

"Unlike task value, self-efficacy did not emerge as a sub-category code in any theme category."

Qualitative Findings

"Patricia (higher belonging group) reported belonging because "[We have] similar interests, it seems like a lot of us are going to be educators, which is why we're taking this class, so I feel a part of where everyone else is coming from.""

"Although Amanda (higher belonging group) was not an education major, she shared her experience as a student minoring in education, "That is probably why I don't feel 100 % belonging, because I'm not going to be a teacher.""

"Unlike students in the higher belonging group, Collin (lower belonging group) explained that he sometimes didn't feel as though he fit in because he was "not pursuing education.""

"Heather (low belonging group) commented, "When we got into discussions, that was when I liked being there [in class] and hearing different people's sides of the story.""

Discussion

"Although specific pedagogical practices were not examined in the current study, our findings indicate that instructor enthusiasm, passion, and the level of interest and caring they show toward their students may play a central role in supporting student motivation and engagement in the classroom social context—an important component in understanding achievement processes."

"The quantitative and qualitative findings from this study illustrated not only the influential role that instructors and peers have in the classroom, but also the role belonging likely plays in college student motivation (via self-efficacy beliefs) as well as relationships with engagement and achievement processes."

"Future research should examine quantitative models of belonging that include both instructor and peer academic and social support in relation to student belonging, motivation, engagement, and achievement."

"The results of this study suggest that college educators should consider the effects of student perceptions of classroom academic and social support—particularly on student feelings of belonging, as these perceptions may affect consequent motivation, engagement, and achievement."

[Section 8]

"Questions also remain about the precise pathways via which college students' belonging perceptions might, in turn, affect their motivation (i.e., self-efficacy and value beliefs for learning tasks), engagement (e.g., attendance and class participation), and achievement; a set of constructs that likely influence students' persistence in higher education."

"We addressed this need by posing two main questions: (1) How do student perceptions of support and belonging relate to student motivation, engagement, and achievement, and (2) In what ways do college students believe classroom contextual characteristics relate to their belonging perceptions?"

"Quantitative data of student perceptions of instructor academic and social support, belonging, self-efficacy, and task value, and their reported class engagement and achievement, and qualitative data detailing student classroom experiences were gathered."

Motivational Engagement and Video Gaming: A Mixed Methods Study [169]

This is a machine-generated summary of:

Hoffman, Bobby; Nadelson, Louis: Motivational engagement and video gaming: a mixed methods study [169]

Published in: Educational technology research and development (2009)

Link to original: https://doi.org/10.1007/s11423-009-9134-9

If you want to cite the papers, please refer to the original.

For technical reasons we could not place the page where the original quote is coming from.

Abstract-Summary
"A mixed methods design was used to identify factors associated with motivational engagement in video gaming."

"Self-report instruments were administered to 189 video game players to assess goal orientations, affect, need for cognition, and perceptions of engagement and flow."

"Regression results revealed motivational engagement for gaming was related to gender, hours of play, task orientation, and socialization."

"Players indicated that gaming was socially captivating, fun, challenging but relaxing, and precipitated positive affect and cognition even when unsuccessful results were achieved."

Motivation and Gaming
"We chose to focus on individuals playing multiple-level video games because these games are characteristically associated with high levels of engagement (Gee [42])."

"These games are usually used for training, and the motivation to engage in simulation games is based on the need to learn more about a situation, or is required for training and non-discretionary (O'Neil and others [43])."

"This is an important distinction because the motivation to engage in simulations used for training may be considered to be more extrinsic, while motivation to engage in games for entertainment is more intrinsic (Malone [44]; Wang and others [45])."

"It is the perceived high levels of intrinsic motivation associated with recreational video games players that render them germane for investigating engagement."

"Several motivational variables have been found to be associated with video gaming."

"In a series of studies assessing psychological need satisfaction (Ryan and others [46]), video-gaming participants reported increased levels of decision making and autonomy when playing games, supporting self-deterministic views of motivation."

Engagement and Video Game Playing

"Academically, engagement is related to achievement, motivation, and task persistence (Shernoff and others [20]), meaningful processing on achievement measures (Walker and others [37]), and is a prerequisite for productive learning (O'Donnell and others [47])."

"Individuals with a conscious and an a priori willingness to approach a task in pursuit of a specific goal can be described as exhibiting motivational engagement."

"How is motivational engagement related to video gaming?"

"Sustained engagement in gaming appears to be related to the relationship between individual skill and game complexity (Malone [44])."

"Gaming is an ideal context to investigate motivational engagement due to existing research that intimates gaming promotes intrinsic motivation, positive affect, and many aspects of the flow experience."

"Given the relationship between the flow experience and positive classroom outcomes such as increased interest and school identification (Shernoff and others [20]), more information regarding motivational engagement should be of significant value to both researchers and practitioners."

Current Study

"We recorded, analyzed, and assessed the perceptions of individuals who played video games for at least five hours per week, with the objective of identifying what factors contributed to motivations and persistence in game playing."

"We administered four self-report measures, analyzed the results, and conducted a regression analysis in conjunction with participant interviews to determine what aspects of gaming and what motivational variables were influential on engagement in video gaming."

"What factors do individuals consider when deciding to engage in video gaming?"

"What are the significant motivational variables associated with video game play?"

Method

"We employed qualitative methods to collect and analyze interview data in order to create a coding structure, taxonomy of domains, and comparative conclusions to uncover the antecedents of gaming motivation and engagement, and understand the persistence to play in the face of obstacles."

"The Phase I coding scheme centered exclusively on the content of the interview protocol, focusing primarily on the following questions: (a) what factors were instrumental in the decision to play video games?; (b) what motivational factors influenced continued game play in the face of difficulty and obstacles?; and (c) what influenced the decision to re-engage or abandon prior gaming objectives."

"We coded data this way for three reasons: (a) participants revealed video game play was motivated by specific goals which were task directed; (b) a social orientation was often communicated by participants as their motivation for playing games; and (c) specific strategies were used to achieve their game-directed goals."

Results and Analysis

"The dependent variable in the regression equation was engagement, based on the total participant score on the Video Game Play measure (Chou and Ting [48])."

"The full sample results from the Video Game Play measure (Chow and Ting [48]) partially supported interview findings as 20% of participants indicated playing video games to "relieve feelings such as anxiety or depression"."

"A loss of a sense of time, another key component of the flow experience was reported by 54% of the full sample and by 89% of the 5 + participants according to the Video Game Play measure (Chow and Ting [48])."

"Based on our interest in transfer of engagement from gaming to other contexts, we asked all participants, "Do you learn anything from playing games?""

"None of the participants indicated that learning was a goal of the video gaming process."

Discussion

"This finding may have resulted from the competing motivations of gaming participants having strong needs for achievement, but equally substantial goals of positive perceptions from peers, which were met through socialization."

"The high positive correlation between task and ego goals, and mastery and performance orientations, supports the contention that engaged players have specific entertainment and gaming objectives, but also have a demonstrative need for recognition and self-valuation as well."

"We believe there is limited transfer of motivational engagement between the context of recreational gaming for entertainment and most current educational domains due to vast contextual differences."

"In light of our research questions, it is important to caution others that the motivational engagement exhibited when game playing is associated with entertainment and will be difficult to transfer to contexts such as a classroom."

[Section 7]

"Several studies have established relationships between gaming and learning (Barab and others [49]; Hopelab [50]; Ito and others [51]; Schrader and McCreery [52]; Squire and others [53]; Warren and others [54]) and results generally indicate knowledge gains associated with video game play."

"Empirical studies investigating video gaming and motivation are less frequent (Ryan and others [46]; Owston [55]; Schrader and McCreery [52])."

"We believe the nature of video gaming represents an ideal context to investigate motivational engagement due to the focused attention and intrinsic motivation exhibited by game players (Wang and others [45])."

"We collected self-report data from a larger sample (which included our interview subjects) to assess goal orientations, the affective factors associated with gaming experiences, need for cognition, and perceptions of engagement and flow to determine the relationship with video game play."

"We discuss research on motivation and video gaming."

Leveraging Learning Experience Design: Digital Media Approaches to Influence Motivational Traits that Support Student Learning Behaviors in Undergraduate Online Courses [170]

This is a machine-generated summary of:

Wong, Joseph T.; Hughes, Bradley S.: Leveraging learning experience design: digital media approaches to influence motivational traits that support student learning behaviors in undergraduate online courses [170]

Published in: Journal of Computing in Higher Education (2022)

Link to original: https://doi.org/10.1007/s12528-022-09342-1

If you want to cite the papers, please refer to the original.

For technical reasons we could not place the page where the original quote is coming from.

Abstract-Summary

"Higher education may benefit from investigating alternative evidence-based methods of online learning to understand students' learning behaviors while considering students' social cognitive motivational traits."

"This mixed-methods study (N = 181) assessed associations of students' social cognitive motivational traits (self-efficacy, task-value, self-regulation) influencing their learning behaviors (engagement, elaboration, critical thinking) resulting from LXD."

"Social cognitive motivational traits were positively predictive of learning behaviors."

"Qualitative content analysis corroborated these findings, explaining how LXD efforts contributed to motivations, learning behaviors, and learning experience."

"Results suggest that mechanisms underpinning LXD and students' learning behaviors are likely the result of dynamically catalyzing social cognitive motivational factors."

"The discussion concludes with the LXD affordances that explain the positive influences in students' social cognitive motivational traits and learning behaviors, while also considering constraints for future iterations."

Introduction

"Higher education may have significant potential to identify practical ways to improve undergraduate online learning experiences through the novel combination of Learner Experience Design (LXD), educational technologies, and testing through design-based research (DBR)."

"Through the sudden immersion into an array of online learning implementation, students and learning experience designers are afforded a unique opportunity to rapidly examine which technology based pedagogical approaches are most effective, based on interventions and observations made within the natural large-scale higher educational settings."

"This combination of low student motivation, poor instructional grounding in learning design, and overlooking the learner's user experience have led to undergraduate learners citing issues of diminished engagement, poor time management, and low levels of confidence with their own abilities to learn online, primarily due to their unfamiliarity and lack of prior experience with online courses (Agarwal & Kaushik [56]; Sun & Rueda [57]; Tullis & Benjamin [58]; Zayapragassarazan [59])."

Theoretical Background

"In order to better support students to develop these skills, actively engage in their coursework, and achieve high completion rates, online courses might be transformed to be grounded in evidence-based teaching pedagogy and learning experience design principles."

"Recent studies have shown that self-efficacy and task-value are linked with students' online learning strategies such as increased engagement, critical thinking (applying previous knowledge to new situations) and use of elaboration (summative aligning of new information) (Ali and others [60]; Artino & McCoach [61])."

"Due to the self-paced nature of the course design, developing students' self-regulation skills throughout the online learning environment was an important factor in improving success for learners."

"By incorporating these LXD choices, the course may better support students' online self-regulation skills in order to further facilitate students' learning behavior strategies such as engagement, elaboration, and critical thinking while distance learning."

Methodology

"From conducting an initial pre-assessment survey, we were able to identify that learners were particularly worried about their confidence in their abilities to learn online, motivations related to distance learning format, and whether or not learners would be able to critically engage with the course materials, as an overwhelming majority of students were first-time distance learners."

"These design choices enable students to actively engage in their own productive metacognitive judgments and motivations to reflect on "how and why" they arrived at their solutions, which has been found to foster learners' critical thinking and use of elaboration skills within an online learning environment (Tullis & Benjamin [58]; Wang and others [15, 62])."

"The course's design intentions were meant to ignite students' motivations and train students to adapt their learning behaviors such as their engagement, critical thinking, and elaboration within the asynchronous self-paced online course."

"This instrument was developed by Artino and McCoach [61] to measure students' self-efficacy and task-value for learning specifically within a self-paced, online course."

Results
"Paired-samples (2-tailed) t-tests were conducted to assess the changes in students' social cognitive motivation variables (self-efficacy, task-value, self-regulation) and learning behaviors (engagement, elaboration, and critical thinking) throughout the 10-week instructional period."

"These results indicate that students' self-efficacy, task-value, and self-regulation within the online learning environment were significantly predictive of their online engagement, accounting for 31.2% of the explained variance."

"These results indicate that students' online self-efficacy, task-value, and self-regulation were significantly predictive of students' critical thinking in the online course $R^2 = 0.471$, $F(3, 178) = 34.3$, $P < 0.001$."

"As students' self-efficacy, task-value beliefs, and self-regulation factors increased while participating in this LXD based online course, on average, their engagement, usage of elaboration, and critical thinking skills increased."

Discussion
"When online learning becomes a dominant model of higher education, methods including asynchronous, synchronous, and blended approaches may benefit from investigations highlighting the affordances and constraints of LXD made during the pandemic time of fully online learning immersion to better prepare instructors, researchers, and designers in how to directly supports students' social cognitive motivational traits and learning behaviors."

"Attuning our understanding of the LXD impacts through the integration of learning design principles with learner-centered user experience heuristics may have helped to discern the resulting increase of students' critical thinking skills as explained by their developing motivations within the course."

"To directly address students' needs, we designed an online course intervention for the purpose of increasing students' social cognitive motivations (self-efficacy, task-value, self-regulation) and learning behaviors (engagement, elaboration, critical thinking) by grounding the online learning environment in practical elements of learning and user experience design."

Conclusion
"This study suggests that implementing asynchronous self-paced online courses with LXD approaches may positively impact students' learning behaviors, potentially through influencing student's social cognitive motivational traits."

"Results also suggest that students' task-value beliefs may be most critical in explaining students' learning behaviors when grounding online courses with LXD, above and beyond self-efficacy and self-regulation."

"We assert an alternative approach to synchronous Zoom teleconferenced lectures by detailing the efforts toward designing an asynchronous self-paced online course, offering a pathway for students to further develop their motivations and their learning behaviors in online environments."

"This study may further support the growing literature on learning experience design in higher education courses by drawing on multiple learning design principles, adopting digital learning tools, and user experience design facets intended to enhance students' online learning experiences through empathy (Kafai [63]), informing designers as well as instructors on how we might effectively improve asynchronous self-paced online teaching and learning of STEM subjects at the R1 institutional level."

The Design of Motivational Agents and Avatars [171]

This is a machine-generated summary of:

Baylor, Amy L.: The design of motivational agents and avatars [171]

Published in: Educational technology research and development (2011)

Link to original: https://doi.org/10.1007/s11423-011-9196-3

If you want to cite the papers, please refer to the original.

For technical reasons we could not place the page where the original quote is coming from.

Abstract-Summary
"While the addition of an anthropomorphic interface agent to a learning system generally has little direct impact on learning, it potentially has a huge impact on learner motivation."

"Agents become increasingly ubiquitous on the Internet, in virtual worlds, and as interfaces for learning and gaming systems, it is important to design them to optimally impact motivation."

"With motivational messages and dialogue (which are not discussed here), the agent's appearance is the most important design feature as it dictates the learner's perception of the agent as a virtual social model, in the Bandurian sense."

Introduction

"An agent social model designed to positively influence inner-city youth to stay in school could be designed to appear as an older "cool" peer, with respect to dress, age, race, gender and socio-economic status, with a persona that represents someone with whom they can identify as part of their in-group (from Baylor [64])."

"While an agent with greater computational functionality may be perceived as more believable and human-like, research indicates that learners perceive, interact socially with, and are influenced by anthropomorphic agents even when their functionality and adaptability are limited (e.g., Baylor and Kim [65, 66]; Guadagno and others [67]; Kim and others [68]; Rosenberg-Kima and others [69, 70]; Ryu and Baylor [71])."

"Appearance and message delivery (e.g., voice, non-verbal communication) together with the motivational message and dialogue are key design considerations for a motivational agent."

"Perhaps the most important aspect of designing a motivational agent is its appearance, as that plays a major role in its effectiveness as a virtual social model in the Bandurian sense (Baylor [64])."

Designing Agents as Virtual Social Models

"Research indicates that for motivational outcomes it is critical that the agent be visibly present, and not just represented as a voice or text box (e.g., Baylor [64]; Baylor and others [72]; Rosenberg-Kima and others [69])."

"Research has shown that sometimes surprising learners with an agent that does not match their expectations or providing a female agent (instead of a male agent) to both male and female younger students may be optimal (e.g., Baylor and Kim [73]; Plant and others [74])."

"Such studies suggest that learners are more influenced by agents that are similar to themselves with respect to appearance-related characteristics (e.g., Bailenson and others [75])."

"The context of the persuasion (e.g., participants' prior knowledge, domain area) as well as the desired motivational outcomes (e.g., interest, attitude change) must be at the forefront when designing an agent's appearance."

Designing Avatars as Ideal Virtual Selves

"Within virtual environments such as web-based chat rooms or gaming environments learners can easily create an avatar to represent their virtual "selves.""

"If learners watch an avatar that looks like themselves exercising and losing weight in a virtual environment, they subsequently exercise more and eat healthier in the real world as compared to a control group (Fox and Bailenson [76])."

"Participants were exposed to either an avatar running on a treadmill that looked like them, another avatar (agent) running, or an avatar loitering that looked like them."

"Within 24 h after the experiment, participants who were exposed to the running avatar that looked like them exercised significantly more than those in other conditions."

"In one study, they manipulated the attractiveness of the avatars and observed participants' behaviors."

"Further, in the case of avatars, designing the avatar appearance is likely to be the only learner-controlled feature."

The Motivational Impact of Agent Message Delivery

"According to Short and others [77], social presence is a function of both verbal cues (e.g., tone of voice, prosody) and nonverbal cues (e.g. facial expression, direction of gaze, posture, and deictic gestures)."

"Researchers have speculated that facial expressions (eye, eyebrow, and mouth movements) and deictic gestures (e.g., pointing with arms and hands, head-nodding) are important for pedagogical agents to promote learning-related outcomes (Atkinson [78]; Johnson and others [79]; Lester and others [80])."

"Based on research in human-to-human communication, gestures can reduce ambiguity by focusing learner attention, and facial expressions can reflect and emphasize agent message, emotion, personality, and other behavioral variables."

"Of particular interest was the interaction effect between type of instruction and agent nonverbal behavior (deictic gestures and facial expression)."

"Results indicated that for enhancing learner attitude, it was best when the agent's facial expression was present but its deictic gestures were absent, thereby allowing the learner to fully focus on the motivational message."

Employing Multiple Agents with Distinct Functionality

"Regardless of the level of agent embodiment, one way to provide learners with both motivational and learning support is to employ multiple agents where each representing distinct agent personas."

"Instead of interacting with multiple anthropomorphic agents, affective support might be enhanced when a student's avatar (i.e., his/her self-representation within the program) co-learns with a computer-driven agent."

"As compared to students in the other two conditions, results indicated that students who worked with the caring co-learner agent had significantly greater feelings of social support (an index comprised of five items: "not alone," "praised," "attended," "appreciated," and "supported"), trust, as well as enhanced recall."

"In an experimental study with 103 undergraduate students, Ebbers [81] manipulated type of anthropomorphic interface agent model (coping vs. mastery) and interaction type (direct vs. vicarious) and found a main effect where the agent as a coping model significantly enhanced motivational outcomes as compared to the agent as a mastery model."

Conclusion

"The most critical design feature for a motivational agent is its appearance."

"It is not necessarily in the best interest of the learners for them to choose the appearance of their agent or avatar."

"In a system designed to promote more positive attitudes toward mathematics for young adults, the agent should reflect what the target learners perceives as a "pro-math" peer, similar in age, gender, ethnicity/race and perceived competency."

"The potential for learners to visualize idealized versions of themselves through their avatars or to work with a motivational coach that reflects an optimal appearance as a social model is particularly powerful."

"More research is needed to define the qualities of voice that are most motivationally appropriate as delivered through an agent or avatar."

"The potential for motivational agents and avatars is here."

Mixed-Methods Research: A New Approach to Evaluating the Motivation and Satisfaction of University Students Using Advanced Visual Technologies [172]

This is a machine-generated summary of:

Fonseca, David; Redondo, Ernest; Villagrasa, Sergi: Mixed-methods research: a new approach to evaluating the motivation and satisfaction of university students using advanced visual technologies [172]

Published in: Universal Access in the Information Society (2014)

Link to original: https://doi.org/10.1007/s10209-014-0361-4

If you want to cite the papers, please refer to the original.

For technical reasons we could not place the page where the original quote is coming from.

Abstract-Summary

"A mixed-methods study evaluating the motivation and satisfaction of Architecture degree students using interactive visualization methods is presented in this paper."

"The advantages and disadvantages of applying mixed evaluation technology are discussed in a case study of the use of interactive and collaborative tools for the visualization of 3D architectonical models."

"The main objective was to evaluate Architecture and Building Science students' the motivation to use and satisfaction with this type of technology and to obtain adequate feedback that allows for the optimization of this type of experiment in future iterations."

Introduction

"The study employs a mixed analysis method to concretely obtain the most relevant aspects of the experience that should be improved both in future interactions and in any new technological implementations within a teaching framework [82]."

"It is common to find studies focused on all teaching types and levels that evaluate the incorporation of technology and technological elements into teaching; the most common examples are the use of computers in the classroom and the use of digital content for online training [83, 84]."

"Teachers design educational experiments based on the technology that is available at their college or that is accessible to their students, assuming (based on their experience) that the use of new IT will be possible and beneficial to students."

"A mixed-methods study evaluating the motivation, satisfaction, and academic performance of degree students is presented."

Mobile Technology, Education, and Their Relationship with Universal Access and Design for All

"In the virtual framework, uncounted resources have been dedicated over the last several decades to improve and generate new models and methods for accessing content (rules and recommendations), thereby adapting those contents to all types of users and devices [85]."

"The visualization and discussion of an architectural project in 3D using mobile devices generate a faster workflow, allow students to adapt their design to the real scale of construction, and allow them to easily modify and customize the project for little or no cost."

"This section, it is reaffirmed that the ideological basis of the project is to evaluate how the student adapts to the use and design of the various visualization methods that are accessible to any user: in person using 2D printed layouts, virtually by posting interactive models on the Internet, or, finally, using AR visualization for a combined interaction."

IT in Education. 3D Models and AR Visualization

"If in the process of designing an educational experience, appropriate individual aspects of the main areas are included, one may be closer to redefining and integrating any type of technology into teaching activities, moving away from classic approaches that have been used in current and past technology integration efforts [86]: Software-focused initiatives Demonstrations of sample resources, lessons, and projects Technology-based educational reform efforts Structured/ standardized professional development workshops or courses Technology-focused teacher education courses These approaches tend to initiate and organize their efforts according to the educational technologies being used (and preferred by the teacher or the institution) rather than the students' learning needs, which is exactly the opposite of the desired approach in which the user is a central element of the experience, due to the user's technological profile, motivation for experiencing new pedagogical methods, and evaluation of both the quantitative and qualitative aspects of the experience."

Mixed-Methods Research and UX in an Educational Framework

"The methodological approach of this work let the end users, in this case students of first course of "Architecture" and "Building Sciences and Technologies" degrees of La Salle Campus Barcelona, Ramon Llull University, participate in the definition of the final product, in this case a pedagogical proposal, through methods that allowed them to be creative during the design process."

"Through qualitative methods, the goal was to explore users' desires, needs, and goals when learning about informatics tools that they would use to present 3D projects in their future work."

"The basic objective of the course was to provide students with basic skills in architectural interpretation and reproduction in both 2D and 3D. The secondary objectives were to enable students to print 2D and 3D reproductions, as well as to explore methods of interactive visualization, primarily through the publication of personal blogs and the display of models with RA at the end of the course."

Results

"The students perceived the usefulness of AR technology in both 3D representations and in future architectural work, evaluating it with positive values ranging between 62.5 and 58.4 %, respectively, although there were still high rates of indecision (neutral responses: 35.4 and 39.6 %, respectively) while waiting for the practical experience of the second phase."

"The objective of this second test was threefold: to compare the efficiency of the two methods by comparing perceptions of the 2D system before and after having worked with methods for 3D viewing, to understand the perceptions students on the use of the technologies for 3D viewing (in particular, the AR), and to assess the degree of usability in general of the content, structure, and methodology of the technology."

"The second objective of this phase focused on student perceptions of the use and usefulness of AR as a system for presenting architectural projects by comparing the system proposed with the method used in the 2D phase."

Discussion

"With the proposed method implemented in the academic year 2012–2013, it was concluded that there is no statistically significant differences ($F = 3.276$, $p = 0.075$), between 2008–2009 (with a mean of 5.45, one can consider this year as a previous indicator, because of in pre-EHEA period, all 2D practice results are between 5.25–5.78) and 2012–2013 ($M = 5.24$), as the statistical significance (two-tailed) is 0.075, which exceeds the threshold of 0.05."

"If the results of the 3D projects are analyzed, which were added to the technological innovation described in this paper, and are compared with those of the previous courses, one notices a significant improvement during the post-EHEA period ($F = 3.48$, $p = 0.05$) and results equal to those of the pre-EHEA period, during which there were no significant differences between courses ($F = 0.30$, $p = 0.57$)."

Conclusions

"The mixed method used has demonstrated its usefulness as a dynamic system for capturing information related to students' experiences with technological elements in education."

"The main advantages demonstrated focused on the identification of aspects related to the design process, and the data were much more specific than those that would have been obtained through quantitative methods only, which is usually the focus of experiments on general questions."

"Focusing on the main objective of the study, student motivation and satisfaction with the proposed system were evaluated, obtaining qualitative feedback for the main items that, according to the students, should be implemented in using RA as a common tool for visualization of and interaction with 3D models in architecture education."

"For future directions of work, two possibilities have been clearly identified: the adaptation of content to people with disabilities and a study of the emotional user response to content displayed according to their profile and the technology used."

The Predictive Effect of Digital Literacy, Self-control and Motivation on the Academic Achievement in the Science, Technology and Society Learning Area [173]

This is a machine-generated summary of:

Pala, Şenol Mail; Başıbüyük, Adem: The Predictive Effect of Digital Literacy, Self-Control and Motivation on the Academic Achievement in the Science, Technology and Society Learning Area [173]

Published in: Technology, Knowledge and Learning (2021)

Link to original: https://doi.org/10.1007/s10758-021-09538-x

If you want to cite the papers, please refer to the original.

For technical reasons we could not place the page where the original quote is coming from.

Abstract-Summary
"Science, Technology and Society (STS) is one of the learning areas that is stated in the Social Studies Course Curriculum."

"Even in the STS learning area of the fifth grade, it is stated that digital literacy and self-control are the skills intended to be gained by students."

"Considering this situation, in addition to the digital literacy and self-control skills, the predictive effects of the STS learning area on the academic achievements of students were investigated in the study."

"The data were collected with 'The Science, Technology and Society Learning Area Achievement Test', The Digital Literacy Scale', 'The Self-control Scale for Children', and 'The Social Studies Course Motivation Scale'."

"According to the research results, it was found that digital literacy, self-control and course motivation predict the academic achievements of the students in the STS learning area."

"It was also determined that the predictive effect of these variables on the academic achievement in the STS learning area was in the order of digital literacy, self-control and course motivation."

Introduction

"Considering the studies conducted, it was seen that no study has been conducted on the status of motivation, digital literacy, and self-control skills for predicting the students' academic achievement in the STS learning area."

"In this direction, the study examined the predictive power of social studies course motivation and the digital literacy and self-control skills that are expected to be acquired by students in the area of STS learning, which are beneficial for their academic achievement."

"In SSCC, which is one of these curricula, it is thought that students should gain digital literacy and self-control skills in order to be successful in the STS learning area."

"This study aimed to reveal to what extent digital literacy and self-control skills and the motivation of students predicted the academic achievements of the STS learning area."

Method

"The sample of the research consisted of 742 fifth-grade students determined with a random method from 18 different secondary schools in the Erzincan province in the spring term of the 2017–2018 academic year."

"Five data collection tools—'Personal Information Form (PIF), 'Science, Technology and Society Learning Area Achievement Test (STSLAA)', 'Digital Literacy Scale for Secondary Students', 'Self-Control Scale for Children (SCS-C)', 'Social Studies Course Motivation Scale (SSCMS)'—were used in the study."

"The 'Basic User', 'Independent User' and 'Proficient User' criteria in the Digital skills Self-assessment Grid were taken into a scale considering the levels of secondary school 5th class students, with literature studies on digital literacy and taking expert opinions."

"The stepwise regression analysis was used in the research to determine the most effective variable related to explaining the students' achievement in the field of STS learning; in other words, to determine the predictive power of predictor variables on the students' achievement in the field of STS learning."

Findings

"Whether or not digital literacy skills, self-control skills and course motivation are significant predictors of the students' achievement in STS learning area was investigated with the stepwise regression analysis method from multiple linear regression analysis."

"One-way ANOVA test results for the significance of the stepwise regression model show that the regression model established by including digital literacy skills, self-control skills and course motivation variables is significant ($F(1738) = 304{,}118$, $p < 0.001$)."

"It was determined that digital literacy skills, self-control skills and course motivation together explained 55.3% of the students' achievement in STS learning (R: 0.744, R^2: 0.553, $F(1,738) = 304,118$, $p < 0.001$)."

"The relative importance of the predictive variables on the students' achievement in the STS learning area was determined in the order of digital literacy skills, self-control skills and course motivation."

Result and Discussion
"In the study that digital literacy skills, self-control skills and course motivation significantly predicted the achievement of the fifth-grade students in the STS learning area."

"The relative significance of the predictive variables on the students' achievement in the STS learning area was determined to be digital literacy skills, self-control skills and course motivation, in this order."

"As the digital literacy skills of the students increased, their academic achievements in the STS learning area also increased."

"Self-control skills were found to be a significant predictor of students' achievement score in the STS learning area."

"As the gains related to the effect of technology on socialization and social relations, complying with the safety rules in the virtual environment and keeping the scientific ethics in studies take place in the STS learning area, self-control skills can influence students' academic achievement in this learning area."

Recommendations
"That digital literacy skills have a positive effect on academic achievements in the STS learning area."

"It was determined that digital literacy and self-control skills positively affect the students' academic achievements in the STS learning area."

"It was observed that gaining skills in the STS learning area is more effective than motivation for the academic success of students."

"The relationship between the STS learning area of the 5th grade and the skill relationship was investigated."

Eye-Tracking and Artificial Intelligence to Enhance Motivation and Learning [174]

This is a machine-generated summary of:

Sharma, Kshitij; Giannakos, Michail; Dillenbourg, Pierre: Eye-tracking and artificial intelligence to enhance motivation and learning [174]

Published in: Smart Learning Environments (2020)

Link to original: https://doi.org/10.1186/s40561-020-00122-x

If you want to cite the papers, please refer to the original.

For technical reasons we could not place the page where the original quote is coming from.

Abstract-Summary

"We then proposed a method to define stimuli-based gaze variables that can be used for any kind of stimulus."

"The proposed stimuli-based gaze variables indicate students' content-coverage (in space and time) and reading processes (area of interest based variables) and attention (i.e., with-me-ness), at the perceptual (following teacher's deictic acts) and conceptual levels (following teacher discourse)."

"We identified a significant mediation effect of the content coverage, reading patterns and the two levels of with-me-ness on the relation between students' motivation and their learning performance."

"Such variables enable common measurements for the different kind of stimuli present in distinct MOOCs."

"Our long-term goal is to create student profiles based on their performance and learning strategy using stimuli-based gaze variables and to provide students gaze-aware feedback to improve overall learning process."

"One key ingredient in the process of achieving a high level of adaptation in providing gaze-aware feedback to the students is to use Artificial Intelligence (AI) algorithms for prediction of student performance from their behaviour."

"We also present a method combining state-of-the-art AI technique with the eye-tracking data to predict student performance."

Introduction

"Besides commonly used variables coming from users' click-streams, keywords and preferences, we can also use eye-tracking variables to accurately measure students' attention during their interaction with learning materials (e.g., MOOC lectures)."

"We address the general question of how gaze-variables (related to students' reading and attention) can help students to watch MOOC videos more efficiently?"

"In this contribution, we address the following two questions: 1.How eye-tracking behaviour mediates the relationship between students' motivation and learning within a MOOC?.2.How well we can predict the learning gain and motivation from the eye-tracking data in its most basic form?"

"In order to answer these questions, we define variables using the stimulus (video lecture) presented to the students."

"To answer the second question, we utilize one of the most basic eye-tracking visualisations, "Heat-maps" (Špakov & Miniotas [87]) to extract features and use state-of-the-art machine learning algorithms to predict the students' learning gains."

Related Work

"Tobagi [88] compared different modalities of video lectures (interactive video, instructional television and television) and preconceptions of difficulty for different modalities and found that there was no significant difference in the learning outcome but there was a significant difference in the level of preconceived difficulty in television and interactive videos."

"Choi and Johnson [89] compared learning outcome and learners motivation (attention, relevance, confidence, satisfaction) in video based learning to traditional textual-instruction based learning and found no difference in learning outcome for the two conditions."

"The students wore more attentive in video based learning condition that the textual-instruction condition."

"We utilize eye-tracking to measure students' attention and then address how students' attention (i.e., "with-me-ne") has the capacity to mediate the relationship between students' motivation and learning within a MOOC video."

Methodology

"The measures used in our study were: content coverage, scanpath (a combination of the fixations and saccades in the order of appearances) based variables, students/teacher co-attention (i.e., with-me-ness) coming from eye-tracking, students' motivation coming from SPQ and students learning (coming from the final test)."

"Conceptual with-me-ness measures how often a student looks at the objects verbally referred to by the teacher during the whole course of time (the complete video)."

"To identify how "with-me-ness" (measured by eye-tracking) mediates the relationship between students' motivation (measured by the questionnaire) and learning (measured by the post quiz) within a MOOC we employ mediation analysis proposed by Baron and Kenny [90]."

"1.Overlay the eye-tracking data on the video to create the heatmap.2.Create the heatmap image from every presentation slide in the video lecture (this step resulted into 15 heatmap image per participant).3.Use the pretrained VGG-19 (Simonyan & Zisserman [91]) deep neural network architecture to extract the 1000 features per image.4.Use a non-overlapping and sliding window of size 10 to reduce the number of features to 100."

Results

"We observe that learning can be significantly predicted by motivation and that content coverage can be predicted by motivation."

"We observe that learning can be significantly predicted by motivation and that perceptual with-me-ness can be predicted by motivation."

"We observe that that learning can be significantly predicted by motivation, and that AOI misses can be predicted by motivation."

"We observe that learning can be significantly predicted by motivation and that perceptual with-me-ness can be predicted by motivation."

"There is a significant prediction of learning by motivation and perceptual with-me-ness, however the coefficient of motivation is not significant anymore."

"We observe that that learning can be significantly predicted by motivation, and that conceptual with-me-ness can be predicted by motivation."

"There is a significant prediction of learning by motivation and conceptual with-me-ness, however the coefficient of motivation is not significant anymore."

Discussions and Conclusions

"The reported study developed and empirically explored two models, where teacher/student co-attention (i.e., with-me-ness) were found to mediate the relationship of motivation and learning in MOOC videos."

"The good performing students having a higher motivation had the highest content coverage (larger areas of the attention) among all the participants, despite having spent the similar time on the slides."

"We could interpret this reading behaviour, based upon the reading time differences, in terms of more attention being paid by the good performing students having a high learning motivation than other student profiles."

"One could use the content coverage for student profiling as well based on the performance and the learning motivation."

"The results also contribute towards our long-term goal of defining the student profiles based on their performance and motivation using the gaze data."

Cyberloafing in IT Classrooms: Exploring the Role of the Psycho-Social Environment in the Classroom, Attitude to Computers and Computing Courses, Motivation and Learning Strategies [175]

This is a machine-generated summary of:

Yılmaz, Ramazan; Yurdugül, Halil: Cyberloafing in IT classrooms: exploring the role of the psycho-social environment in the classroom, attitude to computers and computing courses, motivation and learning strategies [175]

Published in: Journal of Computing in Higher Education (2018)

Link to original: https://doi.org/10.1007/s12528-018-9184-2

If you want to cite the papers, please refer to the original.

For technical reasons we could not place the page where the original quote is coming from.

Abstract-Summary
"The purpose of this study is to find out the level of cyberloafing behaviors of the students in information technologies (IT) classrooms; and to examine the relationship between cyberloafing behaviors and the psycho-social perceptions, attitudes and motivation for the courses and learning strategies."

"The study was carried out with the students who take Computing I course in the IT classroom."

"The results demonstrate that students' cyberloafing behaviors are influenced by their psycho-social perceptions, attitudes and learning strategies."

"The results demonstrate that students' cyberloafing behaviors aren't influenced by their motivation for the courses."

Introduction

"The purpose of this study is to find out the level of cyberloafing behaviors of the students in IT classrooms; and to examine the relationship between cyberloafing behaviors and the psycho-social environment in the classroom perceptions, attitudes towards the course and motivation for the course and learning strategies."

"Besides, within the scope of the study, the structural relationships among psycho-social environment perceptions of the students towards the IT classrooms, their attitudes towards computers and computing classes, their motivation towards the course and learning strategies were also analyzed to define students' perceptions of the courses in the computer classrooms."

"When the psycho-social environment features of IT classrooms are evaluated in terms of cyberloafing behaviors, it can be said that positive perceptions towards IT classrooms will lead students to have increased interest towards the course, learning activities in the course and knowledge sharing related to the course."

Method

"'I explain my ideas to other students' Calculated Cronbach-alpha reliability coefficients of the scale for this study carried out; calculated as .70 for 'student cohesiveness', .78 for 'involvement', .82 for 'investigation', .77 for 'task orientation', .81 for 'cooperation' and .85 for 'equity'."

"Calculated Cronbach-alpha reliability coefficients of the scale for this study carried out; calculated as .77 for 'anxiety', .73 for 'enjoyment', .71 for 'usefulness of computers', .70 for 'usefulness of course'."

"The Calculated Cronbach-alpha reliability coefficients of the MSLQ-M scale for this study carried out; calculated as .73 for 'intrinsic goal orientation', .70 for 'extrinsic goal orientation', .79 for 'task value', .71 for 'control of learning beliefs', .82 for 'self-efficacy for learning and performance' and .77 for 'test anxiety'."

"Calculated Cronbach-alpha reliability coefficients of the MSLQ-LS scale for this study carried out; calculated as .71 for 'rehearsal', .69 for 'organization', .72 for 'elaboration', .77 for 'critical thinking', .66 for 'help seeking', .72 for 'peer learning', .75 for 'self-regulation', .70 for 'effort regulation' and .73 for 'time and study environment'."

Findings

"Looking at these numbers, it can be said that students' scores are average in CLA, WIHIC, ACCC, MSLQ-M, MSLQ-LS scales."

"Based on the findings, it can be said that there is generally a low correlation between CLA and WIHIC, ACCC, MSLQ-M, MSLQ-LS scales."

"When the case is analyzed from the sub-scales of CLA, a low relationship is found between CLA(social)-WIHIC ($r = .341$, $p < .01$), CLA(individual)-MSLQ-LS ($r = .292$, $p < .01$), CLA(social)-MSLQ-LS ($r = .323$, $p < .01$) and CLA(news)-MSLQ-LS ($r = .291$, $p < .01$)."

"A moderate level of relationship is found among the other scales: WIHIC-ACCC ($r = .473$, $p < .01$), WIHIC–MSLQ-M ($r = .524$, $p < .01$), WIHIC–MSLQ-LS ($r = .627$, $p < .01$), ACCC–MSLQ-M ($r = .624$, $p < .01$), ACCC–MSLQ-LS ($r = .479$, $p < .01$), MSLQ-M–MSLQ-LS ($r = .746$, $p < .01$)."

Discussion

"This study was carried out to find out the structural relationships between cyberloafing behaviors of the students in IT classrooms and their perceptions of the psycho-social environment in the classroom, attitudes towards the computer and course and motivation towards the course and learning strategies."

"The results of structural model in the study indicated that there was a path coefficient in a positive direction between the cyberloafing behaviors of the students and their perceptions of the psycho-social environment in the classroom, attitudes towards the computer and course, and motivation towards the course and learning strategies however, this coefficient was low."

"Of the study, whilst the cyberloafing behaviors of the students in IT classroom perceptions of the psycho-social environment, and attitudes towards computer and computing courses, motivation and learning strategies are moderate."

Conclusions

"In the context of the study, the structural relations between cyberloafing behaviors, psycho-social environment perceptions in the classroom, attitudes towards the course and motivation for the course and learning strategies were examined."

"The results of the study, in which the structural relationships between students' psycho-social environment perceptions in the classroom, attitudes towards the course and motivation for the course and learning strategies were examined, offer teachers and researchers suggestions in terms of designing the IT classroom, developing and managing a positive classroom atmosphere, increasing student motivation, developing positive attitudes towards computers and computing classes and the use of learning strategies in these classrooms."

"Development of equality, objectivity and justice beliefs among students will contribute to the positive development of psycho-social environment perceptions in the classroom and this will enable the development of the attitudes and motivation of students towards the class and to effective use of learning strategies."

Context Matters: Differing Implications of Motivation and Help-Seeking in Educational Technology [176]

This is a machine-generated summary of:

Karumbaiah, Shamya; Ocumpaugh, Jaclyn; Baker, Ryan S.: Context Matters. Differing Implications of Motivation and Help-Seeking in Educational Technology [176]

Published in: International Journal of Artificial Intelligence in Education (2021)

Link to original: https://doi.org/10.1007/s40593-021-00272-0

If you want to cite the papers, please refer to the original.

For technical reasons we could not place the page where the original quote is coming from.

Abstract-Summary

"EdTech researchers often have access to larger and more diverse samples of student data to test replication across broad demographic contexts as compared to either the small-scale experiments or the larger convenience samples often seen in experimental psychology studies of learning."

"The source of typical EdTech data (i.e., online learning systems) and concerns related to student privacy often limit the opportunity to collect demographic variables from individual students—the sample is diverse, but the researcher does not know how that diversity is realized in individual learners."

"We use this approach to better understand the role of social factors in students' self-regulated learning and motivation-related behaviors, behaviors whose effectiveness appears to be highly variable between groups."

"We demonstrate that school-level demographics can be significantly associated with the relationships between students' help-seeking behavior, motivation, and outcomes (math performance and math self-concept)."

"By studying the conditions under which these relationships vary across different demographic contexts, we challenge implicit assumptions of generalizability and provide an evidence-based commentary on future research practices in the EdTech community surrounding how we consider diversity in our field's investigations."

Introduction

"Even when EdTech researchers obtain larger sample sizes, the typical source of EdTech data (i.e., intelligent tutoring systems, adaptive learning platforms, educational game websites) often limits the practicality of obtaining demographic variables from individual students."

"Adaptive EdTech with automated decision-making attempts to meet individual student needs, but past research points out that technologies that aim to benefit all students might disproportionately benefit the more advantaged groups."

"It seems important that EdTech designers and researchers make greater efforts to overcome the challenges involved in collecting demographic data in order to ensure population validity, equity-oriented EdTech design, and fairness-aware educational data mining (e.g., Ocumpaugh and others [92])."

"Our broader research goal is to investigate if the current designs of adaptive EdTech lead to inequitable student outcomes across different demographics."

Prior Work

"We also review the past research on the influence of student demographics on help-seeking, motivation, and the two outcome variables—math performance and self-concept."

"We then look at previous research on intrinsic and extrinsic motivation, especially as it relates to social and demographic differences in education, based on evidence that help-seeking is associated with the student's motivation (Butler [93]; Nelson-Le Gall & Resnick [94])."

"Most studies analyzed the effectiveness of hints by focusing on the relationship between help-seeking behavior(s) and student outcome(s), with some researchers emphasizing that the intentionality of help-seeking behavior makes it a good candidate for understanding students' self-regulated learning (SRL) strategies (Aleven and others [95]; Goldin and others [96])."

"We focus on school as the social context and analyze the influence of school demographics on the relationship between student outcomes (math performance and math self-concept) and their help-seeking and motivational behaviors."

Data Collection

"Another automatic outcome of earning more points and streaks in RM Foundations is that doing so opens up more challenging and optional problems (B and C level), which are otherwise locked until students demonstrate mastery in simpler problems (A level)."

"We use the number of items purchased and the number of B and C level problems attempted as proxies for students' extrinsic and intrinsic motivation, respectively."

"Our proxy for differentiating extrinsic motivation from intrinsic assumes that students understand that it is easier to earn points by solving the simpler problems than it is to do so by taking on more challenging material."

"There were 75 students using Reasoning Mind Foundations per school (min = 25; SD = 70) and 364 per school district (SD = 730), with one large urban district in Texas constituting the majority of our data, with 3039 students across 62 schools."

Data Exploration
"From the interaction log data, we operationalize help-seeking behavior as the number of hints used by a student in Reasoning Mind Foundations."

"We operationalize intrinsically motivated behavior as the number of B and C-level problems attempted by a student in Reasoning Mind Foundations."

"For the purposes of this paper, math performance is defined as the accuracy of student responses to A-level problems in Reasoning Mind Foundations, i.e., the ratio of the number of correct answers to the number of problems attempted."

"Students' self-concept in mathematics was measured using a five-item survey adapted from Marsh and others [97]."

"We explored the school-level differences in student outcomes (math performance and self-concept) and hint usage."

"We also considered school-level data on the percentage of students representing major ethnic/racial groups, using the categories provided by the TEA."

Analysis
"We conduct a two-step data analysis to explore how help-seeking and the two motivations might differ based on student demographics, while controlling for performance and math self-concept."

"We determine how closely students' math performance and self-concept measures correlate to their hint usage and motivational behaviors within each school, using Spearman ρ correlations due to non-normality in the data."

"We produce three types of measures for the three behaviors for each student, the correlation between a behavior and performance on A-level problems, the correlation between a behavior and the pre-year survey of self-concept, and the correlation between a behavior and the post-year survey of self-concept."

"We determine whether the differences in these correlations are themselves correlated to school-level demographics."

Results
"When we do not aggregate students into school-level populations (instead treat them all as a single population), there is still a significant relationship between intrinsic motivation and math performance ($\rho = 0.17$, $p < 0.001$) with a correlation

value closer to the average across schools but doing so fails to capture the range and variance in this relationship."

"When we do not aggregate students into school-level populations, there is still a significant relationship between extrinsic motivation and math performance ($\rho =$ 0.04, < 0.001), but the correlation value is very close to zero."

"The association between higher usage of the gamified extrinsic motivations and math performance is positive among students from suburban and rural schools and negative among students from urban schools."

"Another concerning trend among schools with a higher percentage of students who have limited English proficiency is the negative association between extrinsic motivation and math performance and the lack of a strong relationship between intrinsic motivation and math performance."

Discussion

"Through this paper, we hope to contribute to the emerging conversations on fairness and equity in EdTech systems in two ways—(1) by investigating inequitable outcomes across student subpopulations on two of the fundamental psychological constructs in EdTech design, (2) by demonstrating the use of publicly-available, school-level demographics for fairness research where individual student demographics may be difficult or impossible to acquire."

"Although this finding is not novel in itself, this paper demonstrates an approach to assess the generalizability of the EdTech research findings by using publicly-available, school-level demographics when we have access to larger data (e.g., interaction logs), which may be particularly important when access to individual student demographics is restricted."

"Collecting student-level data is not always feasible, and this study has demonstrated that school-level aggregates can still help us understand the role of demographic factors in understanding motivation and help-seeking behaviors."

Conclusion

"We make these recommendations within the context of help-seeking behaviors, which are an important part of self-regulation, and student autonomy more broadly, but which is also a behavior that may be particularly susceptible to cultural differences (i.e., Ogan and others [98])."

"Amidst the mixed results from empirical studies on the effectiveness of hints, Aleven and colleagues (Aleven and others [95]) continue to recommend the use of hints in EdTech systems and suggest making four key methodological distinctions when studying interventions designed to promote help-seeking—(1) effects on learning in the same learning environment versus a new environment; (2) effects on current learning versus future learning; (4) effects on learning in the same domain versus another; (3) effects on SRL processes versus domain-level learning."

"We recommend EdTech designers and researchers consider the role of students' demographic contexts while making design choices to motivationally enhance their systems."

An Investigation of the Influence of Intrinsic Motivation on Students' Intention to use Mobile Devices in Language Learning [177]

This is a machine-generated summary of:

Sun, Yanyan; Gao, Fei: An investigation of the influence of intrinsic motivation on students' intention to use mobile devices in language learning [177]

Published in: Educational technology research and development (2019)

Link to original: https://doi.org/10.1007/s11423-019-09733-9

If you want to cite the papers, please refer to the original.

For technical reasons we could not place the page where the original quote is coming from.

Abstract-Summary
"This study examines the relationships among intrinsic motivation, critical variables related to technology adoption, and students' behavioral intention in mobile-assisted language learning (MALL)."

"The results indicated that although intrinsic motivation did not have a direct influence on students' behavioral intention in MALL, it had a positive influence on students' behavioral intention through the two intervening variables, perceived usefulness and task technology fit."

"Perceived ease of use, however, was not associated with students' behavioral intention directly, nor was it predicted by intrinsic motivation."

Introduction
"Mobile technologies have been used to deliver formal classroom-based or computer-based language learning contents (e.g. Abdous and others [101, 102]; Jolliet [103]) as well as to support new ways of language learning such as collaborative projects (e.g. Murphy and others [104]) and cross-cultural communication (e.g. Comas-Quinn and others [105]; Shao [106])."

"In technology-rich language learning environments, motivation has a positive association with students' learning behavior."

"Given the importance of motivation in language learning, it is considered as an important variable that measures the effects of learning via mobile technologies."

"While the importance of motivation in technology-assisted language learning is well-recognized, in most studies, motivation was examined to determine the effectiveness of MALL or the relationship between motivation and students' performance such as tests scores."

"Whether motivation affects the adoption of mobile technologies in language learning, however, still remains unexplored."

"Though previous literature consistently identified the positive correlation between motivation and student learning in technology-supported environments, the mechanism of how motivation exerted such influence is rarely understood."

Literature Review and Hypothesis Development

"We included both perceived usefulness and perceived ease of use as variables that may be affected by intrinsic motivation and as potential significant predictors for students' behavioral intention in adopting MALL."

"We included the core element of TTF in our research model and assumed that it had influence on students' behavioral intention in adopting mobile devices for language learning as well as acting as an intervening variable between intrinsic motivation and students' behavioral intention."

"Intrinsic motivation positively influences students' perceived usefulness (of mobile devices) and the perceived usefulness is positively associated with their intention to adopt MALL."

"Intrinsic motivation positively influences students' perceived ease of use (of mobile devices) and the perceived ease of use is positively associated with their intention to adopt MALL."

Method

"The survey results indicated that most students spent some time on activities via mobile devices on a daily basis, though there was a small number of students who did not use mobile devices as often."

"All students spent time on social networks via their mobile devices every day and 98.8% ($N = 167$) of them spent between 11–30 min and more than 3 h. The survey instrument contained 7 points Likert-scale items asking students to rate statements on intrinsic motivation, perceived usefulness, task technology fit, and perceived ease of use, and their behavioral intentions to use mobile devices to learn English as a foreign language (7 = Strongly agree, 6 = Agree, 5 = Somewhat Agree, 4 = Neutral, 3 = Somewhat disagree, 2 = Disagree, 1 = Strongly disagree)."

"Definition, we selected three subscales from McAuley and others [107] IMI survey to measure participants' intrinsic motivation, which are interests/enjoyment, efforts and importance, and perceived competence."

Results
"All variables had a Cronbach's alpha value greater than .80, which indicated excellent (intrinsic motivation = .92, perceived usefulness = .91, task technology fit = .92) or good (perceived ease of use = .88, behavioral intention = .86) internal consistency (Nunnaly [108])."

"The results showed intrinsic motivation was positively related to perceived usefulness, task technology fit, perceived ease of use and behavioral intention."

"In the revised model, the determinant of behavioral intention with the largest casual effect was task technology fit (.47), followed by perceived usefulness (.22)."

"Approximately 42% of variance in behavioral intention, 9% of variance in task technology fit, 57% of variance in perceived usefulness, and 14% of variance in perceived ease of use was explained by the model."

Discussion
"Intrinsic motivation positively influences students' perceived usefulness (of mobile devices), and the perceived usefulness is positively associated with their intention to adopt MALL."

"Although intrinsic motivation was not directly associated with students' behavioral intention in MALL, the results indicated that it indirectly affected students' behavioral intention through perceived usefulness and task technology fit."

"The results indicated that students' intrinsic motivation on language learning had positive influence on their attitudes towards the usefulness as well as the task technology fit in MALL."

"For the task technology fit, the results indicated that intrinsically motivated students are more likely to perceive the mobile learning environment as suitable for their language learning task."

"Intrinsic motivation positively influences students' perceived ease of use (of mobile devices), and the perceived ease of use is positively associated with their intention to adopt MALL."

Conclusion
"While most research focuses on exploring specific types of instructional design that may promote students' motivation in MALL, this study contributes to the literature by revealing the relationships among intrinsic motivation, important variables related to technology adoption, and students' behavioral intention towards MALL."

"The results indicated that although intrinsic motivation did not have a direct influence on students' behavioral intention in MALL, it had a positive mediating effect on students' behavioral intention through the two intervening variables, perceived usefulness and task technology fit."

"As an important construct in the TAM, perceived ease of use had been confirmed as a predictor of students' behavioral intention in mobile learning in some studies (e.g. Cheon and others [109]) while in other studies it had no association with students' intention to use mobile technologies in learning (Park and others [110])."

"The results of the present study echoed the conclusion that perceived ease of use was not associated with students' behavioral intention in MALL."

"It is possible that perceived ease of use no longer influence students' intention to adopt MALL when they are already familiar with mobile technologies."

Performance and Motivation in a Middle School Flipped Learning Course [178]

This is a machine-generated summary of:

Winter, Joshua W.: Performance and Motivation in a Middle School Flipped Learning Course [178]

Published in: TechTrends (2017)

Link to original: https://doi.org/10.1007/s11528-017-0228-7

If you want to cite the papers, please refer to the original.

For technical reasons we could not place the page where the original quote is coming from.

Abstract-Summary

"Flipped learning is a teaching approach that promotes collaboration by using technology to 'flip' traditional instruction."

"Research on flipped learning is limited, in that studies are mostly conducted in postsecondary classrooms."

"My purpose was to identify the relationship between student motivation and performance in a flipped learning course."

"Significant differences in survey responses between performance levels revealed that the technology-based content in flipped learning may lead to increased motivation and improved performance."

"My findings suggest flipped learning benefits average achieving students through differentiated instruction."

"Practical implications include designing learning spaces to maximize student engagement and incorporating learner-appropriate strategies with flipped learning."

"Future research should focus on learning space design in different K-12 environments."

Introduction

"Flipped learning is a pedagogical approach designed to improve group learning by using technology to flip traditional instruction."

"Despite its popularity, flipped learning is criticized for its overreliance on lectures, its ineffectiveness for students without technology access, and a lack of comprehensive research."

"Flipped learning is composed of two integral but inherently different learning spaces: an individual learning space that includes instructional content enhanced by technology and a group learning space or collaborative environment."

"Technology-supported learning environments benefit middle school students."

"Another middle school study compared flipped learning to traditional techniques and found little difference in students' conceptual understanding, however, the teacher did note increased engagement during collaborative activities (Kirvan and others [111])."

"Differentiated instruction requires teachers employ different approaches for individual students to meet the various learning needs in a classroom."

"In middle school studies, motivation and engagement have shown to increase within technology-supported learning environments (Godzicki and others [112])."

Purpose

"Educational research is expected to provide practicing educators with knowledge and techniques to achieve learning objectives."

"This study's purpose was to investigate flipped learning in middle school, focusing on the relationship between student performance and motivation."

"By integrating research methods with learner appropriate strategies, the findings are intended to help teachers effectively use flipped learning to benefit students."

"The following were my research objectives: To determine the relationship between students' motivation and learning performance measures in a middle school social studies classroom using flipped learning."

Methods
"Students progressed sequentially, completing approximately one assignment a week during an eight-week unit."

"For each assignment, students had to take notes and complete an online questionnaire (Google Forms) on the content covered in the screencast."

"Students with incomplete assignments had to complete the assignment during class prior to working on collaborative activities in the group space."

"Instructions were provided through the LMS and students progressed at their own pace, working in different groups at each station."

"Stations did not have to be completed sequentially and students were encouraged to choose based on interest."

"Student check-ins at my desk allowed me to assess homework assignments completed in the individual space and progression in the group space."

"While not exclusively reflective of performance, course grades provided a measurement by which students can be grouped for statistical purposes."

Findings
"When asked if collaborative activities were understood in the group space, top performers responded significantly higher than low performers, $F(2,32) = 3.65$, $p = 0.037$."

"When asked how hard they worked, middle performers responded significantly higher than low performers, $F(2,32) = 4.29$, $p = 0.022$."

"When asked if their best effort was put forth, middle performers responded significantly higher than low performers, $F(2,32) = 8.41$, $p = 0.001$."

"When asked to rate their effort in the individual space, top performers responded significantly higher than low performers, $F(2,32) = 8.49$, $p = 0.001$."

"When ask to rate their effort in the group space, top performers responded significantly higher than low performers, $F(2,32) = 4.05$, $p = 0.027$."

Discussion
"Logically, a motivated and engaged student is more likely to perform well and experience meaningful learning."

"Studies in online schools have identified a relationship between self-efficacy and performance, claiming effort regulation increases motivation "when students engage in learning tasks perceived as easy to execute and interesting and enjoyable" (Kim and others [113], p. 263)."

"This suggests technology-based content perceived by students as requiring limited effort can motivate and improve performance."

"This suggests flipped learning provides significant benefits for average achieving students."

"The increased performance is considered a result of the differentiation within flipped learning (Bergmann and Sams [114])."

"Effective flipped learning creates time and space to identify student strengths and weaknesses."

"This face-to-face communication between student and teacher is a result of the flipped learning paradigm."

"Teacher suggestions guide the student and help individualize the learning process."

"Individualized learning then helps foster motivation, increase effort, and ultimately improve performance."

Conclusion

"My findings provide a unique view of flipped learning that adds to limited research in K-12 education."

"My study suggests flipped learning motivates students through technology-based content and is beneficial for average achieving students when effectively implemented with age-appropriate strategies, specifically differentiated instruction."

"My study also suggests flipped learning is a model approach for middle school, where student success depends on differentiation and addresses the acute disparity in middle school students."

"Future studies should investigate flipped learning and learning space design in K-12 environments, especially middle and high school courses."

Studying Technology-Based Strategies for Enhancing Motivation in Mathematics [179]

This is a machine-generated summary of:

Star, Jon R; Chen, Jason A; Taylor, Megan W; Durkin, Kelley; Dede, Chris; Chao, Theodore: Studying technology-based strategies for enhancing motivation in mathematics [179]

Published in: International Journal of STEM Education (2014)

Link to original: https://doi.org/10.1186/2196-7822-1-7

If you want to cite the papers, please refer to the original.

For technical reasons we could not place the page where the original quote is coming from.

Abstract-Summary

"During the middle school years, students frequently show significant declines in motivation toward school in general and mathematics in particular."

"We evaluate the impact of three brief technology-based activities on students' short-term motivation in math."

"16,789 5th to 8th grade students and their teachers in one large school district were randomly assigned to three different technology-based activities, each representing a different framework for motivation and engagement and all designed around an exemplary lesson related to algebraic reasoning."

"We investigated the relationship between specific technology-based activities that embody various motivational constructs and students' engagement in mathematics and perceived competence in pursuing STEM careers."

"Results indicate that the effect of each technology activity on students' motivation was quite modest."

"These effects were modified by students' grade level and not by their demographic variables."

"Teacher-level variables did not have an effect on student outcomes."

"Our results are also encouraging about developers' ability to create instructional interventions and professional development that can be effective when experienced by a wide range of students and teachers."

Background

"To the self-efficacy and implicit theories of ability, value beliefs are also a significant determinant in students' motivation and achievement (Eccles Parsons and others [115])."

"How can the constructs described above be targeted through technology-based educational experiences to support the motivation of students in mathematics and science?"

"To investigate the potential impact of technology-based activities on students' mathematics motivation, we designed three different types of technology activities (or 'inductions'). (We use the term 'induction' to refer to the technology activities, to avoid possible confusion between the technology activities and math lesson activities (described below))."

"Motivationally, Induction 1 was designed to primarily impact students' self-efficacy."

"Students in grades 5 to 8 participated in a four-day classroom-based experience, beginning with a one-day technology activity, followed by a two-day mathematics lesson on algebraic reasoning, and concluding with revisiting the same technology induction on the final day."

Method

"In the elementary schools, the 163 5th grade teachers, who taught all subjects to the same group of students each day, implemented the intervention with their homeroom students."

"The excluded participants had teachers with lower self-efficacy for student engagement and instruction ($p = .002$) and self-efficacy for technology use ($p < .001$) than included participants."

"The teacher pre-test included items that tapped teachers' own teaching self-efficacy for instruction and student engagement (22 items), technology use (7 items), and mathematics (12 items)."

"The first level of the model, the student level, included students' prior knowledge (VA-SOL) scores, pretest math learning scores, pretest self-efficacy scores, pretest implicit theory of ability scores, pretest value scores, and demographic information, including ELL status, grade, gender (male coded as 1 and female coded as 0), and ethnicity."

"The second level of the model, the teacher level, measured the effect of experimental condition, teachers' self-efficacy for student engagement and instruction, teachers' self-efficacy for technology use, teachers' mathematics self-efficacy, and teachers' implicit theory of math ability."

Results

"Comparing Inductions 1 and 2, students in Induction 2 earned similar math learning scores to students in Induction 1, $\beta = 0.003$, $p = .872$."

"Comparing Inductions 1 and 3, students in Induction 3 had similar math learning scores to students in Induction 1, $\beta = -0.01$, $p = .409$."

"There were no significant differences between any of the inductions on the student self-efficacy variable, nor were there any significant interactions between inductions and grade or inductions and prior knowledge (p's > .128)."

"As grade increased, Induction 2 led to higher implicit view of math ability scores than Induction 1, $\beta = 0.12$, $p < .001$."

"There was also not a significant interaction between Induction 3 and grade, $\beta = 0.03$, $p = .271$, nor between Induction 3 and prior knowledge (VA-SOL), $\beta < 0.001$, $p = .371$."

Discussion

"Our first research question concerned the general impact of the four-day intervention on students' motivation in mathematics, particularly self-efficacy, implicit theory of ability, and value."

"We found that induction type and student-level factors had a moderate influence on the motivational impact of the intervention."

"A third possible explanation for this finding is that, although all three inductions did target different aspects of motivation, these inductions were not the only component of the overall four-day intervention that was designed to influence students' motivation."

"A second pattern that emerges from the complex results of our second research question is that the effects of each induction on students' motivation were influenced by students' age, as evidenced by the frequency of significant induction type by grade interactions."

"It is possible that this approach (which did enable teachers to implement the two-day lesson with fidelity) also helps explain the absence of teacher-level effects on student motivation."

Conclusion

"Investigating along a developmental span the relationship between specific technology-based motivational activities and student interest in STEM careers is important, because much potential talent in science, technology, engineering, and mathematics is now lost."

"Our research interweaved alternative motivational activities with effective and authentic mathematics learning, in order to take initial steps toward developing insights about the added value of technology for building confidence in math and science capability, seeing one's abilities in STEM as able to improve over time, and developing a passion or sustained interest in becoming a scientist or engineer."

"Further research is needed to determine the degree, duration of, and type of instructional intervention necessary to substantially impact multi-dimensional, deep-rooted motivational constructs, such as self-efficacy."

Endnotes

"[b]Student and teacher assessments also included additional items assessing several other motivational constructs."

"Students were administered a short assessment immediately after the conclusion of the Day 1 technology-based motivational activities that focused on several of these additional motivational constructs."

"We report only on those student and teacher variables that were explicitly considered in the design of the inductions and that were specifically hypothesized to be related to the effectiveness of the intervention."

[Section 7]

"As a step toward improving our understanding of the potential impact of technology-based activities on students' motivation in mathematics, the goal of this project was to investigate the relationship between (a) specific technology-based activities that exemplify various motivational constructs, (b) students' engagement in mathematics and perceived competence in pursuing STEM careers, and (c) students' mathematics learning from a short algebra lesson."

"As part of a four-day school-based intervention, students in grades 5 to 8 in a large school district were randomly assigned to three different technology-based activities, each representing a different framework for motivation and engagement designed around an exemplary lesson related to the learning of algebra."

"What is the impact of the four-day intervention on students' motivation in mathematics, including interest in pursuing STEM careers?"

"We begin by reviewing evidence on how and why technology-based activities might impact students' motivation in STEM fields."

From Distance Learning to Integrated Digital Learning: A Fuzzy Cognitive Analysis Focused on Engagement, Motivation, and Participation During COVID-19 Pandemic [180]

This is a machine-generated summary of:

Capone, Roberto; Lepore, Mario: From Distance Learning to Integrated Digital Learning: A Fuzzy Cognitive Analysis Focused on Engagement, Motivation, and Participation During COVID-19 Pandemic [180]

Published in: Technology, Knowledge and Learning (2021)

Link to original: https://doi.org/10.1007/s10758-021-09571-w

If you want to cite the papers, please refer to the original.

For technical reasons we could not place the page where the original quote is coming from.

Abstract-Summary

"This work focuses on Distance Learning during the COVID-19 pandemic to improve students' motivation, participation, and engagement, trying to contain the drop-out phenomenon."

"Distance Learning at the time of COVID-19 is an educational methodology and it can be considered the only occasion to keep an educational connection between students and teachers."

"Experimentation in a Mathematics STEM class was carried out evaluating the impact of Distance Learning on students' levels of motivation, participation, and engagement, computed through a Fuzzy Cognitive Map."

"The results, which have been analysed through Technological Pedagogical Content Knowledge and Instrumental Genesis theories, show on one hand that Distance Learning is valid as an additional and support methodology but, on the other hand, they highlight the ineffectiveness of completely remote teaching."

Introduction and Motivations

"Our research is motivated firstly by the need to analyze how the student's status, defined in terms of engagement, motivation and participation, could change in an emergency context, secondly, by how to maintain these levels high not only by transferring the teaching contents from face-to-face teaching to completely remote teaching but adapting them according to appropriate teaching models and techniques."

"The data collected seem to show that, thanks to the use of an adaptive e-learning platform and restructuring of educational content by using technologies, the students have maintained an adequate level of engagement and motivation parameters comparable to those of previous years."

"This study tries to answer some of these questions by structuring the virtual learning environment for university students in order to get the most out of an adaptive e-learning system with the possibility of formally analyzing the current situation of the student, defined by indices of engagement, motivation, and participation, calculated through an FCM."

Related Works

"Some authors refer to the role of motivation and engagement (Kong et al. [116]; Xiong et al. [117]), others to the increase of participation (Wolcott [118]; Li et al. [119]), all elements considered important to foster a student-centered learning environment."

"The engagement can be active (learner actively participates in the learning environment by publishing posts in the forum, asking questions, etc); passive (the learner only answers or vote the posts of other users; she follows the lessons but without asking questions, etc); disengagement, when the learner has poor participation and interest in the course (Ramesh et al. [120])."

"Kim et al. [121] define engagement as cognitive and affective participation in learning activities."

"Motivation and participation are related to the use of technology in learning and teaching Mathematics (Giesbers et al. [122])."

"This study is part of a line of research already undertaken by the authors and concerns the experimentation of teaching methodologies in STEM courses to motivate students to study mathematics and facilitate their learning."

Conceptual Framework

"Teachers are faced with a difficult challenge: they must integrate the knowledge of pedagogical contents or the knowledge of the content that deals with the teaching process, the knowledge of the technological contents that refer to how technology can create new representations for the specific content, technological pedagogical knowledge which refers to the knowledge of how various technologies can be used in teaching."

"The teacher thus makes use of technological artifacts that undergo a triple process of instrumental genesis: from the didactic point of view, because their use is aimed at generating knowledge and enhancing skills; from the pedagogical point of view, because their use is subordinated to suitable teaching methods activated by the teacher and aimed at the construction of mathematical meanings; from a technological point of view, because the use of technologies is not an end but implements an effective mobilization of strategies aimed at learning."

The Case Study

"In order to motivate students, encourage participation and engagement, keep students' degree of attention alive, stimulate their epistemic curiosity, interest, optimism, and passion, it was necessary to readjust the disciplinary methodological knowledge and teaching contents through the use of an adaptive e-learning platform."

"The adaptive e-learning platform was used to provide students with the teaching material of the course both in a textual and interactive format, the assignments to be completed at the end of the topic, check the corrections of the exercises, etc In addition, the integrated feedback system has allowed the teacher personalized teaching, centered on the needs of the individual student."

"The data provided by the system for the different parameters of interest are analyzed through FCM execution to create the student's status; by comparing the parameters of the participants it will be possible to understand if the use of completely remote teaching did not negatively affect the status of students."

Results and Discussion

"To verify if the obtained results are statistically significant, we performed an ANOVA test, comparing two groups: Blended Learning (students who attended classroom lessons and used the supporting e-learning platform) and Distance Learning (students who followed the course completely online)."

"Having the e-learning platform as the only tool available to access the teaching material and to carry out the exercises, the students of the academic year 2019/2020 show, even if slightly, a higher level of interaction than the students of 2018/2019 where the use of the platform was an additional part, used to deepen or to carry out some extra exercise compared to what was done in the presence."

"The engagement is, albeit slightly, greater for students of the academic year 2019/2020 who have massively used the e-learning platform as the only tool to follow the course and access the content made available by the teachers."

Discussion

"The same enthusiasm is not shown by the students of the year 2019/2020 who found themselves using e-learning as the only learning methodology while noting its usefulness as the only means of interaction with teachers."

"From the quantitative analysis conducted on the data from the questionnaires administered to students, two important aspects can be noted: the students tried to make up for the physical presence by trying to interact constantly on the e-learning platform forum."

"Another important practical implication obtained from this analysis is that motivation, participation and engagement are altered by the emergency situation and the distance learning used."

"Engagement increased as a result of massive use of adaptive learning platform and of all the technological tools as the only means to follow the course and feel part of a community; however, by integrating Technological Content Knowledge, Pedagogical Content Knowledge, and Technological Pedagogical Knowledge to the use of an adaptive e-learning platform, students have reached adequate levels of competence."

Conclusions

"This research work focused on distance learning and the virtual environment during the Covid-19 pandemic, highlighting how motivation, participation, and engagement are affected."

"The main findings are: Distance Learning is excellent as an additional and support methodology but highlight the ineffectiveness of completely remote teaching; motivation, participation and engagement are altered by the emergency situation and the distance learning used."

"Engagement increased as a result of massive use of adaptive learning platform and of all the technological tools as the only means to follow the course and feel part of a community; by integrating Technological Content Knowledge, Pedagogical Content Knowledge, and Technological Pedagogical Knowledge to the use of an adaptive e-learning platform, students have reached adequate levels of competence."

"The authors are working on modifying the adaptive e-learning platform used, integrating appropriate teaching methodologies to maintain high levels of engagement participation and motivation even in the long term."

Using Digital Resources for Motivation and Engagement in Learning Mathematics: Reflections from Teachers and Students [181]

This is a machine-generated summary of:

Chao, Theodore; Chen, Jason; Star, Jon R.; Dede, Chris: Using Digital Resources for Motivation and Engagement in Learning Mathematics: Reflections from Teachers and Students [181]

Published in: Digital Experiences in Mathematics Education (2016)

Link to original: https://doi.org/10.1007/s40751-016-0024-6

If you want to cite the papers, please refer to the original.

For technical reasons we could not place the page where the original quote is coming from.

Abstract-Summary

"Students' motivation to learn mathematics often declines during the middle grades."

"Evidence from creating technology-based tasks and resources to motivate students to learn mathematics is mixed, partially because most interventions only loosely incorporate motivational constructs."

"This article is part of a larger research project examining the impact of three digital resources on students' motivation and learning in mathematics."

"Students then engaged with these resources before and after a 2-day mathematical patterns lesson."

"Findings suggest that, even with a minimal encounter over 1 or 2 days, students were able to notice the motivational constructs present within these digital resources."

Background Theoretical Framework

"Four decades of research have shown the importance of self-efficacy to outcomes such as persistence and perseverance in the face of difficulties, academic achievement, and students' choice of college majors and careers (Bandura 123; Brown and Lent 124; for a review, see Pajares and Urdan [125])."

"Students holding an incremental mindset pursue a task or explore a situation simply for the sake of learning."

"Digital resources can support gains in self-efficacy when learning about science, particularly through providing mastery experiences for students, by allowing them to see the fruits of their own labor."

"Liu and others [24] showed that students' self-efficacy for learning science increased after participation in a computer-enhanced, problem-based learning (PBL) environment called Alien Rescue."

"These two studies illustrate how technology can support student self-efficacy by providing relevant mastery experiences – the strongest hypothesized source of self-efficacy."

"Dweck and colleagues have developed Brainology, a web-based series of learning modules designed to teach students about the incremental mindset (termed the 'growth' mindset)."

The Three Digital Resources

"This connection of the class mathematical task to the game allowed us to see whether the game could help the students make use of the scaffolded mathematical exercises within the game – not explicitly labeled as mathematics – as a means to build their self-efficacy in solving similar pattern tasks in the classroom."

"The particular mathematical challenges in the game were directly connected to the 'trains' pattern task."

"Our belief was that students would be engaged in mathematical thinking while interacting with the game, based upon their exploring the various challenges, organizing the permutations of the rods within them, and then generalizing from the pattern."

"Regardless of which resource with which a student engaged, all students explored a 2-day mathematical patterns lesson built around a combinatorics task (the 'trains problem')."

Methods

"While each group had a mix of (teacher-identified) highly motivated and unmotivated students, at the time of the interviews the research team was unaware which students were which."

"For the second research question about teachers' perceptions of their students' interests in and confidence in succeeding in mathematics, we hypothesized that teachers' perceptions would mirror what students reported: that the game would focus on self-efficacy, that Brainology would focus on implicit theory of ability and the video would focus on interest in and enjoyment of learning mathematics."

"For the teacher interviews, we used a two-stage expert analysis protocol (Corbin and Strauss [126]) to explore teacher reflection specific to motivating students using digital resources."

"All four research team members came together for several rounds of theoretical discussions to attempt to uncover emergent themes involving technology, sources of student motivation and the trains lesson, connecting these themes to what we found in the student data."

Results
"Ten of the thirty-two student interviews were about the game, fourteen interviews were about the Brainology modules, and eight interviews were about the video."

"We then report on the main themes (technology, sources of student motivation, and the trains lesson) and sub-themes we found among the ten teacher interviews."

"Only five of the interviews mentioned mastery experiences despite our expecting students to attend to them in the game since it was built around mastery experiences."

"All four of the Brainology® teachers mentioned how their students struggled to engage with the content or connect it to learning mathematics."

"These extracts from the teacher interviews indicate that self-efficacy in beating the game and enjoyment and interest in the video were the only motivational constructs from our framework that teachers seemed to notice in their students."

Discussion
"Vicarious experiences were salient for some students across all three resources, and for more than a third of them in all."

"As a source of self-efficacy, such experiences seemed to resonate with students much more than mastery experiences, implicit theory of ability, or interest and enjoyment within a short time frame, suggesting students are attentive to vicarious experiences within these novel technology experiences, even over such a short time span."

"We saw a glimmer of hope in how even a brief encounter led to students recognizing and responding to the motivational constructs and noticing vicarious experiences everywhere."

"Our effectiveness as designers of educational technology hinges on our ability to understand not only how to embed motivational constructs within the technology, but also how students and teachers translate these constructs into designs and practices that align with how students actually use and experience them."

[Section 6]
"One approach to this problem of low student motivation is to use technology-based resources both to spark students' interest in learning mathematics and to develop greater confidence in mathematical problem solving."

"To explore their impact on student motivation to learn mathematics, we present a qualitative analysis of the views of eighty-eight students (in grades five to eight) and their ten teachers as they participated in a week-long motivational technology intervention, during which students and teachers discussed both the specific resource they had worked with and their motivation for learning mathematics."

"Two research questions guided our investigation. (1) From the standpoint of motivation and engagement, what were students' perceptions of and experiences with the digital resources? (2) How did teachers of mathematics perceive their students' self-efficacy, their implicit theories of ability, and their interest and enjoyment in learning mathematics?"

Examining the Role of Spatial Skills and Mathematics Motivation on Middle School Mathematics Achievement [182]

This is a machine-generated summary of:

Atit, Kinnari; Power, Jason R.; Veurink, Norma; Uttal, David H.; Sorby, Sheryl; Panther, Grace; Msall, Camille; Fiorella, Logan; Carr, Martha: Examining the role of spatial skills and mathematics motivation on middle school mathematics achievement [182]

Published in: International Journal of STEM Education (2020)

Link to original: https://doi.org/10.1186/s40594-020-00234-3

If you want to cite the papers, please refer to the original.

For technical reasons we could not place the page where the original quote is coming from.

Abstract-Summary
"Measures of spatial skills and mathematics motivation were completed by 1056 seventh grade students."

"Using hierarchical linear modeling, spatial skills and math motivation were examined relative to students' performance on a state standardized mathematics subtest."

"Results indicate that spatial skills and motivation interact to significantly predict students' mathematics performance."

"These results suggest that spatial skills in combination with motivation play a significant role in middle school students' mathematics achievement."

Introduction

"We aim to understand the nuances of how a set of cognitive skills important for mathematical reasoning, spatial skills (e.g., Casey, Nuttall, Pezaris, & Benbow, 128; Reuhkala, 129), combine with students' motivation for learning mathematics, to predict middle school students' mathematics performance."

"Researchers have found that the motivational factors encompassed within the model predict student behaviors promoting mathematics learning and achievement (e.g., Berger & Karabenick [129]), but that expectancy and value beliefs towards the domain of mathematics declines as the student progresses through their grade school education."

"We first examined whether spatial skills and mathematics motivation individually predicted middle school students' mathematical achievement and whether these relations were moderated by gender as prior research has found that gender differences are apparent on many tests of spatial skills (e.g., Miller & Halpern, 2014) and also on many motivational constructs relevant to mathematics learning during adolescence (Jacobs and others [130])."

Method

"The MMQ is a self-report questionnaire measuring students' motivation to learn mathematics adapted from Glynn and others's [131] Science Motivation Questionnaire (SMQ)."

"We adapted the SMQ to examine students' motivation to learn mathematics, or the level at which students' strive to learn the mathematical domain, by changing the word "science" to the word "math" in each of the items."

"The M-STEP mathematics subtest is one measure in a battery of standardized assessments given to all Michigan students near the end of the school year in grades three to eight to assess mastery of state standards (Michigan Department of Education [132])."

"We examined sixth grade M-STEP mathematics subtest score to minimize the degree to which students' cognitive profiles may have changed due to academic experiences between the time of assessments."

"Seventh grade mathematics and science teachers administered the spatial measure and the MMQ to their students at the beginning of the academic year."

Results

"Following Cohen's [133] conventions, students' spatial skills were weakly correlated with their MMQ score and their science GPA and strongly correlated with their M-STEP mathematics subtest performance."

"Analyses revealed that after controlling for science GPA, spatial skills (model 1) and MMQ score (model 2) each individually significantly predict students' performance on the M-STEP mathematics subtest (ps < .001)."

"In model 3, we included spatial skills and MMQ score as independent factors in the same model and found that spatial skills and MMQ score independently contribute to predicting students' M-STEP mathematics subtest performance, ps < .001."

"Analyses revealed a significant effect of spatial skills, a significant effect of MMQ score, and a significant interaction between spatial skills and MMQ score on students' M-STEP mathematics subtest performance, ps < .05."

Discussion

"Our findings start to reveal how middle school students' motivation for learning mathematics combines with their spatial skills to help predict mathematics performance."

"Consistent with these findings, our results suggest that spatial skills and students' motivation for learning mathematics interact to predict mathematics performance in middle school students."

"Work by Ganley and Vasilyeva [134] indicates that the mechanism by which spatial skills and motivational factors, specifically self-efficacy and mathematics anxiety, combine to predict mathematics performance in middle school students differs by gender."

"As prior research indicates that strong spatial skills are critical to early mathematics learning (e.g., Gunderson and others [135]; Verdine and others [136]), and results from our study revealed that both spatial skills and motivation to learn mathematics contribute to students' middle school mathematics outcomes, integrating both of these components into the everyday K-12 mathematics curriculum could result in long-term benefits to all students' mathematical achievement."

Conclusions

"Our findings provide insight into how spatial skills and motivation for learning mathematics are intertwined as underlying factors of middle school students' mathematics achievement."

"Our findings highlight that the role of motivation in students' mathematics achievement does not differ for students who demonstrate different levels of spatial skills."

“These results suggest that future interventions may have the strongest impact by targeting both the development of strong spatial skills and positive motivational factors pertinent to mathematics learning for middle school students.”

The Impact of Physics Education Technology (PhET) Interactive Simulation-Based Learning on Motivation and Academic Achievement Among Malawian Physics Students [183]

This is a machine-generated summary of:

Banda, Herbert James; Nzabahimana, Joseph: The Impact of Physics Education Technology (PhET) Interactive Simulation-Based Learning on Motivation and Academic Achievement Among Malawian Physics Students [183]

Published in: Journal of Science Education and Technology (2022)

Link to original: https://doi.org/10.1007/s10956-022-10010-3

If you want to cite the papers, please refer to the original.

For technical reasons we could not place the page where the original quote is coming from.

Abstract-Summary
“The study investigated the impact of PhET simulation-based learning on students’ motivation and academic achievement in learning oscillations and waves among Malawian secondary students.”

“The following research questions guided the study: (i) What were students’ motivation and academic achievement levels at the beginning of the study in oscillation and waves?”

“Independent samples t-test showed a statistical difference between the two groups on post-test of the academic achievement.”

"The ANCOVA test on motivation constructs showed a significant difference with a small effect size between the study groups on self-efficacy, active learning strategies, performance goals, achievement goals, learning environment stimulation, and attitudes towards learning with computer learning."

"The results from the study suggest that PhET simulation-based learning improved the learning of oscillations and waves."

"PhET simulation-based learning provides visualizations and teaching aids that help easily understand content knowledge, hence improving students' academic achievement and motivation levels."

Introduction

"Bozkurt and Ilik [137], Banda and Nzabahimana [138], and Otrel-Cass and others [139, 140] noted that dynamic simulations help educators to facilitate the learning of physics concepts that are challenging and abstract and help students develop skills in virtual laboratory experimentation."

"One technology resource commonly used in physics classes is Physics Education Technology (PhET) interactive simulations because it has many features."

"PhET interactive simulation activities offer more to the way students learn physics in a way that intrigues the mind of learners with mouth-watering visualizations of abstract physics concepts; hence, they develop a deeper understanding of physics concepts (Ganasen & Shamuganathan [141]; Perkins [142]; Wieman and others [143])."

"These features and advantages make PhET simulations ideal for physics in African schools as they can be easily incorporated into mainstream learning."

"Little research has been done to explore the effect of PhET simulation-based learning on academic achievement and motivation in physics education."

Literature Review

"In the context of this study, exposing students to PhET interactive simulations in the teaching and learning of physics enabled students to activate their visual systems, which will more likely enhance their capacity to retain more information, or students' cognitive engagement will increase or decrease; hence, this will impact the academic achievement."

"PhET simulation-based learning impacts the affective and motivation factors that mediate the learning process by increasing or decreasing students' motivation levels."

"Koh and others [144] studied the impact of 3D simulation learning on engineering students' performance and motivation."

"The following research questions guided the study: (i) To what levels do PhET interactive simulation-based learning impact students' motivation and achievement in oscillations and waves? (ii) Is the change in post-test due to the students' characteristics in non-randomized settings or the PhET interactive simulation-based learning?"

Methodology

"Since it is a quantitative approach, the study employed the survey design (questionnaire) to measure the effect of PhET on motivation and the pre-post-test to measure the effect of PhET on students' achievement."

"As recommended by Theobald and Freeman [145], data were also collected on measures to investigate if the differences in post-test scores between the groups are due to students' composition (student characteristics) or the treatment in non-randomized studies."

"The data were collected on measures that reflect students' ability and preparation (students characteristics) which were considered to be: (i) students' prior knowledge about oscillation and waves (pre-test); (ii) students' understanding of the larger discipline (end of term aggregate score); (iii) student work habits and study skills (the previous end of term physics score); and (iv) the intervention (treatment = 1 if the student is in the treatment class, treatment = 0 if the student is in the control class)."

Results

"A t-test of independent samples on the pre-test showed no significant difference between the control and experimental groups with $t\ (278) = -0.066$, $p = 0.883$ with the lowest Cohen's d value of 0.0079."

"The independent sample test of the post-test between the groups showed significant changes between the group, [$t(271.497) = 9.532$, $p < 0.01$]."

"For every point added in the pre-test by the student, the post-test increases by 0.208 or else equal, and the relationship between the pre-test and post-test is not significant (p-value = 0.207)."

"An increase at the end of term aggregate score by one-unit results in an increase in post-test by 0.135, and the relationship between the end of term aggregate score and post-test is not significant (p-value = 0.107)."

"The results of the ANCOVA analysis on post-survey self-efficacy indicated a significant difference in mean [$F(1,271) = 43.744$, $p < 0.01$] between the groups while adjusting for pre-survey self-efficacy."

Discussion and Conclusion

"The study's primary purpose was to determine the impact of PhET simulation-based learning on students' academic achievement and motivation on oscillations and waves among Malawian secondary school students."

"The analysis of the impact of PhET interactive simulation-based learning on students' motivation and academic achievement in oscillations and waves suggests that PhET simulation-based learning can improve academic achievement and motivation among learners."

"PhET simulation-based learning improves students' understanding of content knowledge, impacting their academic achievement."

"Student's performance in the experimental group after the intervention (treatment) underscored the significance of using PhET simulation-based learning to engage learners' minds and efficiently stimulate them to be more receptive to new information."

"Students in the experimental group were more engaged, interested, and motivated, making the learning of oscillations and waves lively as shown by the ANCOVA analysis on the construct of attitude towards learning with PhET simulations which measured items on interest, engagement, and attention span."

The Implication of the Study

"The improvement of academic achievement and motivation towards oscillation and waves by the students in the experimental group explains how PhET simulation-based learning positively impacts the cognitive and affective domain of motivation envisioned in the variables measured in this study."

"The results from the study suggest that PhET simulation-based learning improved the learning of oscillations and waves by laying forth visualizations that helped in quickly understanding content knowledge hence more improvement in academic achievement (Koh and others [144]; Moreno [146])."

"The results from the study suggest that PhET simulation-based learning improved oscillation and wave learning by making the learning process more contextualized, meaningful, and matching to the learners, an environment filled with dynamic visualizations."

Recommendations of the Study

"This study calls on educators in Africa to explore PhET simulations contextualized for their local language and other computer simulations to clarify abstract concepts to enhance understanding of oscillations and waves and other topics in active learning designs."

"The researchers noted the need for further studies on the impact of PhET simulation-based learning on motivation and academic achievement in terms of gender."

"This study did not consider the impact of PhET simulation-based learning on influencing academic achievement and motivation in terms of gender."

"The researchers also noted a need for studies on collaborative learning enhanced by PhET simulation-based learning on motivation and academic achievement."

"A study on the impact of PhET simulation-based learning on cognitive engagement can substantiate the findings on academic achievement."

Serious Games and Blended Learning; Effects on Performance and Motivation in Medical Education [184]

This is a machine-generated summary of:

Dankbaar, Mary: Serious games and blended learning; effects on performance and motivation in medical education [184]

Published in: Perspectives on Medical Education (2016)

Link to original: https://doi.org/10.1007/s40037-016-0320-2

If you want to cite the papers, please refer to the original.

For technical reasons we could not place the page where the original quote is coming from.

Abstract-Summary
"The aims of this thesis were: 1) Compare the effectiveness of blended versus classroom training for the acquisition of knowledge; 2) Investigate the effectiveness and critical design features of serious games for performance improvement and motivation."

"The results of the research studies indicated that: 1) For knowledge acquisition, blended learning is equally effective and attractive for learners as classroom learning; 2) A serious game with realistic, interactive cases improved complex cognitive skills for residents, with limited self-study time."

"Although the same game was motivating for inexperienced medical students and stimulated them to study longer, it did not improve their cognitive skills, compared with what they learned from an instructional e-module."

"This indicates an 'expertise reversal effect', where a rich learning environment is effective for experts, but may be contra-productive for novices (interaction of prior knowledge and complexity of format)."

"A blended design is equally effective and attractive as classroom training."

"More design-based research is needed on the effects of task complexity and other design features on performance improvement, for both novices and experts."

Introduction
"The first aim of this thesis was to compare the effectiveness of a blended training design with classroom training for the acquisition of knowledge."

"One of the promising new online formats to train complex skills in a motivating way is provided by serious games."

"Little is yet known about the effectiveness and optimal design of serious games in health care training [147, 148]."

"The second aim of this thesis was to investigate the effectiveness and critical design features of serious games for performance improvement and motivation."

Methods
"Study 2: Before investigating the effectiveness of a serious game in emergency care (abcdeSIM), we performed a validation study on commonly used formats, both nationally and internationally, in the assessment of emergency care skills for residents."

"Study 4: As the previous study was inconclusive on the design choices that were responsible for the performance improvement, in this study we focused on the question of what the effects are of adding high-fidelity cases (a simulation game) versus low fidelity (text-based) cases to an instructional e-module on the cognitive skills and motivation of fourth-year medical students."

"Study 5: The fifth study focused on the question of whether undergraduate medical students develop better patient safety knowledge and awareness and are more motivated after using a serious game (Air Medic Sky-1), typically highly interactive, or after using an e-module, typically less interactive, on the same patient safety topics."

Results
"A serious game on emergency care with realistic, interactive cases improved complex cognitive skills for residents, despite limited self-study time [149]."

"Although the same game was motivating for inexperienced medical students and stimulated them to study longer, it did not improve their cognitive skills, compared with what they learned from an instructional e-module [150]."

"Our study on a game in patient safety indicates that video-lectures (in a game) and text-based lectures (in an e-module) are equally effective in developing knowledge on patient safety."

"Although serious games with interactive patient cases are strongly engaging for students and stimulate them to study longer, they do not necessarily result in better performance in patient safety issues [151]."

Discussion

"Our combined results on a serious game in emergency care, where the game improved residents' cognitive skills, but did not improve inexperienced students' skills, indicate an 'expertise reversal effect' [152]: a rich learning environment, such as a highly interactive game, may benefit experts, but is counter-productive for novice learners (interaction between prior knowledge and complexity of format)."

"Although serious games with realistic tasks often enhance motivation, they are as such not sufficient for improving performance."

"Games may support skills learning, provided task complexity matches the learner's competency level."

"Design-based research should shed more light on how game design may enhance skills learning."

Advice

"One piece of advice for PhD students on how to successfully complete a PhD project: Choose a subject and research focus which fits your passion and genuine interest."

"You will find PhD research is fun to do, at times challenging, but a rewarding learning experience."

Teachers' Motivating Style and Students' Motivation and Engagement in STEM: the Relationship Between Three Key Educational Concepts [185]

This is a machine-generated summary of:

De Loof, Haydée; Struyf, Annemie; Boeve-de Pauw, Jelle; Van Petegem, Peter: Teachers' Motivating Style and Students' Motivation and Engagement in STEM: the Relationship Between Three Key Educational Concepts [185]

Published in: Research in Science Education (2019)

Link to original: https://doi.org/10.1007/s11165-019-9830-3

Copyright of the summarized publication:

If you want to cite the papers, please refer to the original.

For technical reasons we could not place the page where the original quote is coming from.

Abstract-Summary
"Self-determination theory (SDT) states that social factors in an educational setting, such as teachers' motivating style, can influence students' motivation and engagement."

"This paper investigates the relationship between STEM teachers' motivating style (autonomy support, provision of structure, involvement) and students' motivation and engagement with regard to STEM."

"Thirty classroom observations were conducted in different STEM lessons, to assess teachers' motivating style and students' engagement."

"The results reveal that STEM teachers' provision of structure is positively linked to students' motivation and engagement with regard to STEM subjects."

"The impact of teachers' autonomy support was negatively predictive for students' autonomous motivation, and positively predictive for students' engagement."

"This study suggests that taking teachers' motivating style into account in future educational initiatives regarding STEM is highly relevant as a means of stimulating students' motivation and engagement."

Introduction
"Teachers can influence students' motivation and engagement through their motivating style, which refers to the degree a teacher supports the students' three basic psychological needs (Tessier and others [153])."

"In classes, where teachers successfully improved their teaching style in terms of psychological need support, both students' self-determined motivation and engagement increased."

"In the studies by Reeve and others [154] and Skinner and others [155], although the link between basic psychological need support and collective engagement was tested in a direct manner, they did not connect these concepts with student motivation."

"It is exactly this gap that we aim to address in the current study; we aim to directly link teachers' basic psychological need support with students' motivation and students' engagement."

"We hypothesize that higher teachers' basic psychological need support predicts higher students' self-determined motivation, lower controlled motivation, and higher engagement."

Method

"One of the overall aims of this approach is to increase students' achievement and motivation with regard to studying STEM in order to attract more students to professions that involve the use of STEM."

"We also took into account other meaningful factors that may influence students' motivation with regard to studying STEM subjects."

"The teachers' motivating style was observed, as well as students' engagement."

"The scale consists of 18 items which assessed 4 measures: teachers' autonomy support (4 items), teachers' provision of structure (5 items), teachers' provision of involvement (4 items), and 1 measure of students' engagement, which included both behavioral and emotional engagement (5 items)."

"If we observed that the teacher regularly walked over to students, the score increased."

"To test the hypothesis concerning the effect of STEM teachers' motivating style on students' engagement, a statistical regression model was created, in which class group characteristics were linked with student outcomes."

Results

"Multilevel analysis with class group as a random factor was performed for the prediction of students' motivation for learning STEM subjects due to the teachers' motivating style."

"The model with teachers' autonomy support, involvement, and structure and class group as random effects did not consistently predict students' motivation, as linear regression showed that only structure could positively predict autonomous motivation ($\beta = 0.26$, $p < 0.05$), while autonomy support negatively predicted autonomous motivation ($\beta = -0.22$, $p < 0.05$)."

"Higher levels of teachers' autonomy support were marginally predictive ($\beta = 0.40$, $p = 0.06$), and structure was significantly predictive ($\beta = 0.55$, $p < 0.05$) for students' engagement."

"Multilevel analysis with class group as a random factor was performed for the prediction of students' engagement with motivation for learning STEM subjects."

Discussion

"Conforming to the SDT and other empirical studies (e.g., Black and Deci [156]; Valås and Søvik [157]), we hypothesized that greater teachers' basic psychological need support (provision of autonomy, relatedness, and structure) in STEM lessons predicts higher autonomous class motivation and lower controlled class motivation in terms of studying STEM (hypothesis 1a)."

"In line with our hypothesis (1b), this study confirms that STEM teachers' motivating style positively affects students' collective engagement."

"The results in the current study also confirm the findings of Reeve and others [154], who made use of observational data, and found a clear effect of teachers' motivating style on students' collective engagement."

"Professional development programs could incorporate information and guidance for teachers on how to use a need-supportive motivating style during instruction in all possible STEM learning environments, in order to increase students' engagement in STEM."

Conclusion
"This study showed the importance of teachers' motivating style in a STEM educational context."

"Teachers' provision of structure is significant in terms of increasing students' motivation to study STEM-related courses on the one hand, and students' engagement in STEM classes on the other."

"The direct investigation of the connection between the concepts of teachers' motivating style, students' motivation, and students' engagement in one study is novel."

The Effects of an Afterschool STEM Program on Students' Motivation and Engagement [186]

This is a machine-generated summary of:

Chittum, Jessica R.; Jones, Brett D.; Akalin, Sehmuz; Schram, Ásta B.: The effects of an afterschool STEM program on students' motivation and engagement [186]

Published in: International Journal of STEM Education (2017)

Link to original: https://doi.org/10.1186/s40594-017-0065-4

If you want to cite the papers, please refer to the original.

For technical reasons we could not place the page where the original quote is coming from.

Abstract-Summary

"To reach students at this critical stage, a design-based afterschool STEM program, titled Studio STEM, was implemented to foster motivation and engagement in STEM topics and activities."

"The purpose of this study is twofold: (a) to investigate how Studio STEM affected students' beliefs about science and whether these beliefs differed from their peers who did not participate in the program, and (b) to examine a case study of one Studio STEM implementation to investigate elements of the curriculum that motivated students to engage in the program."

"We also found that students could be motivated in a voluntary afterschool program (Studio STEM) in which they grappled with STEM concepts and activities, and could verbalize specific program elements that motivated them."

"Through this study, we found that students could be motivated in Studio STEM and that the experience had a positive impact on their perceptions about science as a field."

"Studio STEM appeared to halt the decline in these students' motivational beliefs about science that typically occurs during the middle school years, indicating that afterschool programs can be one way to help students maintain their motivation in science."

Background

"To add to these findings, we were interested in the extent to which these types of programs could also affect students' motivation and engagement in STEM, and their beliefs about science."

"Another purpose was to examine a case study of a particular Studio STEM implementation to investigate elements of the curriculum that motivated students to engage in the afterschool program."

"We focused this study on two primary research questions directly related to these two purposes: RQ1: To what extent do students' motivational beliefs about science change as a result of participating in Studio STEM?"

"RQ2: What aspects of the Studio STEM program affect students' motivation to engage in the curriculum?"

Studio STEM Program

"Studio STEM used an inquiry-based approach and an interdisciplinary curriculum to help middle school students learn about energy conservation."

"Classroom teachers and STEM undergraduate college students from a nearby university (referred to as facilitators) worked with students who learned skills and facts as they progressed through the process of solving a problem using a design approach."

"In the Save the Penguins curriculum, students were faced with the problem of saving penguins by designing better-insulated houses to reduce the carbon dioxide emissions that are a result of heating homes with fossil fuels."

"They used this knowledge to work in small groups to design, build, and test penguin houses made out of different materials (e.g., felt, shiny Mylar, aluminum foil) to protect a penguin-shaped ice cube from melting in the warm temperatures of a test oven."

"Students learn about force, motion, and the Law of Conservation of Energy and then use these principles as they design, build, test, and re-test a mini solar-powered car."

Conceptual Frameworks

"The five key principles of the MUSIC model are that students are more motivated when they perceive that they are empowered, they perceive that the content or activities are useful, they believe that they can be successful, they are interested in the topic or activities, and they feel cared for by others in the learning environment (Jones [158, 159])."

"The usefulness component of the MUSIC model includes instructional strategies that encourage students to perceive that their coursework (e.g., assignments, activities) is useful for their short- or long-term goals, or in the real world (Jones [158])."

"Instructional strategies related to the caring component of the MUSIC model are aimed at creating a learning environment in which students feel that their instructors and classmates care about their learning and general well-being (Jones [158])."

Methods

"All fifth-, sixth-, and seventh-grade students (including Studio STEM participants and non-participants) in the two schools who were present when data were collected during school hours completed a Science Questionnaire in December 2011 and May 2013 ($N = 121$)."

"Included in our analyses were a group of Studio STEM students ($n = 19$) who completed the school-wide measure of their motivational beliefs about science at two time points, which provided both "before" Studio STEM (December 2011) and "after" Studio STEM (May 2013) scores."

"Although all present fifth-, sixth-, and seventh-grade students at both schools completed the Science Questionnaire each year, our sample was limited to 102 students because only those who completed the questionnaire fully at the same two time points (December 2011 and May 2013) were included in the study."

Results

"We asked students what made them feel like they could be successful in Studio STEM, and three themes emerged: (a) others were helpful during Studio STEM, (b) they felt successful during specific activities and presentations, and (c) they already felt like they were good at or enjoyed related topics or activities."

"Six students (23% of the coded excerpts) explained that there were moments when they did not feel successful during Studio STEM, including at times when building the model car and when working with motors to lift a basket of cubes."

"Responses (16%) suggested that some students found several structural aspects of Studio STEM to be interesting, such as the teaching methods, the structure of the presentations, and that they had many opportunities to experiment by themselves and participate in hands-on activities (e.g., "The [facilitators] explained a lot of things, [the presentations] were fun to watch, and [the teachers] told us about what we were doing and what we were going to do.")."

Discussion

"Following participation in two Studio STEM programs (Save the Penguins and then Save the Seabirds), the Studio STEM students maintained their increased attainment value, and, furthermore, they reported that they found science more interesting and useful, and were more competent in their science abilities than their peers."

"Although this study is only an initial step in examining students' motivation in this way, our findings may indicate that participation in problem-based after-school science and engineering programs like Studio STEM can serve in assuaging the documented decline in science-based motivation perceptions many students experience during those academic years."

"The high mean values for each of the subscales in the Studio STEM questionnaire, and the high proportion of interview responses that described positive attributes of the program suggest that the students were motivated to engage while participating in Save the Seabirds activities."

Limitations

"For the school-wide science questionnaire (RQ1), there were 19 Studio STEM participants, which included participants in the Studio STEM program at two schools."

"Due to the small program size and inherent limitations on program enrollment, we posit that our Studio STEM participant samples were adequate and representative of the population of study."

"For RQ1, we limited our sample of non-participants to students who completed the science questionnaire at two specific time points."

"Further, the sample represents all students who were present at the school during both data collection events; thus, the sample is representative of the school population and the Studio STEM participants' classmates."

"A second limitation concerns possible selection bias in Studio STEM participation."

Conclusions

"Studio STEM largely had a positive impact on the participants' motivation and engagement in the program-based science and engineering activities, as well as their motivational beliefs about science and intentions to pursue a college education."

"Although students cited some negative aspects of the program and activities, their perceived motivation and interview responses were consistently and mostly positive."

"Through this study, we found that students could be motivated in a voluntary afterschool program in which they grappled with STEM concepts and activities, and that the experience had a positive impact on their perceptions about science as a field."

"This finding indicates that afterschool programs can be one way to help students maintain their motivation in science."

Reflective Writing About the Utility Value of Science as a Tool for Increasing STEM Motivation and Retention—Can AI Help Scale Up? [187]

This is a machine-generated summary of:

Beigman Klebanov, Beata; Burstein, Jill; Harackiewicz, Judith M.; Priniski, Stacy J.; Mulholland, Matthew: Reflective Writing About the Utility Value of Science as a Tool for Increasing STEM Motivation and Retention – Can AI Help Scale Up? [187]

Published in: International Journal of Artificial Intelligence in Education (2017)

Link to original: https://doi.org/10.1007/s40593-017-0141-4

If you want to cite the papers, please refer to the original.

For technical reasons we could not place the page where the original quote is coming from.

Abstract-Summary

"The integration of subject matter learning with reading and writing skills takes place in multiple ways."

"Students learn to read, interpret, and write texts in the discipline-relevant genres."

"We conduct a detailed investigation into the potential of natural language processing technology to support evaluation of such writing at scale: We devise a set of features that characterize UV writing across different genres, present common themes, and evaluate UV scoring models using essays on known and new biology topics."

Introduction

"Hulleman and Harackiewicz [160] found that their Utility Value Intervention (UVI) raised interest and grades for 9th grade science students who had low performance expectations, relative to a control group."

"This would allow disciplinary instructors, for example, to assign UVI to students as homework to be performed using a technological platform; the automatically generated utility value score would then be reported to the instructor."

"The large majority of students would complete the UVI assignment successfully; students who receive a low UV score apparently had difficulty articulating UV and might need additional help in seeing the relevance of course material to their personal and social lives."

"We envision that with the maturation of the NLP technology for analysis and assessment of reflective writing of this kind it would be possible to provide scaffolding while the students are working on the UVI assignment, in the form of pre-writing and interim writing tasks to help students learn to make and clearly articulate UV connections to course content."

Utility Value Intervention (UVI)

"Grounded in Eccles' Expectancy-Value Theory (Eccles and others [161]; Eccles [162]), the UVI, in which students write about the personal relevance of course material, helps students discover connections between course topics and their lives – in their own terms."

"Students were asked to write an essay or a letter posing a question related to the recently studied module and answering it while making sure to incorporate utility value (henceforth, UV), that is, how the biology topic was related to their own life (Essay) or the addressee's (Letter)."

"The utility value and control writing assignments were coded by research assistants for the level of utility value articulated in each essay, on a scale of 0–4, based on how specific and personal the utility value connection was to the individual."

"An example of a trait that is acquired but not heritable is fitness."

Method

"The data for every genre is partitioned into TRAIN, DEV (or development), and TEST sets."

"Since we plan to evaluate multiple models that use different feature combinations, we use the DEV set for preliminary evaluations of a large number of models."

"A small number of best performing models from the preliminary evaluation are subsequently evaluated on the TEST set that contains fresh data, unseen during model training and development."

"The TEST data is sampled by student, namely, if a student was selected to appear in TEST data, all three of her writing samples appeared in TEST sets for the relevant genres."

"We use cross-validation methodology on the TRAIN data to evaluate the extent to which prediction models trained on writing samples addressing a given set of topics would generalize to a writing sample addressing a new topic."

Data

"It is readily apparent that while Essay, Letter, and Summary genres have substantial amounts of data, the Society genre was administered relatively infrequently, and thus has relatively few writing samples."

"The Summary genre is the control genre, where no UV was required; accordingly, 59% of the samples in this genre have the UV score of 0, and an additional 38% have the UV score of 1."

"The Letter genre has 91% of its samples in UV scores of 3 or 4."

"The distribution in Society genre is also more spread out; we keep in mind, however, that there are relatively few essays available in this genre."

Qualitative Analysis of Utility Value in Student Writing

"To gain some insight into the vocabulary supporting the expression of utility value, we calculated a Word Utility Score (WUS) for each word as follows: where T is the set of topics T = {Cell Biology, Ecology, Evolution, Animal Physiology, Plant Physiology, Genetics}, g_t is the genre_topic set for genre g and topic t, and Summary_t is the genre_topic set for the Summary genre for topic t. A word with a high WUS would appear in genre_topic sets for multiple topics for the UVI genres (Essay, Letter, Society), and would tend to not appear in Summary genre_ topic sets for the different topics."

"The words children and friends have a very high WUS of 14, as they are part of the genre_topic set not only in all Letter topics, but in all Essay topics and some Society topics as well – and in none of the Summary topics."

Features for Utility Value Prediction

"For the following grammatical categories, we calculate log frequency per 1,000 words: pro_sg1: First person singular pronouns (e.g., I, mine) pro_pl1: First person plural pronouns (e.g., we, ourselves) pro_2: Second person pronouns (e.g., you) det_pos: Possessive determiners (e.g., their) pro_indef: Indefinite pronouns (e.g., anyone) Since expression of UV is likely to refer to everyday concerns and activities, we expect essays rich in UV to be less technical, on average, than essays that only summarize the technical content of a biology course, and therefore use shorter, more common, and more concrete words, as well as a larger variety of words."

"We define the following: wordln: Average word length (in letters) wf_median: Median word frequency academicWL: Proportion of academic words (Coxhead [163]) in content words in the essay concrete: Log frequency per 1000 words of words from the MRC concreteness database (Coltheart [164]) types: # of different words (types count) We define a feature that captures use of language that is common for the given genre in the given topic, under the assumption that, for example, different personal essays on Ecology might pick similar subtopics in Ecology and also possibly present similar UV statements."

Preliminary Study: Evaluating Features and Feature Combinations

"The single strongest predictor of UV score for each genre is one of the set of pronouns – first person singular for Essay and Society, second person pronouns for Letter, and first person plural pronouns for Summary."

"For Essay genre, there is more variability in the UV scores, so a feature that captures typical Essay words for the given topic is less strongly correlated with UV scores."

"The directions of the significant correlations are stable with a single exception of plural pronouns in Essays vs Summaries; all features apart from Ingestion are significant predictors in at least two different genres; even the magnitudes of the correlations are often quite stable (consider family-level correlations for all but Pronouns family)."

"Feature, the TRAIN+DEV data was used in two ways: (a) to find how this feature can best contribute to the prediction of UV scores, with random forest regressor – this is similar to all features; (b) to find the genre_topic words."

Experiment 1: Predicting UV Scores in Writing on Known Topics

"In order to perform well on these genres, the automated models need to learn to distinguish between adjacent scores on the UV scale (3 vs 4), which could be a more difficult task than in the Essay context, where we observe a larger variability in UV scores."

"It is also possible that, differently from personal essays, where frequent reference to oneself using a first person singular pronoun is a hallmark of high UV writing, the beneficiaries of the utility value statement in Letters are more diverse – both in terms of actual referents (you, family, children) and in linguistic form (children, offspring, your child point to the same type of significant individuals but use different words)."

Experiment 2: Predicting UV in Essays on New Topics

"We therefore ask how well the model generalizes to new topics that were not addressed in the system's training data."

"We expect this model to show robust performance across topics, as these features use functional, rather than content words; it is the latter that are most likely to change with topic, while the former should not depend on topic (much)."

"It seems that for cases where these is little training data available (between 215 and 258 for the different topics in Society set), the model with only the pronoun features and, generally, with fewer features, would be expected to perform in a more robust fashion."

"If a substantial amount of data is available for training (the order of magnitude of 2,500 essays), the Full model is clearly preferable, both in known-topic and in a new-topic setting, but one might want to opt for a simpler model if training data is scarce."

Related Work

"Hypothesizing that UVI writing would be more personal and social and hence contain references to self and other people, they selected the following LIWC categories: personal pronouns (I, us, your), long words (> 6 letters), family words (mom, sister), friend words (friend, neighbor), human words (adult, baby), social processes words (advice, discuss, encourage)."

"These results indicate that when people are asked to write truthfully, they opt for highly personalized writing with language suggesting cognitive engagement."

"Authentic, truthful, and personal engagement being the goal of utility value interventions, perhaps a UVI genre that allows reference to individuals beyond the intimate circle of self, friends, and family should be treated with some caution, since people might find it easier to write without true and authentic engagement about strangers."

Discussion

"The results of our study are, we believe, encouraging for the potential application of UV scoring technology to help scale up the UV intervention, especially in the personal Essay genre."

"Another useful result of the current study is the semi-automatic analysis of the kinds of themes that come up in UV-rich writing; a further development of UV-analysis technology could enable retrieval of UV excerpts using theme index to a database of UV-rich writing, as a preparatory activity for the student when thinking about his own utility."

"From the point of view of computational modeling for prediction of UV scores, the Letter genre is more challenging, due both to the skewness of UV distribution and to a more diverse patterns of writing-to-address-other-people's-problems, as compared to reflecting on one's own plans and goals."

"For the personal Essay genre, the score distribution is more spread out, and the correlations attained by the computational model are generally high, r=.78−.79."

Summary and Conclusion

"This is the first application of NLP technology for predicting utility value expressed in a student's writing sample."

"Automated scoring is particularly important since (a) biology teachers often view this kind of assignment as tangential to their main goal of teaching the subject matter of the STEM course, and (b) human scoring of writing samples for expression of utility value requires significant training."

"Analyses such as the qualitative investigation reported here could also be utilized to design scaffolds for utility-value writing, namely, guided exercises aimed at helping students articulate the utility value they derive from studying the STEM subject."

"The cross-genre analysis of utility-value writing conducted here sheds light on linguistic characteristics of this type of writing that recur across different genres (personal essay, letter, summary) as well as those that are specific to certain genres; these, along with a thematic breakdown of commonly used utility language, could aid teachers when designing and evaluating utility-value writing assignments in biology, as well as serve as a comparison when investigating utility value articulation in other disciplines."

Bibliography

1. Autio, O. (1997). Oppilaiden teknisten valmiuksien kehittyminen peruskoulussa. [Student's development in technical abilities in Finnish comprehensive school], Research Reports No. 117. Helsinki: The University of Helsinki, Department of Teacher Education.
2. Järvinen, T. & Autio, O. (2008). Luokanopettajaksi opiskelevien näkemyksiä teknisen työn opettamisesta. [Classroom teacher students' opinions about teaching technology education]. In A. Kallioniemi, (Ed.) Uudistuva ja kehittyvä ainedidaktiikka Research Reports No. 298 (389–400). Helsinki: University of Helsinki, Department of Teacher Education.
3. Gilman, R., & Anderman, E. M. (2006). Motivation and its relevance to school psychology: An introduction to special issue. Journal on School Psychology, 44, 325–329.

4. Peltonen, M., & Ruohotie, P. (1992). Oppimismotivaatio [Learning motivation]. Keuruu: Otava.
5. Deci, E. L., & Ryan, R. M. (1985). Intrinsic motivation and self-determination in human behavior. New York: Plenum Press.
6. Stipek, D. J. (1996). Motivation and instruction. In D. C. Berliner & R. C. Calfee (Eds.), Handbook of Educational Psychology (pp. 85–113). New York. McMillan.
7. Weber, K., & Custer, R. (2005). Gender-Aods. Journal of Technology Education, 16(2), 55–71.
8. Mammes, I. (2004). Promoting girls' interest in technology through technology education: A research study. International Journal of Technology and Design Education, 14, 89–100.
9. Silverman, S., & Pritchard, A. (1996). Building their future: Girls and technology education in Connecticut. Journal of Technology Education, 7(2), 41–54.
10. Malczyk, B. R. (2018). Multimodal instruction, the new hybrid: A student-centered approach to blended learning. The Journal of Nonprofit Education and Leadership, 8(1), 16. https://doi.org/10.18666/JNEL-2018-V8-I1-8347
11. Nicol, D. J., & Macfarlane-Dick, D. (2006). Formative assessment and self-regulated learning: A model and seven principles of good feedback practice. Studies in Higher Education, 31(2), 199–218.
12. Hung, M. L., Chou, C., Chen, C. H., & Own, Z. Y. (2010). Learner readiness for online learning: Scale development and student perceptions. Computers & Education, 55(3), 1080–1090.
13. Kebritchi, M., Lipschuetz, A., & Santiague, L. (2017). Issues and challenges for teaching successful online courses in higher education: A literature review. Journal of Educational Technology Systems, 46(1), 4–29.
14. Kizilcec, R. F., Pérez-Sanagustín, M., & Maldonado, J. J. (2017). Self-regulated learning strategies predict learner behavior and goal attainment in Massive Open Online Courses. Computers & Education, 104, 18–33. https://doi.org/10.1016/j.compedu.2016.10.001
15. Wang, C. H., Shannon, D. M., & Ross, M. E. (2013). Students' characteristics, self-regulated learning, technology self-efficacy, and course outcomes in online learning. Distance Education, 34(3), 302–323.
16. Rotgans, J. I., & Schmidt, H. G. (2012). The intricate relationship between motivation and achievement: Examining the mediating role of self-regulated learning and achievement-related classroom behaviors. International Journal of Teaching and Learning in Higher Education, 24(2), 197–208.
17. Raes, A., Vanneste, P., Pieters, M., Windey, I., Van Den Noortgate, W., & Depaepe, F. (2020). Learning and instruction in the hybrid virtual classroom: An investigation of students' engagement and the effect of quizzes. Computers & Education, 143, 103682. https://doi.org/10.1016/j.compedu.2019.103682
18. Bower, M., Dalgarno, B., Kennedy, G. E., Lee, M. J., & Kenney, J. (2015). Design and implementation factors in blended synchronous learning environments: Outcomes from a cross-case analysis. Computers & Education, 86, 1–17. https://doi.org/10.1016/j.compedu.2015.03.006
19. Pintrich, P., & Schunk, D. H. (2002). Motivation in education: Theory, research, and applications (2nd ed.). Upper Saddle River, NJ: Prentice Hall.
20. Shernoff, D. J., Csikszentmihalyi, M., Schneider, B., Shernoff, E. S., et al. (2003). Student engagement in high school classrooms from the perspective of Flow Theory. School Psychology Quarterly, 18(2), 158–176.
21. Liu, M., Toprac, P., & Yuen, T. (2009). What factors make a multimedia learning environment engaging: A case study. In R. Zheng (Ed.), Cognitive effects of multimedia learning (pp. 173–192). Hershey, PA: Idea Group Inc.
22. Liu, M., & Bera, S. (2005). An analysis of cognitive tool use patterns in a hypermedia learning environment. Educational Technology Research and Development, 53(1), 5–21.

23. Liu, M., Bera, S., Corliss, S., Svinicki, M., & Beth, A. (2004). Understanding the connection between cognitive tool use and cognitive processes as used by sixth graders in a problem-based hypermedia learning environment. Journal of Educational Computing Research, 31(3), 309–334.
24. Liu, M., Hsieh, P., Cho, Y. J., & Schallert, D. L. (2006). Middle school students' self-efficacy, attitudes, and achievement in a computer-enhanced problem-based learning environment. Journal of Interactive Learning Research, 17(3), 225–242.
25. Barab, S., Thomas, M., Dodge, T., Carteaux, R., & Tuzun, H. (2005). Making learning fun: Quest Atlantis, a game without guns. Educational Technology Research and Development, 53(1), 86–107.
26. Ketelhut, J. D. (2007). The impact of student self-efficacy on scientific inquiry skills: An exploratory investigation in River City, a multi-user virtual environment. Journal of Science Education and Technology, 16(1), 99–111.
27. Dudfield, A. (1999). Literacy and cyberculture. Delaware: International Reading Association, Newark, DE.
28. Judson, E. (2010). Improving technology literacy: Does it open doors to traditional content? Educational Technology Research and Development, 58(3), 271–284.
29. Connell, J. P., & Wellborn, J. G. (1991). Competence, autonomy and relatedness: a motivational analysis of self-system processes. In M. Gunnar & L. A. Sroufe (Eds.), Minnesota symposium on child psychology: self-processes and development. Chicago: University of Chicago Press.
30. Buhs, E. (2005). Peer rejection, negative peer treatment, and school adjustment: Self-concept and classroom engagement as mediating processes. Journal of School Psychology, 43, 407–424.
31. Connell, J. P., Spencer, M. B., & Aber, J. L. (1994). Educational risk and resilience in African American youth: context, self, action, and outcomes in school. Child Development, 65, 493–506.
32. Hamre, B. K., & Pianta, R. C. (2005). Can instructional and emotional support in the first grade classroom make a difference for children at risk of school failure? Child Development, 76, 949–967.
33. Van Ryzin, M., Gravely, A., & Roseth, C. (2009). Autonomy, belongingness, and engagement in school as contributors to adolescent psychological well-being. Journal of Youth and Adolescence, 38, 1–12.
34. Furrer, C., & Skinner, E. (2003). Sense of relatedness as a factor in children's academic engagement and performance. Journal of Educational Psychology, 95, 148–162.
35. Wentzel, K. R. (1997). Student motivation in middle school: the role of perceived pedagogical caring. Journal of Educational Psychology, 89, 411–419.
36. Wentzel, K. R. (1998). Social relationships and motivation in middle school: The role of parents, teachers, and peers. Journal of Educational Psychology, 90, 202–209. https://doi.org/10.1037/0022-0663.90.2.202 .
37. Walker, C. O., Greene, B. A., & Mansell, R. A. (2006). Identification with academics, intrinsic/extrinsic motivation, and self-efficacy as predictors of cognitive engagement. Learning and Individual Differences, 16(1), 1–12. https://doi.org/10.1016/j.lindif.2005.06.004 .
38. Goodenow, C. (1993). Classroom belonging among early adolescent students: Relationship to motivation and achievement. The Journal of Early Adolescence, 13, 21–43.
39. Garcia, T., & Pintrich, P. R. (1996). Assessing students' motivation and learning strategies in the classroom context: the Motivated Strategies for Learning Questionnaire. In M. Birenbaum & F. J. R. C. Dochy (Eds.), Alternatives in assessment of achievements, learning processes, and prior knowledge (pp. 319–339). Boston: Kluwer Academic Press.
40. Barnes, D. C., Engelland, B. T., Matherine, C. F., Martin, W. C., Orgeron, C. P., Ring, J. K., et al. (2008). Developing a psychometrically sound measure of collegiate teaching proficiency. College Student Journal, 42, 199–213.
41. Saldaña, J. (2013). The coding manual for qualitative researchers (2nd ed.). Thousand Oaks: Sage.

42. Gee, J. P. (2003). What video games have to teach us about learning and literacy. New York: Palgrave Macmillan.
43. O'Neil, H. F., Wainess, R., & Baker, E. (2005). Classification of learning outcomes: Evidence from the computer games literature. The Curriculum Journal, 16, 455–474.
44. Malone, T. W. (1981). Toward a theory of intrinsically motivating instruction. Cognitive Science, 4, 333–369.
45. Wang, C., Khoo, A., Liu, W., & Divaharan, S. (2008). Passion and intrinsic motivation in digital gaming. CyberPsychology & Behavior, 11(1), 39–45.
46. Ryan, R. M., Rigby, C. S., & Przybylski, A. (2006). The motivational pull of video games: A self-determination theory approach. Motivation and Emotion, 30, 347–363.
47. O'Donnell, A. M., Reeve, J., & Smith, J. K. (2009). Educational psychology: Reflection for action. Hoboken, NJ: Wiley & Sons.
48. Chou, T., & Ting, C. (2003). The role of flow experience in cyber-game addiction. CyberPsychology & Behavior, 6, 663–675.
49. Barab, S., Sadler, T., Heiselt, C., Hickey, D., & Zuiker, S. (2007). Relating narrative, inquiry, and inscriptions: Supporting consequential play. Journal of Science Education and Technology, 16(1), 59–82.
50. Hopelab. (2006). Our research: Re-mission™ outcomes study. Retrieved June 11, 2008, from http://www.hopelab.org/our-research/re-mission-outcomes-study
51. Ito, M., Horst, H., Bittanti, M., Boyd, D., Herr-Stephenson, B., Lange, P. G., et al. (2008). Living and learning with new media: Summary of the findings from the digital youth project. White paper, the John D. and Catherine T. MacArthur foundation reports on digital media and learning. Retrieved July 14, 2009, from http://digitalyouth.ischool.berkeley.edu/report
52. Schrader, P. G., & McCreery, M. (2007). The acquisition of skill and expertise in massively multiplayer online games. Educational Technology Research & Development. Retrieved October 17, 2008, from http://www.springerlink.com/content/n2496u376825u512/
53. Squire, K. D., DeVane, B., & Durga, S. (2008). Designing centers of expertise for academic learning through video games. Theory into Practice, 47, 240–251.
54. Warren, S. J., Dondlinger, M. J., & Barab, S. A. (2008). A MUVE towards PBL writing: Effects of a digital learning environment designed to improve elementary student writing. Journal of Research on Technology and Education, 41, 113–140.
55. Owston, R. D. (2009). Digital immersion, teacher learning, and games. Educational Researcher, 38(4), 270–273.
56. Agarwal, S., & Kaushik, J. S. (2020). Student's perception of online learning during COVID pandemic. In Indian journal of pediatrics (Vol. 87, Issue 7, p. 554). Springer. https://doi.org/10.1007/s12098-020-03327-7 .
57. Sun, J. C. Y., & Rueda, R. (2012). Situational interest, computer self-efficacy and self-regulation: Their impact on student engagement in distance education. British Journal of Educational Technology, 43(2), 191–204. https://doi.org/10.1111/j.1467-8535.2010.01157.x
58. Tullis, J. G., & Benjamin, A. S. (2011). On the effectiveness of self-paced learning. Journal of Memory and Language, 64(2), 109–118. https://doi.org/10.1016/j.jml.2010.11.002
59. Zayapragassarazan, Z. (2020). Covid-19: Strategies for online engagement of remote learners. Springer. https://doi.org/10.7490/f1000research.1117835.1
60. Ali, L., Hatala, M., Gašević, D., & Winne, P. H. (2014). Leveraging MSLQ data for predicting students achievement goal orientations. Journal of Learning Analytics, 1(3), 157–160. https://doi.org/10.18608/jla.2014.13.11
61. Artino, A. R., & McCoach, D. B. (2008). Development and initial validation of the online learning value and self-efficacy scale. Journal of Educational Computing Research, 38(3), 279–303. https://doi.org/10.2190/EC.38.3.c
62. Wang, C.-H., Shannon, D. M., & Ross, M. E. (2013). Students' characteristics, self-regulated learning, technology self-efficacy, and course outcomes in online learning. Science, 2, 775.
63. Kafai, Y. B. (2005). The classroom as "living laboratory": Design-based research for understanding, comparing, and evaluating learning science through design. Educational Technology, 45(1), 28–34.

64. Baylor, A. L. (2009). Promoting motivation with virtual agents and avatars: Role of visual presence and appearance. Philosophical Transactions of the Royal Society B—Biological Sciences, 364(1535), 3559–3565.
65. Baylor, A. L., & Kim, Y. (2005). Simulating instructional roles through pedagogical agents. International Journal of Artificial Intelligence in Education, 15(1), 95–115.
66. Baylor, A. L., & Kim, S. (2009). Designing nonverbal communication for pedagogical agents: When less is more. Computers in Human Behavior, 25(2), 450–457.
67. Guadagno, R. E., Blascovich, J., Bailenson, J. N., & McCall, C. (2007). Virtual humans and persuasion: The effects of agency and behavioral realism. Media Psychology, 10(1), 1–22.
68. Kim, Y., Baylor, A. L., & Shen, E. (2007). Pedagogical agents as learning companions: The impact of agent emotion and gender. Journal of Computer Assisted Learning, 23(3), 220–234.
69. Rosenberg-Kima, R. B., Baylor, A. L., Plant, E. A., & Doerr, C. E. (2007). The importance of interface agent visual presence: Voice alone is less effective in impacting young women's attitudes toward engineering. Paper presented at the Persuasive 2007, Stanford, CA.
70. Rosenberg-Kima, R. B., Baylor, A. L., Plant, A., & Doerr, C. E. (2008). Interface agents as social models for female students: The effects of agent visual presence and appearance on female students' attitudes and beliefs. Computers in Human Behavior, 24(6), 2741–2756.
71. Ryu, J., & Baylor, A. L. (2005). The psychometric structure of pedagogical agent persona. Technology, Instruction, Cognition and Learning (TICL), 2(4), 291–315.
72. Baylor, A. L., Ryu, J., & Shen, E. (2003). The effects of pedagogical agent voice and animation on learning, motivation and perceived persona. Proceedings of world conference on educational multimedia, hypermedia and telecommunications, Honolulu, Hawaii (pp. 452–458). Norfolk, VA: AACE.
73. Baylor, A. L., & Kim, Y. (2004). Pedagogical agent design: The impact of agent realism, gender, ethnicity, and instructional role. In J. Lester et al. (Eds.), Lecture notes in computer science: Intelligent tutoring systems (Vol. 3220, pp. 592–603). Berlin/Heidelberg: Springer.
74. Plant, E. A., Baylor, A. L., Doerr, C. E., & Rosenberg-Kima, R. B. (2009). Changing middle-school students' attitudes and performance regarding engineering with computer-based social models. Computers and Education, 53(2), 209–215.
75. Bailenson, J. N., Blascovich, J., & Guadagno, R. E. (2008). Self-representations in immersive virtual environments. Journal of Applied Social Psychology, 38(11), 2673–2690.
76. Fox, J., & Bailenson, J. N. (2009). Virtual self-modeling: The effects of vicarious reinforcement and identification on exercise behaviors. Media Psychology, 12(1), 1–25.
77. Short, J., Williams, E., & Christie, B. (1976). The social psychology of telecommunications. London: Wiley.
78. Atkinson, R. K. (2002). Optimizing learning from examples using animated pedagogical agents. Journal of Educational Psychology, 94, 416–427.
79. Johnson, W. L., Rickel, J. W., & Lester, J. C. (2000). Animated pedagogical agents: Face-to-face interaction in interactive learning environments. International Journal of Artificial Intelligence in Education, 11, 47–78.
80. Lester, J. C., Towns, S. G., Callaway, C. B., Voerman, J. L., & FitzGerald, P. J. (2000). Deictic and emotive communication in animated pedagogical agents. In J. Sullivant (Ed.), Embodied conversational agents. Boston, MA: MIT Press.
81. Ebbers, S. J. (2007). The impact of social model agent type (coping, mastery) and social interaction type (vicarious, direct) on learner motivation, attitudes, social comparisons, affect and learning performance (Vol. Ph.D). Tallahassee, FL: Florida State University.
82. Creswell, J.W., Plano-Clark, V.L.: Designing and Conducting Mixed Methods Research. Sage Publications Inc, Thousand Oaks (2011)
83. Law, N., Pelgrum, W.J., Plomp, T.: Pedagogy and ICT Use in Schools Around the World: Findings from the IEA SITES 2006 Study. Springer, Hong Kong (2008)
84. Roca JC, Gagné M (2008) Understanding e-learning continuance intention in the workplace: a self-determination theory perspective. Comput Hum Behav 24:1585–1604

85. Rømen, D., Svanæs, D.: Validating WCAG versions 1.0 and 2.0 through usability testing with disabled users. Univ. Access Inf. Soc. 11(4), 375–385 (2012). https://doi.org/10.1007/s10209-011-0259-3
86. Harris, J., Mishra, P., Koehler, M.: Teachers' technological pedagogical content knowledge and learning activity types: Curriculum-based technology integration reframed. J. Res. Technol. Educ. 41(4), 393–416 (2009)
87. Špakov, O., & Miniotas, D. (2007). Visualization of eye gaze data using heat maps. Elektronika ir elektrotechnika, 74, 55–58.
88. Tobagi, F. A. (1995). Distance learning with digital video. IEEE Multimedia, 2(1), 90–93.
89. Choi, H. J., & Johnson, S. D. (2005). The effect of context-based video instruction on learning and motivation in online courses. American Journal of Distance Education, 19(4), 215–227.
90. Baron, R. M., & Kenny, D. A. (1986). The moderator-mediator variable distinction in social psychological research: Conceptual, strategic and statistical considerations. Journal of Personality and Social Psychology, 51, 1173–1182.
91. Simonyan, K., & Zisserman, A. (2014). Very deep convolutional networks for large-scale image recognition arXiv preprint arXiv:1409.1556.
92. Ocumpaugh, J., Baker, R., Gowda, S., Heffernan, N., & Heffernan, C. (2014). Population validity for educational data mining models: A case study in affect detection. British Journal of Educational Technology, 45(3), 487–501.
93. Butler, R. (2006). An achievement goal perspective on student help seeking and teacher help giving in the classroom: Theory, research, and educational implications. Help seeking in academic settings: Goals, groups, and contexts, 15–44.
94. Nelson-Le Gall, S., & Resnick, L. (1998). Help seeking, achievement motivation, and the social practice of intelligence in school. Strategic help seeking: Implications for learning and teaching (pp. 39–60).
95. Aleven, V., Roll, I., McLaren, B. M., & Koedinger, K. R. (2016). Help helps, but only so much: Research on help seeking with intelligent tutoring systems. International Journal of Artificial Intelligence in Education, 26(1), 205–223.
96. Goldin, I. M., Koedinger, K. R., & Aleven, V. (2012). Learner differences in hint processing. In K. Yacef, O. Zaïane, A. Hershkovitz, M. Yudelson, & J. Stamper (Eds.), Proceedings of the 5th international conference on educational data mining (EDM 2012) (pp. 73–80). Worcester: International Educational Data Mining Society.
97. Marsh, H. W., Trautwein, U., Lüdtke, O., Köller, O., & Baumert, J. (2005). Academic self-concept, interest, grades, and standardized test scores: Reciprocal effects models of causal ordering. Child Development, 76(2), 397–416.
98. Ogan, A., Walker, E., Baker, R., Rodrigo, M. M. T., Soriano, J. C., & Castro, M. J. (2015). Towards understanding how to assess help-seeking behavior across cultures. International Journal of Artificial Intelligence in Education, 25(2), 229–248.
99. Lee, D. Y., & Lehto, M. R. (2013). User acceptance of YouTube for procedural learning: An extension of the technology acceptance model. Computers & Education,61(1), 193–208.
100. Klopping, I. M., & McKinney, E. (2004). Extending the technology acceptance model and the task-technology fit model to consumer e-commerce. Information Technology, Learning, and Performance Journal,22(1), 35–48.
101. Abdous, M., Camarena, M., & Facer, B. (2009). MALL technology: Use of academic podcasting in the foreign language classroom. ReCALL,21(1), 76–95.
102. Abdous, M., Facer, B., & Yen, C.-J. (2012). Academic effectiveness of podcasting: A comparative study of integrated versus supplemental use of podcasting in second language classes. Computers & Education,58(1), 43–52.
103. Jolliet, Y. (2007). M-Learning: A pedagogical and technological model for language learning on mobile phones. In J. Fong & F.-L. Wang (Eds.), Blended learning: Proceeding of workshop on blended learning 2007 (pp. 327–339). Hong Kong: City University of Hongkong.
104. Murphy, P., Bollen, D., & Langdon, C. (2012). Mobile technology, collaborative reading, and elaborative feedback. In J. Díaz-Vera (Ed.), Left to my own devices: Learner autonomy and

mobile-assisted language learning innovation and leadership in English language teaching (pp. 131–159). Bingley: Emerald Group Publishing Limited.
105. Comas-Quinn, A., Mardomingo, R., & Valentine, C. (2009). Mobile blogs in language learning: Making the most of informal and situated learning opportunities. ReCALL,21(1), 96–112.
106. Shao, Y. (2011). Second language learning by exchanging cultural contexts through the mobile group blog. In S. Thouësny & L. Bradley (Eds.), Second language teaching and learning with technology: Views of emergent researchers (pp. 143–168). Dublin: Research-publishing.net.
107. McAuley, E., Duncan, T., & Tammen, V. V. (1989). Psychometric properties of the intrinsic motivation inventory in a competitive sport setting: A confirmatory factor analysis. Research Quarterly for Exercise and Sport,60(1), 48–58.
108. Nunnaly, J. (1978). Psychometric theory. New York: McGraw-Hill.
109. Cheon, J., Lee, S., Crooks, S. M., & Song, J. (2012). An investigation of mobile learning readiness in higher education based on the theory of planned behavior. Computers & Education,59(3), 1054–1064.
110. Park, S. Y., Nam, M. W., & Cha, S. B. (2012). University students' behavioral intention to use mobile learning: Evaluating the technology acceptance model. British Journal of Educational Technology,43(4), 592–605.
111. Kirvan, R., Rakes, C. R., & Zamora, R. (2015). Flipping an algebra classroom: analyzing, modeling, and solving systems of linear equations. Computers in the Schools, 32(3–4), 201–223. https://doi.org/10.1080/07380569.2015.1093902 .
112. Godzicki, L., Godzicki, N., Krofel, M., & Michaels, R. (2013). Increasing motivation and engagement in elementary and middle school students through technology-supported learning environments. (Master's thesis), Saint Xavier University, Chicago. Retrieved from http://files.eric.ed.gov/fulltext/ED541343.pdf .
113. Kim, C., Park, S. W., Cozart, J., & Lee, H. (2015). From motivation to engagement: The role of effort regulation of virtual high school students in mathematics courses. Journal of Educational Technology & Society, 18(4), 261–272.
114. Bergmann, J., & Sams, A. (2012). Flip your classroom: Reach every student in every class every day. Eugene, Oregon: International Society for Technology in Education; Association for Supervision and Curriculum Development.
115. Eccles (Parsons) JS, Adler TF, Futterman R, Goff SB, Kaczala CM, Meece JL, Midgley C: Expectancies, values, and academic behaviors. In Achievement and Achievement Motivation. Edited by: Spence JT. San Francisco: W. H. Freeman; 1983:75–146.
116. Kong, Q. P., Wong, N. Y., & Lam, C. C. (2003). Student engagement in mathematics: Development of instrument and validation of construct. Mathematics Education Research Journal, 15(1), 4–21.
117. Xiong, Y., Li, H., Kornhaber, M. L., Suen, H. K., Pursel, B., & Goins, D. D. (2015). Examining the relations among student motivation, engagement, and retention in a MOOC: A structural equation modeling approach. Global Education Review, 2(3), 23–33.
118. Wolcott, L. L. (2003). Dynamics of faculty participation in distance education: Motivations, incentives, and rewards. Handbook of Distance Education, 549–565.
119. Li, J., Wong, S. C., Yang, X., & Bell, A. (2020). Using feedback to promote student participation in online learning programs: evidence from a quasi-experimental study. Educational Technology Research and Development, 68(1), 485–510.
120. Ramesh, A., Goldwasser, D., Huang, B., Daumé III, H., & Getoor, L. (2013). Modeling learner engagement in MOOCs using probabilistic soft logic. In: NIPS workshop on data driven education (Vol. 21, p. 62).
121. Kim, C., Park, S. W., Cozart, J., & Lee, H. (2015). From motivation to engagement: The role of effort regulation of virtual high school students in mathematics courses. Journal of Educational Technology and Society, 18(4), 261–272.

122. Giesbers, B., Rienties, B., Tempelaar, D., & Gijselaers, W. (2013). Investigating the relations between motivation, tool use, participation, and performance in an e-learning course using web-videoconferencing. Computers in Human Behavior, 29(1), 285–292.
123. Bandura, A. (1997). Self-efficacy: the exercise of control. New York: W. H. Freeman.
124. Brown, S., & Lent, R. (2006). Preparing adolescents to make career decisions. In F. Pajares & T. Urdan (Eds.), Adolescence and education self-efficacy beliefs of adolescents (Vol. 5, pp. 201–223). Greenwich: Information Age.
125. Pajares, F., & Urdan, T. (2006). Adolescence and education: Vol. 5 self-efficacy beliefs of adolescents. Greenwich: Information Age.
126. 127. Corbin, J., & Strauss, A. (2008). Basics of qualitative research: techniques and procedures for developing grounded theory (3rd ed.). Thousand Oaks: Sage.
127. 128. Casey, M. B., Nuttall, R., Pezaris, E., & Benbow, C. P. (1995). The influence of spatial ability on gender differences in mathematics college entrance test scores across diverse samples. Developmental Psychology, 31(4), 697–705. https://doi.org/10.1037/0012-1649.31.4.697 .
128. 129. Reuhkala, M. (2001). Mathematical skills in ninth-graders: Relationship with visuo-spatial abilities and working memory. Educational Psychology, 21(4), 387–399. https://doi.org/10.1080/01443410120090786 .
129. 130. Berger, J.-L., & Karabenick, S. A. (2011). Motivation and students' use of learning strategies: Evidence of unidirectional effects in mathematics classrooms. Learning and Instruction, 21(3), 416–428.
130. 131. Jacobs, J. E., Lanza, S., Osgood, D. W., Eccles, J. S., & Wigfield, A. (2002). Changes in children's self-competence and values: Gender and domain differences across grades one through twelve. Child Development, 73(2), 509–527.
131. 132. Glynn SM, Taasoobshirazi G, Brickman P: Nonscience majors learning science: a theoretical model of motivation. J Res Sci Teach. 2007, 44: 1088-1107. https://doi.org/10.1002/tea.20181.
132. Michigan Department of Education (2017). Michigan Student Test of Educational Progress. Retrieved from http://www.michigan.gov/mde/0,4615,7-140-22709_70117%2D%2D-,00.html .
133. 134. Cohen, J. (1988). Statistical power analysis for the behavioral sciences (1st ed.). New York: Academic Press.
134. 135. Ganley, C. M., & Vasilyeva, M. (2011). Sex differences in the relation between math performance, spatial skills, and attitudes. Journal of Applied Developmental Psychology, 32(4), 235–242. https://doi.org/10.1016/j.appdev.2011.04.001 .
135. 136. Gunderson, E. A., Ramirez, G., Beilock, S. L., & Levine, S. C. (2012). The relation between spatial skill and early number knowledge: The role of the linear number line. Developmental Psychology, 48(5), 1229–1241. https://doi.org/10.1037/a0027433 .
136. 137. Verdine, B. N., Golinkoff, R. M., Hirsch-Pasek, K., & Newcombe, N. S. (2017). Links between spatial and mathematical skills across the preschool years. Monographs of the Society for Research in Child Development, 82(1), 1–150.
137. 138. Bozkurt, E., & Ilik, A. (2010). The effect of computer simulations over students' beliefs on physics and physics success. Procedia Social and Behavioral Sciences, 2, 4587–4591. https://doi.org/10.1016/j.sbspro.2010.03.735
138. 139. Banda, H. J., & Nzabahimana, J. (2021). Effect of integrating physics education technology simulations on students conceptual understanding in physics: A review of literature. Physical Review Physics Education Research, 17(2), 23108. https://doi.org/10.1103/PhysRevPhysEducRes.17.023108
139. Otrel-Cass, K., Girault, I., Renken, M., Peffer, M., & Chiocarriello, A. (2015). Considerations for integrating simulations in the science classroom (pp. 29–34). https://doi.org/10.1007/978-3-319-24615-4_6
140. Otrel-Cass, K., Renken, M., Peffer, M. E., Girault, I., & Chiocarriello, A. (2016). Inquiry-based science education and problem-based learning: Motivations, objectives, and challenges

relevant to computer simulations. In Simulations as scaffolds in science education. SpringerBriefs in Educational Communications and Technology. Springer, Cham. https://doi.org/10.1007/978-3-319-24615-4_4
141. Ganasen, S., & Shamuganathan, S. (2017). The effectiveness of physics education technology (PhET) interactive simulations in enhancing matriculation students' understanding of chemical equilibrium and remediating their misconceptions. In: Karpudewan M., Md Zain A., Chandrasegaran A. (eds).Overco. Springer. https://doi.org/10.1007/978-981-10-3437-4_9
142. 143. Perkins, K. (2020). Transforming STEM learning at scale: PhET interactive simulations. Childhood Education, 96(4), 42–49. https://doi.org/10.1080/00094056.2020.1796451
143. 144. Wieman, C., Adams, W., Loeblein, P., & Perkins, K. (2010). Teaching physics using PhET simulations. The Physics Teacher, 48(4), 225–227. https://doi.org/10.1119/1.3361987
144. 145. Koh, C., Tan, H. S., Tan, K. C., Fang, L., Fong, F. M., Kan, D., Lye, S. L., & Wee, M. L. (2010). Investigating the effect of 3D simulation- based learning on the motivation and. Journal of Engineering Education. https://doi.org/10.1002/j.2168-9830.2010.tb01059.x
145. Theobald, R., & Freeman, S. (2014). Is it the intervention or the students ? Using linear regression to control for student characteristics in undergraduate STEM education research. CBE—Life Sciences Education, 13, 41–48. https://doi.org/10.1187/cbe-13-07-0136
146. 147. Moreno, R. (2007). Optimizing learning from animations by minimizing cognitive load: Cognitive and affective consequences of signaling and segmentation methods. Applied Cognitive Psychology, 21(6), 765–781. https://doi.org/10.1002/acp.1348
147. 148. Connolly TM, Boyle EA, MacArthur E, Hainey T, Boyle JM. A systematic literature review of empirical evidence on computer games and serious games. Comput Educ. 2012;59:661–86.
148. 149. Sitzmann T. A meta-analytic examination of the instructional effectiveness of computer-based simulation games. Pers Psychol. 2011;64:489–528.
149. 150. Dankbaar M, Bakhuys Roozeboom M, et al. Preparing residents effectively in emergency care skills training with a serious game. Simul Healthc. 2016; https://doi.org/10.1097/SIH.0000000000000194 .
150. 151. Dankbaar MEW, Alsma J, Jansen EEH, van Merrienboer JJG, van Saase JLCM, Schuit SCE. An experimental study on the effects of a simulation game on students' clinical cognitive skills and motivation. Adv Health Sci Educ Theory Pract. 2015;21:505–21.
151. Dankbaar M, Richters O, Kalkman C, et al. Comparative effectiveness of a serious game and an emodule on patient safety knowledge and awareness. BMC Med Educ. https://doi.org/10.1186/s12909-016-0836-5. [Accepted for publication].
152. Young JQ, Van Merrienboer J, Durning S, Ten Cate O. Cognitive load theory: implications for medical education: AMEE guide No. 86. Med Teach. 2014;36:371–84.
153. 154. Tessier, D., Sarrazin, P., & Ntoumanis, N. (2010). The effect of an intervention to improve newly qualified teachers' interpersonal style, students' motivation and psychological need satisfaction in sport-based physical education. Contemporary Educational Psychology, 35, 242–253. https://doi.org/10.1016/j.cedpsych.2010.05.005
154. 155. Reeve, J., Jang, H., Carrell, D., Jeon, S., & Barch, J. (2004). Enhancing students' engagement by increasing teachers' autonomy support. Motivation and Emotion, 28, 147–169.
155. 156. Skinner, E., Furrer, C., Marchand, G., & Kindermann, T. (2008). Engagement and disaffection in the classroom: Part of a larger motivational dynamic? Journal of Educational Psychology, 100(4), 765–781.
156. 157. Black AE, Deci EL: The effects of instructors' autonomy support and students'autonomous motivation on learning organic chemistry: a self- determination theory perspective. Sci Edu. 2000, 84: 740-756. https://doi.org/10.1002/1098-237X(200011)84:6<740::AID-SCE4>3.0.CO;2-3.
157. 158. Valås, H., & Søvik, N. (1994). Variables affecting students' intrinsic motivation for school mathematics: Two empirical studies based on Deci and Ryan's theory on motivation. Learning and Instruction, 3(4), 281–298. https://doi.org/10.1016/0959-4752(93)90020-Z .

158. Jones, B. D. (2009). Motivating students to engage in learning: The MUSIC model of academic motivation. International Journal of Teaching and Learning in Higher Education, 21(2), 272–285. Retrieved from http://www.isetl.org/ijtlhe/pdf/IJTLHE774.pdf
159. Jones, B. D. (2015). Motivating students by design: Practical strategies for professors. Charleston, SC: CreateSpace.
160. 161. Hulleman, C. S., & Harackiewicz, J. M. (2009). Promoting interest and performance in high school science classes. Science, 326(5958), 1410–1412.
161. 162. Eccles, J., Adler, T., Futterman, R., Goff, S., Kaczala, C., & Meece, J. (1983). Expectations, values and academic behaviors. In Spence, J. T. (Ed.) Perspective on achievement and achievement motivation (pp. 75–146). San Francisco, CA: W. H. Freeman.
162. 163. Eccles, J. (2009). Who am I and what am I going to do with my life? Personal and collective identities as motivators of action. Educational Psychologist, 44(2), 78–89.
163. 164. Coxhead, A. (2000). A new academic word list. TESOL Quarterly, 34(2), 213–238.
164. 165. Coltheart, M. (1981). The MRC psycholinguistic database. Quarterly Journal of Experimental Psychology, 33A, 497–505.
165. Autio, Ossi; Hietanoro, Jenni; Ruismäki, Heikki Taking part in technology education: elements in students' motivation. International Journal of Technology and Design Education (2010). https://doi.org/10.1007/s10798-010-9124-6
166. Athens, Wendy Self-regulation, motivation, and outcomes in HyFlex classrooms. Educational technology research and development (2023). https://doi.org/10.1007/s11423-023-10243-y
167. Liu, Min; Horton, Lucas; Olmanson, Justin; Toprac, Paul A study of learning and motivation in a new media enriched environment for middle school science. Educational technology research and development (2011). https://doi.org/10.1007/s11423-011-9192-7
168. Zumbrunn, Sharon; McKim, Courtney; Buhs, Eric; Hawley, Leslie R. Support, belonging, motivation, and engagement in the college classroom: a mixed method study. Instructional Science (2014). https://doi.org/10.1007/s11251-014-9310-0
169. Hoffman, Bobby; Nadelson, Louis Motivational engagement and video gaming: a mixed methods study. Educational technology research and development (2009). https://doi.org/10.1007/s11423-009-9134-9
170. Wong, Joseph T.; Hughes, Bradley S. Leveraging learning experience design: digital media approaches to influence motivational traits that support student learning behaviors in undergraduate online courses. Journal of Computing in Higher Education (2022). https://doi.org/10.1007/s12528-022-09342-1
171. Baylor, Amy L. The design of motivational agents and avatars. Educational technology research and development (2011). https://doi.org/10.1007/s11423-011-9196-3
172. Fonseca, David; Redondo, Ernest; Villagrasa, Sergi Mixed-methods research: a new approach to evaluating the motivation and satisfaction of university students using advanced visual technologies. Universal Access in the Information Society (2014). https://doi.org/10.1007/s10209-014-0361-4
173. Pala, Şenol Mail; Başıbüyük, Adem The Predictive Effect of Digital Literacy, Self-Control and Motivation on the Academic Achievement in the Science, Technology and Society Learning Area. Technology, Knowledge and Learning (2021). https://doi.org/10.1007/s10758-021-09538-x
174. Sharma, Kshitij; Giannakos, Michail; Dillenbourg, Pierre Eye-tracking and artificial intelligence to enhance motivation and learning. Smart Learning Environments (2020). https://doi.org/10.1186/s40561-020-00122-x
175. Yılmaz, Ramazan; Yurdugül, Halil Cyberloafing in IT classrooms: exploring the role of the psycho-social environment in the classroom, attitude to computers and computing courses, motivation and learning strategies. Journal of Computing in Higher Education (2018). https://doi.org/10.1007/s12528-018-9184-2
176. Karumbaiah, Shamya; Ocumpaugh, Jaclyn; Baker, Ryan S. Context Matters: Differing Implications of Motivation and Help-Seeking in Educational Technology. International Journal of Artificial Intelligence in Education (2021). https://doi.org/10.1007/s40593-021-00272-0

177. Sun, Yanyan; Gao, Fei An investigation of the influence of intrinsic motivation on students' intention to use mobile devices in language learning. Educational technology research and development (2019). https://doi.org/10.1007/s11423-019-09733-9
178. Winter, Joshua W. Performance and Motivation in a Middle School Flipped Learning Course. TechTrends (2017). https://doi.org/10.1007/s11528-017-0228-7
179. Star, Jon R; Chen, Jason A; Taylor, Megan W; Durkin, Kelley; Dede, Chris; Chao, Theodore Studying technology-based strategies for enhancing motivation in mathematics. International Journal of STEM Education (2014). https://doi.org/10.1186/2196-7822-1-7
180. Capone, Roberto; Lepore, Mario From Distance Learning to Integrated Digital Learning: A Fuzzy Cognitive Analysis Focused on Engagement, Motivation, and Participation During COVID-19 Pandemic. Technology, Knowledge and Learning (2021). https://doi.org/10.1007/s10758-021-09571-w
181. Chao, Theodore; Chen, Jason; Star, Jon R.; Dede, Chris Using Digital Resources for Motivation and Engagement in Learning Mathematics: Reflections from Teachers and Students. Digital Experiences in Mathematics Education (2016). https://doi.org/10.1007/s40751-016-0024-6
182. Atit, Kinnari; Power, Jason R.; Veurink, Norma; Uttal, David H.; Sorby, Sheryl; Panther, Grace; Msall, Camille; Fiorella, Logan; Carr, Martha Examining the role of spatial skills and mathematics motivation on middle school mathematics achievement. International Journal of STEM Education (2020). https://doi.org/10.1186/s40594-020-00234-3
183. Banda, Herbert James; Nzabahimana, Joseph The Impact of Physics Education Technology (PhET) Interactive Simulation-Based Learning on Motivation and Academic Achievement Among Malawian Physics Students. Journal of Science Education and Technology (2022). https://doi.org/10.1007/s10956-022-10010-3
184. Dankbaar, Mary Serious games and blended learning; effects on performance and motivation in medical education. Perspectives on Medical Education (2016). https://doi.org/10.1007/s40037-016-0320-2
185. 186. De Loof, Haydée; Struyf, Annemie; Boeve-de Pauw, Jelle; Van Petegem, Peter Teachers' Motivating Style and Students' Motivation and Engagement in STEM: the Relationship Between Three Key Educational Concepts. Research in Science Education (2019). https://doi.org/10.1007/s11165-019-9830-3
186. Chittum, Jessica R.; Jones, Brett D.; Akalin, Sehmuz; Schram, Ásta B. The effects of an afterschool STEM program on students' motivation and engagement. International Journal of STEM Education (2017). https://doi.org/10.1186/s40594-017-0065-4
187. Beigman Klebanov, Beata; Burstein, Jill; Harackiewicz, Judith M.; Priniski, Stacy J.; Mulholland, Matthew Reflective Writing About the Utility Value of Science as a Tool for Increasing STEM Motivation and Retention – Can AI Help Scale Up?. International Journal of Artificial Intelligence in Education (2017). https://doi.org/10.1007/s40593-017-0141-4

Neuroscience and Motivation

6

Introduction by the Editor

The study of neuroscience and its relation to motivation is a dynamic and rapidly evolving field, merging insights from psychology, biology, and neurology to understand how the brain influences motivation. Motivation, fundamentally, is the process that initiates, guides and sustains goal-directed behaviors. It encompasses the drives that propel humans to act, ranging from basic physiological needs to complex psychological aspirations. The studies in this chapter explore the intricate relationship between neuroscience and motivation, how neural mechanisms underpin motivational processes, and how this understanding can be applied to enhance motivation in various contexts, including education, and personal development. Motivation is traditionally divided into intrinsic and extrinsic categories. The neural substrates of these motivational types involve distinct but overlapping circuits in the brain. Key brain regions implicated in motivation include the prefrontal cortex and various neurotransmitter systems, notably dopamine (Simpson et al, 2022) and serotonin (Yagishita, 2020).

Neuroscience and Motivation

Advances in neuroimaging techniques, such as functional magnetic resonance imaging (fMRI) and positron emission tomography (PET), have allowed researchers to use non-invasive techniques in studying the neural underpinnings of motivation in more detail. These techniques have revealed that motivated behavior involves complex interactions between multiple brain regions and neurotransmitter systems (Lopez-Gamundi et al, 2021). For example, fMRI studies have shown that tasks involving intrinsic motivation, such as solving puzzles for personal satisfaction, activate different neural circuits compared to tasks involving extrinsic motivation, such as working for monetary rewards. These findings suggest that

M. S. Khine (ed.), *Motivation Science*, https://doi.org/10.1007/978-981-97-9247-4_6

while there is overlap in the neural mechanisms underlying different types of motivation, there are also distinct pathways that correspond to specific motivational states (Freed, 2022).

Understanding the neural basis of motivation has important implications for education. By using insights from neuroscience, educators can design learning environments and interventions that enhance student motivation. For example, incorporating elements that stimulate the reward system, such as immediate feedback and recognition, can increase student engagement and motivation. In addition, strategies that promote intrinsic motivation, such as providing autonomy and opportunities for mastery, can activate neural circuits associated with self-directed learning and sustained effort. Educational technologies, such as adaptive learning platforms, can also be designed to personalize learning experiences based on an understanding of individual motivational profiles, thereby optimizing engagement and outcomes. In the field of mental health, insights into the neural mechanisms of motivation can inform the development of treatments for conditions characterized by motivational deficits, such as depression and addiction. Depression is often associated with hypoactivity in the reward circuitry, leading to anhedonia (loss of interest or pleasure in activities) and reduced motivation. Treatments that target the dopaminergic system, such as certain antidepressants and behavioral therapies, can help restore motivation by increasing reward sensitivity and reinforcing positive behaviors. In addition, maladaptive motivation driven by drug-related cues can be addressed by interventions that modify neural responses to these cues, thereby reducing craving and promoting recovery.

Moreover, understanding the neural basis of motivation can inform personal development strategies. Techniques such as mindfulness and cognitive-behavioral therapy can influence neural circuits involved in motivation. Mindfulness practices have been shown to enhance prefrontal cortex function and increase connectivity between the prefrontal cortex and limbic regions, promoting better self-regulation and sustained motivation (Wheeler et al, 2017). Cognitive-behavioral therapy, which involves restructuring negative thought patterns and reinforcing positive behaviors, can modulate neural pathways associated with reward and motivation, leading to improved goal-directed behavior. Even with substantial progress in understanding the neuroscience of motivation, there are still many unanswered questions and challenges. A major challenge is the complexity of motivational processes and their variability across individuals. Motivation is influenced by a variety of factors, including genetics, environment, and personal experience, making it difficult to identify universal neural mechanisms (Wagner, 2021). In addition, while neuroimaging studies provide valuable insights, they often involve artificial tasks and controlled environments that may not fully capture the complexity of real-world motivational dynamics.

Future research in the neuroscience of motivation could be enhanced by integrative approaches that blend neuroimaging with other methods, such as longitudinal studies to better capture the dynamic and context-dependent nature of motivation. Advances in neurotechnology, such as brain-computer interfaces and neurofeedback, also hold promise for enhancing motivation by providing real-time modulation of neural activity (Wolkenstein & Friedrich, 2021). These technologies could potentially be used to develop personalized interventions that optimize motivation based on an individual's unique neural profile. The study of neuroscience and motivation is a rich and diverse field that provides valuable insights into the neural mechanisms underlying motivated behavior. Key brain regions such as the prefrontal cortex, and amygdala (Smith & Torregrossa, 2021), as well as neurotransmitter systems such as dopamine and serotonin, play a critical role in regulating motivation (Salamone & Correa, 2024). Understanding these neural underpinnings has important implications for education, mental health, and personal development, and provides a basis for designing interventions and strategies to enhance motivation. Despite the complexities and challenges, ongoing research in this area holds great promise for advancing our understanding of motivation and improving outcomes in various areas of life.

The chapters in this section include how individual needs influence motivation effects: a neuroscientific study on McClelland's need theory, cross-frequency coupling of brain oscillations in studying motivation and emotion, immersive virtual reality during gait rehabilitation increases walking speed and motivation: a usability evaluation with healthy participants and patients with multiple sclerosis and stroke, and motivational sensitivity of outcome-response priming. Other studies described in this section incorporate motivational intensity and direction into the study of emotions: implications for brain mechanisms of emotion and cognition-emotion interactions, affective motivational direction drives asymmetric frontal hemisphere activation, two types of dopamine neurons distinctly convey positive and negative motivational signals, and in-group biases and oculomotor responses: beyond simple approach motivation.

References

Freed, W. J. (2022). *Motivation and Desire: A New Way to Think About Why We Do Everything and Its Basis in Neuroscience.* Springer Nature.

Lopez-Gamundi, P., Yao, Y. W., Chong, T. T., Heekeren, H. R., Mas-Herrero, E., & Marco-Pallarés, J. (2021). The neural basis of effort valuation: A meta-analysis of functional magnetic resonance imaging studies. *Neuroscience & Biobehavioral Reviews, 131*, 1275–1287.

Salamone, J. D., & Correa, M. (2024). The neurobiology of activational aspects of motivation: exertion of effort, effort-based decision making, and the role of dopamine. *Annual Review of Psychology, 75*, 1–32.

Simpson, E. H., Gallo, E. F., Balsam, P. D., Javitch, J. A., & Kellendonk, C. (2022). How changes in dopamine D2 receptor levels alter striatal circuit function and motivation. *Molecular psychiatry, 27*(1), 436–444.

Smith, D. M., & Torregrossa, M. M. (2021). Valence encoding in the amygdala influences motivated behavior. *Behavioural brain research, 411*, 113370.

Wagner, H. (2021). *The psychobiology of human motivation.* Routledge.

Wheeler, M. S., Arnkoff, D. B., & Glass, C. R. (2017). The neuroscience of mindfulness: How mindfulness alters the brain and facilitates emotion regulation. *Mindfulness*, *8*, 1471–1487.

Wolkenstein, A., & Friedrich, O. (2021). Brain-computer interfaces: current and future investigations in the philosophy and politics of neurotechnology. *Clinical Neurotechnology meets Artificial Intelligence: Philosophical, Ethical, Legal and Social Implications*, 69–80.

Yagishita, S. (2020). Transient and sustained effects of dopamine and serotonin signaling in motivation-related behavior. *Psychiatry and clinical neurosciences*, *74*(2), 91–98.

Machine Generated Summaries

Disclaimer: The summaries in this chapter were generated from Springer Nature publications using extractive AI auto-summarization: An extraction-based summarizer aims to identify the most important sentences of a text using an algorithm and uses those original sentences to create the auto-summary (unlike generative AI). As the constituted sentences are machine selected, they may not fully reflect the body of the work, so we strongly advise that the original content is read and cited. The auto generated summaries were curated by the editor to meet Springer Nature publication standards. To cite this content, please refer to the original papers.

Machine generated keywords: brain, reward, network, neural, region, activation, eeg, response, signal, patient, cortex, stimulus, cortical, action, functional

Motivation from Control [246]

This is a machine-generated summary of:

Eitam, Baruch; Kennedy, Patrick M.; Tory Higgins, E.: Motivation from control [246]

Published in: Experimental Brain Research (2013)

Link to original: https://doi.org/10.1007/s00221-012-3370-7

If you want to cite the papers, please refer to the original.

For technical reasons we could not place the page where the original quote is coming from.

Abstract-Summary

"Human motivation is sensitive to value—to the outcomes of actions."

"People's motivation is sensitive to information about their standing in relation to outcome attainment ('outcome feedback')."

"We argue and present the first evidence for the existence of another motivational sensitivity in humans—a sensitivity to our degree of control on the environment and hence to information about that control ('control feedback')."

"We then show that increased motivation is not because more information about task performance is available and that motivation is increased only in conditions in which control over the effects can be firmly established by the mind."

Introduction

"Effects are, of course, crucial information for knowing whether a purposive, goal-directed action was successful in obtaining a desired outcome (Carver and Scheier [1]; Lewin [2]; Miller [3])."

"In current theories of human action, effects provide information about how we are doing (or what we should do) in relation to obtaining desired outcomes."

"Such theories emphasize the information that effects carry about valued outcomes."

"When functioning as outcome feedback, action effects motivate by providing information about our standing in relation to a desired outcome."

Motivation from Control Effects

"The idea that human motivation is sensitive to one's control over what happens is not new (White 4; for a recent review, see Higgins [5])."

"If 'effectance' is itself a motivator as White [4] has argued, effects as feedback about one's control should be directly motivating."

"In other species, there is some evidence that effects as feedback about one's potential or actual control also affect motivation."

"Empirically establishing that a control effect motivates humans will further challenge the still dominant position that action and cognition are motivated solely by the search for desired outcomes (Eitam and Higgins [6]; Higgins [7, 5]; see also Bromberg-Martin and Hikosaka [8])."

Testing the Control Effect Motivation (CEM) Hypothesis
"In three experiments, we show that humans will be more motivated when their action has a perceivable effect (Experiment 1)."

"If actions do not have such a perceivable effect, regardless of whether the mind knows it is standing in relation to a current goal, the actor, deprived of such control feedback, will be less motivated (Experiment 2)."

"CEM hypothesis breaks from earlier theorizing on the role of effects in human action by arguing that control effects are motivating only when they are clear feedback about one's control (Experiment 2 and 3)."

General Method
"In this task, participants respond to colored circles that appear, one by one, at the top of the screen and rapidly descend in a vertical line to the bottom."

"A fixed time interval had passed from the appearance of the previous trial, regardless of the participants' response speed in the former trial."

"Participants completed 3 blocks of 60–120 trials each."

Experiment 1: Having an Effect Enhances Performance
"The sole difference between the two conditions was that in the Effect condition when participants pressed the correct key—corresponding to the horizontal location of the descending circle—the circle briefly flashed and disappeared."

"In the No Effect condition, the circle continued falling and no perceptual indication that the participant's response had an effect was given."

"Participants in the Effect condition could have clearer/more salient outcome feedback, and clearer outcome feedback produces more certainty about their responses."

"To test this alternative (outcome feedback) account, in Experiment 2, we equated the outcome feedback for participants in the Effect and No Effect conditions, while maintaining the difference between these conditions regarding control feedback."

"We did this by giving participants in both the Effect and the No Effect groups trial-by-trial outcome feedback about their performance."

Experiment 2: Matching Outcome Feedback
"To give all participants full outcome feedback, we added a running score counter at the top center of the screen."

"The task instructions told the participants that there was a score counter that would provide feedback about their performance."

"If outcome feedback was to explain the results, attending only to the score should lead to a response-time advantage over attending only to an effect (and even more to attending to both) because it would both supply full outcome feedback while preparing the participants to readily 'ambush' the circles at their 'source' (i.e., the top of the screen)."

"Crucially, this occurred even when participants in the No Effect condition were given full outcome feedback in the form of a running score."

"The performance benefit from having an effect must have been driven by the difference in control feedback."

Experiment 3: Introducing Uncertainty About Self-authorship
"Participants were run individually and were randomly assigned to one of the four conditions: No Effect, Immediate Effect, a Short Lag Effect condition in which an effect (the flash and disappearance of the circle) occurred 300 ms after a correct response, and a Long Lag Effect condition in which an effect occurred 600 ms after a correct response."

"t tests performed on the relevant means revealed the following findings: First, replicating Experiments 1 and 2, participants in the Immediate Effect condition were faster than those in the No Effect condition."

"The performance results of Experiment 3, beyond providing a second replication of the motivating influence of an immediate effect, supply evidence that this motivation relates to mechanisms of judgment of authorship over managing an effect (i.e., control)."

"The results of Experiment 3 show that the motivational consequence of having an effect is tightly time locked to a commission of a response."

General Discussion
"Three experiments demonstrate that having an effect enhances performance."

"Having an effect did not influence performance through functioning as outcome feedback—specifically, by providing clear feedback that a trial has ended or feedback about being successful in goal attainment (Experiments 2 and 3)."

"While we argue that effects are motivating regardless of the value (and hence, task relevance) of an effect, in the current experiments, the effect was not completely task irrelevant."

"While Experiment 2 clearly shows that an effect does motivate beyond mere outcome feedback (a running score), whether a completely task irrelevant effect will do the same awaits further research."

How Might Having an Effect Motivate?
"At our current level of knowledge, we can only speculate about the mechanism that translates control, and specifically effects, into motivation."

"We stress, however, that such reinforcement comes from the event being judged as feedback about successful control and hence the hypothesized mechanism by which effects motivate is different from both classic and current notions of reinforcement."

"According to that theory, the stimulus itself motivates through the learned relationship between it and a natural reinforcer (a valued outcome), which gives the stimulus its own incentive salience."

"An offshoot of this is that any external or internal stimulus may become a sign for potential or actual control (control feedback) and thus motivate."

"In line with this idea, recent neuroscientific work has demonstrated that a stimulus that comes to indicate the opportunity for choice, another form of control effectiveness (Higgins [5]), activates the brain's 'reward system' and that this happens regardless of the outcome value of that choice (Leotti and Delgado [9])."

"Further research is needed to establish which of these mechanisms underlie how control effects operate motivationally."

Why do Effects Motivate?

"We suggested that the mental representation of stimuli that are perceived to be of high control relevance (e.g., an effect of one's action) would be more accessible and will leave a stronger mark in the mind, for example, such stimuli will be learned better."

"Further speculating, it may be that dopamine both marks the representation of control effects, thereby increasing their relevance and accessibility in memory, and reinforces the action plan that produced them increasing the chance that the action is selected (and the effect reproduced)."

"Such a function would also answer the question of why task (or 'value')-irrelevant effects are learned (i.e., encoded) at all and would explain how the ideomotor principle could operate even without a conscious phase of action–effect contingency learning."

Implications of Control Effect as Motivator

"From the perspective of CEM, if people's involvement in the fate of others was represented by their mind as an instance of control, then it would be internally reinforcing to them."

"Effect as motivator may also have implication for physical and mental health, specifically for Parkinson's disease and schizophrenia."

"Regarding CEM, the question is whether the malfunction concerns the muted benefit that Parkinson's patients get from having control effects (Redgrave and Gurney 10; Redgrave and others [11])."

"Moore and others [12] have recently demonstrated that an implicit measure of having a control effect—intentional binding—was intact for people suffering from Parkinson's while explicit judgments of authorship were not (i.e., they were positively correlated with dopamine agonist medication)."

"One possibility is that the motivational impact of having a control effect will be strengthened due to an increase in dopaminergic availability."

"Using CEM, it is possible to address both Parkinson's and schizophrenia as control-to-motivation disorders."

Imaging the Neural Circuitry and Chemical Control of Aggressive Motivation [247]

This is a machine-generated summary of:

Ferris, Craig F; Stolberg, Tara; Kulkarni, Praveen; Murugavel, Murali; Blanchard, Robert; Blanchard, D Caroline; Febo, Marcelo; Brevard, Mathew; Simon, Neal G: Imaging the neural circuitry and chemical control of aggressive motivation [247]

Published in: BMC Neuroscience (2008)

Link to original: https://doi.org/10.1186/1471-2202-9-111

If you want to cite the papers, please refer to the original.

For technical reasons we could not place the page where the original quote is coming from.

Abstract-Summary

"Synchronized changes in neuronal activity across multiple brain areas can be viewed as functional neuroanatomical circuits coordinating the thoughts, memories and emotions for particular behaviors."

"FMRI in conscious rats combined with 3D computational analysis was used to identifying the putative distributed neural circuit involved in aggressive motivation and how this circuit is affected by drugs that block aggressive behavior."

"Brain areas previously identified as critical in the organization and expression of aggressive behavior were activated, e.g., lateral hypothalamus, medial basal amygdala."

"Oral administration of a selective vasopressin V_{1a} receptor antagonist SRX251 or the selective serotonin reuptake inhibitor fluoxetine, drugs that block aggressive behavior, both caused a general suppression of the distributed neural circuit involved in aggressive motivation."

"The putative neural circuit of aggressive motivation identified with fMRI includes neural substrates contributing to emotional expression (i.e. cortical and medial amygdala, BNST, lateral hypothalamus), emotional experience (i.e. hippocampus, forebrain cortex, anterior cingulate, retrosplenial cortex) and the anterior thalamic nuclei that bridge the motor and cognitive components of aggressive responding."

"Drugs that block vasopressin neurotransmission or enhance serotonin activity suppress activity in this putative neural circuit of aggressive motivation, particularly the anterior thalamic nuclei."

Background

"Normal aggressive behavior and aggression characterized by impulsivity and violence are envisioned to be organized and controlled by a distributed neural circuit, i.e., subsets of interconnected neurons conveying sensory and motor information to and from sites of integration [13]."

"Our present understanding of this putative neural circuit controlling aggressive behavior is based on early studies using chemical and electrical stimulation and lesion techniques in discrete brain areas [14, 15]."

"Immunostaining for immediate early gene proteins as cellular markers of neuronal activity helped identify multiple areas across the brain presumably involved in aggressive behavior [16, 17]."

"Because head restraint is a limitation in any awake animal fMRI study, it is not possible to image the neural circuit involved in the consummatory aspects of aggression like attacks and bites."

"The technique also allows the activity of this neural circuit to be imaged in the presence or absence of different drug treatments that affect aggressive responding."

Results

"This brain activity is reduced with SRX251 treatment in the presence of the intruder, but less so with fluoxetine treatment."

"There appears to be a general decrease in BOLD signal in all major regions with SRX251 but less so for fluoxetine treatment as compared to the activity observed with mate alone or mate/intruder."

"This activation of the primary olfactory system in the presence of SRX251 in response to a receptive female but not during aggressive motivation attests to the specificity of drug action and the stimulus-dependent nature of BOLD activation."

"Both drug treatments reduce the volume of activation in many brain areas identified as part of the putative neural circuitry of aggressive motivation."

"The time course for BOLD activation associated with aggressive motivation (black triangle) is markedly different with SRX251 and fluoxetine treatments."

Discussion

"In these studies, the lateral hypothalamus was particularly sensitive showing both a significant increase in the volume of activation and increase in BOLD signal to aggression provoking stimuli."

"The BOLD signal change did not exceed the 2% baseline threshold over most of the imaging period in the presence of SRX251 and only marginally so for fluoxetine with aggressive motivation but not receptive female."

"The mammillary body, a key area in the "Papez circuit" connecting the hypothalamus to the anterior thalamic nuclei, is not activated with aggressive motivation either by a measure of volume of activation or change in BOLD signal over time (data not shown)."

"SRX251 treatment suppressed activity in these areas in response to aggression-provoking stimuli but not to sexual stimuli, which may explain why sexual behavior is spared with V_{1a} receptor antagonism."

"Imaging showed that both anti-aggressive drugs caused an apparent decrease in general arousal as measured by the reduction in volume of activation and BOLD signal."

Conclusion

"The putative neural circuit of aggressive motivation described in this study using fMRI has the key components of the "Papez circuit" together with the olfactory circuitry and lateral hypothalamus that were missing from his original thesis."

"The distributed neural circuit of aggressive motivation described herein includes neural substrates contributing to emotional expression (i.e. cortical and medial amygdala, BNST, lateral hypothalamus), emotional experience (i.e. hippocampus, forebrain cortex, anterior cingulate, retrosplenial cortex) and the anterior thalamic nuclei that bridge the motor and cognitive components of aggressive responding."

"Oral fluoxetine, known to cause accumulation of 5-HT in hypothalamic intersitium, suppressed aggression and diminished BOLD activation across the putative neural circuit of aggressive motivation."

"Oral SRX251 a V1a receptors antagonist, suppressed aggression and produced a general reduction in BOLD activation in the neural circuitry of aggression similar to that seen with fluoxetine."

Methods
"On the day of an imaging session, male residents were tested for aggressive motivation by placing a novel, adult male intruder into their homecage for 5 min."

"Of each imaging session a high resolution anatomical data set was collected using a RARE pulse sequence (12 slice; 1.2 mm; FOV 3.0 cm; 256 × 256; TR 2.1 sec; TE 12.4 msec; NEX 6; 7 min acquisition time)."

"To control for this issue a phantom positioned in the rat restrainer was imaged for a duration of 10 min consisting of 5 min of baseline data followed by 5 min following the introduction of two adult male rats into the vivarium (n = 4)."

"To assess false activation due to subject motion we collected fMRI data from awake rats (n = 8) over a 10 min scanning session in the absence of any stimulation."

Motivation and Movement: The Effect of Monetary Incentive on Performance Speed [248]

This is a machine-generated summary of:

Mir, Pablo; Trender-Gerhard, Iris; Edwards, Mark J.; Schneider, Susanne A.; Bhatia, Kailash P.; Jahanshahi, Marjan: Motivation and movement: the effect of monetary incentive on performance speed [248]

Published in: Experimental Brain Research (2011)

Link to original: https://doi.org/10.1007/s00221-011-2583-5

If you want to cite the papers, please refer to the original.

For technical reasons we could not place the page where the original quote is coming from.

Abstract-Summary
"The effect of monetary incentive on total time (TT, (RT + MT)) was assessed in 16 healthy participants using four paradigms: warned simple RT (wSRT),

unwarned simple RT (uSRT), uncued choice RT (uCRT), and precued choice RT (pCRT)."

"wSRT, uSRT, and pCRT tasks all allow advance preparation and preprogramming of the movement, whereas uCRT does not."

"We found a significant effect of monetary incentive in shortening TTs in wSRT, uSRT, and pCRT tasks, but no effect on the uCRT task."

"These results demonstrate that monetary incentive can speed up movement initiation and execution in human participants, but only in tasks where preprogramming of the response is possible."

"This suggests that in reaction time tasks such as these, monetary incentive is having its effect by enhancing preparation of preprogrammed movement, but has little effect when movements cannot be specified in advance."

"These "RT and reward" tasks provide a useful paradigm for investigation into the effects of monetary incentive on reaction times in man and to study motivational modulation of movement speed in health and disease."

Introduction

"From observation of human behavior, we know that speed of movement initiation and execution can be influenced by motivational factors, for example walking faster when in a hurry (sense of urgency) or writing faster during an exam (potential reward of good results)."

"Can monetary reward be used as an incentive to speed up movement in RT tasks in humans?"

"As speeding up of RTs in response to reward-predicting stimuli in animal studies is obtained in precued RT tasks where advance preparation is possible (Hollerman and others [18]; Hassani and others [19]), we predicted that monetary incentive would be most likely to have an effect when preprogramming or advance preparation of movements was possible and that speed improvements would be seen for SRT and pCRT tasks but that no effect of monetary incentive would be observed in uCRT tasks."

Methods

"Participants were then required to release the home key as quickly as possible and press the response key."

"On 50% of trials (unwarned SRT, uSRT), there was no S1 or warning signal and the fixation cross was followed by a filled square immediately, in which case, the participant was required to press the response key straight away."

"For uncued choice reaction time (uCRT, 50% of trials), the fixation cross was followed after a delay of 1–4 s by a filled square without presentation of an empty square (i.e., no precue or S1) requiring the participant to press the appropriate response key straight away."

"Because of the nature of the task, requiring participants to lift their index finger from the home key on stimulus presentation (RT) and moving to the response key (MT), it is possible that monetary incentive may influence both movement initiation and execution."

Results

"We performed a three-way ANOVA with TASK (uSRT, wSRT, uCRT, pCRT), BLOCK (2–4), and REWARD (rewarded session, unrewarded session) as main factors."

"Corrected pairwise comparisons showed this interaction to be due to a shortening of TTs in blocks 3 and 4 compared with block 2 in the uSRT, wSRT, and pCRT tasks ($P < 0.01$ for all), but no significant change in TTs in blocks 3 and 4 compared with block 2 for the uCRT task."

"While a significant TASK × BLOCK × REWARD would have been the clearest indication of the impact of reward on the RT tasks, nevertheless, the above significant BLOCK × REWARD and TASK × REWARD interactions and subsequent pairwise comparisons indicated that provision of reward shortened TTs relative to the unrewarded block and produced a differential effect on TT across tasks, speeding up TTs for uSRT, wSRT, and pCRT but not in uCRT."

Discussion

"It appears that monetary incentive speeds up movement in RT tasks (uSRT, wSRT, pCRT) where volitional preparation and preprogramming of a response are possible."

"This suggestion is supported in our study by the lack of effect of reward on a task where movement preparation is not possible (uCRT)."

"While the lack of a reward effect in the uCRT task argues against such general effects, nevertheless, in future studies, the potential contributions of attention or learning can be dissociated from a purely motivational effect by, for example, introducing further blocks to determine whether movement speed returns to baseline levels when reward is no longer provided."

"Our results indicate that in relation to reaction time tasks, reward has its major effect on task performance by influencing aspects of movement preparation, allowing faster movement initiation in tasks where preprogramming of the movement is possible."

A Microbiome-Dependent Gut–Brain Pathway Regulates Motivation for Exercise [249]

This is a machine-generated summary of:

Dohnalová, Lenka; Lundgren, Patrick; Carty, Jamie R. E.; Goldstein, Nitsan; Wenski, Sebastian L.; Nanudorn, Pakjira; Thiengmag, Sirinthra; Huang, Kuei-Pin; Litichevskiy, Lev; Descamps, Hélène C.; Chellappa, Karthikeyani; Glassman, Ana; Kessler, Susanne; Kim, Jihee; Cox, Timothy O.; Dmitrieva-Posocco, Oxana; Wong, Andrea C.; Allman, Erik L.; Ghosh, Soumita; Sharma, Nitika; Sengupta, Kasturi; Cornes, Belinda; Dean, Nitai; Churchill, Gary A.; Khurana, Tejvir S.; Sellmyer, Mark A.; FitzGerald, Garret A.; Patterson, Andrew D.; Baur, Joseph A.; Alhadeff, Amber L.; Helfrich, Eric J. N.; Levy, Maayan; Betley, J. Nicholas; Thaiss, Christoph A.: A microbiome-dependent gut–brain pathway regulates motivation for exercise [249]

Published in: Nature (2022)

Link to original: https://doi.org/10.1038/s41586-022-05525-z

If you want to cite the papers, please refer to the original.

For technical reasons we could not place the page where the original quote is coming from.

Abstract-Summary
"An important factor stimulating the engagement in both competitive and recreational exercise is the motivating pleasure derived from prolonged physical activity, which is triggered by exercise-induced neurochemical changes in the brain."

"We report on the discovery of a gut–brain connection in mice that enhances exercise performance by augmenting dopamine signalling during physical activity."

"We find that microbiome-dependent production of endocannabinoid metabolites in the gut stimulates the activity of TRPV1-expressing sensory neurons and thereby elevates dopamine levels in the ventral striatum during exercise."

"Our study also suggests that interoceptomimetic molecules that stimulate the transmission of gut-derived signals to the brain may enhance the motivation for exercise."

Main

"Exercise is possibly the single most important and accessible lifestyle component that offers protection from a large range of diseases [20]."

"The ability to exercise is the result of a complex interplay between the musculoskeletal system, cardiovascular and respiratory factors, and the motivational state [21]."

"We identified a metabolite-triggered gut–brain connection that regulates the motivation for exercise, linking the intestinal microbiome to midbrain dopamine signalling via afferent sensory neurons."

Metagenomic Impact on Exercise Performance

"We used a machine-learning approach to identify variables with a strong predictive contribution to exercise performance, again by using treadmill parameters as the endpoint."

"These results suggest a strong functional contribution of the microbiome to exercise capacity in DO mice."

"To identify taxonomic elements of the microbiome that drive its contribution to physical performance, we turned to isogenic C57BL/6J mice, which allowed us to vary only microbiome parameters, while keeping all other factors constant."

"We inoculated germ-free mice with a member of the Erysipelotrichaceae family contributing to the prediction of exercise performance (Eubacterium rectale), a member of the Lachnospiraceae family likewise involved in the prediction of exercise (Coprococcus eutactus), a member of the Lachnospiraceae family with no predictive value for exercise (Blautia producta) and a taxonomically distinct member of the microbiota with high predictive power (Ruminococcus gnavus)."

"These data demonstrate the causative role of the microbiome in regulating exercise performance and highlight certain members of the Erysipelotrichaceae and Lachnospiraceae families as possible taxonomic contributors."

Microbiome Impact on Striatal Dopamine

"To explore the impact of the microbiome on the striatal response to exercise, we performed single-nucleus RNA-sequencing (RNA-seq) of the striatum of antibiotic-treated mice before and after treadmill exercise."

"We next evaluated whether the exercise-triggered dopamine response was mechanistically involved in the performance-stimulating effect of the microbiome."

"These experiments suggest that striatal dopamine signalling is an important mediator of the microbiome impact on exercise activity."

"We thus sought to explore the mechanisms whereby microbiome ablation impaired exercise-induced dopamine levels in the striatum."

"We thus hypothesized that sustained MAO levels are responsible for the blunted dopamine response to exercise in microbiome-depleted mice and that blockade of MAO would normalize dopamine signalling."

"These results suggest that intestinal microbial colonization enhances exercise-induced dopamine signalling in the striatum by limiting MAO-driven dopamine turnover."

Exercise Regulation by Sensory Neurons
"TRPV1 is broadly expressed by sensory neurons, we sought to determine whether spinal or vagal afferents were the primary mediator of the microbiome impact on exercise."

"To explore whether stimulation of sensory neurons affected the MAO-dopamine pathway, we measured Maoa transcripts and post-exercise dopamine levels in the striatum."

"We then co-treated microbiome-depleted and control mice with capsaicin and sulpiride, to inhibit dopamine signalling while stimulating afferent neurons."

"These findings demonstrate that the impact of the microbiome on dopamine levels in the striatum and exercise performance requires $TRPV1^{+}$ neurons."

Exercise Regulation by Gut Metabolites
"To explore whether DRG neuron stimulation by microbiome products was associated with striatal dopamine levels and exercise performance, we used stool extracts from the DO cohort and tested them for DRG neuron-stimulatory capacity."

"To determine whether FAAs are sufficient to restore exercise capacity in the absence of the microbiome, we pursued a dietary supplementation approach."

"OEA has been shown to stimulate dopamine release in the context of dietary reward [22], we asked whether gastric infusion of FAAs was sufficient to recapitulate the effects of exercise on dopamine levels in the striatum."

"We investigated whether FAA production by the microbiome can account for its performance-stimulating effect."

"We explored the mechanisms by which intestinal FAAs may stimulate sensory neurons and their impact on exercise motivation."

"These observations thus raised the possibility that stimulation of CB1 by FAAs may drive sensory neuron activity, striatal dopamine release and exercise performance."

Discussion
"We demonstrate that the brain circuitry involved in regulating the motivation for physical activity is not strictly central nervous system autonomous but is shaped by peripheral influences that originate in the intestinal microbial community, suggesting a possible mechanistic basis for understanding interindividual variability in exercise motivation and performance."

"Our results show that the gut microbiome contributes to the generation of intestinal FAA metabolites that trigger CB1-expressing $TRPV1^+$ sensory neurons, which in turn send an exercise-induced afferent signal to the brain and promote the downregulation of MAO expression in the striatum."

"Evolutionarily, the regulation of exercise-induced reward and motivation circuits by gut metabolites may serve to couple nutrient availability and the state of intestinal microbial colonization to the readiness and capacity to engage in prolonged physical activity."

"It also shows that exercise-induced activity of spinal sensory neurons drives striatal dopamine signalling through the regulation of MAO."

Methods
"In which pharmacological modulators were administered, the baseline for each mouse was recorded for three consecutive days and then, based on the experiment, mice were intraperitoneally injected daily between 5 and 7 p.m. with pharmacological modulators or vehicle, and data were acquired for another seven days following the specific treatments below."

"Targeting CB1, mice were injected daily intraperitoneally with CB1 agonist CP55940 (Sigma Aldrich, catalogue No."

"Brains were homogenized in $PBS^{-/-}$ and centrifuged at 5,000g for 10 min at + 4 °C, and then 50 μl of supernatant was used for quantification of dopamine in brain tissue following the manufacturer's instructions (BioVision, K4219-100)."

"Brains were homogenized in $PBS^{-/-}$ and centrifuged at 5,000g for 10 min at +4 °C, and then 50 μl of supernatant was mixed with 50 μl of sample dilution buffer and the quantification of total acetylcholine and glutamic acid were performed according to manufacturers' protocols (Amplex Acetylcholine/ Acetlycholinesterase Assay Kit, catalogue No."

Motivation(s) from Control: Response-Effect Contingency and Confirmation of Sensorimotor Predictions Reinforce Different Levels of Selection [250]

This is a machine-generated summary of:

Hemed, Eitan; Karsh, Noam; Mark-Tavger, Ilya; Eitam, Baruch: Motivation(s) from control: response-effect contingency and confirmation of sensorimotor predictions reinforce different levels of selection [250]

Published in: Experimental Brain Research (2022)

Link to original: https://doi.org/10.1007/s00221-022-06345-3

If you want to cite the papers, please refer to the original.

For technical reasons we could not place the page where the original quote is coming from.

Abstract-Summary

"To reliably manipulate the environment and attain their goals they would benefit from a constant modification of motor-responding based on responses' current effect on the current environment."

"We further show that evaluations of a response's effectiveness can occur simultaneously, driven by at least two different processes, each relying on different statistical properties of the feedback and affecting a different level of responding."

"We show the continuous effect of (a) a sensorimotor process sensitive only to the conditional probability of effects given that the agent acted on the environment (i.e., action-effects) and of (b) a more abstract judgement or inference that is also sensitive to the conditional probabilities of occurrence of feedback given no action by the agent (i.e., inaction-effects)."

"This study is the first to show that different evaluation processes of a response's effectiveness influence different levels of responding."

Introduction

"Further studies provided additional evidence that the facilitation of response speed due to immediate action-effects does not depend on performance feedback, but most likely on sensorimotor SOA driving reinforcement of actions."

"Given the above we predicted the following: (a) action-effects trigger the sequence that generates a sensorimotor SOA and consequently credits (i.e., reinforces) the effective motor-programs. (b) This form of reinforcement is argued to operate on a non-conceptual level; therefore, (b1) it will be sensitive to the conditional probability P(E|R) rather than to Delta-P, that also involves P(E|¬R) and (b2) as this form of SOA credits a specific motor program it will affect only the very specific manner of execution here measured as response speed and not the 'volitional aspects' of the response (Brass and Haggard [23]), here measured as response frequency. (c) A conceptual, regularity-detecting, mechanism of SOA will consider both action and inaction-effects and hence will be captured by Delta-P. The conceptual mechanism is argued to be detached from the motor-system (Wen and Haggard [24]; but see Synofzik and others [25]), so it can modulate actions through inputting to action-selection processes—the "whether and what to do" measured here as response frequency."

Experiment 1
"Compared with the action-effects present, inaction-effects present condition we found that response frequency was nominally higher on the baseline condition, with an insignificant difference (Bayesian: insensitive result) [Lower-tail—t(28) = − 1.43, p = 0.082, Cohen's d = − 0.52, 95% CI (− 1.28, 0.24), BF 1:0 = 1.31]."

"We found that response frequency was significantly higher on the baseline condition compared with the action-effects absent, inaction-effects present condition [Lower-tail—t(16) = − 2.61, p = 0.010, Cohen's d = − 1.11, 95% CI (− 1.99, − 0.23), BF 1:0 = 10.05], but not significantly different than the condition in which both action-effects and inaction-effects were absent [Lower-tail—t(31) = − 0.53, p = 0.299, Cohen's d = − 0.18, 95% CI (− 0.9, 0.53), BF 1:0 = 0.50]."

Experiment 2a
"With our prediction, in this experiment we did not find that RTs were faster in this condition compared with action-effects absent, inaction-effects present [Upper-tail—t(17) = 0.93, p = 0.182, Cohen's d = 0.17, 95% CI (− 0.200, 0.54), BF 1:0 = 0.57]."

"As predicted, we found lower response frequency when only inaction-effects were present [Lower-tail—t(17) = − 2.93, p = 0.005, Cohen's d = − 0.94, 95% CI (− 1.73, − 0.16), BF 1:0 = 11.08] or when both action- or inaction-effects were absent [Lower-tail—t(17) = − 3.73, p = 0.001, Cohen's d = − 1.27, 95% CI (− 2.2, − 0.34), BF 1:0 = 47.64]."

Experiment 2b
"The timepoint for determining if the current trial is a No-Response trial was sampled from a pre-generated normal distribution of 200 values (μ = 650 ms, σ = 40; based on the results of Experiment 2a and designed to ensure that the timing of inaction-effects was (a) not late in the trial and (b) would not truncate the response window)."

"We compared the action-effects present, inaction-effects absent condition to all other conditions and found that as predicted, response frequency was significantly lower when both action-effect and inaction-effects were present (yet Bayesian analysis shows that the current data is insensitive) [Lower-tail—t(35) = – 1.87, p = 0.035, Cohen's d = – 0.41, 95% CI (– 0.86, 0.04), BF 1:0 = 1.65]."

"We also found lower response frequency for the contrast with action-effects absent, inaction-effects present [Lower-tail—t(35) = – 1.96, p = 0.029, Cohen's d = – 0.46, 95% CI (– 0.96, 0.03), BF 1:0 = 1.90]."

Experiment 3a

"Participants were familiarized with the auditory cues and completed a short training block (32 trials) during which the experimenter gave them vocal feedback on their performance, but they did not receive any action-effects or inaction-effects regardless of the experimental condition."

"With our prediction, we found that when both action-effects and inaction-effects were present, response frequency was significantly higher [Lower-tail—t(38) = – 1.96, p = 0.029, Cohen's d = – 0.52, 95% CI (– 1.05, 0.02), BF 1:0 = 2.500]."

"We did not find the predicted lower response frequency when action-effects were absent, regardless of whether inaction-effects were present [Lower-tail—t(52) = 0.50, p = 0.689, Cohen's d = 0.13, 95% CI (– 0.4, 0.66), BF 1:0 = 0.19] or absent [Lower-tail—t(53) = 1.17, p = 0.876, Cohen's d = 0.31, 95% CI (– 0.23, 0.85), BF 1:0 = 0.14], both with Bayesian support for the null hypothesis."

Experiment 3b

"It is possible that the tightly controlled nature of the task (using auditory cues, rather than free operant procedure) minimized the (albeit, limited) ability of action-effects and inaction-effects to influence response frequency, even though participants still displayed a similar pattern of RT to that seen in all previous experiments."

"With our predictions, we did not find significantly lower response frequency when both inaction-effects and action-effects were present [Lower-tail—t(80) = – 0.48, p = 0.318, Cohen's d = – 0.06, 95% CI (– 0.32, 0.19), BF 1:0 = 0.19]."

"We did confirm our prediction that response frequency is higher on the baseline condition (action-effects present, inaction-effects absent) compared with the condition where only inaction-effects are present [Lower-tail—t(80) = – 2.72, p = 0.004, Cohen's d = – 0.35, 95% CI (– 0.62, – 0.09), BF 1:0 = 7.58] or when both action-effects and inaction-effects are absent [Lower-tail—t(80) = – 4.85, p < 0.001, Cohen's d = – 0.65, 95% CI (– 0.93, – 0.36), BF 1:0 = 5525.65]."

General Discussion
"If this sort of prediction influenced their response speed, this comes as evidence against our point that response speed in our task is mostly sensitive to sensorimotor processes and that if our participants were only aware of the probability of inaction-effects our pattern of results would have changed (e.g., slower RTs for the action-effect present, inaction-effect present condition)."

"Exploring several other conditions where Delta-P is 0 (e.g., like one where both action-effect probability and inaction-effect probability are 0.3) could have answered the question of whether the facilitation we observed for response speed on the action-effect present, inaction-effect present occurred from sheer frequency of feedback, as Elsner and Hommel's study suggests or from high action-effect probability as we argue."

The Influence of Action–Outcome Contingency on Motivation from Control [251]

This is a machine-generated summary of:

Penton, Tegan; Wang, Xingquan; Coll, Michel-Pierre; Catmur, Caroline; Bird, Geoffrey: The influence of action–outcome contingency on motivation from control [251]

Published in: Experimental Brain Research (2018)

Link to original: https://doi.org/10.1007/s00221-018-5374-4

If you want to cite the papers, please refer to the original.

For technical reasons we could not place the page where the original quote is coming from.

Abstract-Summary
"The sense of agency, Oxford University Press, New York, 26), suggests that actions associated with a high sense of agency are intrinsically rewarding and thus motivate response selection."

"Previous studies support this theory by demonstrating that factors impacting on sense of agency (e.g. probability of an outcome following an action) also motivate selection of actions."

"We report a novel test of the control-based response selection framework in the domain of action–outcome contingency."

"The contingency between actions and their outcome has previously been demonstrated to impact the sense of agency, but its impact on the motivation to perform actions has not yet been examined."

"A contingency manipulation was employed where the probability of an outcome occurring in the absence of a button press was also varied in blocks throughout the experiment."

"Results demonstrated a significant influence of contingency on response speed, and a significant effect of probability on response selection, consistent with predictions from the control-based response selection framework."

"The current findings support the idea that actions associated with an increased sense of agency are intrinsically rewarding, and identify how individual differences may impact on this process."

Introduction

"To these data, several studies have demonstrated a relationship between action–outcome contingency and implicit sense of agency (e.g. Moore and Haggard [27]; Moore and others [28]; Sidarus and others [29]), demonstrating that action–outcome contingency, as well as the probability of outcome, influences the implicit sense of agency."

"This framework has been tested in a series of studies by Karsh, Eitam and colleagues which show that those factors previously demonstrated to impact the sense of agency, such as probability of an outcome, and temporal contiguity and predictability of an outcome, also influence the likelihood of an action being selected (e.g. Karsh and Eitam [30]; Karsh and others [31]; Eitam and others [32])."

"Predictions from the Control-Based Response Selection Framework are clear; contingency manipulations should affect action responses as they have previously been demonstrated to impact upon the sense of agency (Moore and Haggard [27]; Moore and others [28]; Sidarus and others [29])."

Methods

"On a given trial, the likelihood of an internally generated event (caused by the participant making a button press) and an externally generated event (outside of the participant's control) was varied to manipulate action–outcome contingency."

"300 ms was subtracted from each of the fast RTs to ensure that externally generated cues presented at these times would occur prior to a participant's button press (early externally generated event)."

"Participants are asked to write the most likely cause of 12 (6 positive and 6 negative) hypothetical events (e.g. "A friend compliments you on your appearance") and then rate the cause along 3, 7-point Likert scales reflecting the extent to which the participant feels the cause of the event was due to an internal factor (the self) or an external factor (internality factor), the stability of the cause of the event across time (stability factor) and the extent to which the cause of the event is unique to one domain or universal (globality factor)."

Results

"For each participant, for the first level of contingency (0% likelihood of an external flash), median RTs from each button (buttons with 0%, 30%, 60%, and 90% chance of an internally generated flash) were multiplied by relative weightings (− 1.5, − 0.5, 0.5 and 1.5) and then summed to create an overall score of the influence of probability in the highest action–outcome contingency condition."

"For the first level of contingency (0% likelihood of an external flash), percentage of responses for each button (buttons with 0%, 30%, 60%, and 90% chance of an internally generated flash when pressed) was multiplied by relative weightings (− 1.5, − 0.5, 0.5 and 1.5) and then summed to create an overall score of the influence of probability in the highest action–outcome contingency condition."

Discussion

"The response selection results are also consistent with predictions from the Control-Based Response Selection Framework if the contingency manipulation affects implicit, but not explicit sense of agency."

"The current finding that action–outcome contingency affects one index of motivation to act (response speed), but not another (response selection), may provide additional insight into a growing literature on differential processes involved in the experience of implicit and explicit sense of agency."

"However, that under the Control-Based Response Selection Framework the predictive ability of autistic traits could arise from an association between autistic traits, and (1) the ability to determine contingency, (2) the degree to which action–outcome contingency impacts sense of agency, and/or (3), the degree to which sense of agency influences response speed."

"Results are in accordance with the Control-Based Response Selection Framework (Karsh and Eitam [26]), suggesting that individuals are motivated to perform actions that afford a sense of agency."

Normalization by Valence and Motivational Intensity in the Sensorimotor Cortices (PMd, M1, and S1) [252]

This is a machine-generated summary of:

Yao, Zhao; Hessburg, John P.; Francis, Joseph Thachil: Normalization by valence and motivational intensity in the sensorimotor cortices (PMd, M1, and S1) [252]

Published in: Scientific Reports (2021)

Link to original: https://doi.org/10.1038/s41598-021-03200-3

If you want to cite the papers, please refer to the original.

For technical reasons we could not place the page where the original quote is coming from.

Abstract-Summary
"We ask if psychological affect causes normalization of the sensorimotor cortices by modulating valence and motivational intensity."

"We utilized several normalization models for valence and motivational intensity in all three regions."

"We found three types of divisive normalized relationships between neural activity and the representation of valence and motivation, linear, sigmodal, and hyperbolic."

"The hyperbolic relationships resemble receptive fields in psychological affect space, where a unit is susceptible to a small range of the valence/motivational space."

"We found that these cortical regions have both strong valence and motivational intensity representations."

Introduction

"Neural networks within the primary sensorimotor cortices need to function within a limited dynamic range while encoding multiple forms of information such as kinematics and kinetics [33–36]."

"To valence, the interaction between reward and punishment can also be represented as motivational intensity or motivational salience."

"We characterized neural activity to cued possible reward and simultaneously cued possible punishment in cS1, rM1, and PMd."

"By cuing both reward and punishment, in this manner, we modulated valence and motivational intensity [37]."

"We demonstrate that neural modulation associated with cued reward and punishment in an operant conditioning task, where reward or punishment was delivered according to an individual's performance, is encoded in cS1, rM1, and PMd."

"We obtained similar results using either a divisive term based on a combined variable comprised of the cued reward level and a scaled version of the cued punishment level or simply a function of the brain region's population activity."

Methods

"Each trial consisted of 6 stages: cue display, autonomous robot reaching, manual isometric grip force output, autonomous robot transporting, manual isometric releasing, and feedback, which could be reward delivery if it was a successful trial or a timeout period if not."

"The number of green squares indicated the level of fruit juice reward that the NHP would receive upon successful completion of the task, with the number present (0–3) corresponding to the number of 0.5 s juice delivery periods."

"For post-feedback analysis, firing rates were averaged over 500 ms following the indication of the result (reward or punishment); thus, there were no temporal confounds due to the different reward and punishment delivery periods."

"The post-feedback period was after the animal had successfully completed or failed the trial, and the animal at that point may only be encoding reward for successfully completed trials or punishment for unsuccessfully completed trials; at least, this was our hypothesis."

Results

"We found approximately 25–48% of the units from the full population of a given brain region were significantly modulated by reward, punishment, or both."

"Our hypothesis for this post feedback period was that the units would be encoding the reward or punishment that the NHP would receive compared to during the post cue period when the outcome was not yet known, and thus there was still uncertainty."

"As an example, on successful trials in the post feedback period, we expected a unit to have the same response when the reward and punishment levels were R = 1, P = 0 or R = 1, P = 1 since the NHP would have obtained the same reward in either case."

"NHP P had slightly more units with their best model from group 1 (considering reward and punishment together for all trials)."

Discussion

"We modulated the level of cued valence, spanning from negative (punishment) to positive (reward), and cued motivational intensity, which was simultaneously modulated by cueing the level of reward they would receive if successful, and level of timeout punishment if unsuccessful."

"There were 4 levels of cued reward and cued punishment, leading to 16 levels of valence and motivational intensity termed affective stimuli."

"Model 4 had the most significant units as their best model, where model 4 only looked at successful trials and reward levels (R = 0–3) while also utilizing the task R levels in the divisive term."

"We found units modulated only by reward or punishment in both NHPs and all cortical regions under study."

"For BCI control signals to be accurately decoded from neural activity, it is important to determine if these units utilize a divisive normalization scheme for affective stimuli or naturally changing moods and emotional states."

The Influence of Motivational Salience on Saccade Latencies [253]

This is a machine-generated summary of:

Rothkirch, Marcus; Ostendorf, Florian; Sax, Anne-Lene; Sterzer, Philipp: The influence of motivational salience on saccade latencies [253]

Published in: Experimental Brain Research (2012)

Link to original: https://doi.org/10.1007/s00221-012-3284-4

If you want to cite the papers, please refer to the original.

For technical reasons we could not place the page where the original quote is coming from.

Abstract-Summary

"The initiation of saccades towards visual stimuli is known to be influenced by the bottom-up salience of stimuli as well as the motivational context of the task."

"We asked whether the initiation of saccades is also influenced by the intrinsic motivational salience of a stimulus."

"The same faces served as target stimuli in a subsequent saccade task, in which their motivational salience was no longer task-relevant."

"We found a specific effect of learned positive stimulus value on the latencies of voluntary saccades: For faces with high versus low positive motivational salience, saccadic latencies were significantly reduced."

"Our findings show for the first time that saccadic preparation is susceptible to the acquired intrinsic motivational salience of visual stimuli."

"Based on the observation that only voluntary saccades but not reactive saccades were modulated, we conclude that the recruitment of neural processes for target identification is required to allow for an influence of motivational stimulus salience on saccadic preparation."

Introduction

"Attentional selection of a certain stimulus in the visual field is often followed by a saccade, that is, a rapid eye movement towards the target of attention."

"The use of both a voluntary and a reactive saccade task allowed us to further specify the level of visual target processing that is required for motivational stimulus salience to have an effect on saccade initiation: If the appearance of a motivationally salient stimulus alone already influenced saccade initiation, then an effect would be expected both on reactive and on voluntary saccades (i.e. in experiments 1 and 2)."

"If an influence of motivational salience on saccade initiation required minimal degree of object-specific visual processing (i.e. the selection of the target object among two stimuli), then we would predict to find an effect on voluntary saccades but not on reactive saccades."

Experiment 1: The Influence of Motivational Salience on Voluntary Saccades

"2 × 2 repeated measures ANOVAs with the factors motivational salience (high vs. low salience) and valence (positive vs. negative valence) were used to probe the influence of previously learned values on saccade latencies in the subsequent voluntary saccade task."

"A repeated measures ANOVA with the factors valence (positive vs. negative) and motivational salience (high vs. low outcome probability in the learning task) did not show a significant main effect (valence $F(1,22) = 0.30$, $p = .59$; motivational salience $F(1,22) = 2.19$, $p = .15$)."

"For positive valence, latencies of voluntary saccades towards faces with high motivational salience were significantly shorter compared to faces with low motivational salience ($t(22) = -3.39$, $p = .005$)."

Experiment 2: The Influence of Motivational Salience on Reactive Saccades

"The statistical analysis of the influence of previously learned stimulus values on saccade latencies in the subsequent task followed the same approach as in experiment 1; 25.22 % (SEM 2.05 %) of all trials had to be discarded using the same exclusion criteria as in the first experiment."

"Participants directed the first saccade towards the optimal stimulus in 50.1 % (SEM: 2.6 %) of the trials in which a reward-related face pair was presented."

"For face pairs associated with punishment, the first saccade was directed towards the optimal stimulus in 52.1 % of the trials (SEM 2.6 %), which again was not significantly different from an equal proportion between the optimal and non-optimal stimulus ($p > .4$)."

General Discussion

"For reactive saccades, we did not find a modulating effect of the motivational salience or valence of the stimulus on latencies."

"However, in our study, the learned motivational salience of the stimuli was entirely irrelevant for the saccade task, that is, participants were not required to generate fast saccades in order to gain a reward or avoid a punishment."

"While larger effects of reward on saccade latencies have been previously reported (e.g. Blaukopf and DiGirolamo 38; Milstein and Dorris 39; Ross and others 40; but see also Jazbec and others 41 for a negative finding), in contrast to these earlier studies, faces' motivational salience was not at stake during our saccade task."

"The finding that only voluntary saccades, but not reactive saccades, were found to be modulated indicates that some degree of object-specific neural processing for target identification is required to allow for an influence of motivational stimulus salience."

How Individual Needs Influence Motivation Effects: A Neuroscientific Study on McClelland's Need Theory [254]

This is a machine-generated summary of:

Rybnicek, Robert; Bergner, Sabine; Gutschelhofer, Alfred: How individual needs influence motivation effects: a neuroscientific study on McClelland's need theory [254]

Published in: Review of Managerial Science (2017)

Link to original: https://doi.org/10.1007/s11846-017-0252-1

If you want to cite the papers, please refer to the original.

For technical reasons we could not place the page where the original quote is coming from.

Abstract-Summary

"This study validates the long-standing need theory of McClelland (Am Psychol 40(7):812–825, 42."

"It neurally tests the assumptions that (1) heterogeneous rewards may result in similarly rewarding effects and (2) that these effects are enhanced if a reward closely matches an employee's need."

"Results show that the heterogeneous types of rewards lead to overlapping neural activations in parts of the brain's reward circuitry, such as the putamen or caudate."

"Each of these rewards uniquely activates brain areas not stimulated by other reward types."

"A closer matching between the type of reward and the participants' individual needs results in stronger neural activations in the reward circuitry."

"These findings support and enhance key assumptions of need theory on a neural level and further promote a personality-based approach to work motivation."

"They suggest need-tailored reward systems for organizations and an increased use of rewards other than money."

Introduction

"To work environments, changes can also be seen regarding the methodological approaches that may be used to validate motivation theory."

"Need theory claims that employees are motivated when their need for power, affiliation and achievement are activated."

"In light of the new developments regarding work environments and methodological approaches, the overall aim of this study is to validate need theory with neuroscientific methods."

"We study whether the effect of these distinct rewards is enhanced if they closely match an employee's need."

"It offers a neuroscientific perspective to McClelland's [42] need theory and uniquely validates it by showing that neural reward activations are stronger when a specific reward closely matches the individual need level of a person."

"The practical implications we draw from our results refer to an individual approach to work motivation in which the varying needs of employees have to be adequately considered."

Theoretical Background and Hypotheses Development

"The work environment offers external factors (e.g. rewards) that promise need-fulfilment."

"When external factors (rewards) interact with internal factors (needs) work motivation is observed (Beckmann and Heckhausen [44]; Schneider and Schmalt [45])."

"Work motivation results from the interaction of an employee's needs and organizational rewards."

"The second hypothesis refers to empirical findings suggesting that work motivation is based on an interaction of needs (internal factors) and rewards (external factors; e.g. Cerasoli and others [46])."

"Building on these findings we test a key assumption of the need theory and examine on a neural level whether a closer match between an employee's needs and the rewards that offer need fulfilment result in more rewarding effects."

"According to McClelland's [42] need theory, higher work motivation should be observed when a reward meets a corresponding need."

Method

"According to the cover story, the more distraction tasks a participant solved correctly in the MR-scanner the higher was his/her chance for (a) a high income in the internship, (b) a prestigious company car, and (c) a respectful leader."

"Following the logic of the money treatment, participants simultaneously saw two pictures in the MR-scanner after completing a task, one represented a car they had rated high regarding prestige and social status (e.g. BMW; Car+) and another picture showing a car, which they had rated low in terms of prestige and social status (e.g., Hyundai; Car−)."

"If the task was solved correctly, the particular reward of each treatment (i.e. the picture for a high income, prestigious car, respectful leader, upward arrow) was framed for 3000 ms."

Results

"Regarding H2, a region of interest analysis focuses on specific reward-related brain areas and offers an answer to the question of whether rewards that closely match an employee's needs result in stronger neural activations of the reward circuitry than rewards that match these needs to a lesser extent."

"In order to test whether the participants differed regarding their number of correctly solved tasks and whether this number of correctly solved tasks varies across the three rewards, a repeated-measures ANOVA with rewards (Leader, Car, Money, and Arrow as control) and scores (Correct vs. Incorrect) was conducted."

"With respect to H2, we studied whether rewards that closely match an employee's need result in stronger neural activations of the reward circuitry than rewards that match this need to a lesser extent."

"To these results, the participants' need for achievement and power was not significantly related to the neural activations that occurred when they anticipated working for a respectful leader."

Discussion and Conclusion

"It appears as if the rewards high income, respectful leadership, and prestigious company car are to some extent interchangeable on the neural level despite the circumstance that they trigger different needs of McClelland's [47] theory."

"If a person with a high need for achievement was rewarded with high income then this person showed stronger activation of the key parts in the reward circuitry compared to receiving the reward prestigious company car or respectful leadership."

"Our region of interest analysis showed that three brain areas of the reward circuitry—the nucleus accumbens, putamen and caudate nucleus—are more activated when rewards closely match a person's need level."

"Our study provides findings that could be referred to when comparing the neural activations of rewards matched with conscious needs to the activations of the same rewards matched with unconscious needs."

Employee Engagement as Human Motivation: Implications for Theory, Methods, and Practice [255]

This is a machine-generated summary of:

Pincus, J. David: Employee Engagement as Human Motivation: Implications for Theory, Methods, and Practice [255]

Published in: Integrative Psychological and Behavioral Science (2022)

Link to original: https://doi.org/10.1007/s12124-022-09737-w

If you want to cite the papers, please refer to the original.

For technical reasons we could not place the page where the original quote is coming from.

Abstract-Summary
"This paper argues for a more sophisticated approach to the engagement concept, grounding it in the vast psychological literature on human motivation."

"We review the leading definitions of employee engagement in the literature and find that they are reducible to a core set of human motives, each backed by full literatures of their own, which populate a comprehensive model of twelve human motivations."

"We propose that there is substantial value in adopting a comprehensive motivational taxonomy over current approaches, which have the effect of "snowballing" ever more constructs adopted from a variety of fields and theoretical traditions."

"We consider the impact of rooting engagement concepts in existing motivational constructs for each of the following: (a) theory, especially the development of engagement systems; (b) methods, including the value of applying a comprehensive, structural approach; and (c) practice, where we emphasize the practical advantages of clear operational definitions."

Introduction

"These 102 concepts also vary broadly in terms of their content, encompassing job characteristics (e.g., variety, challenge, enrichment); individual traits (e.g., conscientiousness, autotelic personality, locus of control); intrapsychic concerns related to the self (e.g., psychological safety, authenticity, opportunities for personal growth); relations with the material world of work (e.g., autonomy, absorption, opportunity to meaningfully contribute); social cognitions, emotions, and motivations (e.g., sense of belonging, demonstrations of caring, opportunities for recognition); and concerns with higher-order, abstract principles (e.g., justice, values, purpose)."

"It's no coincidence that the major definitions of the employee engagement construct, despite their widely ranging theoretical origins, happen to fall perfectly in line with the definition of motivation, given by Pincus [48] as an individual-level, unobservable state of emotion or desire operating on the will and, as a psychological mediator, causing it to act."

Conclusion

"This paper responds to repeated, urgent calls for integration of the diverse and proliferating concepts related to employee engagement."

"As a final example of the utility of our model, we return to one of the many laments over the state of engagement theory and measurement."

"As indicated in the parentheses, our model easily accommodates all of these perspectives, mini-theories, and concept within a single model, providing a kind of "unified field theory" of employee engagement."

"We contend that the secret to unlocking a meta-theory to encompass all of these perspectives, and all of the dimensions they propose, has always been hidden in plain sight within the very first descriptions of employee engagement."

Cross-frequency Coupling of Brain Oscillations in Studying Motivation and Emotion [256]

This is a machine-generated summary of:

Schutter, Dennis J. L. G.; Knyazev, Gennady G.: Cross-frequency coupling of brain oscillations in studying motivation and emotion [256]

Published in: Motivation and Emotion (2011)

Link to original: https://doi.org/10.1007/s11031-011-9237-6

If you want to cite the papers, please refer to the original.

For technical reasons we could not place the page where the original quote is coming from.

Abstract-Summary

"We provide evidence that studying interdependencies between brain oscillations may be a valuable approach to study the electrophysiological processes associated with motivation and emotional states."

"Studies will be presented showing that amplitude-amplitude coupling between delta-alpha and delta-beta oscillations varies as a function of state anxiety and approach-avoidance-related motivation, and that changes in the association between delta-beta oscillations can be observed following successful psychotherapy."

"These studies suggest that cross-frequency coupling of brain oscillations may contribute to expanding our understanding of the neural processes underlying motivation and emotion."

Introduction

"Ever since the successful recording of 10 Hz brain oscillations over the intact scalp of human subjects by Berger in 49, the study of brain oscillations in the electroencephalogram (EEG) has become an increasingly important focus of interest in studying brain-behavior relationships."

"Brain oscillations are the result of the continue waxing and waning of electric field potentials generated simultaneously in a large number of vertically oriented apical dendrites of pyramidal nerve cells in the cerebral cortex (Niedermeyer and Lopes da Silva [50])."

"In spite of having a cortical read-out brain oscillations come about by direct or indirect interactions with other local and distal structures subsequently forming complex networks."

Brain Oscillations

"It has been proposed that more complex networks operate using higher frequency ranges."

"In line, slow wave activity corresponding to delta (1–3 Hz) and theta (4–7 Hz) frequency range stems from the 'evolutionary older' subcortical structures (including the brainstem and septo-hippocampal complex)."

"Fast wave activity in the alpha (8–12 Hz) and beta (13–30 Hz) frequency range, on the other hand, is suggested to find its origin in thalamo-cortical and cortico-cortical circuits respectively (Knyazev and Slobodskaya [51]; Knyazev and others [52]; Robinson [53, 54, 55, 56])."

"Dominant rhythms shift from low frequency to higher frequency with increasing complexity of the brain, yet lower frequencies are suggested to be preserved and arguably reflect neural systems that are evolutionary more ancient."

Indices of Cross-frequency Coupling

"This measure mixes phase and amplitude relations between signals or the phase-amplitude relationship between different frequencies."

"In order to study a pure phase–phase relationship, that is phase synchronizations, across frequencies, a method to estimate time varying quadratic phase coupling of electrophysiological signals, the bi-phase locking value was introduced (Darvas and others [57])."

"This method is useful for examining amplitude-independent phase-locking, but is limited to coupling among integer-multiple frequency bands (e.g., 10, 20, 40 Hz)."

"Both phase-phase and phase-amplitude cross-frequency coupling have been shown to relate to specific cortical areas and specific stages of information processing and have received much attention in recent works."

"Amplitude-amplitude associations between different frequency EEG bands could be studied both in the within- (i.e., across a number of states within each individual) and between- (i.e., across a sample of subjects) subject domain."

Interrelations with Motivation, Emotion and Personality

"Apart from these factor-analytic studies, a large body of evidence coming from different experimental manipulations and observations in psychopathological and juvenile populations confirms a general tendency for alpha and slower (delta and theta) frequencies to be inversely related to each other, with a prevalence of slow frequencies in some pathological conditions and early developmental stages being associated with impulsivity and disinhibition (for a review see Knyazev [58])."

"An association study yielded comparable results as males who were high in baseline salivary testosterone exhibited significantly reduced frontal delta-beta coupling compared to those low in testosterone (Miskovic and Schmidt [59])."

"Schutter and van Honk [60] demonstrated that higher levels of cortisol were associated with increased frontal delta-beta coupling in healthy subjects."

"A study by Miskovic and others [61] provides evidence that individuals diagnosed with social anxiety disorder (generalized subtype) exhibit high levels of EEG delta-beta coupling during the anticipation of a self-presentation task, whereas cognitive behavioural therapy significantly reduces the anticipatory delta-beta coupling."

Discussion and Future Directions

"Empirical studies have demonstrated that delta-alpha and delta-beta frequency coupling in the human EEG are associated with phenomenology, behavior, and hormones."

"It should be emphasized that although we review works studying the delta-alpha and the delta-beta interactions together under the concept of cross-frequency coupling, they probably reflect quite different phenomena."

"The observed increases in delta oscillations paralleled by a reduction in the coupling between delta and beta oscillations after a single administration of testosterone suggest a modulation of the cross-frequency coupling driven by a change in delta wave activity (Schutter and van Honk [62])."

"This pattern of results suggests that the increase of delta power in medial frontal and orbitofrontal cortices during an anxiogenic situation in high state anxiety subjects, might be the driving force behind the (between-subject) cross-frequency coupling and that this coupling is closely related to the within-subject cross-frequency amplitude-amplitude and phase-amplitude coupling."

Conclusion

"Research has shown that brain functions are realized by simultaneous oscillations in various frequency bands (Basar [63])."

"To examining oscillations in pre-specified bands, interactions and relations between the different frequency bandwidths is another important aspect that needs to be considered in unraveling the workings of the human brain and its functions."

"The first series of studies have provided evidence that the cross-frequency coupling between slow and fast wave oscillations is sensitive to steroid hormones and associated with varying motivational states in which normal and pathological forms of anxiety have received most attention."

"In spite of the remaining issues surrounding the cross-frequency approach, studying interrelations between different brain wave oscillations may yield important new insights into the organization and dynamics of human brain functions."

Immersive Virtual Reality During Gait Rehabilitation Increases Walking Speed and Motivation: A Usability Evaluation with Healthy Participants and Patients with Multiple Sclerosis and Stroke [257]

This is a machine-generated summary of:

Winter, Carla; Kern, Florian; Gall, Dominik; Latoschik, Marc Erich; Pauli, Paul; Käthner, Ivo: Immersive virtual reality during gait rehabilitation increases walking speed and motivation: a usability evaluation with healthy participants and patients with multiple sclerosis and stroke [257]

Published in: Journal of NeuroEngineering and Rehabilitation (2021)

Link to original: https://doi.org/10.1186/s12984-021-00848-w

If you want to cite the papers, please refer to the original.

For technical reasons we could not place the page where the original quote is coming from.

Abstract-Summary
"The rehabilitation of gait disorders in patients with multiple sclerosis (MS) and stroke is often based on conventional treadmill training."

"Virtual reality (VR)-based treadmill training can increase motivation and improve therapy outcomes."

"The present study evaluated an immersive virtual reality application (using a head-mounted display, HMD) for gait rehabilitation with patients to (1) demonstrate its feasibility and acceptance and to (2) compare its short-term effects to a semi-immersive presentation (using a monitor) and a conventional treadmill training without VR to assess the usability of both systems and estimate the effects on walking speed and motivation."

"In a within-subjects study design, 36 healthy participants and 14 persons with MS or stroke participated in each of the three experimental conditions (VR via HMD, VR via monitor, treadmill training without VR)."

"For both groups, the walking speed in the HMD condition was higher than in treadmill training without VR and in the monitor condition."

"Most of the healthy study participants (89%) and patients (71%) preferred the HMD-based training among the three conditions and most patients could imagine using it more frequently."

"The present study evaluated the usability of an immersive VR system for gait rehabilitation in a direct comparison with a semi-immersive system and a conventional training without VR with healthy participants and patients."

"The study demonstrated the feasibility of combining a treadmill training with immersive VR."

"Due to its high usability and low side effects, it might be particularly suited for patients to improve training motivation and training outcome e. g. the walking speed compared with treadmill training using no or only semi-immersive VR."

Background

"Treadmill training can be combined with virtual reality (VR) to increase its efficacy and the patients' motivation, as demonstrated in several studies [64–69]."

"In a randomized controlled trial with patients with MS comparing conventional with VR-based treadmill training, both groups improved walking endurance and speed [66]."

"No study has conducted a direct comparison of an immersive and a semi-immersive VR based treadmill training with patients with stroke and MS."

"For the current study, a novel VR-based treadmill training was implemented and its feasibility tested with healthy participants and patients with stroke and MS with gait disorders."

"The aim of the study was to evaluate an immersive VR application for supervised gait rehabilitation of patients with MS or stroke, to test its feasibility and acceptance and to compare its effects to those of a semi-immersive application and to a conventional treadmill training."

Methods

"Patients filled in the questionnaire before and after each VR-condition, while healthy participants filled in the questionnaire only after each condition."

"To avoid frustration and ensure that all patients reach the end of the path, the length of the path was individually adjusted based on their average walking speed in the condition without VR."

"Treadmill training without VR was the first treadmill condition to be completed for both healthy participants and patients to become familiar with the treadmill and to establish a baseline walking speed for the VR conditions."

"To determine whether the type of treadmill training (no VR, semi-immersive VR, immersive VR) or the presentation order of VR in the VR conditions had an effect on the dependent variables such as walking speed and heart rate, repeated measures analyses of variance (rmANOVAs) were calculated."

Results

"The mood of the healthy participants was better after the HMD condition than after the monitor condition and higher than after the condition without VR (both $p < 0.001$)."

"The healthy participants reported higher motivation after the HMD condition than after the monitor condition ($p < 0.001$) and after the condition without VR ($p < 0.001$)."

"was higher in the HMD condition than in the monitor condition for healthy participants ($p < 0.001$) and for patients ($p = 0.032$)."

"The mental load was higher in the HMD condition than in the condition without VR for patients ($p = 0.039$) and, in the healthy participants, higher than in the condition without VR ($p < 0.001$) and higher than in the monitor condition ($p = 0.043$)."

Discussion

"The present study demonstrated the feasibility and acceptance of an immersive virtual reality-based treadmill training as a rehabilitation method for neurological patients with gait disorders."

"The study demonstrated that the immersive presentation of a virtual scenario via an HMD leads to a higher walking speed of patients with MS and stroke than a semi-immersive VR presentation or a treadmill training without VR."

"And in line with many previous studies [70, 71], healthy participants and patients alike reported a higher sense of presence in the virtual environment presented via an HMD compared to a semi-immersive presentation via a monitor."

"For patients, however, mental demands were significantly higher in the HMD condition compared with the treadmill training without VR."

"This study revealed the feasibility and acceptance of an immersive VR-based treadmill training among a group of healthy users and patients with MS and stroke."

Conclusions

"The study demonstrated the feasibility and patients' acceptance of a treadmill training with immersive VR via an HMD."

"The increased walking speed demonstrated in the current study suggests that it could lead to an improved training outcome compared with treadmill training using no or only semi-immersive VR."

"Despite these promising results, the decision, whether the advantages of immersive VR outweigh the risks associated with wearing an HMD should be based on a patient-by-patient basis and first and foremost on the will of the individual patient."

Relationship Between Metacognitive Awareness and Motivation to Learn in Medical Students [258]

This is a machine-generated summary of:

Siqueira, Marina Alves Martins; Gonçalves, Johnatan Padovez; Mendonça, Vitor Silva; Kobayasi, Renata; Arantes-Costa, Fernanda Magalhães; Tempski, Patricia Zen; Martins, Mílton de Arruda: Relationship between metacognitive awareness and motivation to learn in medical students [258]

Published in: BMC Medical Education (2020)

Link to original: https://doi.org/10.1186/s12909-020-02318-8

If you want to cite the papers, please refer to the original.

For technical reasons we could not place the page where the original quote is coming from.

Abstract-Summary

"In self-regulated learning, learning is defined as metacognitively guided, intrinsically motivated and strategic."

"In the context of medical education, the development of self-regulated learning can be associated with better academic and clinical performance."

"This report focuses on demonstrating the association between metacognitive awareness and motivation to learn among medical students in the clinical sciences portion of their education (3rd and 4th years of the medical programme) and characterizing medical students' motivational factors."

"We have selected validated self-report questionnaires for the evaluation of metacognition (the Schraw and Dennison Metacognitive Awareness Inventory - MAI) and motivation to learn (the Baranik, Barron and Finney Achievement Goals for a Work Domain - AGWD)."

"We also performed a qualitative analysis based on an open-ended question: "What motivates me the most in medical training?""

"Only the knowledge about cognition domain of MAI was significantly associated with motivation to learn."

"We found that higher scores on the knowledge about cognition domain of MAI was associated with the mastery approach goal orientation ($p = 0.003$, median 0.71, IQR 0.23) and that lower scores on this same domain was associated with a mastery avoidance goal orientation ($p = 0.034$, median 0.65, IQR 0.14)."

"Metacognitive awareness and motivation to learn are closely related."

"This association may represent a potential target for the educational process, as deans and faculty can adopt strategies focused on promoting self-regulated learning concerning students' motivational factors."

Background

"SRL suggests that integration among metacognition, strategic learning and motivation is useful for explaining whether some students engage purposefully in learning processes and goal-directed activities [74]."

"SRL theory incorporates cognitive, affective and social factors, representing an inclusive perspective of students' learning and their motivations [75]."

"In the context of medical education, SRL theory can be represented by students who are active participants in their learning and are guided through key processes as strategies to attain goals, adapt behaviours and optimize learning and performance [72, 73]."

"Students who develop metacognitive strategies can plan, monitor and regulate their cognition process."

"Based on social-cognitive learning theory, individuals are motivated to achieve personal goals through the self-regulation of thoughts, actions and environmental factors [72, 76]."

"This study focuses on demonstrating a possible association between metacognitive awareness and motivation to learn in medical students in the clinical sciences portion of their education."

Methods

"We performed a cross-sectional study involving medical students enrolled in the clinical sciences portion of their medical programme at the University of Sao Paulo (USP), Brazil."

"The study combined a quantitative analysis based on validated self-report questionnaires for the evaluation of metacognition (the Schraw and Dennison Metacognitive Awareness Inventory (MAI) [77]) and motivation to learn (the Baranik, Barron and Finney Achievement Goals for a Work Domain (AGWD) [78–82]) and qualitative analysis based on an open-ended question: "What motivates me the most in medical training?""

"The translated and validated Brazilian Portuguese version comprises 18 items, each specifically associated with one of 4 achievement goals: mastery approach, mastery avoidance, performance approach, and performance avoidance [83]."

"We included an open question to assess the students' motivations: "What motivates me the most in medical training?""

"Descriptive statistics were used to analyse demographic data and students' metacognitive and motivational scores."

Results

"We identify this trait in the following examples: "What motivates me the most in medical training is the conviction that in the future, I will participate in patients' histories and make efforts to make their lives better through the knowledge I acquire.""

"I am motivated by the challenge of learning and practising medicine, as long as it involves a high impact on people's lives.""

"Another sample of responses converged to practical curricular activities, participation in student organizations and social relationships as components of medical students' experience that motivated them during medical training.""

"I feel motivated when faculty promote activities in which I can talk to patients."

"My friends are who motivate me the most in academic environments."

"I feel motivated to be part of a sports team in the University Athletic Association."

Discussion

"Our study sought to investigate the relationship between metacognitive awareness and motivation to learn within the context of Brazilian undergraduate medical students."

"We found that mastery approach-oriented students, whose priority is the development of competencies and skills, deeper learning and personal satisfaction, presented higher scores on one of the MAI domains (knowledge about cognition, $p = 0.003$, median 0.71, IQR 0.23)."

"According to previous studies, medical students' motivation to learn is influenced by interests in science/medicine, social interests, altruism, flexible work hours, prestige, and financial security [78]."

"Our results can be used to make deans and faculty aware of the necessity of adopting strategies focused on promoting self-regulated learning concerning students' motivational factors."

"This study was designed to better understand a complex and multivariable picture: medical students' potentialities and weaknesses in learning, thus encouraging an important discussion about the metacognitive, behavioural, motivational, and affective aspects of learning."

Conclusion

"The majority of students endorsed a mastery avoidance goal orientation, which is significantly associated with lower metacognitive awareness."

"Students reported being motivated by aspects related to personal satisfaction, altruism and the medical curriculum."

"Personal satisfaction was the aspect with the highest number of responses, while financial and social status had less importance."

"Understanding the metacognitive awareness and motivations of our students will help us to support them during medical training."

Multimodal Data Indicators for Capturing Cognitive, Motivational, and Emotional Learning Processes: A Systematic Literature Review [259]

This is a machine-generated summary of:

Noroozi, Omid; Pijeira-Díaz, Héctor J.; Sobocinski, Marta; Dindar, Muhterem; Järvelä, Sanna; Kirschner, Paul A.: Multimodal data indicators for capturing cognitive, motivational, and emotional learning processes: A systematic literature review [259]

Published in: Education and Information Technologies (2020)

Link to original: https://doi.org/10.1007/s10639-020-10229-w

If you want to cite the papers, please refer to the original.

For technical reasons we could not place the page where the original quote is coming from.

Abstract-Summary

"This systematic review on data modalities synthesises the research findings in terms of how to optimally use and combine such modalities when investigating cognitive, motivational, and emotional learning processes."

"We provide findings in terms of target journal, country, subject, participant characteristics, educational level, foci, type of data modality, research method, type of learning, learning setting, and modalities used to study the different foci."

"From the 207 publications, 98 focused exclusively on the cognitive aspects of learning, followed by 27 publications that only focused on motivation, while only five publications exclusively focused on emotional aspects."

"Only 10 publications focused on a combination of cognitive, motivational, and emotional aspects of learning."

"Our results plea for the increased use of objective measures, highlight the need for triangulation of objective and subjective data, and demand for more research on combining various aspects of learning."

"Further, rather than researching cognitive, motivational, and emotional aspects of learning separately, we encourage scholars to tap into multiple learning processes with multimodal data to derive a more comprehensive view on the phenomenon of learning."

Introduction

"Multimodal research, which is studying individuals' learning endeavours by gathering and exploring different multimodal data, might help to explain the complex interplay of cognitive, motivational, and emotional processes during the learning process."

"Digital technologies and advanced educational tools can provide researchers with the possibility to combine subjective and objective data, to trace various cognitive and affective learning processes of the learners, to make micro-level environmental interactions and their responses of the body and brain visible (see Reimann and others [84]), and to analyse different multimodal data (see Noroozi and others [85])."

"To our knowledge, no systematic review has been conducted to provide an overview of the affordances of multimodal data in terms of understanding the cognitive, motivational, and emotional processes during learning."

"We summarise, analyse, and interpret a comprehensive set of data modalities that are used in combination for measuring various aspects of cognitive, motivational, and emotional learning processes."

"What and how are data modalities used to capture cognitive, motivational, and emotional learning processes?"

Method

"Publications that were excluded from the further analysis did not: 1) include evidence related to the learning sciences and report at least one of the aspects of learning processes and/or outcomes (i.e., learning cognition, motivation, emotion); 2) use at least two modalities of data (i.e., studies focused only on one data modality); 3) belong to the formal educational levels such as primary, secondary, high school, college, or higher education (studies conducted in summer-schools, second language courses, distance learning, online courses, and other extra-curricular activities were also included); 4) report empirical findings on the topic such as conceptual, methodological, and theoretical publications."

Results

"The majority of multimodal research studies (116 publications) used mixed methods to analyse various aspects of cognitive, motivational, and emotional learning processes; only eight studies exclusively used qualitative methods (e.g., interviews and observations) and 83 exclusively used quantitative methods (e.g., students' products, surveys and performance tests)."

"Out of 721 occurrences of modalities in the reviewed publications, 437 occurrences focused on measuring the cognitive aspect of learning, followed by 203 related to motivation."

"The focus of each publication in terms of data modalities was paired with the type of learning setting (i.e., regular, online, and extra-curricular) of the multimodal publications."

"Modalities such as interviews and observation were the most typical type of data collection for the extra-curricular learning setting in the reviewed publications."

"The focus of each publication in terms of data modalities was paired with the type of learning (i.e., individual, group, and mixed) of the publications."

Discussion

"Such a multimodal journal might act as a venue to develop standard procedures and tools for processing and combining different data modalities in learning research."

"Multimodal data collection from multiple cultures might be particularly necessary when investigating the motivational and emotional aspects of learning."

"The findings showed that utilising multimodal data in collaborative learning settings constituted only a small portion of the publications, whereas the majority of the studies focused on individual learning."

"Our findings showed that few multimodal studies (four papers) in online learning environments used log data."

"Multimodal approaches in learning research can help to tackle the constraints of typical single-channel data (e.g., subjective, objective, or physiological data), and help to draw more valid and reliable inferences about the learning processes (Harley and others [86]; Pantic and Rothkrantz [87])."

"Future research should make use of objective data modalities to extend current knowledge on the emotional aspects of learning."

Conclusion, Limitations and Future Work

"Our findings led to the conclusion that multimodal data is a vast area of research across various learning domains."

"This would help to understand the affordances and limitations of multimodal data in the whole spectrum of learning domains."

"Studies, collecting multimodal data from collaborative learning settings might help to capture the complex social, motivational, and emotional processes that arise during collaborative learning."

"Although multimodal data were mostly gathered from on-campus settings, a significant portion of studies collected multimodal data from online or blended learning settings."

"This highlights the future potential of multimodal data in learning."

"Further, rather than researching cognitive, motivational, and emotional aspects of learning separately, we encourage scholars to tap into multiple learning processes with multimodal data to derive a more comprehensive view on the phenomenon of learning."

"This study has considered those papers studying learning processes with multimodal data, which uses the terms "multimodal" or "multichannel"."

Motivational Sensitivity of Outcome-Response Priming: Experimental Research and Theoretical Models [260]

This is a machine-generated summary of:

Watson, Poppy; Wiers, Reinout W.; Hommel, Bernhard; de Wit, Sanne: Motivational sensitivity of outcome-response priming: Experimental research and theoretical models [260]

Published in: Psychonomic Bulletin & Review (2018)

Link to original: https://doi.org/10.3758/s13423-018-1449-2

If you want to cite the papers, please refer to the original.

For technical reasons we could not place the page where the original quote is coming from.

Abstract-Summary

"Outcome-response (O-R) priming is at the core of various associative theories of human intentional action."

"While there is a wealth of evidence that both the perceptual and motivational properties of an outcome can be encoded in the O-R association and mediate O-R priming, we critically examine the integration of these mechanisms and the conditions under which motivational factors constrain the sensory O-R priming effect."

"We discuss the clinical relevance of this O-R priming mechanism, whether it can satisfactorily account for human goal-directed behaviour, and the implications for theories of human action control."

Theories of Action Control

"Work (the findings of which are discussed in more detail below; see section Modulation of O-R Priming by Changes in Outcome Value) some theories of action control, such as recent formulations of the associative-cybernetic model (S. de Wit & Dickinson [88]), include an O→R mechanism as one path to action but supplement this with a forward R→O pathway to fully capture goal-directed action control."

"Many human studies have been conducted with the aim of shedding light on the role of sensory and motivational outcomes in O-R priming."

"The bidirectional hypothesis assumes that bidirectional R-O associations are formed during training as a consequence of the causal relationship between the instrumental response and the outcome, allowing for later 'backwards' response priming in the O-R direction (Elsner & Hommel [89]; Pavlov [90]; Rescorla [91])."

Outcome Anticipation and O-R Priming

"We also review evidence for the O-R priming mechanism from various strands of research utilizing instrumental discrimination paradigms and response-priming tasks in which outcomes are presented either directly to participants or are signalled indirectly (via Pavlovian cues)."

"This interference comes about because the response signalled by the discriminative stimulus (S-R) conflicts with the response triggered by the outcome anticipation (O-R priming)."

"It is argued that in the latter condition, anticipation of the instrumental outcome activates both associated responses via O-R associations, regardless of which response is signalled to be correct by the discriminative stimulus."

"We should note that a related group of studies have used a simpler version of the PIT paradigm, in which only a single response was trained (e.g. S1-O followed by R1-O) to show the motivating (and inhibitory) effects of Pavlovian cues on ongoing appetitive (and avoidance) responses towards either monetary or chocolate rewards (in humans; Colagiuri & Lovibond [92]; Garbusow and others [93]; Garofalo & di Pellegrino [94]; Guitart-Masip and others [95]; Lovibond & Colagiuri [96]; Talmi, Seymour, Dayan, & Dolan [97])."

Motivational Modulation of O-R Priming

"Using food outcomes, Watson and others [98] examined both direct O-R priming (with pictures of food outcomes that had been associated during the training phase with particular responses) and indirect S-O-R priming (using Pavlovian stimuli that had previously been associated with those food pictures, but never with a response)."

"The authors found that incongruent responses were carried out slower on trials that were primed by the high-reward colour, suggesting that the presentation of the outcome in Phase 2 triggered the previously learned response (via an O-R mechanism) and that this priming effect was more difficult to overcome in the high-value condition."

"A unique feature of the O-R priming studies that did show insensitivity to outcome devaluation (Hogarth [99]; Hogarth & Chase [100]; Watson and others [101]) is that participants were instructed during the instrumental (and test) phases that, whilst they would not be told which reward was available, only one reward was available on each trial."

Further Points of Discussion

"The insensitivity to outcome devaluation displayed by both (S-)O-R priming and S-R habits that are triggered by specific contexts is problematic for many current approaches to treatment that rely on explicitly devaluing outcome value (e.g. by health warnings), as the data reviewed above suggests that this approach will have little effect on reducing cue-elicited responding for signalled rewards (Boutelle & Bouton, [102]; Verhoeven and others, [103])."

"Through this discriminative extinction training, participants may have learned explicitly that the CS functioned as a hierarchical cue signalling that the instrumental response for the alternative outcome (rather than the signalled outcome) would be reinforced, thereby allowing an explicit strategy to override the associative O-R priming effect."

"These studies provide convincing evidence that the indirect O-R priming effect (in which cue-elicited anticipation of outcomes triggers associated responses) is constrained by additional factors such as action valence."

Implications for Theories of Action Control

"Seabrooke and others [104] argue that the standard PIT paradigm is highly sensitive to O-R priming effects for the devalued outcome (as measured in reference to a baseline condition, where participants tend to respond rarely for the devalued outcome)."

"A final possibility is that O-R priming may simply be an inflexible mechanism that is based purely on learned associations between responses and sensory/affective properties of outcomes that is not at some stage integrated with motivational processes that allow for adjustments on the basis of changes in outcome value."

"The current paradigms may isolate the sensory O-R priming mechanism and thereby prevent the integration with mechanisms that allow for modulation of behaviour on the basis of outcome value to become visible."

"To further assess the validity of models that include the integration of the specific O-R priming effect with a more general motivational mechanism, future studies should adopt more ecologically valid paradigms with multiple cues, responses, and outcomes."

[Section 6]

"Many different associative theories are based upon the central idea that in the course of exploration and learning, associative links between responses (R) and outcome (O) representations are formed (Asratyan [105]; Gormezano & Tait [106]; Hommel, Müsseler, Aschersleben, & Prinz [107]; James [108])."

"Activation of the outcome representation (thinking about the light coming on) leads to activation of the associated motor patterns required to achieve it (pushing the light switch)."

"Evidence for such O-R associative links comes from multiple converging strands of research showing that presentation (or anticipation) of outcomes activates associated motor responses and that preparing motor responses activates anticipation of outcomes."

"In the current manuscript, we review O-R priming effects, focusing on the integration of sensory and motivational aspects of action control."

Ambient Lighting Alters Motivational Responses to Advertisements for Foods of Different Energetic Value [261]

This is a machine-generated summary of:

Bailey, Rachel L.; Wang, Grace Tianjiao; Liu, Jiawei: Ambient lighting alters motivational responses to advertisements for foods of different energetic value [261]

Published in: Motivation and Emotion (2021)

Link to original: https://doi.org/10.1007/s11031-021-09891-5

If you want to cite the papers, please refer to the original.

For technical reasons we could not place the page where the original quote is coming from.

Abstract-Summary
"Ambient lighting has been identified as a motivationally relevant context that affects cognition, emotion, and behavior, including food consumption and choice."

"This study adds to this body of work utilizing optimal foraging theory, which predicts individuals will prefer foods that deliver the most energy while costing the least energy to obtain."

"An experiment was performed in which young adults viewed food advertisements that varied in energy density level in either dim or bright contexts."

"Individuals exhibited greater motivational intensity and preference when exposed to more energy dense foods in light contexts, but preference dissipated in dark contexts."

"Less energy dense foods elicited more motivational preference in dark contexts compared to light."

Introduction
"Further, because motivation can be seen as dispositional (e.g., organisms seek food to satisfy biological imperatives) as well as contextualized (e.g., organisms seek food under particular circumstances of higher opportunity and lower risk), understanding how contexts such as ambient lighting can influence more generalized, dispositional behavior is of crucial importance in understanding and theorizing about motivation (Baumeister [114])."

"Because this study is interested in how contexts create different motivational responses generally, and how ambient lighting contexts create different food preferences specifically, we focus on a theoretical framework that seeks to explain and predict the primary motivation of food-seeking behavior: obtaining energy."

"We expected individuals to exhibit greater motivational preference (indexed via corrugator supercilii and orbicularis oculi activation) when exposed to stimuli that contained more energy dense foods in a light context and these differences to be less pronounced in the dark."

Methods

"A 2 (Food energy density: more energy dense, less energy dense) × 12 (Repetition) × 34 (Time of exposure in half seconds) × 2 (Environmental context: light, dark) mixed factorial design experiment was used."

"Attitude toward the ad was obtained using a three 7-point item scale ($\alpha = 0.95$) originally developed by Kamp and MacInnis [115]."

"One participant elected not to participate in the psychophysiological data collection portion of the experiment."

"The participants viewed and rated each ad for ad and brand attitudes and purchase intentions."

"Data were submitted to a 2 (Food energy density: more energy dense, less energy dense) × 12 (Repetition) × 2 (Environmental circumstances: light, dark) repeated-measures ANCOVA."

"SCL and OO change data were computed using the half-second of ad onset as the reference point, as they were used to assess in change from onset of appetitive stimuli in the aversive/neutral contexts."

Results

"We predicted that in a light context, individuals would exhibit greater motivational preference across exposure to advertisements for foods that were more energy dense compared to foods that were less energy dense."

"This prediction also called for an interaction of environmental circumstances and energy density of foods advertised with time of exposure on corrugator supercilii (CS) activation, which was not statistically significant, $F(33, 1188) = 1.415$, $p = 0.061$, $\eta_p^2 = 0.038$."

"In dark contexts, individuals exhibited more CS activation overall but this was moderated by energy density of the food ads."

"We also predicted individuals would exhibit greater motivational intensity [indexed via skin conductivity level (SCL)] when exposed to stimuli that contained more energy dense foods in general, but especially when in a dark context."

Discussion

"As predicted, individuals exhibited less difference in motivational preference (OO and CS) for more and less energy dense foods in a dark context."

"We predicted that motivational intensity and sympathetic arousal would be greater for more energy dense foods viewed in a dark context (where foraging would be more costly, thus requiring more resources to make it worthwhile), we found that more energy density foods elicited greater sustained motivational intensity across viewing and darkness elicited less intensity overall."

"Dark contexts inhibited sympathetic arousal here, if the foraging were not perceived as costly even in a dark environment (i.e., the energy dense food was already prepared and at hand), individuals would be more likely to select more energy dense foods over other choices that were either more costly to obtain or less energy dense."

Conclusion

"This paper is in line with previous work suggesting that ambient lighting is a motivationally-relevant context for humans that influences eating behavior."

"Dark contexts inhibited sympathetic arousal and motivational preference for foods of greater energy density, which suggests that in circumstances of costlier foraging, individuals may be less bound to energy dense choices."

"If the foraging were not perceived as costly even in a known aversive context like darkness (i.e., energy dense food was already prepared and at hand), individuals would be more likely to eat more energy dense foods over other choices that were either more costly to obtain or less energy dense."

"Motivational conflict introduced by appetitive stimuli in an aversive context inhibits approach responses in general and levels the playing field on preferences for higher and lower energy density foods."

Working Memory Training in Relation to Anxiety, Stress, and Motivation [262]

This is a machine-generated summary of:

Sari, Berna A.; Tarman, Güliz Zeynep; Ozdogan, Busra; Metin, Baris; Derakshan, Nazanin: Working Memory Training in Relation to Anxiety, Stress, and Motivation [262]

Published in: Journal of Cognitive Enhancement (2020)

Link to original: https://doi.org/10.1007/s41465-020-00176-2

If you want to cite the papers, please refer to the original.

For technical reasons we could not place the page where the original quote is coming from.

Abstract-Summary

"We examine how WM training can impact trait anxiety and test anxiety for upcoming examinations, and the role of motivation and the sustainability of training effects were also examined."

"Participants completed measures of trait anxiety, test anxiety, and attentional control before and after the intervention and at follow-up times (1 week and 7 weeks)."

"Results found that adaptive training compared with the non-adaptive control training resulted in reductions in trait anxiety across time."

"Exploratory analyses showed that perceived success in the training task was associated with pre-to-post-intervention reductions in test anxiety."

"Our findings suggest that WM training has the potential to reduce anxiety and test anxiety-related symptoms under stress and emphasize the role of perceived success."

Introduction

"WM training using the adaptive dual n-back task was shown to result in sustainable reductions in anxiety-related symptomatology in survivors of breast cancer (Swainston and Derakshan [116]) and anxiety and depressive vulnerability in adolescents (Beloe and Derakshan [117])."

"Despite the accumulating evidence behind the efficacy of WM training in influencing anxiety-related symptomatology, it is unclear how motivational factors, situational stress, and the perceived success of training can influence transfer-related outcomes from training."

"One of the aims of the current investigation is to examine the influence of motivation on training-related gains."

"Self-report measures of anxiety and attentional control were collected prior to the intervention, at 1-week follow-up as well as at 7-week follow-up times to examine the sustainability and trajectory of training-related outcomes over time (see Green and others [118])."

"We could examine if motivation was associated with training-related improvements."

"We predicted that participants in the adaptive training group compared with the active control group will have greater reductions in anxiety-related symptoms post-intervention including at follow-up periods."

Method

"In the training task, level of task difficulty (n) changed depending on participant performance."

"If accuracy was less than 75%, task difficulty decreased by one level of n. Participants were informed about the difficulty of the level in the beginning of each block."

"In the control task, regardless of the accuracy rates, participants performed 1-back level of this task for 20 blocks in every session."

"During the intervention, after each daily session of training, participants were asked to respond to questions on a scale ranging from 0 (lowest possible rating) to 10 (highest possible rating) with regard to their anxiety levels, perceived success, and motivation."

"At the end of each session, participants were asked to rate how successful they think they were during the task, how motivated they were to perform the task, and how anxious they felt."

Result

"Groups did not differ in daily self-report scores of anxiety, motivation, and perceived success in training (indexed by the slope of change) either (all ts < 1.08, NS)."

"A 2 (group: training, control) × 4 (Time: pre-intervention, post-intervention, first follow-up, second follow-up) linear mixed effects model (LMM) analysis with the maximum likelihood method was utilized to assess changes on self-reported anxiety (STAI), attentional control (ACS), and test anxiety (WTAS) scores."

"This showed that participants in the adaptive training group showed a significant linear trend for reductions in STAI scores across time from pre-intervention to 2nd follow-up, indicating sustained reductions across time, compared with the control group."

"The relationship between the slope of motivation and perceived success scores during the training intervention period and pre- to post-changes in trait questionnaires scores were assessed via correlational analyses for the training and control groups separately."

Discussion

"Training-related improvements on reductions in trait anxiety were observed such that the adaptive training group compared with the active control group showed sustainable reductions across time in trait anxiety from pre-intervention to 2nd follow-up."

"Although the results indicated no direct benefits of WM training on test anxiety, reductions in test anxiety correlated with increased perceived success during the training intervention for the participants in the training group."

"Perceived success is also associated with self-efficacy (Lane and others [119]), the results of the current study encourage future research assessing the role of self-efficacy in cognitive training, as well as the effects of cognitive training on self-efficacy.. In the current study, we also attempted to investigate whether training and control groups differed in motivation and whether motivation was associated with training-related improvements."

The 20-min Trial of the Maintenance of Wakefulness Test is Profoundly Affected by Motivation [263]

This is a machine-generated summary of:

Shreter, Roni; Peled, Ron; Pillar, Giora: The 20-min trial of the Maintenance of Wakefulness Test is profoundly affected by motivation [263]

Published in: Sleep and Breathing (2006)

Link to original: https://doi.org/10.1007/s11325-006-0072-z

If you want to cite the papers, please refer to the original.

For technical reasons we could not place the page where the original quote is coming from.

Abstract-Summary

"The current gold standard is considered the Maintenance of Wakefulness Test (MWT), although there are two different potential protocols to perform it and the normal reference range has been determined on selected populations."

"The effect of potential "penalty" on failing the test (i.e., presenting sleepiness) has not been seriously studied."

"We therefore planned this study to examine the effect of motivation on the MWT results in a potentially sleepy population."

"We hypothesized that with the knowledge that participants may lose their driving license if they fail the test, the results would indicate significantly less sleepiness than reported for other populations."

"Fifty-four consecutive subjects at high risk of sleepiness referred to the sleep laboratory for evaluation of their ability to maintain wakefulness were studied."

"Only 5 of the participants fell asleep in any of the MWT trials (one to five trials), which could not be predicted by severity of OSA, age, BMI, or variables derived from the previous night PSG."

"Of the 21 patients with severe untreated OSA, only 1 patient fell asleep during the MWT20."

"We conclude that motivation profoundly affects the MWT results, raising the question what should be considered normal or abnormal when sleepy results may cause penalty or other practical implications."

"We believe that these results suggest that for renewal of driving license purposes, the MWT20 is insufficient."

Introduction

"The test measures the daytime sleep latency (time it takes the patient to fall asleep) for each session."

"Most studies on determining the normative values for MWT have used randomly selected healthy subjects, and the threshold to define abnormal findings was set at 1 standard deviation (SD) or 2 SD from mean daytime sleep latency [120, 121]."

"For the MWT20 test, Doghramji and others [122] found the average sleep latency to be 18.7 ± 2.6 min and the cutoff for abnormal results was determined as 13.5 min."

"A later study [123] found average sleep latency in healthy subjects of $19.4 + 2.1$ min and determined a cutoff at 15.2 min."

"In cases where subjects are at risk of loosing their driver's licenses and/or their jobs if they fail the MWT, they are highly motivated to succeed (maintain wakefulness) and their mean daytime sleep latency may be significantly different from the other study populations."

Methods

"The results of the MWT and the clinical recommendations of the sleep laboratory's physicians largely affect the eligibility of said drivers to retain or renew their driving license."

"The study included all 54 subjects who were referred to the Technion sleep laboratory between the years 2002 and 2004."

"The subjects included professional drivers (bus, taxi, and truck) who were known to be suffering from breathing disorders during sleep or having documented chronic daytime sleepiness (six of our subjects had been previously diagnosed as suffering from EDS base on MSLT results)."

"All subjects were aware that the results of the MWT would affect their eligibility to retain their driver's license."

Test Procedure
"All subjects were given five 20-min trials of Maintenance of Wakefulness Test (MWT) preformed at 0800, 1000, 1200, 1400, and 1600 hours."

"Subjects were instructed to lie in bed and maintain wakefulness during the test."

"Subjects were instructed not to perform any unusual actions to maintain wakefulness (including standing, repetitive movements in bed, talking, etc)."

"Recordings during this test were similar to those during the night and included two-channel EEG (C3-A2, C4-O1), two-channel EOG, submental EMG, quantitative snoring intensity, and closed circuit video monitoring by a trained technician."

"During intervals between each MWT, subjects were instructed not to sleep and were closely monitored by closed circuit monitors."

"Our protocol did deviate slightly from the Practice Parameters recommendations for the MWT [124], in that our subjects were instructed to lie rather than sit in bed during the MWT sessions."

Results
"We should note that 33 (61%) subjects whom we diagnosed as suffering from OSA were untreated at the time of the MWT, of those 23 patients had a previous diagnosis of OSA, but only 13 were treated (most by Continuous Positive Airway Pressure, CPAP)."

"Of the 54 subjects who underwent the MWT, only 5 (9.3%) fell asleep at any stage of the procedure and only 3 (5.6%) had a sleep onset latency of less than 13.5 min."

"Ten subjects had less than 4.5 h of real nocturnal sleep time, but only 1 of them fell asleep during the MWT."

"One important finding was that all treated patients with OSA (even if severe) could maintain normal alertness during the test and did not fall asleep in any of the MWT episodes."

Discussion
"Because all subjects knew the exact reason for their being tested and the results expected of them, we could not rely on their subjective sleepiness report (Epworth score) which was below 6 in all patients, including those with previous sleep-related accidents and pathological MSLT (even though theoretically as in the MWT, the low Epworth score's could be a reflection of real alertness)."

"Although most of our subjects would be expected to be sleepy (based on past medical and driving history and on previous nights PSG/MSLT results), most of them were able to maintain wakefulness for 20 min in each trial."

"In the Doghramji study [122], the average sleep latency for the MWT20 test was 18.1 ± 3.6 min, and they proposed a cutoff for abnormality at 13.5 min."

"It is yet to be determined whether with the MWT40 protocol, more patients will indeed fall asleep on these occasions, or perhaps even 60-min trial length may be required."

Event-Related Potentials Indicate Motivational Relevance of Cocaine Cues in Abstinent Cocaine Addicts [264]

This is a machine-generated summary of:

van de Laar, Maria C.; Licht, Robert; Franken, Ingmar H. A.; Hendriks, Vincent M.: Event-related potentials indicate motivational relevance of cocaine cues in abstinent cocaine addicts [264]

Published in: Psychopharmacology (2004)

Link to original: https://doi.org/10.1007/s00213-004-1928-1

If you want to cite the papers, please refer to the original.

For technical reasons we could not place the page where the original quote is coming from.

Abstract-Summary
"The present study aimed to investigate the motivational relevance of cocaine cues and whether or not an appetitive emotional system is involved employing event-related potential (ERP) measurements."

"Cocaine-addicted subjects and healthy controls were exposed to neutral and cocaine-related pictures whilst ERPs were recorded simultaneously over frontal, parietal and midline sites."

"Patients exhibited ERP amplitude discrepancies between neutral and cocaine-related pictures for N300, late slow positive wave (LSPW) and sustained slow positive wave (SSPW), whilst this effect was absent in control subjects."

"Differences in neutral and cocaine cue-evoked ERP waves were also found at left frontal sites for LSPW and SSPW in the patient group only."

"The present study provides evidence that the sensitivity of ERP correlates for cocaine cues may be an indicator of motivational and emotional processes in drug-dependent individuals."

Introduction

"The contribution of left frontal sites could be related to appetitive sensation of drug cues since left frontal brain areas are involved when subjects are experiencing increased approach tendencies and right frontal brain areas when experiencing increased avoidance tendencies (Watson and Tellegen [127]; Davidson and others [128]; Fox [129]; Fox and others [130]; Davidson [131])."

"Another technique, which is useful to study the relationship between brain activity and the motivational relevance of drug cues in addicts, is a specific application of electroencephalogram (EEG), known as event-related potentials (ERPs)."

"Various imaging studies have been conducted with the aim to find indications of the involvement of prefrontal brain areas in craving experiences and its relationship to motivational and emotional processes."

"ERP responses appear to offer a useful technique to assess motivational relevance and the emotional value of drug-related stimuli and may provide further understanding of craving and motivational drive to seek drugs in substance dependency."

Method

"Twenty-six male cocaine-dependent patients (mean age 35.4 years, SD=7.6 years) and 20 healthy male control subjects (mean age 40.6 years, SD=9.9 years) voluntarily participated in the present study."

"For the patient group, self-reported craving questionnaires [Desire for Drug Questionnaire (DDQ) and Obsessive Compulsive Drug Use Scale (OCDUS)] were completed prior to the preparation of physiological recordings."

"The two-item Self-Assessment Manikin (Bradley and Lang [132]) was used to obtain subjective reports concerning valence and arousal properties of the cocaine-related pictures presented during ERP recordings in controls and patients."

"In the ANOVAs on frontal ERP components, group (control versus patients) served as the between-subjects factor, and cue type (neutral and cocaine) and laterality (left, F3 and F7, versus right, F4 and F8) as within-subjects factors."

"It was decided to only calculate correlations between cocaine-cue-evoked ERP amplitudes found in significant group interaction effects and self-reported craving levels."

Results

"For N300, the overall analysis showed a significant main effect for cue type at F3 and F4 ($F_{1,44}$=23.20, P<0.0001), indicating that neutral pictures elicited larger negative amplitudes than cocaine pictures."

"Further analysis revealed a significant effect of cue type ($F_{1,25}$=14.62, P=0.001) in the patient group only, showing that neutral pictures evoked larger negative amplitudes than cocaine pictures."

"For SSPW, a main cue-type effect was found at F3 and F4 ($F_{1,44}$=5.27, P=0.027), indicating larger positive amplitudes for cocaine pictures than neutral pictures."

"For P600 and SPW, significant cue-type effects were found (all $F_{1,44}$ values>29.99, all P values<0.0001), indicating larger positive amplitudes for cocaine pictures than for neutral pictures."

"For SSPW, a significant cue-type effect was found ($F_{1,44}$=9.17, P=0.004), indicating larger positive amplitudes for cocaine pictures than for neutral pictures."

Discussion

"This observed difference in neutral and cocaine cue-evoked ERP waves at left frontal sites could indicate the possible involvement of an emotional system, since frontal brain areas are differentially involved in processing emotional laden stimuli (Silberman and Weingartner [133]; Davidson [126]) and appear to play a fundamental role in approach and avoidance motivational tendencies (Watson and Tellegen [127]; Davidson and others [128]; Davidson [131])."

"The present study hypothesised that cocaine-addicted subjects would process cocaine cues differently than neutral cues and that this effect would be manifested in changes in late positive waves at midline sites."

"The absence of group-specific cue-evoked ERP responses at parietal and midline sites could indicate the unique function of the lateral frontal lobes in processing emotional and motivational relevant cues (drug-related pictures) in substance-dependent individuals."

"There were contradictory findings of similar valence and arousal reports on cocaine-related pictures by cocaine-addicted subjects and healthy controls and the observed patient group-specific cocaine cue-evoked ERP amplitudes."

Incorporating Motivational Intensity and Direction Into the Study of Emotions: Implications for Brain Mechanisms of Emotion and Cognition-Emotion Interactions [265]

This is a machine-generated summary of:

Harmon-Jones, Eddie; Gable, Philip A.: Incorporating motivational intensity and direction into the study of emotions: implications for brain mechanisms of emotion and cognition-emotion interactions [265]

Published in: Netherlands journal of psychology (2008)

Link to original: https://doi.org/10.1007/bf03076416

If you want to cite the papers, please refer to the original.

For technical reasons we could not place the page where the original quote is coming from.

Abstract-Summary

"Contemporary dimensional models of emotion regard the positive to negative valence dimension as an important organising principle."

"This affective valence organising principle has also been used in research concerned with how emotions affect cognition, and much research has suggested that positive affects have different effects on cognition than negative affects."

"We review recent research that questions the utility of the affective valence dimension in understanding the functions of asymmetrical frontal cortical activity and in understanding the effects of emotions on cognition."

Affective Valence, Motivational Direction, and Asymmetrical Frontal Cortical Activity

"Past research had essentially confounded emotional valence (positive, negative affect) with motivational direction (approach, withdrawal motivation), and researchers were claiming that relatively greater left than right frontal cortical activity reflected greater approach motivation and positive affect, whereas relatively greater right than left frontal cortical activity reflected greater withdrawal motivation and negative affect."

"Given previous research showing that approach-motivated positive affect is associated with increased left prefrontal cortical activation (Gable & Harmon-Jones, 2008b), we investigated whether greater left frontal activation associated with high approach-motivated positive affect would relate to attentional narrowing (Harmon-Jones & Gable, in press)."

"We examined whether neural activations associated with approach motivation would relate to the effect of approach-motivated positive affect on narrowed attention."

Conclusion

"Approach-related emotions such as anger or desire involve several brain regions, but the reviewed research establishes the importance of the left prefrontal cortex in approach motivation independent of affective valence."

"Damage to these regions of frontal cortex cause depression vs. mania (Robinson & Downhill, 1995), and rTMS manipulations of left vs. right cortical regions affect mood and attentional processing in manners consistent with the idea that asymmetrical frontal cortical activity is involved in motivational direction (d'Alfonso and others, 2000; van Honk & Schutter, 2006)."

"It is possible that sub-cortical structures are lateralised for approach and withdrawal motivational processes in amphibians, reptiles, and birds but that these lateralisations are preserved and elaborated into the frontal cortices of primates."

"Greater left than right frontal cortical activity is associated with approach motivation and not positive affect per se."

[**Section 6.3**]

"Contemporary dimensional models of emotion regard the positive to negative valence dimension as an important organising principle (Lang, 1995; Watson, 2000)."

"This affective valence organising principle has also been used in research concerned with how emotions affect cognition, and much research has suggested that positive affects have different effects on cognition than negative affects."

"We will review research that questions the utility of the affective valence dimension in understanding the functions of asymmetrical frontal cortical activity and in understanding the effects of emotions on cognition."

Impact of Motivation on Selected Aspects of Attention in Children with ADHD [266]

This is a machine-generated summary of:

Skalski, Sebastian; Pochwatko, Grzegorz; Balas, Robert: Impact of Motivation on Selected Aspects of Attention in Children with ADHD [266]

Published in: Child Psychiatry & Human Development (2020)

Link to original: https://doi.org/10.1007/s10578-020-01042-0

If you want to cite the papers, please refer to the original.

For technical reasons we could not place the page where the original quote is coming from.

Abstract-Summary

"The purpose of this study was to assess the impact of extrinsic motivation on selected aspects of attention in children with ADHD, as well as to measure cortical activity and dimensions of motivation as per the self-determination theory."

"Children with ADHD were more likely to undertake activity under the pressure of external stimuli and exhibited attention deficits regarding vigilance, visual search and divided attention."

"Differences between groups regarding attention decreased in conditions of increased motivation, indicating that motivation can reduce cognitive deficits in children with ADHD."

Introduction

"Based on the described behavior samples, it seems that children with ADHD may exhibit a higher level of extrinsic motivation, introjected motivation and amotivation, as well as a lower level of intrinsic motivation and identified regulation; however, there are no studies describing the activity of children with ADHD as per the self-determination theory."

"Based on the aforementioned articles, we assumed that: (H1) children with ADHD may exhibit lower levels of cortical activity as measured using (a) TBR in the midline and the (b) HEG ratio in prefrontal areas than TD children; (H2) children with ADHD may display lower levels of (a) intrinsic motivation and (b) identified regulation as well as higher levels of (c) external regulation, (d) introjected regulation and (e) amotivation to learning compared to TD children; (H3) children with ADHD may exhibit deficits regarding: (a) vigilance, (b) visual search and (c) divided attention; (H4) increasing the motivation may reduce attention deficits in children with ADHD with regard to: (a) vigilance, (b) visual search and (c) divided attention."

Methods

"The measurement of cortical activity, fluid intelligence and motivation to learning was conducted at the beginning of the study."

"The assessment of the impact of motivation on selected aspects of attention was carried out in a quasi-experiment model with repeated measurement."

"The study used psychological tools to determine the level of intelligence, motivation and attention: Raven's colored progressive matrices (CPM) in Polish standardization [134] for measuring educational capacity (fluid intelligence) in children."

"The task of the examined person is to detect the double (irregular) jumps by pressing a button."

"The task consists of 48 boards with 5–20 elements on the screen."

"In order to determine the significance of differences, the Student's t-test, Mann–Whitney's U test and the repeated measures multivariate analysis of variance ANOVA were used."

"The repeated measures ANOVA did not include the sphericity test due to two levels of independent variable within each effect."

Results

"No significant differences were observed between ADHD and TD children in conditions of increased motivation for the following dependent variables: omission errors ($t = 1.78$, $p = 0.081$, $r = 0.23$), RT in visual search of 5–20 elements ($t = 1.65$, $p = 0.105$, $r = 0.21$), RT slope in visual search ($t = 0.38$, $p = 0.702$, $r = 0.05$) and RT in single-tasking ($t = 0.67$, $p = 0.505$, $r = 0.09$)."

"In the case of TD children, all of the controlled attention indicators were similar in normal and increased motivation conditions: omission errors ($t = 1.61$, $p = 0.118$, $r = 0.21$), commission errors ($t = 0.19$, $p = 0.854$, $r = 0.02$), RT in visual search of 5–20 elements ($t = 1.50$, $p = 0.145$, $r = 0.19$), RT slope in visual search ($t = 0.46$, $p = 0.651$, $r = 0.06$), RT in single-tasking ($t = 0.51$, $p = 0.613$, $r = 0.07$) and RT in multi-tasking ($t = 0.15$, $p = 0.886$, $r = 0.02$)."

Discussion

"We looked at the differences in the dimensions of motivation to learning between ADHD and TD children as per the self-determination theory (Hypothesis 2a–e)."

"In the multitasking test for divided attention (Hypothesis 3c), children with ADHD showed longer RT in single-tasking (medium effect: $r = 0.41$) and in multitasking (large effect: $r = 0.50$), which is consistent with many previous reports in this respect [135–137]."

"Our study confirmed the positive impact of extrinsic motivation on selected aspects of attention in children with ADHD, who have improved their capacity regarding most controlled aspects of attention."

"The results of children with ADHD improved under the influence of motivation, but they still differed from those of the control group (children without disorders did not improve their performance under conditions of increased motivation)."

"In the Reijnen and Opwis study [138] in conditions of increased motivation, children with ADHD exhibited comparable performance in visual search to those in the control group."

Summary
"The study confirmed significant differences between children with ADHD and TD children in terms of TBR and HEG ratio, which indicate different patterns of cortical activity in prefrontal and central areas."

"The study also confirmed the occurrence of attention deficits in children with ADHD."

"Cognitive deficits in children with ADHD are not permanent neuropsychological impairments."

"Evoking interest and more frequent strengthening help reduce cognitive deficits in children with ADHD."

Increasing Motivation in Robot-Aided Arm Rehabilitation with Competitive and Cooperative Gameplay [267]

This is a machine-generated summary of:

Novak, Domen; Nagle, Aniket; Keller, Urs; Riener, Robert: Increasing motivation in robot-aided arm rehabilitation with competitive and cooperative gameplay [267]

Published in: Journal of NeuroEngineering and Rehabilitation (2014)

Link to original: https://doi.org/10.1186/1743-0003-11-64

If you want to cite the papers, please refer to the original.

For technical reasons we could not place the page where the original quote is coming from.

Abstract-Summary

"In order to explore the potential of such games, we designed a two-player game played using two ARMin arm rehabilitation robots."

"Three game modes were tested: single-player (competing against computer), competitive (competing against human), and cooperative (cooperating with human against computer)."

"All modes were played by 30 unimpaired subjects and 8 impaired chronic stroke subjects."

"The subjects filled out the Intrinsic Motivation Inventory questionnaire after each game mode, as well as a final questionnaire about game preferences and their personality."

"Nearly all unimpaired subjects preferred playing the two-player game modes to the single-player one, as they enjoyed talking and interacting with another person."

"Unimpaired subjects who liked the competitive mode also put significantly more effort into it than into the other modes."

"The subjects' personalities roughly predicted which mode they would prefer, which was especially evident in a poorly-matched impaired pair that preferred the single-player mode."

"Results indicate great potential for two-player rehabilitation games, in the form of greater enjoyment as well as potentially more intensive exercise compared to single-player games."

"The right game type needs to be chosen for each subject depending on skill and personality, along with selecting an appropriate co-player."

Introduction

"Virtual reality may be able to increase patient motivation, which has been described by professionals as an important determinant of rehabilitation outcome [139]."

"The general agreement in the stroke rehabilitation literature is that virtual reality should serve as a tool to increase motivation and exercise intensity, whether directly using difficulty adaptation or indirectly through elements that make patients willing to exercise for longer periods of time."

"We thus see that, while the basic potential of multiplayer games for stroke rehabilitation has been demonstrated, it is clearly necessary to evaluate several different game types to find the optimal game with regard to motivation and exercise intensity."

"This paper takes the first step by evaluating subjectively felt motivation in an arm rehabilitation game with three modes: competing against a computer opponent, competing against a human opponent, and cooperating with a human teammate."

Materials and Methods

"There are three game modes: In the single-player mode, the subject controls the bottom mallet while the top one is controlled by the computer."

"The subjects were then grouped according to their favorite game mode, and separate Friedman tests were performed on the four IMI scales for only subjects who favored the competitive mode and for only subjects who favored the cooperative mode in order to identify differences between these two groups."

"After grouping the subjects according to favorite game mode, the groups' differences in score and the six personality scales were also evaluated using the Mann–Whitney U test."

"To determine if subjects' favorite game mode can be predicted from personality scores, a stepwise linear discriminant classifier was trained with seven possible inputs (age and the six personality scores) and one output (competitive or cooperative)."

Results

"Two competitive subjects stated that the human opponent was more challenging, even though one of the two had a better game score in the competitive mode than in the single-player mode."

"When the co-player's age and personality scores were included, the classifier was able to predict the favorite game mode for 25 of 28 subjects (89.3%) in crossvalidation."

"For subject 1, the preference toward the competitive mode was clear from her behavior during the game, as she showed clear pleasure at scoring a point and displeasure at the opponent scoring a point."

"Subject 8 chose the single-player mode as his favorite, stating that he did not enjoy competition in any setting and did not like the cooperative mode since he was worried that he might cause the pair to lose."

Discussion

"The subject's personality allowed the favorite game mode to be predicted in 82% of cases, with competitive players having higher emotional stability and competitiveness."

"Subjects of all ages enjoyed the two-player gameplay, and half of them chose the competitive mode as their favorite - including two subjects over 60 years old, despite expectations that older people would be less competitive [140] and that impaired subjects may find competitive gameplay uncomfortable [141]."

"The finding that rehabilitation games should be chosen according to the personality of each subject in fact applies not only to two-player games, but also to virtual environments for rehabilitation in general."

"The cooperative mode was often chosen as the most enjoyable in our study even though we subjectively believe that it requires the least physical activity; as the playing field is split in half between players, each player only needs to block the puck half of the time."

Conclusions

"Our study showed that most subjects do prefer playing a two-player rehabilitation game to a single-player one, as they enjoy being able to talk and otherwise interact with the other person."

"The preferred game mode depends strongly on each subject as well as their co-player."

"Subjects who enjoyed competitive gameplay indicated higher self-reported effort/importance in the competitive game mode, suggesting that such gameplay may lead to more intensive exercise."

"These findings were mostly confirmed in impaired chronic stroke subjects, who enjoyed both competitive and cooperative gameplay, sometimes expressing a strong preference for one of the two modes."

"In an instructive example, two impaired subjects preferred playing alone; one because he found competitive situations very unpleasant in general and was worried about disappointing his co-player, the other because he liked competition but did not enjoy losing."

Affective Motivational Direction Drives Asymmetric Frontal Hemisphere Activation [268]

This is a machine-generated summary of:

Poole, Bryan D.; Gable, Philip A.: Affective motivational direction drives asymmetric frontal hemisphere activation [268]

Published in: Experimental Brain Research (2014)

Link to original: https://doi.org/10.1007/s00221-014-3902-4

If you want to cite the papers, please refer to the original.

For technical reasons we could not place the page where the original quote is coming from.

Abstract-Summary

"This past work has demonstrated that greater left-frontal activation is related to positive-approach, and greater right-frontal activation is related to negative-withdrawal."

"The current experiment sought to illuminate whether frontal asymmetry is related to motivation or affective valence by examining frontal-lateralized late positive potentials (f-LPPs) and frontal cortical alpha power activation to approach-positive, approach-negative, and withdrawal-negative affects in the same participants."

"Midline LPPs to approach-positive and approach-negative pictures related to greater left-frontal cortical activation."

"These results suggest that greater relative left-frontal activation is associated with positive and negative approach-motivated states in the same participants."

The Valence Model of Hemispheric Activity

"Much of this past work has demonstrated that greater left hemisphere activity is associated with positive affect, whereas greater right hemisphere activity is associated with negative affect."

"Early support for this view emerged from studies in which hemispheric activity in the frontal cortex was shown to be asymmetrically related to affective processing."

"Greater right hemisphere activation was related to withdrawal-motivated states and traits (e.g., depression), whereas greater left hemisphere activation was related to approach-motivated states and traits (e.g., euphoria; Terzian and Cecotto [142])."

"This research demonstrated that relative left-frontal activation is associated with approach-positive affects (Ahern and Schwartz 143; Tomarken and others [144]), whereas relative right-frontal activation is associated with withdrawal-negative affects (Buss and others [145]; Davidson and others [146]; Silva and others [147]; Tomarken and others [148]; Wheeler and others [149])."

Support for the Motivational Model of Hemispheric Activity

"Greater left- than right-frontal activation is related to approach motivation (Schutter and others [150]), regardless of affective valence."

"In relation to fontal asymmetry, greater trait and state anger relate to increased left-frontal activation and decreased right-frontal activation (Harmon-Jones [151, 152]; Harmon-Jones and others [153]; Harmon-Jones and Sigelman [154]; Harmon-Jones and others [155])."

"If anger is an approach-motivated affective state, then it should evoke similar neurophysiological activity as approach-motivated positive states."

"Using EEG-derived event-related potentials, early neural measures of approach motivation have also supported the motivational direction model of frontal asymmetry."

"Approach-motivated positive (vs. neutral) pictures evoke greater left- than right-frontal LPP amplitudes, and approach-motivated negative (anger) pictures evoke greater left-frontal LPP amplitudes than neutral pictures (Gable and Harmon-Jones [156])."

"Midline LPP amplitudes to anger pictures predicted greater left-frontal activation toward anger pictures, but not neutral pictures."

"This finding suggests that the LPP to approach-motivated stimuli is related to state neurophysiological measures of approach motivation."

The Current Experiment

"We designed the current experiment to address this gap in past research on affect and frontal asymmetry by testing whether approach-positive (e.g., desire), approach-negative (e.g., anger), and withdrawal-negative (e.g., disgust) affects enhance asymmetric cortical activity as measured by the LPP and alpha frequency band power."

"Based on the pervading perspective that frontal activation is associated with emotional valence, arousal was synonymous with the intensity of affective valence: All high arousing/high intensity positive emotions should evoke greater left-frontal activation, whereas all high arousing/high intensity negative emotions should evoke greater right-frontal activation."

"High arousing emotion states associated with approach (vs. withdrawal) motivation should evoke greater left (vs. right) frontal activation regardless of the intensity of emotional valence."

"Because the LPP and frontal asymmetry are related to motivational processes (Gable and Poole [157]), we predicted that midline LPPs to approach-positive and approach-negative pictures, but not withdrawal-negative pictures, should relate to greater left-frontal cortical activity."

Method

"Approach-positive pictures consisted of appetitive images (e.g., delicious desserts), which have been used in previous research to evoke approach-motivated positive affect (Gable and Harmon-Jones [158, 159, 156, 160]; Harmon-Jones and Gable [161])."

"Withdrawal-negative pictures were scenes of aversive objects (e.g., mutilated bodies) and have reliably evoked withdrawal-motivated negative affect (Gable and Harmon-Jones 156; Gable and Poole 162)."

"Each block consisted of 64 pictures: 32 pictures of one affective picture type and 32 matching neutral pictures."

"LPP amplitudes to affective pictures were aggregated across picture type within each block, and LPP amplitudes to neutral pictures were aggregated to form a single LPP index to neutral pictures."

"Consistent with past studies measuring state frontal cortical activation to affective pictures using alpha band power (see Coan and Allen [163], for a review; Harmon-Jones and others [164]; Hewig and others [165]; Jackson and others [166]), power spectra epochs 1.024 s in duration were extracted through a Hamming window (50 % taper of distal ends)."

Results

"There was no difference between f-LPP asymmetry to approach-positive and approach-negative pictures ($p = .37$) or withdrawal-negative pictures ($p = .49$)."

"The difference between f-LPP asymmetry to approach-negative and withdrawal-negative pictures was also not significant, $p = .14$."

"Consistent with past research examining the influence of affective pictures on frontal asymmetry (Elgavish and others [167]; Gable and Harmon-Jones [159]; Hagemann and others [168]; Harmon-Jones [152]), approach-positive pictures [$t(47) = .04$, $p = .97$], approach-negative pictures [$t(47) = .06$, $p = .95$], or withdrawal-negative pictures [$t(47) = 1.50$, $p = .14$] did not evoke differences in frontal asymmetry as compared to neutral pictures."

"Frontal asymmetry to approach-positive pictures was not related to frontal asymmetry to withdrawal-negative pictures, $r = -.02$, $p = .87$."

"Cp-LPP amplitudes to approach-negative pictures predicted greater relative left-frontal activity to the approach-negative pictures, $r(46) = .36$, $p = .011$."

Discussion

"These results refute models that continue to associate all positive affects with approach and all negative affects with withdrawal, and demonstrate that motivational direction, rather than affective valence or general arousal, is related to asymmetric hemisphere activation."

"Past research has demonstrated that withdrawal-motivated affects such as disgust and fear are associated with relatively less left-frontal activation than approach-motivated affects such as anger and desire (Harmon-Jones and Allen [169, 170]; Sutton and Davidson [171])."

"Critical support for our hypotheses that approach motivational states evoke similar neurophysiological activity comes from evidence that approach-motivated affects related to greater left LPP amplitudes and left-frontal activation, whereas withdrawal-motivated affects related to relatively less left LPP amplitudes and left-frontal activation."

"Our results contribute to a growing body of research linking positive and negative approach-motivated affects with left hemispheric activation (for review see Harmon-Jones and others [172])."

Two Types of Dopamine Neuron Distinctly Convey Positive and Negative Motivational Signals [269]

This is a machine-generated summary of:

Matsumoto, Masayuki; Hikosaka, Okihide: Two types of dopamine neuron distinctly convey positive and negative motivational signals [269]

Published in: Nature (2009)

Link to original: https://doi.org/10.1038/nature08028

If you want to cite the papers, please refer to the original.

For technical reasons we could not place the page where the original quote is coming from.

Abstract-Summary

"The activity of dopamine neurons is not related to body movements, but to reward events."

"Recent theories suggest that dopamine neurons, as a homogeneous functional group, guide motor learning via reward-related signals."

"This suggests that dopamine neurons have a more complex involvement in learning control, encoding more subtle signals than simple reward."

Main

"Neurons showing a significant inhibition and excitation were classified as airpuff conditioned stimulus (ACS)-inhibited type ($n = 24$) and ACS-excited type ($n = 38$), respectively ($P < 0.05$, Wilcoxon signed-rank test)."

"These results suggest that the ACS-inhibited neurons encode motivational value on a single scale, and are most strongly excited in response to the most positive stimulus (100% reward conditioned stimulus) and most strongly inhibited in response to the most negative stimulus (100% airpuff conditioned stimulus)."

"These results raise the possibility that the responses of the airpuff-excited dopamine neurons to a conditioned stimulus reflect the motivational salience of the conditioned stimulus."

"Using the same Pavlovian procedure, we have shown that lateral habenula neurons are excited by the airpuff-predicting conditioned stimulus and inhibited by the reward-predicting conditioned stimulus, indicating that they encode motivational value similarly to the airpuff-inhibited dopamine neurons, but in the opposite manner [173]."

Methods Summary

"The head holder, the recording chamber and the eye coil connectors were all embedded in dental acrylic that covered the top of the skull, and were connected to the skull using acrylic screws."

"We conditioned two monkeys using a Pavlovian procedure with an appetitive unconditioned stimulus (liquid reward) and an aversive unconditioned stimulus (airpuff)."

"During the Pavlovian procedure, we recorded the activity of dopamine neurons in and around the SNc and VTA."

"We analysed anticipatory licking, anticipatory blinking and neuronal responses during the Pavlovian procedure."

"Details of the Pavlovian procedure, identification of dopamine neurons, analysis methods, and histological procedure can be found in the full Methods."

Online Methods

"In the appetitive block, three conditioned stimuli (red circle, green cross and blue square for monkey N; yellow ring, cyan triangle and blue square for monkey D) were associated with a liquid reward (apple juice) as an unconditioned stimulus with 100%, 50% and 0% probability, respectively."

"In analyses of neuronal activity, responses to each conditioned stimulus were defined as the discharge rate during the interval 150 to 325 ms after conditioned stimulus onset minus the background discharge rate during the 250 ms before conditioned stimulus onset."

"Response to reward omission was defined as the discharge rate during the interval 200 to 500 ms after the conditioned stimulus ended minus the background discharge rate during the 250 ms before the conditioned stimulus ended."

"Response to airpuff omission was defined as the discharge rate during the interval 150 to 350 ms after the conditioned stimulus ended minus the background discharge rate during the 250 ms before the conditioned stimulus ended."

EEG

Machine generated keywords: eeg, brain, network, functional, spatial, dynamic, source, activation, technique, signal, region, target, fmri, stimulus, prior

Optimizing EEG Source Reconstruction with Concurrent fMRI-Derived Spatial Priors [270]

This is a machine-generated summary of:

Abreu, Rodolfo; Soares, Júlia F.; Lima, Ana Cláudia; Sousa, Lívia; Batista, Sónia; Castelo-Branco, Miguel; Duarte, João Valente: Optimizing EEG Source Reconstruction with Concurrent fMRI-Derived Spatial Priors [270]

Published in: Brain Topography (2022)

Link to original: https://doi.org/10.1007/s10548-022-00891-3

If you want to cite the papers, please refer to the original.

For technical reasons we could not place the page where the original quote is coming from.

Abstract-Summary

"The spatial resolution of fMRI makes fMRI derived spatial priors very convenient for EEG reconstruction, however, only task activation maps and resting-state networks (RSNs) have been explored so far, overlooking the recent, but already accepted, notion that brain networks exhibit dynamic functional connectivity fluctuations."

"The lack of a systematic comparison between different source reconstruction algorithms, considering potentially more brain-informative priors such as fMRI, motivates the search for better reconstruction models."

"Using simultaneous EEG-fMRI data, here we compared four different inversion algorithms (minimum norm, MN; low resolution electromagnetic tomography, LORETA; empirical Bayes beamformer, EBB; and multiple sparse priors, MSP) under a Bayesian framework (as implemented in SPM), each with three different sets of priors consisting of: (1) those specific to the algorithm; (2) those specific to the algorithm plus fMRI task activation maps and RSNs; and (3) those specific to the algorithm plus fMRI task activation maps and RSNs and network modules of task-related dFC states estimated from the dFC fluctuations."

"Optimal overlap/proportion values were found using EBB and priors specific to this algorithm and fMRI task activation maps and RSNs or MSP and considering all the priors (algorithm priors, fMRI task activation maps and RSNs and dFC state modules), respectively, indicating that fMRI spatial priors, including dFC state modules, might contain useful information to recover EEG source components reflecting neuronal activity of interest."

"Our main results show that providing fMRI spatial derived priors that reflect the dynamics of the brain might be useful to map neuronal activity more accurately from EEG-fMRI."

Introduction

"The inverse problem relates with determining the sources in the brain that generate a given scalp distribution of electrical potential differences (i.e., EEG topography)."

"From dFC fluctuations, a limited, but variable, number of dFC states have been recurrently identified in the literature as the building blocks of brain functional connectivity (dynamics) (Preti and others [174]), which are hypothesized to be associated with different cognitive, vigilance or pathological brain states (Thompson [175]); however, they have not been considered as potential spatial priors for EEG source reconstruction so far."

"Considering the present limitations described in this section, here we compared four different inversion algorithms (MN, LORETA, empirical Bayes beamformer, EBB; and MSP), each with two different sets of additional fMRI-derived spatial priors (activation maps and RSNs, with and without including dFC states) on EEG data collected concurrently with fMRI at 3 T from 6 multiple sclerosis (MS) patients and 7 healthy subjects performing visual perception tasks and during rest."

Materials and Methods

"From the pre-processed fMRI data, several potential priors for EEG source reconstruction were subsequently extracted (procedures described next), namely:

resting-state networks for all runs, and task-related activity maps and dynamic functional connectivity (dFC) states for the task runs only."

"The dFC analysis described here was only performed on the fMRI data collected during the task runs (localizer and two BM runs), since its purpose was to objectively identify a small set of dFC states associated with the tasks, and to use them as spatial priors in the subsequent reconstruction of EEG sources."

"To those specific to a given algorithm, other CCs estimated from the fMRI-derived spatial priors (RSNs, activation maps and task-based dFC states) were considered (the procedures for their estimation are briefly described below)."

Results

"The quality of EEG source reconstruction provided by the different combinations of (four) inversion algorithms and (three) sets of CCs, was first evaluated in terms of the posterior P(model|data) and VE of the associated models, which are commonly considered in PEB frameworks."

"The combination MSP + S_1 (with S_1 containing only CCs specifically associated with the inversion algorithm) yielded the highest P(model|data) and the highest VE, only followed by LORETA + S_2 and LORETA + S_3."

"The ANOVA of the VE values revealed significant main effects of the inversion algorithms and sets of CCs, as well as a significant interaction."

Discussion

"We aimed at optimizing the reconstruction of EEG sources by considering spatial priors derived from concurrently acquired fMRI data when solving the inverse problem, coupled with a systematic comparison of different inversion algorithms and sets of covariance components (CCs) reflecting those spatial priors, on a parametric empirical Bayesian (PEB) framework."

"We systematically compared the quality of the source reconstruction of EEG data performed using different combinations of four inversion algorithms and three sets of covariance components incorporating different types of spatial priors derived from concurrently acquired fMRI data."

"We found that according to the quality metrics reflecting the presence of neuronal activity, combining the EBB or MSP algorithms with CC sets including fMRI task activation maps and RSNs yields the overall best source reconstruction, and that by further including dFC state modules as spatial priors, the quality of EEG sources from the task runs is optimal."

Music of Brain and Music on Brain: A Novel EEG Sonification Approach [271]

This is a machine-generated summary of:

Sanyal, Shankha; Nag, Sayan; Banerjee, Archi; Sengupta, Ranjan; Ghosh, Dipak: Music of brain and music on brain: a novel EEG sonification approach [271]

Published in: Cognitive Neurodynamics (2018)

Link to original: https://doi.org/10.1007/s11571-018-9502-4

If you want to cite the papers, please refer to the original.

For technical reasons we could not place the page where the original quote is coming from.

Abstract-Summary
"We present a novel method with which we can sonify the electroencephalogram (EEG) data recorded in "control" state as well as under the influence of a simple acoustical stimuli—a tanpura drone."

"For this pilot project we chose to study the nonlinear correlations between musical stimulus (tanpura drone as well as music clips) and sonified EEG data."

"Till date, there have been no study which deals with the direct correlation between a bio-signal and its acoustic counterpart and also tries to see how that correlation varies under the influence of different types of stimuli."

"This study tries to bridge this gap and looks for a direct correlation between music signal and EEG data using a robust mathematical microscope called Multifractal Detrended Cross Correlation Analysis (MFDXA)."

"We took EEG data of 10 participants in 2 min "control condition" (i.e. with white noise) and in 2 min 'tanpura drone' (musical stimulus) listening condition."

"The EEG signals from different electrodes were sonified and MFDXA technique was used to assess the degree of correlation (or the cross correlation coefficient γ_x) between the EEG signals and the music clips."

Introduction
"The first thing to be addressed while dealing with sonification is what is meant by 'sonification'."

"As per the definition of International Conference on Auditory Display (ICAD) "the use of non-speech audio to convey information; more specifically sonification is the transformation of data relations into perceived relations in an acoustic signal for the purposes of facilitating communication or interpretation" (Kramer and others [176])."

"Hermann [177] gives a more classic definition for sonification as"a technique that uses data as input, and generates sound signals (eventually in response to optional additional excitation or triggering)"."

History of Sonifying EEG Waves

"Sonification is, however, not an exclusively scientific endeavor; the technique has been commonly applied within the domain of experimental music."

"The interplay between science and art is beautifully demonstrated by the sonification of the EEG, a practice that was first described in the early 1930s and subsequently gave rise to a variety of medical and artistic applications."

"In neurophysiology, sonification was used to complement visual EEG analysis, which had become increasingly complex by the middle of the 20th century."

"By the end of the century, advances in EEG and sound technology ultimately gave rise to brain–computer music interfaces (BCMIs), a multidisciplinary achievement that has enhanced expressive abilities of both patients and artists (Miranda and Castet [178])."

"In 1934, the renowned neurophysiologist and Nobel laureate Edgar Adrian (1889–1977) first described the transformation of EEG data into sound (Adrian and Matthews [179])."

Applications of EEG Sonification

"American psychologist Joe Kamiya (b. 1925) had first demonstrated that subjects could operantly learn to control their alpha activity when real-time auditory feedback was provided, but the technique only became popular after his accessible report in Psychology Today (1968)."

"Real time EEG sonification enjoys a wide number of applications including diagnostic purposes like epileptic seizure detection, different sleep states etc (Glen 180; Olivan and others 181; Khamis and others [182]), neuro-feedback applications (Hinterberger and others [183]; McCreadie and others [184]), brain-controlled musical instruments (Arslan and others [185]) while a special case involves converting brain signals directly into meaningful musical compositions (Miranda and Brouse 186; Miranda and others [187])."

"We have a number of studies which deal with the emotional appraisal in human brain corresponding to a wide variety of scale-free musical clips using EEG/fMRI techniques (Sammler and others [188]; Koelsch and others [189]; Lin and others [190]; Lu and others [191]; Wu and others [192]), but none of them provides a direct correlation between the music sample used and the EEG signal generated using the music as a stimulus, although both are essentially complex time series variations."

"The EEG signals are lobe specific and characterized with a lot of fluctuations corresponding to different musical and other types of stimulus."

The Role of Tanpura in Hindustani Classical Music

"We chose to study the EEG response corresponding to the simplest (and yet very complex) musical stimuli—the Tanpura drone."

"The Tanpura drone primarily establishes the "Sa" or the scale in which the musical piece is going to be sung/played."

"It will be quite interesting to study the response of brain simultaneously to a simple drone sound using different non-linear techniques."

"Because there is a felt resonance in perception, psycho-acoustics of Tanpura drone may provide a unique window into the human psyche and cognition of musicality in human brain."

""Fractal Analysis" technique has been used to study the non linear nature of the tanpura signals (Sengupta and others [193])."

"It demands robust non-linear methods to assess the complex nature of tanpura signals."

Emotions and Indian Classical Music

"In Indian classical music (ICM), each raga has a well defined structure consisting of a series of four/five or more musical notes upon which its melody is constructed."

"From the time of Bharata's Natyashastra (Ghosh [194]), there have been a number of treatises which speak in favor of the various rasas (emotional experiences) that are conveyed by the different forms of musical performances."

"A number of works tried to harvest this immense potential by studying objectively the emotional experiences attributed to the different ragas of Indian classical music (Balkwill and Thompson [195]; Chordia and Rae [196]; Wieczorkowska and others [197]; Mathur and others [198])."

"The studies have revealed unlike Western Music pieces the emotions evoked by Indian Classical Music pieces are often more ambiguous and far more subdued."

Fractals and Multifractals in EEG Study

"In recent past, the Detrended Fluctuation Analysis (DFA) has become a very useful technique to determine the fractal scaling properties and long-range correlations in noisy, non-stationary time-series (Hardstone and others [199])."

"Multifractal Detrended Fluctuation Analysis (MFDFA) (Kantelhardt and others [200]) technique analyzes the musical signal in different scales and gives a multifractal spectral width which is the measure of complexity of the signal."

"A robust technique called Multifractal Detrended Cross correlation Analysis (MFDXA) (Zhou [201]) has been used to analyze the multifractal behaviors in the power-law cross-correlations between any two non-linear time series data (in this case music and sonified EEG signals)."

"With this technique, all segments of the music clips and the sonified EEG signals were analyzed to find out a cross correlation coefficient (γ_x) which gives the degree of correlation between these two categories of signals."

Overview of the Present Work

"The main constraint in establishing a direct correlation between EEG signals and the stimulus sound signal is the disparity in sampling frequency of the two; while an EEG signal is generally sampled at up to 512 samples/sec (in our case it is 256 samples/sec), the sampling frequency of a normal recorded audio signal is 44100 samples/sec."

"A human response psychological data was taken initially to standardize the emotional appraisal corresponding to the acoustic clips chosen for this study, followed by EEG response using the same musical clips as stimuli."

"The alpha frequency wave is extracted from the entire EEG signal using Wavelet Transform technique as elaborated in Banerjee and others (2016) as alpha waves are mainly associated with emotional activities corresponding to music clips (Sammler and others [188]; Maity and others [202]; Babiloni and others [203])."

"The MFDXA technique has been applied to assess the degree of correlation between the two non-linear, non-stationary signals (in this case output EEG signals and the input acoustic musical signals)."

Materials and Methods

"For the human response data, 50 participants ($M = 33$, $F = 17$) were asked to rate the music clips in a sheet on the basis of what emotion they felt for each clip."

"The EEG experiment was conducted in the afternoon (around 2 PM) in a room with the subjects sitting relaxed in a comfortable chair with their eyes closed during the experiment."

"The EEG experiments were conducted in the afternoon (around 2 PM) in an air conditioned room with the subjects sitting in a comfortable chair in a normal diet condition."

"Sequence 2–3 was repeated With Music (Darbari Kannada) In between the application of stimuli, the subjects were asked to rate the clips on the basis of emotional appraisal."

"The listening test experiment was performed following the same protocol with each subject being asked to mark the emotional appraisal corresponding to each music clip."

Methodology

"We used a robust non-linear cross correlation technique called Multi Fractal Detrended Cross Correlation Analysis (MFDXA) (Zhou [201]) in this case, taking the tanpura drone/music signal as the first input and a music induced resampled EEG signal (electrode wise) as the second input."

"Scaling exponent $\lambda(q)$ represents the degree of the cross-correlation between the two time series."

"If scaling exponent $\lambda(q)$ is independent of q, the cross-correlations between two time series are monofractal."

"If scaling exponent $\lambda(q)$ is dependent on q, the cross-correlations between two time series are multifractal."

"Scaling exponent $\lambda(q)$ represents the degree of the cross-correlation between the two time series x(i) and y(i)."

"We want to point out how two non-linear signals (in this case music and EEG) are cross-correlated in various time scales."

Results and Discussion

"The same analysis was repeated for all other electrodes and music clips as well as between the EEG electrodes which revealed the amount cross-correlation between the different electrodes of human brain."

"Apart from these, F3-FP2 correlation also increase consistently again in the 2nd and 3rd segments of the audio clip while inter lobe frontal correlation, i.e. between F3 and F4 electrodes show the highest rise in the last two parts of the audio signal."

"Experiment, EEG data corresponding to emotional music stimuli were sonified and degree of cross-correlation was evaluated for the audio clip and the sonified EEG signals."

"The two fronto-polar electrodes, FP1/FP2 and the frontal midline electrode Fz are the ones which show significant increase in cross-correlation coefficient for both the music clips."

Conclusion

"A direct evidence of correlation existing between audio signal and the sonified EEG signals obtained from different lobes of human brain is obtained."

"The rise in correlation is different in scale in different electrodes, but in general we have found a stipulated time period wherein the effect of music on human brain is the maximum."

"The degree of correlation increases significantly under the effect of even a neutral musical stimulus like tanpura drone, which does not elicit any emotional appraisal as is evident from the psychological study."

"We have direct quantitative evidence of correlation between music and EEG signals as well as the level of arousal in among the electrodes in regard to emotional music stimuli."

"The inter lobe correlation between frontal and fronto-polar electrodes appear to increase most in case of the sad music clips."

Mean-Field Modeling of Brain-Scale Dynamics for the Evaluation of EEG Source-Space Networks [272]

This is a machine-generated summary of:

Allouch, Sahar; Yochum, Maxime; Kabbara, Aya; Duprez, Joan; Khalil, Mohamad; Wendling, Fabrice; Hassan, Mahmoud; Modolo, Julien: Mean-Field Modeling of Brain-Scale Dynamics for the Evaluation of EEG Source-Space Networks [272]

Published in: Brain Topography (2021)

Link to original: https://doi.org/10.1007/s10548-021-00859-9

If you want to cite the papers, please refer to the original.

For technical reasons we could not place the page where the original quote is coming from.

Abstract-Summary
"To estimate these large-scale brain networks, the emergent method called "electroencephalography (EEG) source connectivity" has generated increasing interest in the network neuroscience community, due to its ability to identify cortical brain networks with satisfactory spatio-temporal resolution, while reducing mixing and volume conduction effects."

"A validation toolbox that provides flexible "ground truth" models is needed for an objective methods/parameters evaluation and, thereby an optimization of the EEG source connectivity pipeline."

"We show how a recently developed large-scale model of brain-scale activity, named COALIA, can provide to some extent such ground truth by providing realistic simulations of source-level and scalp-level activity."

"Using a bottom-up approach, the model bridges cortical micro-circuitry and large-scale network dynamics."

"We provide an example of the potential use of COALIA to analyze, in the context of epileptiform activity, the effect of three key factors involved in the "EEG source connectivity" pipeline: (i) EEG sensors density, (ii) algorithm used to solve the inverse problem, and (iii) functional connectivity measure."

"We aim at presenting a proof-of-concept of the interest of COALIA in the network neuroscience field, and its potential use in optimizing the EEG source-space network estimation pipeline."

Introduction

"We highlight the implications of this model in enhancing our interpretation of the reconstructed brain networks and in evaluating the key factors of the EEG source connectivity pipeline, such as (1) EEG sensor density, (2) solution of the EEG inverse problem, and (3) functional connectivity measure."

"We generate epileptiform, cortical activity (confined to the left hemisphere) and present a (not exhaustive) comparative study to evaluate, in this specific context, the effect of: (1) five different electrode densities [256, 128, 64, 32, 19]; (2) two inverse solution algorithms, the weighted minimum norm estimate (wMNE) and the exact low resolution electromagnetic tomography (eLORETA); and (3) two functional connectivity measures, the phase locking value (PLV) and the weighted phase lag index (wPLI) as they represent one of the most used combination of methods in the context of EEG source-space network estimation."

Materials and Methods

"All sources belonging to a single patch were synchronized at a zero lag, while a delay of 30 ms was introduced between the two subnetworks to reflect the propagation of spikes between relatively distant regions in the brain."

"We used two methods widely used in EEG source connectivity analysis: the weighted minimum norm estimate (wMNE) and the exact low-resolution electromagnetic tomography (eLORETA)."

"We used linear mixed model analyses to investigate the effects of electrode number, inverse solution method, and connectivity measure on the accuracy of the estimated networks."

"We used the lmer function of the {lme4} package (Bates and others [204]) with the following model that includes, epoch, electrode number, inverse solution method and connectivity measures as interacting fixed effects, and also a random intercept for epochs: We applied a square root transform to the data, since this led to a better compliance of the model with the assumptions of normality and homoscedasticity of model's residuals than for raw data."

Results
"wMNE/PLV and eLORETA/PLV performed similarly and were slightly less accurate than wMNE/wPLI at a high sensor density."

"The influence of the sensor density was confirmed by the statistical analysis with a significant sensor density effect ($F_{(4,532)} = 333.53$, $p < 0.001$, conditional $R^2 = 0.84$, marginal $R^2 = 0.82$)."

"Regarding the inverse solution, wMNE significantly outperformed eLORETA ($F_{(1,532)} = 281.75$, $p < 0.001$, conditional $R^2 = 0.84$, marginal $R^2 = 0.82$)."

"The combination inverse solution/connectivity measure combination had also a significant effect on the network estimation accuracy ($F_{(1,532)} = 478.91$, $p < 0.001$, conditional $R^2 = 0.84$, marginal $R^2 = 0.82$)."

"The highest network estimation accuracy was reached using wMNE/wPLI, while the worst performance was obtained with eLORETA/wPLI."

"Differences between eLORERA/PLV and wMNE/PLV and between eLORETA/wPLI and wMNE/wPLI were not statistically significant."

Discussion

"Our study showed that a more crucial parameter is the combination of inverse method and connectivity measure."

"Although wMNE/wPLI performed better than eLORETA/wPLI, the PLV connectivity measure performed similarly with both eLORETA and wMNE methods."

"It is therefore worth noting that the results obtained in this study are specific to the analyzed condition, i.e., epileptic spikes: therefore, we cannot be certain that the same combination of methods will provide the best network estimation accuracy when analyzing networks related to cognitive tasks or resting state, such as the alpha/beta DMN for instance (i.e., physiological activity), which is the main objective of the future steps of this work."

"These observations highlight the importance of applying an effective preprocessing of EEG signals before reconstructing the cortical networks, as well as a robust source connectivity method to correctly estimate functional connectivity."

Methodological Considerations
"Our objective was to provide a typical example of the use of the COALIA model to investigate the effect of different pipeline-related parameters on EEG source-space network analysis."

"With our approach, we aimed at promoting the use of computational modeling as a ground-truth to evaluate parameters of EEG source connectivity methods."

"It would be even more realistic for the network neuroscience field to use the model to simulate different rhythms of resting-state data (alpha/beta-band activity in the DMN network, for instance) and then evaluate the desired techniques in such context, rather than restricting the study to spikes/background activity scenarios."

"The connectivity assessed here using PLV and wPLI is bidirectional, however, since we introduced a time delay between the two subnetworks, we propose that studying directional connectivity metrics (e.g., Granger causality) may also lead to additional insights specifically related to Bartolomei's model, where epileptic activity is transferred from the epileptogenic network to the propagation network (Bartolomei and others [205])."

Conclusion

"We have provided evidence that COALIA, a recently developed, physiologically-inspired computational model can provide a ground-truth for comparative studies aiming at optimizing the EEG-source connectivity pipeline."

"Using this model-based approach, several methodological questions can be addressed."

"We assessed the effect of the number of EEG electrodes, as well as the inverse solution/connectivity measure combination in the context of simulated epileptic activity."

"Our results suggest that a higher network estimation accuracy requires a high number of EEG electrodes, and suggest a careful choice of an efficient inverse solution/connectivity measure combination."

In-group Biases and Oculomotor Responses: Beyond Simple Approach Motivation [273]

This is a machine-generated summary of:

Moradi, Zahra Zargol; Manohar, Sanjay; Duta, Mihaela; Enock, Florence; Humphreys, Glyn W.: In-group biases and oculomotor responses: beyond simple approach motivation [273]

Published in: Experimental Brain Research (2018)

Link to original: https://doi.org/10.1007/s00221-018-5221-7

If you want to cite the papers, please refer to the original.

For technical reasons we could not place the page where the original quote is coming from.

Abstract-Summary

"Previous studies suggest that in-group bias facilitates approach motor responses, but disrupts avoidance ones."

"Such motor biases are shown to be more robust when the out-group is threatening."

"We investigated whether, under controlled visual familiarity and complexity, in-group biases still promote pro-saccade and hinder anti-saccades oculomotor responses."

"They were then instructed to listen to the group-relevant auditory cue (name of own and a rival university) followed by one of the shapes."

"Half of the participants were instructed to look towards the visual target if it matched the preceding group-relevant auditory cue and to look away from it if it did not match."

"This design allowed us to orthogonally manipulate the effect of in-group bias and cognitive control demand on oculomotor responses."

"Both pro- and anti-saccades were faster and more accurate following the in-group auditory cue."

"Our findings indicate that under higher cognitive control demands individuals' oculomotor responses improved following the motivationally salient cue (in-group)."

"Future studies could extend our findings using non-threatening out-groups instead."

Introduction

"In tasks involving oculomotor responses, participants are usually required to make an eye movement towards (pro-saccade) or away from (anti-saccade) an in-group or out-group target (Liuzza and others [206]; Wu and others [207])."

"Familiarity could drive the salience effects of in-group observed for pro- (approach) vs. anti-saccade (avoidance) oculomotor responses."

"The complexity and familiarity of the stimulus may fundamentally mask the actual effect of in-group biases on pro- and anti-saccade oculomotor responses."

"We aimed to test whether in-group biases facilitate pro-saccade over anti-saccade oculomotor responses by controlling for familiarity and visual complexity of the targets."

"We orthogonally manipulated the motivational salience of the auditory cue (in-group vs. out-group) and cognitive control demands required for each movement type (pro- vs. anti-saccade) while controlling for the familiarity and complexity of the visual target."

Method

"In the eye-movement task, participants were presented with a spoken audio label of in- and out-group universities ("Oxford" or "Cambridge") followed by a visual target (the same geometrical shapes that were used in the social associative learning task)."

"In the low-demand version of the task, participants were instructed to make a pro-saccade to the visual target if it matched the preceding auditory cue and to make an anti-saccade away from it if it did not match the preceding auditory cue."

"In the high-demand version of the task, participants were instructed to do the opposite and make a pro-saccade toward the visual target if it did not match the preceding auditory cue and make an anti-saccade away from the visual target if it did match the preceding auditory cue."

Results

"For all analyses, a mixed-effects $2 \times 2 \times 2$ ANOVA was performed, with the within-subject independent variables of the required movement type (pro-saccade vs. anti-saccade) and group relevance of the auditory cue (in-group vs. out-group) and the between-subject variable of the task type (High vs. Low demand)."

"There was a significant interaction between movement type and task type, with larger pro- vs. anti-saccade differences evident in the low-demand version of the task (match-approach) [$t(26) = 10.45$, $p < 0.001$, $d = 4.09$; mean difference $\pm$ SD $= 0.417 \pm 0.039$] compared to the high demand (match-avoid) version of the task [$t(25) = 4.72$, $p < 0.001$, $d = 1.70$; mean difference $\pm$ SD $= 0.207 \pm 0.043$]."

Discussion

"Previous findings suggest that in-group biases facilitate pro-saccades while hindering or slowing down anti-saccade oculomotor responses (Ciardo and others [208]; Liuzza and others [206]; Moradi and others [209])."

"To reduce these confounds, we examined the relationship between in-group biases and pro- vs. anti-saccade oculomotor responses, controlling for the familiarity and visual complexity of the targets."

"Our study provides evidence for the effects of an in-group bias on both pro- and anti-saccade oculomotor responses."

"Our results indicate that when controlled for familiarity and visual complexity, the in-group bias does not merely enhance performance in pro-saccades and nor does it hinder anti-saccades."

"Previous studies have shown that social and motivational factors including group identification may affect the directional errors in anti-saccade oculomotor responses (Manohar and others 210; Moradi and others [209])."

"Controlling for familiarity and visual complexity, our study provides further evidence indicating that previous findings might be affected by the higher level of familiarity of in-group targets, and that anti-saccade visual motor responses require greater cognitive control."

Limitations

"Future studies might start to investigate how in-group biases affect oculomotor responses in the absence of intergroup rivalry."

"We would also like to indicate that, in our study, the group information was highly relevant to the eye-movement task."

"It would be interesting for the future studies to test whether making the group information irrelevant to the eye-movement task can still have an impact on participants' looking behaviour."

"This association was a unimodal association (visual), while in the eye-movement task, participants were required to translate the group-relevant cue from visual modality to auditory modality (cue) and then judge whether the following visual target matched this auditory cue."

"Such cross-modal translation might have made the information processing in the eye-movement task more difficult and therefore resulted in a smaller effect of in-group biases on pro- and anti-saccades."

An Integrated TMS-EEG and MRI Approach to Explore the Interregional Connectivity of the Default Mode Network [274]

This is a machine-generated summary of:

Esposito, Romina; Bortoletto, Marta; Zacà, Domenico; Avesani, Paolo; Miniussi, Carlo: An integrated TMS-EEG and MRI approach to explore the interregional connectivity of the default mode network [274]

Published in: Brain Structure and Function (2022)

Link to original: https://doi.org/10.1007/s00429-022-02453-6

If you want to cite the papers, please refer to the original.

For technical reasons we could not place the page where the original quote is coming from.

Abstract-Summary

"By means of an integrative approach, we examined both the structural and functional connections of the default mode network (DMN) in a group of sixteen healthy subjects."

"The target network was structurally explored by diffusion-weighted imaging, tested by neurophysiological means, and retested by means of concurrent transcranial magnetic stimulation and electroencephalography (TMS-EEG)."

"A series of correlational analyses were performed to explore the relationship between the amplitude of early-latency TMS-evoked potentials and the indexes of structural connectivity (weighted number of fibres and fractional anisotropy)."

"Structural features of the corpus callosum predicted the activation of the directly connected nodes, i.e., parietal-parietal nodes, and of the broader DMN network, i.e., parietal-frontal nodes, as identified with functional magnetic resonance imaging."

"Our results suggested that the proposed integrated approach would allow us to describe the contributory causal relationship between structural connectivity and functional connectivity of the DMN."

Introduction
"Accumulating evidence points to a large number of brain regions, for instance, the default mode network (DMN), that are functionally connected even at rest (Greicius and others 212; Raichle [213])."

"We used an innovative approach to study the relationship between structural connectivity and functional connectivity in the DMN."

"Our project was initiated with the idea that with the proposed integrative approach combining TMS-EEG guided by MRI, it would be feasible to obtain a detailed view of the spatial (MRI reveals the structural pathways), temporal (EEG enables measurements of the time course of the activity of the cortical areas) and effective (TMS provides information about directionality) features of the DMN at a macroscopic level."

"The proposed methodologically integrated MRI-TMS-EEG approach represents a tool that can be used to describe the relationship among the structural, functional, and effective connectivity of brain networks (Voineskos and others [214])."

Materials and Methods
"Even though the gap between the MRI and the two TMS-EEG sessions was quite long, there is evidence that the reliability of the resting-state networks, over time, is higher than that of other brain networks (see Noble and others [215])."

"During the TMS-EEG sessions, TMS was applied over four different areas of the brain, the right and left prefrontal and parietal areas, while EEG was recorded concurrently (see below for details)."

"For each subject, each TMS-EEG session was divided into six blocks: two 5-min adaptation blocks of resting EEG recorded at the beginning and the end of the session (data not analysed) and four TMS-EEG blocks."

"The ground electrode was placed at FPz, the online reference was placed at TP9, and an electrode at TP10 recorded an online signal to "balance" data, in terms of electrode location, in the offline re reference process (see the TMS-EEG session)."

Data Preprocessing and Statistical Analysis
"The pipeline steps consisted of slice-timing correction, rigid-body realignment, and removal of movement-susceptibility interactions, subtraction of baseline fluctuations by fitting a fourth-order polynomial, low-pass filtering using a second-order Butterworth filter having $f_{-3\,dB} = 0.09$ Hz, removal of covariance with the six first-order head movement vectors (translations and rotations) and with average white matter and cerebrospinal fluids derived from individual tissue masks derived from the structural scan."

"Physiological and TMS-related artifact components were detected using INFOMAX-ICA and removed based on their scalp distribution, frequency, timing, and amplitude."

"We assessed the Pearson correlation between the TEPs for all the electrodes at each time point, from 7 to 60 ms, with the two indexes of structural connectivity of the forceps, i.e., the weighted number of fibres and FA (i.e., connectivity), in which t tests were run to evaluate the hypothesis that the correlation coefficient R is different from zero."

Results
"When the TEP voltage was positive, as was the case for the electrodes close to stimulation, this correlation indicated that the TEP amplitude decreased for higher FA in the forceps major."

"This correlation indicated a decrease in signal amplitude with higher FA values of the forceps major."

"The pattern that emerged from the significant clusters demonstrated a general reduction in the voltage recorded from electrodes close to the stimulation site and over frontocentral areas and an increase in the voltage in posterior electrodes contralateral to the TMS target in correlation with the FA values of the forceps major."

"In the bilateral parieto-occipital electrodes (negative cluster), the positive voltage on electrodes ipsilateral to stimulation decreased as FA increased."

"Voltage on electrodes contralateral to stimulation increased with higher FA values for the forceps major."

Discussion
"The reported results were derived from a methodologically integrated approach, with which we obtained a dynamic view of DMN connectivity."

"The TMS pulse is able to induce energy, detected via EEG, which dynamically diffuses within the main functional nodes of the network through its related structural pathways."

"The reported results support the notion that the proposed approach is feasible to use to investigate the relation between the structural connectivity and the neurophysiological responses of the DMN."

"The results showed an extended relation of every single node within networks since the stimulation of each of the cortical nodes of the DMN correlated with its underlying pathway and returned a dynamically integrated outcome."

"A significant advantage of this method highlighted by this study is that the correlation between the temporal dynamics and the underlying spatial pathways reverberates within the cortical nodes of the network."

Reward and Neuroscience

Machine generated keywords: reward, food, neutral, brain, representation, processing, procedure, signal, stimulation, intensity, region, female, behavioral, stimulus, participant

Two Drosophila Neuropeptide Y-like Neurons Define a Reward Module for Transforming Appetitive Odor Representations to Motivation [275]

This is a machine-generated summary of:

Pu, Yuhan; Zhang, Yiwen; Zhang, Yan; Shen, Ping: Two Drosophila Neuropeptide Y-like Neurons Define a Reward Module for Transforming Appetitive Odor Representations to Motivation [275]

Published in: Scientific Reports (2018)

Link to original: https://doi.org/10.1038/s41598-018-30113-5

If you want to cite the papers, please refer to the original.

For technical reasons we could not place the page where the original quote is coming from.

Abstract-Summary
"This innate appetitive response may require coordinated signaling activities of dopamine (DA) and neuropeptide F (NPF; a fly homolog of neuropeptide Y)."

"Laser lesioning of these NPF neurons abolished odor induction of appetitive arousal, while their genetic activation mimicked the behavioral effect of appetitive odors."

"The NPF neurons discriminatively responded to appetitive odor stimuli, and their odor responses were blocked by targeted lesioning of a pair of dopaminergic third-order olfactory neurons that appear to be presynaptic to the NPF neurons."

"The two NPF neurons contribute to appetitive odor induction of impulsive-like feeding by selectively decoding DA-encoded ascending olfactory inputs and relaying NPF-encoded descending motivational outputs for behavioral execution."

Introduction

"How such dopaminergic third-order olfactory neurons contribute to a reward circuit for selective recognition of appetitive odor representations by the larval brain remains incompletely understood."

"We have also found that Dop1R1 is likely expressed in NPF neurons, and this Dop1R1 activity may mediate a gating mechanism that tunes NPF signaling in response to appetitive odor-evoked DA inputs [217]."

"The odor responses of the two NPF neurons were blocked by targeted lesioning of a pair of dopaminergic third-order olfactory neurons, and these NPF neurons appear to be postsynaptically connected with such DA neurons in a brain region named the lateral horn."

"In combination, our findings suggest that the two NPF neurons contribute to appetitive odor induction of impulsive-like feeding by selectively converting DA-encoded ascending olfactory inputs to NPF-encoded descending motivational outputs for behavioral execution."

Results

"These results indicate that DL and DM pairs of NPF neurons have distinct functions, and the latter is essential for odor-aroused appetite."

"We found that fed larvae with lesioned DL-NPF neurons showed significantly increased feeding response, while those with lesioned DM-NPF neurons failed to do so."

"These results suggest that activation of the two DM-NPF neurons is also sufficient to induce appetitive arousal in fed larvae."

"These observations suggest that each DM-NPF neuron provides a link between food odor-responsive DL2 neurons and its target neurons in the SEZ, and Dop1R1 may tune the response of NPF neurons to odor-evoked DA signals that are not too strong or too weak [217]."

"In a reciprocal test, we found that the odor responses of DL2 neurons were similar regardless whether two DM-NPF neurons were lesioned or not."

"In combination, our results strongly suggest that DM-NPF neurons act downstream to the food odor-responsive DL2 neurons."

Discussion

"In each pathway, the DM-NPF neuron defines the highest-order circuit module for food odor processing."

"Through combined use of targeted laser microsurgery and dTrpA1-mediated neuronal activation, we also show that remote activation of two DM-NPF neurons in behaving fed larvae appears to be sufficient to mimic the appetizing effect of food odor stimulation."

"These findings point to the presence of a Dop1R1-mediated gating mechanism that tunes the NPF neuronal response to odor-evoked DA signals, and thereby selectively assigns appetitive significance to DA signals that are otherwise meaningless behaviorally."

"We have provided several lines of evidence suggesting that in each brain hemisphere, the DM-NPF neuron receives ascending DA signals from an assembly of four dopaminergic DL2 neurons in the lateral horn."

"When stimulated by a stream of appetitive odor vapor, the DL2 neurons responded more rapidly than the NPF neuron, and lesions in the DM-NPF neuron had no effect on the odor response of the DL2 neurons."

Materials and Methods

"Larvae were reared at 25 °C, and early third instars (~74 hr after egg laying, AEL) were fed before behavioral experiments as previously described [218]."

"Synchronized early third instars, fed on yeast paste, were stimulated for 5 minutes in a sealed 1.5L-chamber that was fumigated with an appetitive dose (5 or 7.5 μl) of pentyl acetate (PA; Sigma-Aldrich, 628-63-7)."

"UAS-dTrpA1 was expressed by allowing larvae to feed in pre-warmed yeast paste in a 31 °C incubator for defined periods, followed by rinsing with 31 °C water prior to feeding assays."

"The neural tissues of larvae were processed as previously described [218]."

"A Ca^{2+} indicator, GCaMP6s, was used for imaging odor responses by DA and NPF neurons."

"The method for making larval CNS preparation is the same as previously described for calcium imaging preparation [218]."

Modulatory Effects of Positive Mood and Approach Motivation on Reward Processing: Two Sides of the Same Coin? [276]

This is a machine-generated summary of:

Paul, Katharina; Pourtois, Gilles; Harmon-Jones, Eddie: Modulatory effects of positive mood and approach motivation on reward processing: Two sides of the same coin? [276]

Published in: Cognitive, Affective, & Behavioral Neuroscience (2020)

Link to original: https://doi.org/10.3758/s13415-019-00764-6

If you want to cite the papers, please refer to the original.

For technical reasons we could not place the page where the original quote is coming from.

Abstract-Summary

"Under positive mood, the Reward Positivity (RewP) component and fronto-medial theta activity (FMθ) in response to monetary feedback were both changed compared with neutral mood."

"To address this question, we combined a mindset manipulation with an imagery procedure to create and maintain three different affective states using a between-subjects design: a neutral mood, and positive mood with either high or low motivational intensity."

"FMθ activity results showed that irrespective of motivational intensity, positive compared with neutral mood altered reward expectancy."

"RewP was not affected by positive mood nor motivational intensity."

"These results suggest that positive mood, rather than motivational intensity, is likely driving the change in reward expectation during gambling, which could reflect the presence of an optimistic bias."

Methods

"In the low approach positive mood condition, participants were instructed to recall a situation that made them feel very good about themselves."

"A mixed model ANOVA with Mood (three groups: neutral, low approach positive, and high approach positive) as between-subjects factor and Time (4 assessments:

baseline and 3 manipulations) as within-subject factor was used, separately for each scale and the frontal alpha-asymmetry index."

"These accuracy data were analyzed using separate mixed-model ANOVAs with the within-subject factors feedback Valence (reward or no-reward) and feedback Expectancy (expected or unexpected), and the between-subjects factor Mood (neutral, low approach positive, or high approach positive)."

"The amplitude of the RewP and FMθ activity were analyzed using separate mied- model ANOVAs with the within-subject factors feedback Valence (reward or no-reward) and feedback Expectancy (expected or unexpected), and the between-subjects factor Mood (neutral, low approach positive, high approach positive)."

Results

"The ANOVA run on the data for the happiness subscale of the DEQ showed significant effects of Time, $F(2.6, 408) = 7.83$, $p < 0.001$, $\eta^2 = 0.043$, Mood, $F(2,158) = 7.54$, $p = 0.001$, $\eta^2 = 0.087$, and the interaction between them, $F(5.2,408) = 8.31$, $p < 0.001$, $\eta^2 = 0.091$."

"The data for the relaxation subscale showed significant main effects of Time, $F(2.6,407) = 4.86$, $p = 0.002$, $\eta^2 = 0.027$, and Mood, $F(2,158) = 9.09$, $p < 0.001$, $\eta^2 = 0.10$, as well as a significant interaction between them, $F(5.1,407) = 8.91$, $p < 0.001$, $\eta^2 = 0.10$."

"Participants in the high approach positive mood reported (trend-significant) higher levels of desire compared with the neutral group after the first and second mood inductions ($p = 0.032$ and 0.055), whereas there was no difference between the groups for any other time point ($ps \geq 0.28$)."

Discussion

"EEG results showed that under positive mood and irrespective of approach motivation intensity, FMθ activity did not differentiate if reward feedback was expected or not, while FMθ activity was clearly larger for unexpected compared with expected no-reward feedback."

"Because this effect was equally strong in both positive mood conditions (compared with a neutral mood control condition), and hence occurred irrespective of the approach motivation intensity, this result suggested that positive affect (but not approach motivation) likely changed reward expectation in a mood congruent way, thereby replicating and extending our previous results (Paul & Pourtois, [219])."

"When brought together with our previous EEG results (Paul & Pourtois, [219]), these new findings confirm that positive mood, but not approach motivation, was likely the underlying factor responsible for this specific change during reward processing."

[Section 4]

"Because positive affect and approach motivation often are confounded (in nature and in experiments), it was therefore unsettled based on our previous study (Paul & Pourtois, [219]) whether positive mood per se, or instead approach motivation, actually drove the modulation of reward processing seen at the EEG level."

"It seems plausible to assume that approach motivation rather than positive mood might have caused the change in the RewP, as well as FMθ, seen in our previous EEG study (Paul & Pourtois, [219])."

"If positive mood accounted for the changes in reward processing seen previously at the EEG level (Paul & Pourtois, [219]), then we surmised that a larger RewP and blunted FMθ activity for unexpected reward should be observed for both conditions with positive mood (regardless of motivational intensity) compared to neutral mood in the current study."

Intrinsic/Extrinsic Motivation

Machine generated keywords: curiosity, intrinsic motivation, intrinsic, reward, region, network, child, extrinsic, extrinsic motivation, memory, intrinsic extrinsic, picture, neural, patient, brain

Efficiency of Structural Connectivity Networks Relates to Intrinsic Motivation in Children Born Extremely Preterm [277]

This is a machine-generated summary of:

Pascoe, Leona; Thompson, Deanne; Spencer-Smith, Megan; Beare, Richard; Adamson, Chris; Lee, Katherine J.; Kelly, Claire; Georgiou-Karistianis, Nellie; Nosarti, Chiara; Josev, Elisha; Roberts, Gehan; Doyle, Lex W.; Seal, Marc L.; Anderson, Peter J.: Efficiency of structural connectivity networks relates to intrinsic motivation in children born extremely preterm [277]

Published in: Brain Imaging and Behavior (2018)

Link to original: https://doi.org/10.1007/s11682-018-9918-9

If you want to cite the papers, please refer to the original.

For technical reasons we could not place the page where the original quote is coming from.

Abstract-Summary

"Intrinsic motivation is essential for academic success and cognitive growth, but limited work has examined the neuroanatomical underpinnings of intrinsic motivation from a network perspective, particularly in early childhood."

"Using graph theoretical analysis, this study investigated global and local properties of structural connectivity networks in relation to intrinsic motivation within a vulnerable group of children at early school age."

"Global and node-specific network measures were analyzed in association with self-reported aspects of intrinsic motivation for school learning (Mastery, Challenge and Curiosity) using linear regression."

"These findings suggest that both integrated and segregated network communication support aspects of intrinsic motivation in childhood, and shed new light on structural network properties important for intrinsic motivation orientations in extremely preterm children at early school age."

Introduction

"The only research to date using graph theory to investigate the association between motivational processing and functional network connectivity was an adult study, which found that global efficiency (i.e., integration within the brain's network) increased during reward conditions (Kinnison and others [220])."

"While this study further implicated a role for the nucleus accumbens, caudate, and the ACC for reward processing (Kinnison and others [220]), to date no study has applied graph theory to investigate structural connectivity networks associated with intrinsic motivation in childhood."

"Given the recent call for attention to the neuroscience of intrinsic motivation (Di Domenico and Ryan [221]), we propose an initial neurobiological account of intrinsic motivation by examining whole-brain structural connectivity networks in relation to mastery, challenge and curiosity, and exploring a priori regions of interest that may be important for intrinsic motivational processes."

Materials and Methods

"Analyses were based on undirected, weighted connectivity graphs, where edges in the graph were weighted according to streamline counts within fiber tracts (i.e., number of streamlines per unit of volume of gray matter node) and normalized by streamline length and average node volume to reflect their relative strength within the network (Hagmann and others [222])."

"In order to compare network measures across individuals, a uniform density across individuals was achieved by applying a density threshold to networks where the top 30% of the maximum number of possible edges in the graph were retained, which has been successfully performed in previous studies examining structural connectivity in preterm children (Thompson and others [223])."

"We explored whole-brain as well as nodal structural connectivity using standard graph theory metrics of weighted-edge graphs (Beare and others [224]; Thompson and others [223])."

Results

"In terms of local efficiency, greater Curiosity was associated with reduced local efficiency ($f^2 = 0.06$), although the association weakened considerably in the secondary analysis, and was not evident for Mastery or Challenge."

"There was weak evidence that greater Mastery was associated with higher global efficiency, which strengthened in the secondary analysis ($f^2 = 0.30$), although a similar association was not evident for Challenge or Curiosity."

"There was evidence that greater Challenge was associated with higher small worldness ($f^2 = 0.10$), although this association weakened slightly in the secondary analyses, and was not evident for Mastery or Curiosity."

"There was evidence that increased connection strength of the right nucleus accumbens region was associated with greater Challenge, though this weakened slightly in the secondary analysis ($f^2 = 0.26$)."

Discussion

"We demonstrated that greater information transfer and small-world organization of whole-brain structural networks, alongside greater connectivity of key brain nodes, are associated with greater desire for mastery, challenge and curiosity in school learning at age 7 years."

"Our finding that greater mastery in school learning is associated with shorter path length and increased global efficiency, indicative of greater information transfer across the network, is consistent with our understanding that intrinsic motivation is a complex cognitive and affective state important for goal-directed behavior, and likely supported by multiple neural structures and processes (Di Domenico and Ryan 221)."

"This study was the first to investigate whole-brain structural connectivity networks in relation to intrinsic motivation, using cutting-edge network science and examining three key components of intrinsic motivation important for school learning in childhood."

Intrinsic Motivation in Patients with Parkinson's Disease: A Neuropsychological Investigation of Curiosity Using Dopamine Transporter Imaging [278]

This is a machine-generated summary of:

Shigemune, Yayoi; Kawasaki, Iori; Midorikawa, Akira; Baba, Toru; Takeda, Atsushi; Abe, Nobuhito: Intrinsic motivation in patients with Parkinson's disease: a neuropsychological investigation of curiosity using dopamine transporter imaging [278]

Published in: Neurological Sciences (2021)

Link to original: https://doi.org/10.1007/s10072-020-04968-4

If you want to cite the papers, please refer to the original.

For technical reasons we could not place the page where the original quote is coming from.

Abstract-Summary
"We investigated intrinsic motivation in PD patients using a task designed to assess the "Pandora effect," which constitutes a curiosity for resolving uncertainty, even if this curiosity is likely to result in negative consequences."

"The results revealed that PD patients, relative to the HCs, viewed the pictures less frequently under both the certain and uncertain conditions."

"Both the PD patients and HCs viewed the pictures at a higher frequency under the uncertain condition than under the certain condition."

"In the PD patients, the proportion of pictures viewed under the certain condition was positively correlated with the distribution of dopamine transporters in the striatum."

"These results suggest that despite the overall decreasing level of interest in viewing negative pictures, the motivation to resolve uncertainty is relatively intact in PD patients."

Introduction

"Previous neuroimaging studies suggested that intrinsic motivation, including curiosity, and extrinsic motivation are involved in the dopaminergic reward pathway [221, 225–227]."

"Despite accumulated evidence indicating that intrinsic and extrinsic motivation share a common neural basis, it remains unclear whether intrinsic motivation deteriorates in PD patients accompanied by the loss of striatal dopaminergic transmission."

"An earlier study indicated that participants were motivated to view even negative pictures, and more chose to view the negative pictures when the contents of the pictures were uncertain [228]."

"This task therefore enabled us to investigate curiosity characterized by an overall interest in viewing negative pictures and to further determine whether the motivation to resolve uncertainty affects motivated behavior."

"We hypothesized that PD patients, relative to healthy controls (HCs), would view negative pictures at a reduced rate and exhibit no effect of stimulus uncertainty owing to a decrease in the distribution of dopamine transporters in the striatum."

Materials and Methods

"To confirm whether curiosity was impaired in the PD patients, the proportions of pictures viewed were analyzed using two-way ANOVA with group (PD patients vs. HCs) and condition (certain vs. uncertain) as factors."

"To further confirm whether the differences identified by the t tests (i.e., TMT, BAS, anxiety, and depression) affected curiosity, we examined whether these measures worked as potential mediators between a factors of group (PD patients vs. HCs) as an independent variable and the proportion of pictures viewed as a dependent variable."

"To investigate the relationship between curiosity and dopamine neurotransmission in the striatum, correlation analysis between the proportion of pictures viewed and the distribution volume ratio in the right and left striata were performed for the PD patients."

Results

"The mean proportions of pictures viewed under certain and uncertain conditions were 42.8% (SD: 31.8) and 66.3% (SD: 31.4) for PD patients and 66.5% (SD: 32.8) and 80.2% (SD: 27.5) for HCs, respectively."

"These group differences might have contributed to the differences in the proportions of pictures viewed between PD patients and HCs."

"In the mediation analysis, the bias-corrected 95% confidence interval for the indirect effect size included 0 for all factors (− 5.1 to 7.4 for TMT score, − 9.5 to 2.9 for BAS score, − 3.1 to 7.9 for anxiety score, and − 9.8 to 2.1 for depression score), indicating TMT, BAS, anxiety, and depression scores had a non-significant indirect effect on the group difference in proportion of pictures viewed."

"The proportions of pictures viewed were collapsed across the certain and uncertain conditions because the PD patients viewed fewer pictures than did the HCs, regardless of the condition."

Discussion

"The PD patients viewed fewer pictures than did the HCs regardless of whether the pictures' contents were specified (i.e., certain condition) or not (i.e., uncertain condition), indicating that an overall interest in seeking information was diminished in the PD patients."

"Consistent with this interpretation, earlier studies of personality traits have suggested that novelty seeking is diminished in PD patients [229, 230]."

"One might assume that the lower proportion of negative pictures viewed by PD patients was caused by increased harm avoidance, a key feature of depression [231, 232]."

"The PD patients, who were associated with higher depression scores in the present study, might have tried to avoid viewing negative pictures due to increased harm avoidance, rather than decreased novelty seeking."

"Both the PD patients and HCs viewed more pictures under the uncertain condition than under the certain condition."

The Effect of Intrinsic and Extrinsic Motivation on Memory Formation: Insight from Behavioral and Imaging Study [279]

This is a machine-generated summary of:

Duan, Hongxia; Fernández, Guillén; van Dongen, Eelco; Kohn, Nils: The effect of intrinsic and extrinsic motivation on memory formation: insight from behavioral and imaging study [279]

Published in: Brain Structure and Function (2020)

Link to original: https://doi.org/10.1007/s00429-020-02074-x

If you want to cite the papers, please refer to the original.

For technical reasons we could not place the page where the original quote is coming from.

Abstract-Summary

"Effects of extrinsic and intrinsic motivation on memory formation have not been studied in combination and thus, it is unknown whether they interact and how such interplay is neurally implemented."

"Both extrinsic monetary reward and intrinsic curiosity enhanced memory performance, without evidence for an interaction."

"Curiosity might enhance memory performance by allocation of attentional resources and reward-related processes; while, monetary reward does so by suppression of task-irrelevant processing."

Introduction

"A common interpretation is that curiosity, as anticipation of rewarding information, enhances memory by sharing common neural mechanism with extrinsic motivation, but this proposal has not been tested directly."

"This undermining effect was partially supported by a behavioral study in which monetary reward improved memory performance but only to uninteresting questions and had no effect on interesting questions in a one-week delayed surprise memory test."

"Murayama and others [236] investigated the neural mechanism of this undermining effect in which performance-based monetary reward impaired intrinsic motivation as indicated by less amount of voluntary engagements in the free-choice task."

"We hypothesized that monetary reward would reduce interest/engagement for high curiosity questions and, therefore, impair memory performance with high rather than low curiosity ratings, which is in line with the undermining effect (Deci and others [237])."

Materials and Methods

"Participants underwent a three-stage paradigm with (1) a screening phase in which they classified trivia questions according to individual prior knowledge and curiosity, (2) a study phase in which they learned a series of 270 individually selected trivia answers while being scanned and (3) a test phase on the next day during which their memory for the studied trivia answers was tested."

"To manipulate intrinsic and extrinsic motivation individually, trivia information was binned during the study phase according to a full factorial design with two factors, each having three levels: Curiosity (Low/Middle/High) and Monetary reward (None/Low/High = 0/1/3 Euros) using data obtained in the screening phase from each particular participant."

"For each participant, motivation-related brain activation was conducted in the question presentation phase and modeled with two regressors (monetary reward and curiosity) whose magnitudes scaled linearly with curiosity bins and monetary reward levels separately."

Results

"The effect of reward on memory was also significant: the higher the potential monetary reward, the better the participants recalled the answer 1 day later (Reward (None/Low/High): 52/55/57% correct; Wald $\chi^2 = 11.656$; $p = 0.003$), which is also consistent with earlier findings (Adcock and others [238])."

"Interaction between monetary reward and curiosity on memory was not significant (Low Curiosity combined with three levels of reward (None/Low/High): 38/43/46% correct; Middle curiosity combined with three levels of reward (None/Low/High): 51/54/56% correct; High curiosity combined with three levels of reward (None/Low/High): 66/67/68% correct; Wald $\chi^2 = 2.79$, $p = 0.59$)."

"The effect of reward on prospective curiosity was also significant: the higher the potential monetary reward on the trivia question, the more the participants want to know more about the topic in the future (Reward (None/Low/High): −0.0136/0.0358/0.0937; Wald $\chi^2 = 6.816$; $p = 0.033$)."

Discussion

"We measured brain activity during memory encoding which was modulated by intrinsic (monetary reward) and extrinsic (curiosity) motivation during an intentional learning task."

"We did not find any interaction effect in behavioral performance, we nevertheless explored the specific activation pattern that pertains in the high reward combined with high curiosity condition, i.e., the facilitation effect of high extrinsic and intrinsic motivation on memory."

"Unlike the curiosity-modulated subsequent memory effect, the monetary reward-modulated subsequent memory effect was associated with decreased activity in posterior midline structures, i.e., greater activity was found during encoding of subsequently forgotten trivia items compared to subsequently remembered ones (negative subsequent memory effect) (Kim 239; Weis and others [240])."

"The direct comparison of the curiosity-modulated subsequent memory effect and the monetary reward-modulated subsequent memory effect showed that these fronto-parietal regions were more strongly activated in the context of curiosity-modulated memory formation."

Conclusions

"By operationalizing extrinsic motivation as potential monetary reward and intrinsic motivation as curiosity, the present study provides the first demonstration of the neural correlates of how intrinsic motivation in the context of extrinsic motivation influences memory."

"Our behavioral results replicate recent findings that extrinsic motivation and intrinsic motivation separately promote memory formation."

"Neuroimaging data showed different neural mechanisms underlying extrinsic and intrinsic motivation and involvement during memory formation."

"These findings suggest that both intrinsic and extrinsic motivation can benefit memory as tested here, although by different mechanisms and, thus, in an additive way."

Development of Goal-Directed Action Selection Guided by Intrinsic Motivations: An Experiment with Children [280]

This is a machine-generated summary of:

Taffoni, Fabrizio; Tamilia, Eleonora; Focaroli, Valentina; Formica, Domenico; Ricci, Luca; Di Pino, Giovanni; Baldassarre, Gianluca; Mirolli, Marco; Guglielmelli, Eugenio; Keller, Flavio: Development of goal-directed action selection guided by intrinsic motivations: an experiment with children [280]

Published in: Experimental Brain Research (2014)

Link to original: https://doi.org/10.1007/s00221-014-3907-z

If you want to cite the papers, please refer to the original.

For technical reasons we could not place the page where the original quote is coming from.

Abstract-Summary
"The spontaneous exploration of the world plays a fundamental role in the development of this capacity, providing subjects with an increasingly diverse set of opportunities to acquire, practice and refine the understanding of action–outcome connection."

"This work presents a new experimental paradigm to study the effect of intrinsic motivation on action–outcome relation learning and action selection during free exploration of the world."

"Three- and four-year-old children were observed during the free exploration of a new toy: half of them were allowed to develop the knowledge concerning its functioning; the other half were not allowed to learn anything."

"The knowledge acquired during the free exploration of the toy was subsequently assessed and compared."

Introduction
"In the computational models of competence-based intrinsic motivations, the agent is typically rewarded when its ability to accomplish a goal improves, independently from the origin of the goal (Chentanez and others [242]; Schembri and others [243]; Baranes and Oudeyer [244]; Santucci and others [245])."

"This work proposes a new experimental paradigm to study the effect of intrinsic motivations on action–outcome relation learning and action selection in absence of a reward or of a predefined task goal."

"The goal of this study is to verify if: (1) intrinsic motivations toward the board may be triggered by the novelty and by the surprising features of the produced stimuli; (2) motivations may be kept alive in absence of an external reward or goal, by experiencing of action–outcome contingency alone; (3) action–outcome contingency may promote the acquisition of action–outcome relations; (4) learning may be split into sub-components related to some features of the action (i.e. where the action should be performed; what should be done)."

Methods
"During the Test phase, the exploration behavior and the performance of each EXP and CTRL subject were assessed separately for each trial, adding the following indices to the ones described above: Spatial correctness index (SC): the difference between the number of correct button pushes (CP) and the number of wrong button

pushes (WP), divided by the NP; its values belong to the range [−1, 1], where −1 indicates the complete absence of spatial correctness in the subject's pushes (he/she never presses the right button), 0 indicates the subject performing an equal number of wrong and right pushes, and 1 indicates the subject understanding the button–box relation (she/he only presses the correct button); Time to reward (TtR): the time (s) used by the subject to retrieve the reward (the value is considered as 'NaN', i.e. not a number, if the reward was not retrieved); Trials to criterion (TtC): the number of WP before the first right push (RP)."

Results

"Of the Test phase, EXP subjects had not understood the spatial relations between boxes and buttons yet: in the first triplet, the difference between SC and zero only showed a small positive trend that was not statistically significant, meaning that the number of correct and wrong pushes was statistically the same."

"This index positively differed from zero starting from the second triplet in the EXP group, confirming that EXP subjects refined the spatial relation between buttons and boxes during the Test trials, and not during the Learning phase."

"In this group the index was always statistically equal to zero, demonstrating that CTRL subjects, who were not allowed to learn anything during the Learning phase, did not have enough time to acquire the spatial relation during the nine trials of the Test phase."

Discussion

"The present experiment consisted of allowing children to discover the relevant feedback stimuli of the mechatronic board, and to learn the action–outcome relations to recall a specific action when, at a later stage, the related outcome becomes desirable (i.e., it becomes an actively pursued goal)."

"The possibility to experience action–outcome contingencies during the Learning phase guided EXP children's exploration: our findings suggest that children adopt an exploration strategy that allows them to learn the effects of specific actions (action knowledge) and the spatial relations (spatial knowledge) separately."

"Our findings also suggest that action–outcomes contingency is not sufficient to acquire spatial relations when an external goal is missing: only by forcing a spatial focusing with the use of a reward (in our experiment a sticker), this knowledge has been acquired."

Conclusions

"The main results of this study can be summarized into four points: (1) the novelty and the surprising features of the mechatronic board, in particular the transient effects that actions can cause on it (e.g., lights and sounds), triggered exploratory behaviors toward the board; (2) the purely novel and surprising aspect of the stimuli is far more exciting than the fact that the contingencies may allow a causal mapping for younger children; (3) action contingencies, and in particular outcomes lasting long enough to allow the performance of new actions and explorations (e.g., box opening), produced an additional motivating effect allowing subjects to acquire new knowledge on the possible outcomes that their actions caused on the board and enhancing their capabilities to accomplish more effective, efficient, and desirable (extrinsic) goals; (4) actions are actually formed by sub-components (e.g., what, where, when, how) possibly relying on different cognitive/brain processes."

Bibliography

1. Carver, D. S., & Scheier, M. F. (1990). Origins and functions of positive and negative affect: A control-process view. Psychological Review, 97(1), 19–35.
2. Lewin K (1935) A dynamic theory of personality: selected papers. MacGraw-Hill, New York
3. Miller GA (1960) Plans and the structure of behavior. Holt, New York
4. White, R. W. (1959). Motivation reconsidered: The concept of competence. Psychological Review, 66, 297–333. https://doi.org/10.1037/h0040934 .
5. Higgins ET (2012) Beyond pleasure and pain: how motivation works. Oxford University Press, New York
6. Eitam B, Higgins ET (2010) Motivation in Mental accessibility: relevance of a representation (ROAR) as a new framework. Soc Personal Psychol Compass 4(10):951–967. https://doi.org/10.1111/j.1751-9004.2010.00309.x
7. Higgins ET (1997) Beyond pleasure and pain. Am Psychol 52(12):1280
8. Bromberg-Martin ES, Hikosaka O (2009) Midbrain dopamine neurons signal preference for advance information about upcoming rewards. Neuron 63(1):119
9. Leotti LA, Delgado MR (2011) The inherent reward of choice. Psychol Sci. https://doi.org/10.1177/0956797611417005
10. Redgrave P, Gurney K (2006) The short-latency dopamine signal: a role in discovering novel actions? Nat Rev Neurosci 7(12):967–975. https://doi.org/10.1938/Nrn2022
11. Redgrave P, Gurney K, Reynolds J (2008) What is reinforced by phasic dopamine signals? Brain Res Rev 58(2):322–339. https://doi.org/10.1016/J.Brainresrev.2007.10.007
12. Moore JW, Schneider SA, Schwingenschuh P, Moretto G, Bhatia KP, Haggard P (2010) Dopaminergic medication boosts action-effect binding in Parkinson's disease. Neuropsychologia 48(4):1125–1132. https://doi.org/10.1016/J.Neuropsychologia.2009.12.014
13. Delgado JMR: Neural constellations in aggressive behavior. Aggression and Violence: A Psychobiological and Clinical Approach. Edited by: Valzelli I, Morgese I. 1980, Milan: Edizione Saaint Vincent, 82-97.
14. Moyer KE: Kinds of aggression and their physiological basis. Comm Behav Biol. 1968, 2: 65-87.
15. Albert DJ, Walsh ML: Neural systems and the inhibitory modulation of agonistic behavior: a comparison of mammalian species. Neurosci Biobehav Rev. 1984, 8 (1): 5-24.
16. Ricci LA, Grimes JM, Melloni RH: Lasting changes in neuronal activation patterns in select forebrain regions of aggressive, adolescent anabolic/androgenic steroid-treated hamsters. Behav Brain Res. 2006, 176: 344-352.
17. Knyshevski I, Connor DF, Harrison RJ, Ricci LA, Melloni RHJ: Persistent activation of select forebrain regions in aggressive, adolescent cocaine-treated hamsters. Behav Brain Res. 2005, 159: 277-286.
18. Hollerman JR, Tremblay L, Schultz W (1998) Influence of reward expectation on behavior-related neuronal activity in primate striatum. J Neurophysiol 80:947–963
19. Hassani OK, Cromwell HC, Schultz W (2001) Influence of expectation of different rewards on behavior-related neuronal activity in the striatum. J Neurophysiol 85:2477–2489
20. Neufer, P. D. et al. Understanding the cellular and molecular mechanisms of physical activity-induced health benefits. Cell Metab. 22, 4–11 (2015).

21. Hawley, J. A., Hargreaves, M., Joyner, M. J. & Zierath, J. R. Integrative biology of exercise. Cell 159, 738–749 (2014).
22. Tellez, L. A. et al. A gut lipid messenger links excess dietary fat to dopamine deficiency. Science 341, 800–802 (2013).
23. Brass M, Haggard P (2008) The what, when, whether model of intentional action. Neuroscientist 14(4):319–325. https://doi.org/10.1177/1073858408317417
24. Wen W, Haggard P (2020) Prediction error and regularity detection underlie two dissociable mechanisms for computing the sense of agency. Cognition. https://doi.org/10.1016/j.cognition.2019.104074
25. Synofzik M, Vosgerau G, Newen A (2008) Beyond the comparator model: a multifactorial two-step account of agency. Conscious Cogn 17(1):219–239. https://doi.org/10.1016/j.concog.2007.03.010
26. Karsh N, Eitam B (2015a) Motivation from control: a response selection framework. The sense of agency. Oxford University Press, New York, 12. https://doi.org/10.1093/acprof:oso/9780190267278.003.0012
27. Moore J, Haggard P (2008) Awareness of action: Inference and prediction. Consci Cognit 17(1):136–144
28. Moore JW, Lagnado D, Deal DC, Haggard P (2009a) Feelings of control: contingency determines experience of action. Cognition 110(2):279–283. https://www.sciencedirect.com/science/article/pii/S0010027708002771?casa_token=PQqmKJMSD_8AAAAA:Lxh2F7APr3-6IKn_t6DB3QFp274UvI37BklVJ6ggrHMVlUh04onORztxuMJQRYHlkAf83BRjMe4
29. Sidarus N, Chambon V, Haggard P (2013) Priming of actions increases sense of control over unexpected outcomes. Consci Cognit 22(4):1403–1411
30. Karsh N, Eitam B (2015a) I control therefore I do: judgments of agency influence action selection. Cognition. https://doi.org/10.1016/j.cognition.2015.02.002
31. Karsh N, Eitam B, Mark I, Higgins ET (2016) Bootstrapping agency: how control-relevant information affects motivation. J Exp Psychol Gen 145(10):1333–1350. https://doi.org/10.1037/xge0000212
32. Eitam B, Kennedy PM, Higgins ET (2013) Motivation from control. Exp Brain Res. https://doi.org/10.1007/s00221-012-3370-7
33. Georgopoulos, A. P., Kalaska, J. F., Caminiti, R. & Massey, J. T. On the relations between the direction of two-dimensional arm movements and cell discharge in primate motor cortex. J. Neurosci. 2, 1527–1537 (1982).
34. Georgopoulos, A. P., Ashe, J., Smyrnis, N. & Taira, M. The motor cortex and the coding of force. Science 256, 1692–1695 (1992).
35. Li, C. S., Padoa-Schioppa, C. & Bizzi, E. Neuronal correlates of motor performance and motor learning in the primary motor cortex of monkeys adapting to an external force field. Neuron 30, 593–607 (2001).
36. Chhatbar, P. Y. & Francis, J. T. Towards a naturalistic brain-machine interface: hybrid torque and position control allows generalization to novel dynamics. PLoS ONE 8, e52286 (2013).
37. Roesch, M. R. & Olson, C. R. Neuronal activity related to reward value and motivation in primate frontal cortex. Science 304, 307–310 (2004).
38. Blaukopf CL, DiGirolamo DJ (2006) Differential effects of reward and punishment on conscious and unconscious eye movements. Exp Brain Res 174:786–792
39. Milstein DM, Dorris MC (2007) The influence of expected value on saccadic preparation. J Neurosci 27:4810–4818
40. Ross M, Lanyon LJ, Viswanathan J, Manoach DS, Barton JJ (2011) Human prosaccades and antisaccades under risk: effects of penalties and rewards on visual selection and the value of actions. Neuroscience 196:168–177
41. Jazbec S, Hardin MG, Schroth E, McClure E, Pine DS, Ernst M (2006) Age-related influence of contingencies on a saccade task. Exp Brain Res 174:754–762
42. McCelland, D. C. (1985). How motives, skills, and values determine what people do. American Psychologist, 40(7), 812–825. https://doi.org/10.1037/0003-066X.40.7.812 .
43. Maslow, A. H. (1943). A theory of human motivation. Psychological Review, 50(4), 370–96.

44. Beckmann J, Heckhausen H (2008) Motivation as a function of expectancy and incentive. In: Heckhausen J, Heckhausen H (eds) Motivation and action. Cambridge University Press, New York, pp 99–136
45. Schneider K, Schmalt H-D (2000) Motivation, 3, überarb. und erw. Aufl. Kohlhammer Standards Psychologie. Basisbuch. Kohlhammer, Stuttgart
46. Cerasoli, C. P., Nicklin, J. M., & Ford, M. T. (2014). Intrinsic motivation and extrinsic incentives jointly predict performance: A 40-year Meta-analysis. Psychological Bulletin. https://doi.org/10.1037/a0035661
47. McClelland DC (1957) Community development and the nature of human motivation: some implications of recent research. C, 57-35. Center for International Studies, Massachusetts Institute of Technology, Cambridge
48. Pincus, J. (2004). The consequences of unmet needs: the evolving role of motivation in consumer research. Journal of Consumer Behaviour: An International Research Review, 3(4), 375–387.
49. Berger, H. (1929). Über das elecktroenzephalogramm des menschen I. Archive Psychiatrie (Nervenkrankheiten), 87, 527–570.
50. Niedermeyer, E., & Lopes da Silva, F. (1999). Electroencephalography: Basic principles, clinical applications and related fields (fourth edition ed.). Baltimore: Lippincott Williams & Wilkins.
51. Knyazev, G. G., & Slobodskaya, H. R. (2003). Personality trait of behavioral inhibition is associated with oscillatory systems reciprocal relationships. International Journal Psychophysiology, 48, 247–261.
52. Knyazev, G. G., Slobodskaya, H. R., & Wilson, G. D. (2004). Personality and brain oscillations: Developmental aspects. In S. P. Shohov (Ed.), Advances in psychology research (Vol. 29, pp. 3–34). New York: Nova Science Publishers.
53. Robinson, D. L. (1982). Properties of the diffuse thalamocortical system and human personality: A direct test of Pavlovian/Eysenckian theory. Personality and Individual Differences, 3, 1–16.
54. Robinson, D. L. (1999). The technical, neurological and psychological significance of 'alpha', 'delta' and 'theta' waves confounded in EEG evoked potentials: A study of peak latencies. Clinical Neurophysiology, 110, 1427–1434.
55. Robinson, D. L. (2000). The technical, neurological and psychological significance of 'alpha', 'delta' and 'theta' waves confounded in EEG evoked potentials: A study of peak amplitudes. Personality and Individual Differences, 28, 673–693.
56. Robinson, D. L. (2001). How brain arousal systems determine different temperament types and the major dimensions of personality. Personality and Individual Differences, 31, 1233–1259.
57. Darvas, F., Miller, K. J., Rao, R. P. N., & Ojemann, J. G. (2009). Nonlinear phase–phase cross-frequency coupling mediates communication between distant sites in human neocortex. Journal of Neuroscience, 29, 426–435.
58. Knyazev, G. G. (2007). Motivation, emotion, and their inhibitory control mirrored in brain oscillations. Neuroscience and Biobehavioral Reviews, 31, 377–395.
59. Miskovic, V., & Schmidt, L. A. (2009). Frontal brain oscillatory coupling among men who vary in salivary testosterone levels. Neuroscience Letters, 464, 239–242.
60. Schutter, D. J. L. G., & van Honk, J. (2005). Salivary cortisol levels and the coupling of midfrontal delta-beta oscillations. International Journal of Psychophysiology, 55, 127–129.
61. Miskovic, V., Moscovitch, D. A., Senn, J., McCabe, R. E., Antony, M. M., & Schmidt, L. A. (2009). Neural correlates of cognitive behavioral therapy in social anxiety disorder: Evidence from frontal brain oscillatory coupling [Abstract]. Society for Neuroscience Abstracts.
62. Schutter, D. J. L. G., & van Honk, J. (2004). Decoupling of midfrontal delta-beta oscillations after testosterone administration. International Journal of Psychophysiology, 53, 71–73.
63. Basar, E. (2006). The theory of the whole-brain-work. International Journal of Psychophysiology, 60, 133–138.

64. Massetti T, Trevizan IL, Arab C, Favero FM, Ribeiro-Papa DC, de Mello Monteiro CB. Virtual reality in multiple sclerosis—A systematic review. Mult Scler Relat Disord. 2016;8:107–12. https://doi.org/10.1016/j.msard.2016.05.014 .
65. Calabro RS, Russo M, Naro A, De Luca R, Leo A, Tomasello P, et al. Robotic gait training in multiple sclerosis rehabilitation: Can virtual reality make the difference? Findings from a randomized controlled trial. J Neurol Sci. 2017;15(377):25–30. https://doi.org/10.1016/j.jns.2017.03.047 .
66. Peruzzi A, Cereatti A, Croce UD, Mirelman A. Effects of a virtual reality and treadmill training on gait of subjects with multiple sclerosis: a pilot study. Mult Scler Relat Disord. 2016;5:91–6. https://doi.org/10.1016/j.msard.2015.11.002 .
67. Yang Y-R, Tsai M-P, Chuang T-Y, Sung W-H, Wang R-Y. Virtual reality-based training improves community ambulation in individuals with stroke: a randomized controlled trial. Gait Posture. 2008;28(2):201–6. https://doi.org/10.1016/j.gaitpost.2007.11.007 .
68. Sveistrup H. Motor rehabilitation using virtual reality. J NeuroEngineering Rehabil. 2004;1(1):10. https://doi.org/10.1186/1743-0003-1-10 .
69. Massetti T, da Silva TD, Crocetta TB, Guarnieri R, de Freitas BL, Bianchi Lopes P, et al. The clinical utility of virtual reality in neurorehabilitation: a systematic review. J Cent Nerv Syst Dis. 2018;1(10):1179573518813541. https://doi.org/10.1177/1179573518813541 .
70. Cummings JJ, Bailenson JN. How immersive is enough? A meta-analysis of the effect of immersive technology on user presence. Media Psychol. 2016;19(2):272–309. https://doi.org/10.1080/15213269.2015.1015740
71. Slater M, Usoh M, Steed A. Depth of presence in virtual environments. Presence Teleoperators Virtual Environ. 1994;3(2):130–44. https://doi.org/10.1162/pres.1994.3.2.130
72. Sandars J, Cleary TJ. Self-regulation theory: applications to medical education: AMEE guide no. 58. Med Teach. 2011;33:875–86.
73. Pintrich, P. R. (2004). A conceptual framework for assessing motivation and self-regulated learning in college students. Educational Psychology Review, 16(4), 385–407.
74. Cleary TJ, Durning SJ, Artino AR. Microanalytic assessment of self-regulated learning during clinical reasoning tasks: recent developments and next steps. Acad Med. 2016;91:1516–21.
75. Pintrich, P. R. (2000). The role of goal orientation in self-regulated learning. In M. Boekaerts, M. Zeidner, & P. R. Pintrich (Eds.), Handbook of Self-Regulation (pp. 451–502). San Diego, CA: Academic Press.
76. Schunk DH, DiBenedetto MK. Motivation and social cognitive theory. Contemp Educ Psychol. 2020;60 December 2019:101832. https://doi.org/10.1016/j.cedpsych.2019.101832
77. Schraw G, Dennison RS. Assessing Metacognitive Awareness; 1994. p. 460–75.
78. Baranik LE, Barron KE, Finney SJ. Measuring goal orientation in a work domain: construct validity evidence for the 2×2 framework. Educ Psychol Meas. 2007;67:697–718.
79. Midgley, C., Kaplan, A., Middleton, M., Maehr, M. L., Urdan, T., Anderman, L. H. et al. (1998). The development and validation of scales assessing students' achievement goal orientations. Contemporary Educational Psychology, 23, 113–131.
80. Baranik LE, Stanley LJ, Bynum BH, Lance CE. Examining the construct validity of mastery-avoidance achievement goals: a meta-analysis. Hum Perform. 2010;23:265–82.
81. Finney SJ, Pieper SL, Barron KE. Examining the psychometric properties of the achievement goal questionnaire in a general academic context. Educ Psychol Meas. 2004;64:365–82.
82. Elliot, A. J., & Murayama, K. (2008). On the measurement of achievement goals: Critique, illustration, and application. Journal of Educational Psychology, 100(3), 613–628.
83. Oliveira-Silva LC. Escala de Metas de Realização no Trabalho: adaptação e evidências de validade. Rev Avaliação Psicológica. 2017;15:1–10.
84. Reimann, P., Markauskaite, L., & Bannert, M. (2014). E-research and learning theory: What do sequence and process mining methods contribute? British Journal of Educational Technology, 45, 528–540.

85. Noroozi, O., Alikhani, I., Järvelä, S., Kirschner, P. A., Juuso, I., & Seppänen, T. (2019). Multimodal data to design visual learning analytics for understanding regulation of learning. Computers in Human Behavior, 100, 298–304.
86. Harley, J. M., Bouchet, F., Hussain, M. S., Azevedo, R., & Calvo, R. (2015). A multi-componential analysis of emotions during complex learning with an intelligent multi-agent system. Computers in Human Behavior, 48, 615–625.
87. Pantic, M., & Rothkrantz, L. J. (2003). Toward an affect-sensitive multimodal human-computer interaction. Proceedings of the IEEE, 91, 1370–1390.
88. de Wit, S., & Dickinson, A. (2009). Associative theories of goal-directed behaviour: A case for animal–human translational models. Psychological Research, 73(4), 463–476. https://doi.org/10.1007/s00426-009-0230-6
89. Elsner B, Hommel B (2004) Contiguity and contingency in action-effect learning. Psychol Res 68(2–3):138–154. https://doi.org/10.1007/s00426-003-0151-8
90. Pavlov, I. P. (1932). The reply of a physiologist to psychologists. Psychological Review, 39(2), 91–127.
91. Rescorla, R. A. (1992). Response-outcome versus outcome-response associations in instrumental learning. Animal Learning & Behavior, 20(3), 223–232. https://doi.org/10.3758/BF03213376
92. Colagiuri, B., & Lovibond, P. F. (2015). How food cues scan enhance and inhibit motivation to obtain and consume food. Appetite, 84, 79–87. https://doi.org/10.1016/j.appet.2014.09.023
93. Garbusow, M., Schad, D. J., Sebold, M., Friedel, E., Bernhardt, N., Koch, S. P., ... Heinz, A. (2015). Pavlovian-to-instrumental transfer effects in the nucleus accumbens relate to relapse in alcohol dependence. Addiction Biology. https://doi.org/10.1111/adb.12243
94. Garofalo, S., & di Pellegrino, G. (2015). Individual differences in the influence of task-irrelevant Pavlovian cues on human behaviour. Frontiers in Behavioral Neuroscience, 9. https://doi.org/10.3389/fnbeh.2015.00163
95. Guitart-Masip, M., Fuentemilla, L., Bach, D. R., Huys, Q. J. M., Dayan, P., Dolan, R. J., & Duzel, E. (2011). Action dominates valence in anticipatory representations in the human striatum and dopaminergic midbrain. The Journal of Neuroscience: The Official Journal of the Society for Neuroscience, 31(21), 7867–7875. https://doi.org/10.1523/JNEUROSCI.6376-10.2011
96. Lovibond, P. F., & Colagiuri, B. (2013). Facilitation of voluntary goal-directed action by reward cues. Psychological Science, 24(10), 2030–2037. https://doi.org/10.1177/0956797613484043
97. Talmi D, Seymour B, Dayan P, Dolan RJ (2008) Human pavlovian-instrumental transfer. J Neurosci 28:360–368
98. Watson, P., Wiers, R. W., Hommel, B., Ridderinkhof, K. R., & de Wit, S. (2016). An associative account of how the obesogenic environment biases adolescents' food choices. Appetite, 96, 560–571. https://doi.org/10.1016/j.appet.2015.10.008
99. Hogarth, L. (2012). Goal-directed and transfer-cue-elicited drug-seeking are dissociated by pharmacotherapy: Evidence for independent additive controllers. Journal of Experimental Psychology. Animal Behavior Processes, 38(3), 266–278. https://doi.org/10.1037/a0028914
100. Hogarth, L., & Chase, H. W. (2011). Parallel goal-directed and habitual control of human drug-seeking: Implications for dependence vulnerability. Journal of Experimental Psychology: Animal Behavior Processes, 37(3), 261–276. https://doi.org/10.1037/a0022913
101. Watson, P., Wiers, R. W., Hommel, B., & de Wit, S. (2014). Working for food you don't desire: Cues interfere with goal-directed food-seeking. Appetite, 79C, 139–148. https://doi.org/10.1016/j.appet.2014.04.005
102. Boutelle, K. N., & Bouton, M. E. (2015). Implications of learning theory for developing programs to decrease overeating. Appetite, 93, 62–74. https://doi.org/10.1016/j.appet.2015.05.013
103. Verhoeven, A. A. C., Watson, P., & de Wit, S. (2018). Failing to pay heed to health warnings in a food-associated environment. Appetite, 120, 616–626. https://doi.org/10.1016/j.appet.2017.10.020

104. Seabrooke, T., Le Pelley, M. E., Hogarth, L., & Mitchell, C. J. (2017). Evidence of a goal-directed process in human Pavlovian-instrumental transfer. Journal of Experimental Psychology: Animal Learning and Cognition. https://doi.org/10.1037/xan0000147
105. Asratyan, E. A. (1974). Conditional reflex theory and motivational behaviour. Acta Neurobiologiae Experimentalis, 34(1), 15–31.
106. Gormezano, I., & Tait, R. W. (1976). The Pavlovian analysis of instrumental conditioning. The Pavlovian Journal of Biological Science, 11(1), 37–55.
107. Hommel B, Musseler J, Aschersleben G, Prinz W (2001) The theory of event coding (TEC): a framework for perception and action planning. Behav Brain Sci 24(5):849–937
108. James, W. (1890). The principles of psychology. New York, NY: Dover Publications.
109. Biswas, D., Szocs, C., Chacko, R., & Wansink, B. (2017). Shining light on atmospherics: How ambient light influences food choices. Journal of Marketing Research, 54(1), 111–123. https://doi.org/10.1509/jmr.14.0115
110. Kasof, J. (2002). Indoor lighting preferences and bulimic behavior: An individual differences approach. Personality and Individual Differences, 32(3), 383–400. https://doi.org/10.1016/S0191-8869(01)00023-X
111. Scheibehenne, B., Todd, P. M., & Wansink, B. (2010). Dining in the dark: The importance of visual cues for food consumption and satiety. Appetite, 55(3), 710–713. https://doi.org/10.1016/j.appet.2010.08.002
112. Wansink, B., & van Ittersum, K. (2012). Fast food restaurant lighting and music can reduce calorie intake and increase satisfaction. Psychology Reports: Human Resources & Marketing, 111(1), 228–232. https://doi.org/10.2466/01.PR0.111.4.228-232
113. Norris, C. J., Gollan, J., Berntson, G. G., & Cacioppo, J. T. (2010). The current status of research on the structure of evaluative space. Biological Psychology, 84(3), 422–436. https://doi.org/10.1016/j.biopsycho.2010.03.011
114. Baumeister, R. F. (2016). Toward a general theory of motivation: Problems, challenges, opportunities, and the big picture. Motivation and Emotion, 40, 1–10.
115. Kamp, E., & MacInnis, D. J. (1995). Characteristics of portrayed emotions in commercials: When does what is shown in ads affect viewers? Journal of Advertising Research, 35(6), 19–28.
116. Swainston, J., & Derakshan, N. (2018). Training cognitive control to reduce emotional vulnerability in breast cancer. Psycho-Oncology, 27(7), 1780–1786.
117. Beloe, P., & Derakshan, N. (2019). Adaptive working memory training can reduce anxiety and depression vulnerability in adolescents. Developmental Science, e12831.
118. Green, C. S., Bavelier, D., Kramer, A. F., Vinogradov, S., Ansorge, U., Ball, K. K., ... & Facoetti, A. (2019). Improving methodological standards in behavioral interventions for cognitive enhancement. Journal of Cognitive Enhancement, 3(1), 2–29.
119. Lane, J., Lane, A. M., & Kyprianou, A. (2004). Self-efficacy, self-esteem and their impact on academic performance. Social Behavior and Personality: an international journal, 32(3), 247–256.
120. Sangal RB, Thomas L, Mitler MM (1992) Maintenance of wakefulness test and multiple sleep latency test. Measurement of different abilities in patients with sleep disorders [see comments]. Chest 101(4):898–902
121. Harma M, Suvanto S, Popkin S, Pulli K, Mulder M, Hirvonen K (1998) A dose–response study of total sleep time and the ability to maintain wakefulness. J Sleep Res 7(3):167–174
122. Doghramji K, Mitler MM, Sangal RB, Shapiro C, Taylor S, Walsleben J, Belisle C, Erman MK, Hayduk R, Hosn R, O'Malley EB, Sangal JM, Schutte SL, Youakim JM (1997) A normative study of the maintenance of wakefulness test (MWT). Electroencephalogr Clin Neurophysiol 103(5):554–562
123. Banks S, Barnes M, Tarquinio N, Pierce RJ, Lack LC, McEvoy RD (2004) The Maintenance of Wakefulness Test in normal healthy subjects. Sleep 27:799–802

124. Littner MR, Kushida C, Wise M, Davila DG, Morgenthaler T, Lee-Chiong T, Hirshkowitz M, Daniel LL, Bailey D, Berry RB, Kapen S, Kramer M (2005) Standards of Practice Committee of the American Academy of Sleep Medicine. Practice parameters for clinical use of the multiple sleep latency test and the maintenance of wakefulness test. Sleep 28:113–121
125. London ED, Ernst M, Grant S, Bonson K, Weinstein A (2000) Orbitofrontal cortex and human drug abuse: functional imaging. Cereb Cortex 10:334–342
126. Davidson RJ (1992) Anterior cerebral asymmetry and the nature of emotion. Brain Cogn 20:125–251
127. Watson D, Tellegen A (1985) Toward a consensual structure of mood. Psychol Bull 98:219–235
128. Davidson RJ, Ekman P, Saron CD, Senulis JA, Friesen WV (1990) Approach-withdrawal and cerebral asymmetry: emotional expression and brain physiology. I. J Pers Soc Psychol 58:330–341
129. Fox NA (1991) If it's not left, it's right. Electroencephalograph asymmetry and the development of emotion. Am Psychol 46:863–872
130. Fox NA, Rubin KH, Calkins SD, Marshall TR, Coplan RJ, Porges SW, Long JM, Stewart S (1995) Frontal activation asymmetry and social competence at four years of age. Child Dev 66:1770–1784
131. Davidson RJ (2000) Cognitive neuroscience needs affective neuroscience (and vice versa). Brain Cogn 42:89–92
132. Bradley MM, Lang PJ (1994) Measuring emotion: the Self-Assessment Manikin and the Semantic Differential. J Behav Ther Exp Psychiatry 25:49–59
133. Silberman EK, Weingartner H (1986) Hemispheric lateralization of functions related to emotion. Brain Cogn 5:322–353
134. Szustrowa T, Jaworowska A (2003) Test Matryc Ravena w Wersji Kolorowej [Raven's Colored Progressive Matrices]. Pracownia Testów Psychologicznych PTP, Warsaw, Poland
135. Chan RCK, Gou M, Zou X, Li D, Hu Z, Yang B (2006) Multitasking performance of Chinese children with ADHD. J Int Neuropsychol Soc 12(4):575–579. https://doi.org/10.1017/S1355617706060693
136. Siklos S, Kerns KA (2004) Assessing multitasking in children with ADHD using a modified six elements test. Arch Clin Neuropsychol 19(3):347–361. https://doi.org/10.1016/S0887-6177(03)00071-4
137. Gawrilow C, Merkt J, Goossens-Merkt H, Bodenburg S, Wendt M (2011) Multitasking in adults with ADHD. ADHD Atten Deficit Hyperact Disord 3(3):253–264. https://doi.org/10.1007/s12402-011-0056-0
138. Reijnen E, Opwis K (2008) Visual search in children with ADHD: the influence of feedback on selective attention. J Vis 8(6):774–774
139. Maclean N, Pound P, Wolfe C, Rudd A: A critical review of the concept of patient motivation in the literature on physical rehabilitation. Soc Sci Med 2000, 50: 495-506.
140. Gajadhar BJ, Nap HH, De Kort YAW, Ijsselsteijn WA: Out of sight, out of mind: co-player effects on seniors' player experience. In Proceedings of the 3rd International Conference on Fun and Games. Leuven, Belgium: ACM Press; 2010:74-83.
141. Alankus G, Lazar A, May M, Kelleher C: Towards customizable games for stroke rehabilitation. In Proceedings of the 2010 International Conference on Human Factors in Computing systems - CHI '10. Atlanta, USA: ACM Press; 2010:2113-2122.
142. Terzian H, Cecotto C (1959) Determination and study of hemisphere dominance by means of intracarotid sodium amytal injection in man: II. Electroencephalographic effects. Bolletino della Societa Ztaliana Sperimentale 35:1626–1630
143. Ahern GL, Schwartz GE (1985) Differential lateralization for positive and negative emotion in the human brain: EEG spectral analysis. Neuropsychologia 23:745–755
144. Tomarken AJ, Davidson RJ, Wheeler RE, Doss RC (1992) Individual differences in anterior brain asymmetry and fundamental dimensions of emotion. J Personal Soc Psychol 62:676–687

145. Buss KA, Malmstadt Schumacher JR, Dolski I, Kalin NH, Goldsmith HH, Davidson RJ (2003) Right frontal brain activity, cortisol, and withdrawal behavior in 6-month-old infants. Behav Neurosci 117:11–20
146. Davidson RJ, Ekman P, Saron CD, Senulis JA, Friesen WV (1990b) Approach-withdrawal and cerebral asymmetry: emotional expression and brain physiology I. J Personal Soc Psychol 58:330–341
147. Silva JR, Pizzagalli DA, Larson CL, Jackson DC, Davidson RJ (2002) Frontal brain asymmetry in restrained eaters. J Abnorm Psychol 111:676–681
148. Tomarken AJ, Davidson RJ, Henriques JB (1990) Resting frontal brain asymmetry predicts affective responses to films. J Personal Soc Psychol 59:791–801
149. Wheeler RE, Davidson RJ, Tomarken AJ (1993) Frontal brain asymmetry and emotional reactivity: a biological substrate of affective style. Psychophysiology 30:82–89
150. Schutter DJLG, de Weijer AD, Meuwese JDI, Morgan B, van Honk J (2008) Interrelations between motivational stance, cortical excitability, and the frontal electroencephalogram asymmetry of emotion: a transcranial magnetic stimulation study. Hum Brain Mapp 29:574–580
151. Harmon-Jones, E. (2004). On the relationship of anterior brain activity and anger: Examining the role of attitude toward anger. Cognition and Emotion, 18, 337-361.
152. Harmon-Jones, E. (2007). Trait anger predicts relative left frontal cortical activation to anger-inducing stimuli. International Journal of Psychophysiology, 66, 154-160.
153. Harmon-Jones, E., Harmon-Jones, C., Fearn, M., Sigelman, J. D., & Johnson, P. (2008). Action orientation, relative left frontal cortical activation, and spreading of alternatives: A test of the action-based model of dissonance. Journal of Personality and Social Psychology, 94, 1-15.
154. Harmon-Jones, E., & Sigelman, J. (2001). State anger and prefrontal brain activity: Evidence that insult-related relative left prefrontal activation is associated with experienced anger and aggression. Journal of Personality and Social Psychology, 80, 797-803.
155. Harmon-Jones E, Vaughn-Scott K, Mohr S, Sigelman J, Harmon-Jones C (2004) The effect of manipulated sympathy and anger on left and right frontal cortical activity. Emotion 4:1–7
156. Gable PA, Harmon-Jones E (2010) Late positive potential to appetitive stimuli and local attentional bias. Emotion 10:441–446
157. Gable PA, Poole BD (2012a) Influence of trait behavioral inhibition and behavioral approach motivation systems on the LPP and frontal asymmetry to anger pictures. Soc Cogn Affect Neurosci. https://doi.org/10.1093/scan/nss130
158. Gable, P. A., & Harmon-Jones, E. (2008). Approach-motivated positive affect reduces breadth of attention. Psychological Science, 19, 476–482.
159. Gable, P. A., & Harmon-Jones, E. (2008b). Relative left frontal activation to appetitive stimuli: Considering the role of individual differences. Psychophysiology, 45, 275-278.
160. Gable PA, Harmon-Jones E (2011) Attentional states influence early neural responses associated with motivational processes: local vs. global attentional scope and N1 amplitude to appetitive stimuli. Biol Psychol 87:303–305
161. Harmon-Jones E, Gable PA (2009) Neural activity underlying the effect of approach-motivated positive affect on narrowed attention. Psychol Sci 20:406–409
162. Gable PA, Poole BD (2012b) Time flies when you're having approach-motivated fun: effects on motivational intensity on time perception. Psychol Sci 23:879–886
163. Coan JA, Allen JJB (2004) Frontal EEG asymmetry as a moderator and mediator of emotion. Biol Psychol 67:7–49
164. Harmon-Jones, E., Lueck, L., Fearn, M., & Harmon-Jones, C. (2006). The effect of personal relevance and approach-related action expectation on relative left frontal cortical activity. Psychological Science, 17, 434-440.
165. Hewig J, Hagemann D, Seifert J, Naumann E, Bartussek D (2004) On the selective relation of frontal cortical asymmetry and anger-out versus anger-control. J Personal Soc Psychol 87:926–939

166. Jackson DC, Mueller CJ, Dolski I, Dalton KM, Nitschke JB, Urry HL, Rosenkranz MA, Ryff CD, Singer BH, Davidson RJ (2003) Now you feel it, now you don't: frontal brain electrical asymmetry and individual differences in emotion regulation. Psychol Sci 14:612–617
167. Elgavish E, Halpern D, Dikman Z, Allen JJB (2003) Does frontal EEG asymmetry moderate or mediate responses to the international affective picture system (IAPS)? Psychophysiology 40:s38
168. Hagemann D, Ewald N, Becker G, Maier S, Bartussek D (1998) Frontal brain asymmetry and affective style: a conceptual replication. Psychophysiology 35:372–388
169. Harmon-Jones, E., & Allen, J. J. B. (1997). Behavioral activation sensitivity and resting frontal EEG asymmetry: Covariation of putative indicators related to risk for mood disorders. Journal of Abnormal Psychology, 106, 159-163.
170. Harmon-Jones, E., & Allen, J. J. (1998). Anger and frontal brain activity: EEG asymmetry consistent with approach motivation despite negative affective valence. Journal of Personality and Social Psychology, 74, 1310–1315.
171. Sutton, S. K., & Davidson, R. J. (1997). Prefrontal brain asymmetry: A biological substrate of the behavioral approach and inhibition systems. Psychological Science, 8, 204-210.
172. Harmon-Jones E, Gable PA, Peterson CK (2010) The role of asymmetric frontal cortical activity in emotion-related phenomena: a review and update. Biol Psychol 84:451–462
173. Matsumoto, M. & Hikosaka, O. Representation of negative motivational value in the primate lateral habenula. Nature Neurosci. 12, 77–84 (2009)
174. Preti MG, Bolton TA, Van De Ville D (2017) The dynamic functional connectome: state-of-the-art and perspectives. Neuroimage 160:41–54. https://doi.org/10.1016/J.NEUROIMAGE.2016.12.061
175. Thompson GJ (2018) Neural and metabolic basis of dynamic resting state fMRI. Neuroimage 180:448–462. https://doi.org/10.1016/J.NEUROIMAGE.2017.09.010
176. Kramer G, Walker B, Bonebright T, Cook P, Flowers JH, Miner N, Neuhoff J (2010) Sonification report: status of the field and research agenda
177. Hermann T (2008) Taxonomy and definitions for sonification and auditory display. In: Proceedings of the 14th international conference on auditory display (ICAD 2008)
178. Miranda ER, Castet J (eds) (2014) Guide to brain-computer music interfacing. Springer
179. Adrian ED, Matthews BH (1934) The Berger rhythm: potential changes from the occipital lobes in man. Brain 57(4):355–385
180. Glen J (2010) Use of audio signals derived from electroencephalographic recordings as a novel 'depth of anaesthesia'monitor. Med Hypotheses 75(6):547–549
181. Olivan J, Kemp B, Roessen M (2004) Easy listening to sleep recordings: tools and examples. Sleep Med 5(6):601–603
182. Khamis H, Mohamed A, Simpson S, McEwan A (2012) Detection of temporal lobe seizures and identification of lateralisation from audified EEG. Clin Neurophysiol 123(9):1714–1720
183. Hinterberger T, Hill J, Birbaumer N (2013) An auditory brain-computer communication device. In: proceedings IEEE BIOCAS'04, 2004. The 19th international conference on auditory display (ICAD-2013), Lodz, Poland
184. McCreadie KA, Coyle DH, Prasad G (2013) Sensorimotor learning with stereo auditory feedback for a brain–computer interface. Med Biol Eng Compu 51(3):285–293
185. Arslan B, Brouse A, Castet J, Filatriau JJ, Lehembre R, Noirhomme Q, Simon C (2005) Biologically-driven musical instrument. In: Proceedings of the summer workshop on multimodal interfaces (eNTERFACE'05). Faculté Polytechnique de Mons, Mons, BL
186. Miranda ER, Brouse A (2005) Interfacing the brain directly with musical systems: on developing systems for making music with brain signals. Leonardo 38(4):331–336
187. Miranda ER, Magee WL, Wilson JJ, Eaton J, Palaniappan R (2011) Brain-computer music interfacing (BCMI): from basic research to the real world of special needs. Music Med 3(3):134–140
188. Sammler D, Grigutsch M, Fritz T, Koelsch S (2007) Music and emotion: electrophysiological correlates of the processing of pleasant and unpleasant music. Psychophysiology 44(2):293–304

189. Koelsch S, Fritz T, Müller K, Friederici AD (2006) Investigating emotion with music: an fMRI study. Hum Brain Mapp 27(3):239–250
190. Lin YP, Wang CH, Jung TP, Wu TL, Jeng SK, Duann JR, Chen JH (2010) EEG-based emotion recognition in music listening. IEEE Trans Biomed Eng 57(7):1798–1806
191. Lu J, Wu D, Yang H, Luo C, Li C, Yao D (2012) Scale-free brain-wave music from simultaneously EEG and fMRI recordings. PLoS ONE 7(11):e49773
192. Wu D, Li CY, Yao DZ (2009) Scale-free music of the brain. PLoS ONE 4(6):e5915
193. Sengupta R, Dey N, Datta AK, Ghosh D (2005) Assessment of musical quality of tanpura by fractal-dimensional analysis. Fractals 13(03):245–252
194. Ghosh M (2002) Natyashastra (ascribed to Bharata Muni). Chowkhamba Sanskrit Series Office, Varanasi
195. Balkwill LL, Thompson WF (1999) A cross-cultural investigation of the perception of emotion in music: psychophysical and cultural cues. Music Percept Interdiscip J 17(1):43–64
196. Chordia P, Rae A (2007) Understanding emotion in raag: an empirical study of listener responses. In: International symposium on computer music modeling and retrieval. Springer, Berlin, pp 110–124
197. Wieczorkowska AA, Datta AK, Sengupta R, Dey N, Mukherjee B (2010) On search for emotion in Hindusthani vocal music. In: Raś ZW, Wieczorkowska AA (eds) Advances in music information retrieval. Springer, Berlin, pp 285–304
198. Mathur A, Vijayakumar SH, Chakrabarti B, Singh NC (2015) Emotional responses to Hindustani raga music: the role of musical structure. Front Psychol 6:513
199. Hardstone R, Poil SS, Schiavone G, Jansen R, Nikulin VV, Mansvelder HD, Linkenkaer-Hansen K (2012) Detrended fluctuation analysis: a scale-free view on neuronal oscillations. Front Physiol 3:450
200. Kantelhardt JW, Zschiegner SA, Koscielny-Bunde E, Havlin S, Bunde A, Stanley HE (2002) Multifractal detrended fluctuation analysis of nonstationary time series. Physica A 316(1):87–114
201. Zhou WX (2008) Multifractal detrended cross-correlation analysis for two non-stationary signals. Phys Rev E 77(6):066211
202. Maity AK, Pratihar R, Mitra A, Dey S, Agrawal V, Sanyal S, Banerjee A, Ghosh D, Sengupta R (2015) Multifractal detrended fluctuation analysis of alpha and theta EEG rhythms with musical stimuli. Chaos Solitons Fract 81:52–67
203. Babiloni C, Buffo P, Vecchio F, Marzano N, Del Percio C, Spada D, Rossi S, Bruni I, Rossini PM, Perani D (2012) Brains "in concert": frontal oscillatory alpha rhythms and empathy in professional musicians. Neuroimage 60(1):105–116
204. Bates, D., Maechler, M., Bolker, B., & Walker, S. (2015). Fitting linear mixed-effects models using lme4. Journal of Statistical Software, 67, 1–48.
205. Bartolomei F, Guye M, Wendling F (2013) Abnormal binding and disruption in large scale networks involved in human partial seizures. EPJ Nonlinear Biomed Phys 1(4):1–16. https://doi.org/10.1140/epjnbp11
206. Liuzza MT, Cazzato V, Vecchione M, Crostella F, Caprara GV, Aglioti SM (2011). Follow my eyes: the gaze of politicians reflexively captures the gaze of ingroup voters. PLoS One, 6(9), e25117
207. Wu EXW, Laeng B, Magnussen S (2012) Through the eyes of the own-race bias: eye-tracking and pupillometry during face recognition. Social neuroscience, 7(2), 202–216. Chicago
208. Ciardo F, Marino BF, Rossetti A, Actis-Grosso R, Ricciardelli P (2013). Face age and social status exert different modulatory effects on gaze following behaviour. CogSci
209. Moradi Z, Duta M, Hewstone M, Yankouskaya A, Enock F, Humphreys GW (2017). The rival doesn't catch my eyes: In-group relevance modulates inhibitory control over anti-saccades. Visual Cognit 1–15
210. Manohar SG, Chong TTJ, Apps MA, Batla A, Stamelou M, Jarman PR, … Husain M (2015) Reward pays the cost of noise reduction in motor and cognitive control. Curr Biol 25(13):1707–1716

211. Momi D, Ozdemir RA, Tadayon E, Boucher P, Shafi MM, Pascual-Leone A, Santarnecchi E (2021) Network-level macroscale structural connectivity predicts propagation of transcranial magnetic stimulation. Neuroimage 229:117698
212. Greicius MD, Supekar K, Menon V, Dougherty RF (2009) Resting-state functional connectivity reflects structural connectivity in the default mode network. Cereb Cortex 19(1):72–78. https://doi.org/10.1093/cercor/bhn059
213. Raichle ME (2015) (2015) The Brain' s default mode network. Annu Rev Neurosci 38(1):433–447. https://doi.org/10.1146/annurev-neuro-071013-014030
214. Voineskos AN, Farzan F, Barr MS, Lobaugh NJ, Mulsant BH, Chen R, Fitzgerald PB, Daskalakis ZJ (2010) The role of the corpus callosum in transcranial magnetic stimulation induced interhemispheric signal propagation. Biol Psychiat 68(9):825–831
215. Noble S, Scheinost D, Constable RT (2019) A decade of test-retest reliability of functional connectivity: a systematic review and meta-analysis. Neuroimage 203:116157
216. Rohwedder, A., Selcho, M., Chassot, B. & Thum, A. S. Neuropeptide F neurons modulate sugar reward during associative olfactory learning of Drosophila larvae. J. Comp. Neurol. 523, 2637–2664 (2015).
217. Pu, Y., Palombo, M. M. M. & Shen, P. Contribution of DA Signaling to Appetitive Odor Perception in a Drosophila Model. Sci Rep 8, 5469 (2018).
218. Wang, Y., Pu, Y. & Shen, P. Neuropeptide-gated perception of appetitive olfactory inputs in Drosophila larvae. Cell Rep 3, 820–830 (2013).
219. Paul, K., & Pourtois, G. (2017). Mood congruent tuning of reward expectation in positive mood: evidence from FRN and theta modulations. Social Cognitive and Affective Neuroscience, 12(5), 765–774. https://doi.org/10.1093/scan/nsx010
220. Kinnison, J., Padmala, S., Choi, J. M., & Pessoa, L. (2012). Network analysis reveals increased integration during emotional and motivational processing. The Journal of Neuroscience, 32(24), 8361–8372. https://doi.org/10.1523/JNEUROSCI.0821-12.2012 .
221. Di Domenico, S. I., & Ryan, R. M. (2017). The emerging neuroscience of intrinsic motivation: A new frontier in self-determination research. Frontiers in Human Neuroscience,11, 145.
222. Hagmann, P., Commoun, L., Gigandet, X., Meuli, R., Honey, C. J., Wedeen, V. J., & Sporns, O. (2008). Mapping the structural core of human cerebal cortex. PLoS Biology, 6(7), 1479–1493. https://doi.org/10.1371/journal.pbio.0060159 .
223. Thompson, D. K., Chen, J., Beare, R., Adamson, C. L., Ellis, R., Ahmadzai, Z. M., et al. (2016). Structural connectivity relates to perinatal factors and functional impairment at 7years in children born very preterm. Neuroimage, 134, 328–337. https://doi.org/10.1016/j.neuroimage.2016.03.070 .
224. Beare, R., Adamson, C., Bellgrove, M. A., Vilgis, V., Vance, A., Seal, M. L., & Silk, T. J. (2017). Altered structural connectivity in ADHD: A network based analysis. Brain Imaging and Behavior, 11(3), 846–858. https://doi.org/10.1007/s11682-016-9559-9 .
225. Sakaki M, Yagi A, Murayama K (2018) Curiosity in old age: a possible key to achieving adaptive aging. Neurosci Biobehav Rev 88:106–116
226. Kidd C, Hayden BY (2015) The psychology and neuroscience of curiosity. Neuron 88:449–460
227. Gruber MJ, Gelman BD, Ranganath C (2014) States of curiosity modulate hippocampus-dependent learning via the dopaminergic circuit. Neuron 84:486–496
228. Hsee CK, Ruan B (2016) The Pandora effect: the power and peril of curiosity. Psychol Sci 27:659–666
229. Santangelo G, Garramone F, Baiano C, D'Iorio A, Piscopo F, Raimo S, Vitale C (2018) Parkinsonism and related disorders personality and Parkinson's disease: a meta-analysis. Parkinsonism Relat Disord 49:67–74
230. Santangelo G, Piscopo F, Barone P, Vitale C (2017) Personality in Parkinson's disease: clinical, behavioural and cognitive correlates. J Neurol Sci 374:17–25
231. Nemes B, Cozman D (2016) The relevance of personality assessment in estimating the risk of onset and the outcome of major depressive disorder. Clujul Med 89:212–215

232. Kampman O, Poutanen O (2011) Can onset and recovery in depression be predicted by temperament? A systematic review and meta-analysis. J Affect Disord 135:20–27
233. Benware CA, Deci EL (1984) Quality of learning with an active versus passive motivational set. Am Educ Res J 21:755–765
234. Kuhbandner C, Aslan A, Emmerdinger K, Murayama K (2016) Providing extrinsic reward for test performance undermines long term memory acquisition. Front Psychol 7. 79. http:// doi.org/ https://doi.org/10.3389/fpsyg.2016.00079
235. Vansteenkiste, M., Simons, J., Lens, W., Sheldon, K. M., & Deci, E. L. (2004). Motivating learning, performance, and persistence: The synergistic effects of intrinsic goal contents and autonomy-supportive contexts. Journal of Personality and Social Psychology, 87, 246–260.
236. Murayama K, Matsumoto M, Izuma K, Matsumoto K (2010) Neural basis of the undermining effect of monetary reward on intrinsic motivation. Proc Natl Acad Sci USA 107:20911–20916. https://doi.org/10.1073/pnas
237. Deci, E., Koestner, R., & Ryan, R. (1999). A meta-analytic review of experiments examining the effects of extrinsic rewards on intrinsic motivation. Psychological Bulletin, 125, 627–668. https://doi.org/10.1037/0033-2909.125.6.627 .
238. Adcock RA, Thangavel A, Whitfield-Gabrieli S, Knutson B, Gabrieli JD (2006) Reward motivated learning: mesolimbic activation precedes memory formation. Neuron 50:507–517. https://doi.org/10.1016/j.neuron.2006.03.036
239. Kim H (2011) Neural activity that predicts subsequent memory and forgetting: a meta-analysis of 74 fMRI studies. Neuroimage 54:2446–2461. https://doi.org/10.1016/j.neuroimage.2010.09.045
240. Weis S, Klaver P, Reul J, Elger CE, Fernández G (2004) Temporal and cerebellar brain regions that support both declarative memory formation and retrieval. Cereb Cortex 14:256–267
241. Ryan, R. M., & Deci, E. L. (2000). Intrinsic and extrinsic motivations: Classic definitions and new directions. Contemporary Educational Psychology, 25(1), 54–67. https://doi.org/10.1006/ceps.1999.1020
242. Chentanez N, Barto AG, Singh SP (2004) Intrinsically motivated reinforcement learning. In: Saul LK, Weiss Y, Bottou L (eds) Advances in neural information processing systems 17. MIT Press, Cambridge, pp 1281–1288
243. Schembri M, Mirolli M, Baldassarre G (2007) Evolving internal reinforcers for an intrinsically motivated reinforcement-learning robot. In: IEEE 6th international conference on development and learning (ICDL 2007), pp 282–287
244. Baranes A, Oudeyer PY (2013) Active learning of inverse models with intrinsically motivated goal exploration in robots. Robot Auton Syst 61(1):49–73
245. Santucci VG, Baldassarre G, Mirolli M (2013) Which is the best intrinsic motivation signal for learning multiple skills? Front Neurorobot 7:22. https://doi.org/10.3389/fnbot.2013.00022
246. Eitam, Baruch; Kennedy, Patrick M.; Tory Higgins, E. Motivation from control. Experimental Brain Research (2013). https://doi.org/10.1007/s00221-012-3370-7
247. Ferris, Craig F; Stolberg, Tara; Kulkarni, Praveen; Murugavel, Murali; Blanchard, Robert; Blanchard, D Caroline; Febo, Marcelo; Brevard, Mathew; Simon, Neal G Imaging the neural circuitry and chemical control of aggressive motivation. BMC Neuroscience (2008). https://doi.org/10.1186/1471-2202-9-111
248. Mir, Pablo; Trender-Gerhard, Iris; Edwards, Mark J.; Schneider, Susanne A.; Bhatia, Kailash P.; Jahanshahi, Marjan Motivation and movement: the effect of monetary incentive on performance speed. Experimental Brain Research (2011). https://doi.org/10.1007/s00221-011-2583-5
249. Dohnalová, Lenka; Lundgren, Patrick; Carty, Jamie R. E.; Goldstein, Nitsan; Wenski, Sebastian L.; Nanudorn, Pakjira; Thiengmag, Sirinthra; Huang, Kuei-Pin; Litichevskiy, Lev; Descamps, Hélène C.; Chellappa, Karthikeyani; Glassman, Ana; Kessler, Susanne; Kim, Jihee; Cox, Timothy O.; Dmitrieva-Posocco, Oxana; Wong, Andrea C.; Allman, Erik L.; Ghosh, Soumita; Sharma, Nitika; Sengupta, Kasturi; Cornes, Belinda; Dean, Nitai; Churchill,

Gary A.; Khurana, Tejvir S.; Sellmyer, Mark A.; FitzGerald, Garret A.; Patterson, Andrew D.; Baur, Joseph A.; Alhadeff, Amber L.; Helfrich, Eric J. N.; Levy, Maayan; Betley, J. Nicholas; Thaiss, Christoph A. A microbiome-dependent gut–brain pathway regulates motivation for exercise. Nature (2022). https://doi.org/10.1038/s41586-022-05525-z

250. Hemed, Eitan; Karsh, Noam; Mark-Tavger, Ilya; Eitam, Baruch Motivation(s) from control: response-effect contingency and confirmation of sensorimotor predictions reinforce different levels of selection. Experimental Brain Research (2022). https://doi.org/10.1007/s00221-022-06345-3
251. Penton, Tegan; Wang, Xingquan; Coll, Michel-Pierre; Catmur, Caroline; Bird, Geoffrey The influence of action–outcome contingency on motivation from control. Experimental Brain Research (2018). https://doi.org/10.1007/s00221-018-5374-4
252. Yao, Zhao; Hessburg, John P.; Francis, Joseph Thachil Normalization by valence and motivational intensity in the sensorimotor cortices (PMd, M1, and S1). Scientific Reports (2021). https://doi.org/10.1038/s41598-021-03200-3
253. Rothkirch, Marcus; Ostendorf, Florian; Sax, Anne-Lene; Sterzer, Philipp The influence of motivational salience on saccade latencies. Experimental Brain Research (2012). https://doi.org/10.1007/s00221-012-3284-4
254. Rybnicek, Robert; Bergner, Sabine; Gutschelhofer, Alfred How individual needs influence motivation effects: a neuroscientific study on McClelland's need theory. Review of Managerial Science (2017). https://doi.org/10.1007/s11846-017-0252-1
255. Pincus, J. David Employee Engagement as Human Motivation: Implications for Theory, Methods, and Practice. Integrative Psychological and Behavioral Science (2022). https://doi.org/10.1007/s12124-022-09737-w
256. Schutter, Dennis J. L. G.; Knyazev, Gennady G. Cross-frequency coupling of brain oscillations in studying motivation and emotion. Motivation and Emotion (2011). https://doi.org/10.1007/s11031-011-9237-6
257. Winter, Carla; Kern, Florian; Gall, Dominik; Latoschik, Marc Erich; Pauli, Paul; Käthner, Ivo Immersive virtual reality during gait rehabilitation increases walking speed and motivation: a usability evaluation with healthy participants and patients with multiple sclerosis and stroke. Journal of NeuroEngineering and Rehabilitation (2021). https://doi.org/10.1186/s12984-021-00848-w
258. Siqueira, Marina Alves Martins; Gonçalves, Johnatan Padovez; Mendonça, Vitor Silva; Kobayasi, Renata; Arantes-Costa, Fernanda Magalhães; Tempski, Patricia Zen; Martins, Mílton de Arruda Relationship between metacognitive awareness and motivation to learn in medical students. BMC Medical Education (2020). https://doi.org/10.1186/s12909-020-02318-8
259. Noroozi, Omid; Pijeira-Díaz, Héctor J.; Sobocinski, Marta; Dindar, Muhterem; Järvelä, Sanna; Kirschner, Paul A. Multimodal data indicators for capturing cognitive, motivational, and emotional learning processes: A systematic literature review. Education and Information Technologies (2020). https://doi.org/10.1007/s10639-020-10229-w
260. Watson, Poppy; Wiers, Reinout W.; Hommel, Bernhard; de Wit, Sanne Motivational sensitivity of outcome-response priming: Experimental research and theoretical models. Psychonomic Bulletin & Review (2018). https://doi.org/10.3758/s13423-018-1449-2
261. Bailey, Rachel L.; Wang, Grace Tianjiao; Liu, Jiawei Ambient lighting alters motivational responses to advertisements for foods of different energetic value. Motivation and Emotion (2021). https://doi.org/10.1007/s11031-021-09891-5
262. Sari, Berna A.; Tarman, Güliz Zeynep; Ozdogan, Busra; Metin, Baris; Derakshan, Nazanin Working Memory Training in Relation to Anxiety, Stress, and Motivation. Journal of Cognitive Enhancement (2020). https://doi.org/10.1007/s41465-020-00176-2
263. Shreter, Roni; Peled, Ron; Pillar, Giora The 20-min trial of the Maintenance of Wakefulness Test is profoundly affected by motivation. Sleep and Breathing (2006). https://doi.org/10.1007/s11325-006-0072-z

264. van de Laar, Maria C.; Licht, Robert; Franken, Ingmar H. A.; Hendriks, Vincent M. Event-related potentials indicate motivational relevance of cocaine cues in abstinent cocaine addicts. Psychopharmacology (2004). https://doi.org/10.1007/s00213-004-1928-1
265. Harmon-Jones, Eddie; Gable, Philip A. Incorporating motivational intensity and direction into the study of emotions: implications for brain mechanisms of emotion and cognition-emotion interactions. Netherlands journal of psychology (2008). https://doi.org/10.1007/bf03076416
266. Skalski, Sebastian; Pochwatko, Grzegorz; Balas, Robert Impact of Motivation on Selected Aspects of Attention in Children with ADHD. Child Psychiatry & Human Development (2020). https://doi.org/10.1007/s10578-020-01042-0
267. Novak, Domen; Nagle, Aniket; Keller, Urs; Riener, Robert Increasing motivation in robot-aided arm rehabilitation with competitive and cooperative gameplay. Journal of NeuroEngineering and Rehabilitation (2014). https://doi.org/10.1186/1743-0003-11-64
268. Poole, Bryan D.; Gable, Philip A. Affective motivational direction drives asymmetric frontal hemisphere activation. Experimental Brain Research (2014). https://doi.org/10.1007/s00221-014-3902-4
269. Matsumoto, Masayuki; Hikosaka, Okihide Two types of dopamine neuron distinctly convey positive and negative motivational signals. Nature (2009). https://doi.org/10.1038/nature08028
270. Abreu, Rodolfo; Soares, Júlia F.; Lima, Ana Cláudia; Sousa, Lívia; Batista, Sónia; Castelo-Branco, Miguel; Duarte, João Valente Optimizing EEG Source Reconstruction with Concurrent fMRI-Derived Spatial Priors. Brain Topography (2022). https://doi.org/10.1007/s10548-022-00891-3
271. Sanyal, Shankha; Nag, Sayan; Banerjee, Archi; Sengupta, Ranjan; Ghosh, Dipak Music of brain and music on brain: a novel EEG sonification approach. Cognitive Neurodynamics (2018). https://doi.org/10.1007/s11571-018-9502-4
272. Allouch, Sahar; Yochum, Maxime; Kabbara, Aya; Duprez, Joan; Khalil, Mohamad; Wendling, Fabrice; Hassan, Mahmoud; Modolo, Julien Mean-Field Modeling of Brain-Scale Dynamics for the Evaluation of EEG Source-Space Networks. Brain Topography (2021). https://doi.org/10.1007/s10548-021-00859-9
273. Moradi, Zahra Zargol; Manohar, Sanjay; Duta, Mihaela; Enock, Florence; Humphreys, Glyn W. In-group biases and oculomotor responses: beyond simple approach motivation. Experimental Brain Research (2018). https://doi.org/10.1007/s00221-018-5221-7
274. Esposito, Romina; Bortoletto, Marta; Zacà, Domenico; Avesani, Paolo; Miniussi, Carlo An integrated TMS-EEG and MRI approach to explore the interregional connectivity of the default mode network. Brain Structure and Function (2022). https://doi.org/10.1007/s00429-022-02453-6
275. Pu, Yuhan; Zhang, Yiwen; Zhang, Yan; Shen, Ping Two Drosophila Neuropeptide Y-like Neurons Define a Reward Module for Transforming Appetitive Odor Representations to Motivation. Scientific Reports (2018). https://doi.org/10.1038/s41598-018-30113-5
276. Paul, Katharina; Pourtois, Gilles; Harmon-Jones, Eddie Modulatory effects of positive mood and approach motivation on reward processing: Two sides of the same coin?. Cognitive, Affective, & Behavioral Neuroscience (2020). https://doi.org/10.3758/s13415-019-00764-6
277. Pascoe, Leona; Thompson, Deanne; Spencer-Smith, Megan; Beare, Richard; Adamson, Chris; Lee, Katherine J.; Kelly, Claire; Georgiou-Karistianis, Nellie; Nosarti, Chiara; Josev, Elisha; Roberts, Gehan; Doyle, Lex W.; Seal, Marc L.; Anderson, Peter J. Efficiency of structural connectivity networks relates to intrinsic motivation in children born extremely preterm. Brain Imaging and Behavior (2018). https://doi.org/10.1007/s11682-018-9918-9
278. Shigemune, Yayoi; Kawasaki, Iori; Midorikawa, Akira; Baba, Toru; Takeda, Atsushi; Abe, Nobuhito Intrinsic motivation in patients with Parkinson's disease: a neuropsychological investigation of curiosity using dopamine transporter imaging. Neurological Sciences (2021). https://doi.org/10.1007/s10072-020-04968-4

279. Duan, Hongxia; Fernández, Guillén; van Dongen, Eelco; Kohn, Nils The effect of intrinsic and extrinsic motivation on memory formation: insight from behavioral and imaging study. Brain Structure and Function (2020). https://doi.org/10.1007/s00429-020-02074-x
280. Taffoni, Fabrizio; Tamilia, Eleonora; Focaroli, Valentina; Formica, Domenico; Ricci, Luca; Di Pino, Giovanni; Baldassarre, Gianluca; Mirolli, Marco; Guglielmelli, Eugenio; Keller, Flavio Development of goal-directed action selection guided by intrinsic motivations: an experiment with children. Experimental Brain Research (2014). https://doi.org/10.1007/s00221-014-3907-z

The manufacturer's authorised representative in the EU is Springer
Nature Customer Service Centre GmbH, Europaplatz 3, 69115 Heidelberg,
Germany. If you have any concerns regarding our products, please
contact ProductSafety@springernature.com

Printed and bound by CPI Group (UK) Ltd, Croydon, CR0 4YY
08/10/2026
02229819-0004